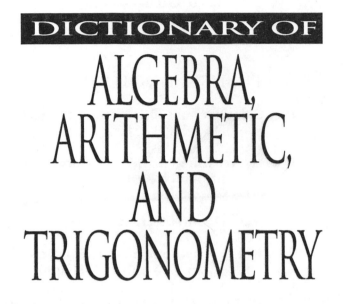

DICTIONARY OF

ALGEBRA, ARITHMETIC, AND TRIGONOMETRY

COMPREHENSIVE DICTIONARY
OF MATHEMATICS

Douglas N. Clark
Editor-in-Chief

Stan Gibilisco
Editorial Advisor

PUBLISHED VOLUMES

Analysis, Calculus, and Differential Equations
Douglas N. Clark

Algebra, Arithmetic, and Trigonometry
Steven G. Krantz

FORTHCOMING VOLUMES

Classical & Theoretical Mathematics
Catherine Cavagnaro and Will Haight

Applied Mathematics for Engineers and Scientists
Emma Previato

The Comprehensive Dictionary of Mathematics
Douglas N. Clark

A VOLUME IN THE
COMPREHENSIVE DICTIONARY
OF MATHEMATICS

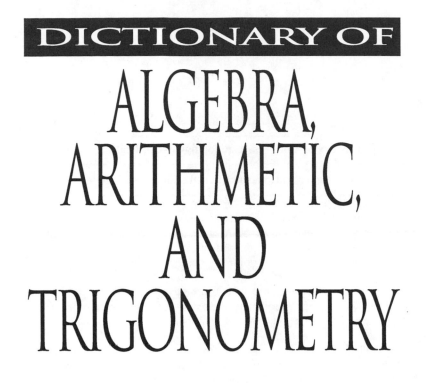

DICTIONARY OF
ALGEBRA, ARITHMETIC, AND TRIGONOMETRY

Edited by
Steven G. Krantz

CRC Press
Boca Raton   London   New York   Washington, D.C.

Library of Congress Cataloging-in-Publication Data

Catalog record is available from the Library of Congress.

No claim to original U.S. Government works
International Standard Book Number 1-58488-052-X
Printed in the United States of America 1 2 3 4 5 6 7 8 9 0
Printed on acid-free paper

PREFACE

The second volume of the CRC Press Comprehensive Dictionary of Mathematics covers algebra, arithmetic and trigonometry broadly, with an overlap into differential geometry, algebraic geometry, topology and other related fields. The authorship is by well over 30 mathematicians, active in teaching and research, including the editor.

Because it is a dictionary and not an encyclopedia, definitions are only occasionally accompanied by a discussion or example. In a dictionary of mathematics, the primary goal is to define each term rigorously. The derivation of a term is almost never attempted.

The dictionary is written to be a useful reference for a readership that includes students, scientists, and engineers with a wide range of backgrounds, as well as specialists in areas of analysis and differential equations and mathematicians in related fields. Therefore, the definitions are intended to be accessible, as well as rigorous. To be sure, the degree of accessibility may depend upon the individual term, in a dictionary with terms ranging from Abelian cohomology to z intercept.

Occasionally a term must be omitted because it is archaic. Care was taken when such circumstances arose to ensure that the term was obsolete. An example of an archaic term deemed to be obsolete, and hence not included, is "right line". This term was used throughout a turn-of-the-century analytic geometry textbook we needed to consult, but it was not defined there. Finally, reference to a contemporary English language dictionary yielded "straight line" as a synonym for "right line".

The authors are grateful to the series editor, Stanley Gibilisco, for dealing with our seemingly endless procedural questions and to Nora Konopka, for always acting efficiently and cheerfully with CRC Press liaison matters.

Douglas N. Clark
Editor-in-Chief

CONTRIBUTORS

Edward Aboufadel
Grand Valley State University
Allendale, Michigan

Gerardo Aladro
Florida International University
Miami, Florida

Mohammad Azarian
University of Evansville
Evansville, Indiana

Susan Barton
West Virginia Institute of Technology
Montgomery, West Virginia

Albert Boggess
Texas A&M University
College Station, Texas

Robert Borrelli
Harvey Mudd College
Claremont, California

Stephen W. Brady
Wichita State University
Wichita, Kansas

Der Chen Chang
Georgetown University
Washington, D.C.

Stephen A. Chiappari
Santa Clara University
Santa Clara, California

Joseph A. Cima
The University of North Carolina at Chapel Hill
Chapel Hill, North Carolina

Courtney S. Coleman
Harvey Mudd College
Claremont, California

John B. Conway
University of Tennessee
Knoxville, Tennessee

Neil K. Dickson
University of Glasgow
Glasgow, United Kingdom

David E. Dobbs
University of Tennessee
Knoxville, Tennessee

Marcus Feldman
Washington University
St. Louis, Missouri

Stephen Humphries
Brigham Young University
Provo, Utah

Shanyu Ji
University of Houston
Houston, Texas

Kenneth D. Johnson
University of Georgia
Athens, Georgia

Bao Qin Li
Florida International University
Miami, Florida

Robert E. MacRae
University of Colorado
Boulder, Colorado

Charles N. Moore
Kansas State University
Manhattan, Kansas

Hossein Movahedi-Lankarani
Pennsylvania State University
Altoona, Pennsylvania

Shashikant B. Mulay
University of Tennessee
Knoxville, Tennessee

Judy Kenney Munshower
Avila College
Kansas City, Missouri

Charles W. Neville
CWN Research
Berlin, Connecticut

Daniel E. Otero
Xavier University
Cincinnati, Ohio

Josef Paldus
University of Waterloo
Waterloo, Ontario, Canada

Harold R. Parks
Oregon State University
Corvallis, Oregon

Gunnar Stefansson
Pennsylvania State University
Altoona, Pennsylvania

Anthony D. Thomas
University of Wisconsin
Platteville, Wisconsin

Michael Tsatsomeros
University of Regina
Regina, Saskatchewan, Canada

James S. Walker
University of Wisconsin at Eau Claire
Eau Claire, Wisconsin

C. Eugene Wayne
Boston University
Boston, Massachusetts

Kehe Zhu
State University of New York at Albany
Albany, New York

A

A-balanced mapping   Let M be a right module over the ring A, and let N be a left module over the same ring A. A mapping ϕ from $M \times N$ to an Abelian group G is said to be A-balanced if $\phi(x, \cdot)$ is a group homomorphism from N to G for each $x \in M$, if $\phi(\cdot, y)$ is a group homomorphism from M to G for each $y \in N$, and if

$$\phi(xa, y) = \phi(x, ay)$$

holds for all $x \in M$, $y \in N$, and $a \in A$.

A-B-bimodule   An Abelian group G that is a left module over the ring A and a right module over the ring B and satisfies the associative law $(ax)b = a(xb)$ for all $a \in A$, $b \in B$, and all $x \in G$.

Abelian cohomology   The usual cohomology with coefficients in an Abelian group; used if the context requires one to distinguish between the usual cohomology and the more exotic non-Abelian cohomology. *See* cohomology.

Abelian differential of the first kind   A holomorphic differential on a closed Riemann surface; that is, a differential of the form $\omega = a(z)\,dz$, where $a(z)$ is a holomorphic function.

Abelian differential of the second kind   A meromorphic differential on a closed Riemann surface, the singularities of which are all of order greater than or equal to 2; that is, a differential of the form $\omega = a(z)\,dz$ where $a(z)$ is a meromorphic function with only 0 residues.

Abelian differential of the third kind   A differential on a closed Riemann surface that is not an Abelian differential of the first or second kind; that is, a differential of the form $\omega = a(z)\,dz$ where $a(z)$ is meromorphic and has at least one non-zero residue.

Abelian equation   A polynomial equation $f(X) = 0$ is said to be an *Abelian equation* if

its Galois group is an Abelian group. *See* Galois group. *See also* Abelian group.

Abelian extension   A Galois extension of a field is called an Abelian extension if its Galois group is Abelian. *See* Galois extension. *See also* Abelian group.

Abelian function   A function $f(z_1, z_2, z_3, \ldots, z_n)$ meromorphic on $\mathbf{C}^n$ for which there exist $2n$ vectors $\omega_k \in \mathbf{C}^n$, $k = 1, 2, 3, \ldots, 2n$, called *period vectors,* that are linearly independent over $\mathbf{R}$ and are such that

$$f(z + \omega_k) = f(z)$$

holds for $k = 1, 2, 3, \ldots, 2n$ and $z \in \mathbf{C}^n$.

Abelian function field   The set of Abelian functions on $\mathbf{C}^n$ corresponding to a given set of period vectors forms a field called an *Abelian function field.*

Abelian group   Briefly, a commutative group. More completely, a set G, together with a binary operation, usually denoted "+," a unary operation usually denoted "−," and a distinguished element usually denoted "0" satisfying the following axioms:

(i.) $a + (b + c) = (a + b) + c$ for all $a, b, c \in G$,

(ii.) $a + 0 = a$ for all $a \in G$,

(iii.) $a + (-a) = 0$ for all $a \in G$,

(iv.) $a + b = b + a$ for all $a, b \in G$.

The element 0 is called the *identity,* $-a$ is called the *inverse* of a, axiom (i.) is called the *associative* axiom, and axiom (iv.) is called the *commutative* axiom.

Abelian ideal   An ideal in a Lie algebra which forms a commutative subalgebra.

Abelian integral of the first kind   An indefinite integral $W(p) = \int_{p_0}^{p} a(z)\,dz$ on a closed Riemann surface in which the function $a(z)$ is holomorphic (the differential $\omega(z) = a(z)\,dz$ is said to be an Abelian differential of the first kind).

Abelian integral of the second kind   An indefinite integral $W(p) = \int_{p_0}^{p} a(z)\,dz$ on a closed Riemann surface in which the function $a(z)$ is

1-58488-052-X/01/$0.00+$.50

meromorphic with all its singularities of order at least 2 (the differential $a(z)\,dz$ is said to be an Abelian differential of the second kind).

Abelian integral of the third kind An indefinite integral $W(p) = \int_{p_0}^{p} a(z)\,dz$ on a closed Riemann surface in which the function $a(z)$ is meromorphic and has at least one non-zero residue (the differential $a(z)\,dz$ is said to be an Abelian differential of the third kind).

Abelian Lie group A Lie group for which the associated Lie algebra is Abelian. *See also* Lie algebra.

Abelian projection operator A non-zero projection operator E in a von Neumann algebra $\mathcal{M}$ such that the reduced von Neumann algebra $\mathcal{M}_E = E\mathcal{M}E$ is Abelian.

Abelian subvariety A subvariety of an Abelian variety that is also a subgroup. *See also* Abelian variety.

Abelian surface A two-dimensional Abelian variety. *See also* Abelian variety.

Abelian variety A complete algebraic variety G that also forms a commutative algebraic group. That is, G is a group under group operations that are regular functions. The fact that an algebraic group is complete as an algebraic variety implies that the group is commutative. *See also* regular function.

Abel's Theorem Niels Henrik Abel (1802-1829) proved several results now known as "Abel's Theorem," but perhaps preeminent among these is Abel's proof that the general quintic equation cannot be solved algebraically. Other theorems that may be found under the heading "Abel's Theorem" concern power series, Dirichlet series, and divisors on Riemann surfaces.

absolute class field Let k be an algebraic number field. A Galois extension K of k is an *absolute class field* if it satisfies the following property regarding prime ideals of k: A prime ideal $\mathbf{p}$ of k of absolute degree 1 decomposes

as the product of prime ideals of K of absolute degree 1 if and only if $\mathbf{p}$ is a principal ideal.

The term "absolute class field" is used to distinguish the Galois extensions described above, which were introduced by Hilbert, from a more general concept of "class field" defined by Tagaki. *See also* class field.

absolute covariant A covariant of weight 0. *See also* covariant.

absolute inequality An inequality involving variables that is valid for all possible substitutions of real numbers for the variables.

absolute invariant Any quantity or property of an algebraic variety that is preserved under birational transformations.

absolutely irreducible character The character of an absolutely irreducible representation. A representation is absolutely irreducible if it is irreducible and if the representation obtained by making an extension of the ground field remains irreducible.

absolutely irreducible representation A representation is absolutely irreducible if it is irreducible and if the representation obtained by making an extension of the ground field remains irreducible.

absolutely simple group A group that contains no serial subgroup. The notion of an absolutely simple group is a strengthening of the concept of a simple group that is appropriate for infinite groups. *See* serial subgroup.

absolutely uniserial algebra Let A be an algebra over the field K, and let L be an extension field of K. Then $L \otimes_K A$ can be regarded as an algebra over L. If, for every choice of L, $L \otimes_K A$ can be decomposed into a direct sum of ideals which are primary rings, then A is an absolutely uniserial algebra.

absolute multiple covariant A multiple covariant of weight 0. *See also* multiple covariants.

absolute number   A specific number represented by numerals such as 2, $\frac{3}{4}$, or 5.67 in contrast with a literal number which is a number represented by a letter.

absolute value of a complex number   More commonly called the *modulus,* the absolute value of the complex number $z = a + ib$, where a and b are real, is denoted by $|z|$ and equals the non-negative real number $\sqrt{a^2 + b^2}$.

absolute value of a vector   More commonly called the *magnitude,* the absolute value of the vector
$$\vec{v} = (v_1, v_2, \ldots, v_n)$$
is denoted by $|\vec{v}|$ and equals the non-negative real number $\sqrt{v_1^2 + v_2^2 + \cdots + v_n^2}$.

absolute value of real number   For a real number r, the nonnegative real number $|r|$, given by
$$|r| = \left\{ \begin{array}{ll} r & \text{if} \quad r \geq 0 \\ -r & \text{if} \quad r < 0. \end{array} \right\}$$

abstract algebraic variety   A set that is analogous to an ordinary algebraic variety, but defined only locally and without an imbedding.

abstract function   (1) In the theory of generalized almost-periodic functions, a function mapping **R** to a Banach space other than the complex numbers.

(2) A function from one Banach space to another Banach space that is everywhere differentiable in the sense of Fréchet.

abstract variety   A generalization of the notion of an algebraic variety introduced by Weil, in analogy with the definition of a differentiable manifold. An *abstract variety* (also called an *abstract algebraic variety*) consists of (i.) a family $\{V_\alpha\}_{\alpha \in A}$ of affine algebraic sets over a given field k, (ii.) for each $\alpha \in A$ a family of open subsets $\{W_{\alpha\beta}\}_{\beta \in A}$ of V_α, and (iii.) for each pair α and β in A a birational transformation between $W_{\alpha\beta}$ and $W_{\alpha\beta}$ such that the composition of the birational transformations between subsets of V_α and V_β and between subsets of V_β and V_γ are consistent with those between subsets of V_α and V_γ.

acceleration parameter   A parameter chosen in applying successive over-relaxation (which is an accelerated version of the Gauss-Seidel method) to solve a system of linear equations numerically. More specifically, one solves $Ax = b$ iteratively by setting
$$x_{n+1} = x_n + R(b - Ax_n) ,$$
where
$$R = \left(L + \omega^{-1} D \right)^{-1}$$
with L the lower triangular submatrix of A, D the diagonal of A, and $0 < \omega < 2$. Here, ω is the *acceleration parameter,* also called the *relaxation parameter.* Analysis is required to choose an appropriate value of ω.

acyclic chain complex   An augmented, positive chain complex
$$\cdots \xrightarrow{\partial_{n+1}} X_n \xrightarrow{\partial_n} X_{n-1} \xrightarrow{\partial_{n-1}} \cdots$$
$$\cdots \xrightarrow{\partial_2} X_1 \xrightarrow{\partial_1} X_0 \xrightarrow{\epsilon} A \to 0$$
forming an exact sequence. This in turn means that the kernel of ∂_n equals the image of ∂_{n+1} for $n \geq 1$, the kernel of ϵ equals the image of ∂_1, and ϵ is surjective. Here the X_i and A are modules over a commutative unitary ring.

addend   In arithmetic, a number that is to be added to another number. In general, one of the operands of an operation of addition. *See also* addition.

addition   (1) A basic arithmetic operation that expresses the relationship between the number of elements in each of two disjoint sets and the number of elements in the union of those two sets.

(2) The name of the binary operation in an Abelian group, when the notation "+" is used for that operation. *See also* Abelian group.

(3) The name of the binary operation in a ring, under which the elements form an Abelian group. *See also* Abelian group.

(4) Sometimes, the name of one of the operations in a multi-operator group, even though the operation is not commutative.

addition formulas in trigonometry   The formulas

$$\cos(\phi + \theta) = \cos\phi\,\cos\theta - \sin\phi\,\sin\theta,$$
$$\sin(\phi + \theta) = \cos\phi\,\sin\theta + \sin\phi\,\cos\theta,$$
$$\tan(\phi + \theta) = \frac{\tan\phi + \tan\theta}{1 - \tan\phi\,\tan\theta}.$$

addition of algebraic expressions   One of the fundamental ways of forming new algebraic expressions from existing algebraic expressions; the other methods of forming new expressions from old being subtraction, multiplication, division, and root extraction.

addition of angles   In elementary geometry or trigonometry, the angle resulting from the process of following rotation through one angle about a center by rotation through another angle about the same center.

addition of complex numbers   One of the fundamental operations under which the complex numbers $\mathbf{C}$ form a field. If $w = a + ib$, $z = c + id \in \mathbf{C}$, with a, b, c, and d real, then $w + z = (a + c) + i(b + d)$ is the result of addition, or the sum, of those two complex numbers.

addition of vectors   One of the fundamental operations in a vector space, under which the set of vectors form an Abelian group. For vectors in $\mathbf{R}^n$ or $\mathbf{C}^n$, if $x = (x_1, x_2, \ldots, x_n)$ and $y = (y_1, y_2, \ldots, y_n)$, then $x + y = (x_1 + y_1, x_2 + y_2, \ldots, x_n + y_n)$.

additive group   (1) Any group, usually Abelian, where the operation is denoted $+$. *See* group, Abelian group.

(2) In discussing a ring R, the commutative group formed by the elements of R under the addition operation.

additive identity   In an Abelian group G, the unique element (usually denoted 0) such that $g + 0 = g$ for all $g \in G$.

additive identity   a binary operation that is called addition and is denoted by "+." In this situation, an additive identity is an element $i \in S$ that satisfies the equation

$$i + s = s + i = s$$

for all $s \in S$. Such an additive identity is necessarily unique and usually is denoted by "0."

In ordinary arithmetic, the number 0 is the *additive identity* because $0 + n = n + 0 = n$ holds for all numbers n.

additive inverse   In any algebraic structure with a commutative operation referred to as addition and denoted by "+," for which there is an additive identity 0, the *additive inverse* of an element a is the element b for which $a + b = b + a = 0$. The *additive inverse* of a is usually denoted by $-a$. In arithmetic, the *additive inverse* of a number is also called its opposite. *See* additive identity.

additive set function   Let X be a set and let $\mathcal{A}$ be a collection of subsets of X that is closed under the union operation. Let $\phi : \mathcal{A} \to F$, where F is a field of scalars. We say that ϕ is *finitely additive* if, whenever $S_1, \ldots, S_k \in \mathcal{A}$ are pairwise disjoint then $\phi(\cup_{j=1}^k S_j) = \sum_{j=1}^k \phi(S_j)$. We say that ϕ is *countably additive* if, whenever $S_1, S_2, \cdots \in \mathcal{A}$ are pairwise disjoint then $\phi(\cup_{j=1}^\infty S_j) = \sum_{j=1}^\infty \phi(S_j)$.

additive valuation   Let F be a field and G be a totally ordered additive group. An additive valuation is a function $v : F \to G \cup \{\infty\}$ satisfying

(i.) $v(a) = \infty$ if and only if $a = 0$,

(ii.) $v(ab) = v(a) + v(b)$,

(iii.) $v(a + b) \geq \min\{v(a), v(b)\}$.

adele   Following Weil, let k be either a finite algebraic extension of $\mathbf{Q}$ or a finitely generated extension of a finite prime field of transcendency degree 1 over that field. By a place of k is meant the completion of the image of an isomorphic embedding of k into a local field (actually the equivalence class of such completions under the equivalence relation induced by isomorphisms of the local fields). A place is infinite if the local field is $\mathbf{R}$ or $\mathbf{C}$, otherwise the place is finite. For a place v, k_v will denote the completion, and if v is a finite place, r_v will denote the maximal compact subring of k_v. An *adele* is an element of

$$\prod_{v \in P} k_v \times \prod_{v \notin P} r_v,$$

where P is a finite set of places containing the infinite places.

adele group   Let V be the set of valuations on the global field k. For $v \in V$, let k_v be the completion of k with respect to v, and let O_v be the ring of integer elements in k_v. The *adele group* of the linear algebraic group G is the restricted direct product

$$\prod_{v \in V} G_{k_v}\left(G_{O_v}\right)$$

which, as a set, consists of all sequences of elements of G_{k_v}, indexed by $v \in V$, with all but finitely many terms in each sequence being elements of G_{O_v}.

adele ring   Following Weil, let k be either a finite algebraic extension of $\mathbf{Q}$ or a finitely generated extension of a finite prime field of transcendency degree 1 over that field. Set

$$k_A(P) = \prod_{v \in P} k_v \times \prod_{v \notin P} r_v \, ,$$

where P is a finite set of places of k containing the infinite places. A ring structure is put on $k_A(P)$ defining addition and multiplication componentwise. The adele ring is

$$k_A = \bigcup_P k_A(P) \, .$$

A locally compact topology is defined on k_A by requiring each $k_A(P)$ to be an open subring and using the product topology on $k_A(P)$.

adjoining   (1) Assuming K is a field extension of k and $S \subset K$, the field obtained by *adjoining* S to k is the smallest field F satisfying $k \subset F \subset K$ and containing S.

(2) If R is a commutative ring, then the ring of polynomials $R[X]$ is said to be obtained by *adjoining* X to R.

adjoint group   The image of a Lie group G, under the adjoint representation into the space of linear endomorphisms of the associated Lie algebra $\mathbf{g}$. *See also* adjoint representation.

adjoint Lie algebra   Let $\mathbf{g}$ be a Lie algebra. The *adjoint Lie algebra* is the image of $\mathbf{g}$ under

the adjoint representation into the space of linear endomorphisms of $\mathbf{g}$. *See also* adjoint representation.

adjoint matrix   For a matrix M with complex entries, the adjoint of M is denoted by M^* and is the complex conjugate of the transpose of M; so if $M = \left(m_{ij}\right)$, then M^* has $\bar{m}_{ji}$ as the entry in its ith row and jth column.

adjoint representation   (1) In the context of Lie algebras, the *adjoint representation* is the mapping sending X to $[X, \cdot]$.

(2) In the context of Lie groups, the *adjoint representation* is the mapping sending σ to the differential of the automorphism $\alpha_\sigma : G \to G$ defined by $\alpha_\sigma(\tau) = \sigma\tau\sigma^{-1}$.

(3) In the context of representations of an algebra over a field, the term *adjoint representation* is a synonym for *dual representation*. *See* dual representation.

adjoint system   Let D be a curve on a nonsingular surface S. The adjoint system of D is $|D + K|$, where K is a canonical divisor on S.

adjunction formula   The formula

$$2g - 2 = C.(C + K)$$

relating the genus g of a non-singular curve C on a surface S with the intersection pairing of C and $C + K$, where K is a canonical divisor on S.

admissible homomorphism   For a group G with a set of operators Ω, a group homomorphism from G to a group G' on which the same operators act, such that

$$\omega(ab) = (\omega a)(\omega b)$$

holds for all $a, b \in G$ and all $\omega \in \Omega$. Also called an Ω-*homomorphism* or an *operator homomorphism*.

admissible isomorphism   For a group G with a set of operators Ω, a group isomorphism from G onto a group G', on which the same operators act, such that

$$\omega(ab) = (\omega a)(\omega b)$$

holds for all $a, b \in G$ and all $\omega \in \Omega$. Also called an Ω-*isomorphism* or an *operator isomorphism*.

admissible normal subgroup   Let G be a group. It is easily seen that a subset N of G is a normal subgroup if and only if there is some equivalence relation $\sim$ on G such that $\sim$ is compatible with the multiplication on G, meaning

$$a \sim b,\ c \sim d \Rightarrow (ac) \sim (bd) \,,$$

and N is the equivalence class of the identity. In case G also has an operator domain Ω, an *admissible normal subgroup* is defined to be the equivalence class of the identity for an equivalence relation $\sim$ that is compatible with the multiplication as above and that also satisfies

$$a \sim b \Rightarrow (\omega a) \sim (\omega b) \text{ for all } \omega \in \Omega \,.$$

admissible representation   Let π be a unitary representation of the group G in a Hilbert space, and let M be the von Neumann algebra generated by $\pi(G)$. The representation π is said to be an *admissible representation* or a *trace admissible representation* if there exists a trace on M^+ which is a character for π.

Ado-Iwasawa Theorem   The theorem that every finite dimensional Lie algebra (over a field of characteristic p) has a faithful finite dimensional representation. The characteristic $p = 0$ case of this is Ado's Theorem and the characteristic $p \neq 0$ case is Iwasawa's Theorem. *See also* Lie algebra.

Ado's Theorem   A finite dimensional Lie algebra $\mathbf{g}$ has a representation of finite degree ρ such that $\mathbf{g} \cong \rho(\mathbf{g})$.

   While originally proved for Lie algebras over fields of characteristic 0, the result was extended to characteristic p by Iwasawa. *See* Ado-Iwasawa Theorem.

affect   For a polynomial equation $P(X) = 0$, the Galois group of the equation can be considered as a group of permutations of the roots of the equation. The *affect* of the equation is the index of the Galois group in the group of all permutations of the roots of the equation.

affectless equation   A polynomial equation for which the Galois group consists of all permutations. *See also* affect.

affine algebraic group   *See* linear algebraic group.

affine morphism of schemes   Let X and Y be schemes and $f : X \to Y$ be a morphism. If there is an open affine cover $\{V_i\}$ of the scheme Y for which $f^{-1}(V_i)$ is affine for each i, then f is an *affine morphism of schemes*.

affine scheme   Let A be a commutative ring, and let $\mathrm{Spec}(A) = X$ be the set of all prime ideals of A, equipped with the spectral or Zariski topology. Let $\mathcal{O}_X$ be a sheaf of local rings on X. The ringed space $(X, \mathcal{O}_X)$ is called the *affine scheme* of the ring A.

affine space   Let V be a real, linear n-dimensional space. Let A be a set of points, which are denoted P, Q. Define a relation between points in A and vectors in V as follows:
(i.) To every ordered pair $(P, Q) \in A \times A$, there is associated a "difference vector" $\overrightarrow{PQ} \in V$.
(ii.) To every point $P \in A$ and every vector $v \in V$ there is associated precisely one point $Q \in A$ such that $\overrightarrow{PQ} = v$.
(iii.) If $P, Q, R \in A$ then

$$\overrightarrow{PQ} + \overrightarrow{QR} = \overrightarrow{PR} \,.$$

In this circumstance, we call A an n-dimensional *affine space*.

affine variety   A variety (common zero set of a finite collection of functions) defined in an affine space.

A-homomorphism   For A-modules M and N, a group homomorphism $f : M \to N$ is called an A-homomorphism if

$$f(am) = af(m) \text{ for all } a \in A,\ m \in M \,.$$

Albanese variety   For V a variety, the *Albanese variety* of V is an Abelian variety $A = \mathrm{Alb}(V)$ such that there exists a rational $f : V \to A$ which generates A and has the universal mapping property that for any rational

$g : V \rightarrow B$, where B is an Abelian variety, there exist a homomorphism $h : A \rightarrow B$ and a constant $c \in B$ such that $g = h f + c$.

Alexander Duality   If A is a compact subset of $\mathbf{R}^n$, then for all indices q and all R-modules G,

$$\overline{H}_q(\mathbf{R}^n, \mathbf{R}^n \setminus A; G) = \overline{H}^{n-q-1}(A; G) .$$

algebra   (1) The system of symbolic manipulation formalized by François Viéte (1540–1603), which today is known as elementary algebra.

   (2) The entire area of mathematics in which one studies groups, rings, fields, etc.

   (3) A vector space (over a field) on which is also defined an operation of multiplication.

   (4) A synonym for universal algebra, which includes structures such as Boolean algebras.

algebra class   An equivalence class of central simple algebras under the relation that relates a pair of algebras if they are both isomorphic to full matrix rings over the same division algebra. Algebras in the same algebra class are said to be "similar." *See also* central simple algebra.

algebra class group   Let K be a field. Two central simple algebras over K are said to be similar if they are isomorphic to full matrix rings over the same division algebra. Similarity is an equivalence relation, and the equivalence classes are called algebra classes. The product of a pair of algebra classes is defined by choosing an algebra from each class, say A and B, and letting the product of the classes be the algebra class containing $A \otimes_K B$. This product is well defined, and the algebra classes form a group under this multiplication, called the algebra class group or Brauer group.

algebra extension   Let A be an algebra over the commutative ring R. Then by an *algebra extension* of A is meant either

   (i.) an algebra over R that contains A; or

   (ii.) an algebra A' containing a two-sided R-module M which is a two-sided ideal in A' and is such that

$$A'/M = A .$$

In this case, M is called the kernel of the extension because it is the kernel of the canonical homomorphism.

algebra homomorphism   Suppose A and B are algebras of the same type, meaning that for each n-ary operation f_A on A there is a corresponding n-ary operation f_B on B. A mapping $\phi : A \rightarrow B$ is called a homomorphism from A to B if, for each pair of corresponding operations f_A and f_B,

$$\phi (f_A (a_1, a_2, \ldots, a_n))$$
$$= f_B (\phi (a_1), \phi (a_2), \ldots, \phi (a_n))$$

holds for all $a_1, a_2, \ldots, a_n \in A$.

   Typically, an algebra A is a ring that also has the structure of a module over another ring R, so that an *algebra homomorphism* ϕ must satisfy
   (i.) $\phi(a_1 + a_2) = \phi(a_1) + \phi(a_2)$ for $a_1, a_2 \in A$,
   (ii.) $\phi(a_1 a_2) = \phi(a_1)\phi(a_2)$ for $a_1, a_2 \in A$,
   (iii.) $\phi(ra) = r\phi(a)$, for $r \in R$ and $a \in A$.

algebraic   (1) An adjective referring to an object, structure, or theory that occurs in algebra or arises through application of the processes used in algebra.

   (2) An adverb meaning a process that involves only the operations of algebra, which are addition, subtraction, multiplication, division, and root extraction.

algebraic addition   In elementary algebra, the addition of algebraic expressions which extends the operation of addition of numbers in arithmetic.

algebraic addition formula   For an Abelian function f, an equation that expresses $f(a + b)$ rationally, in terms of the values of a certain $(p + 1)$-tuple of Abelian functions, evaluated at the points $a, b \in \mathbf{C}$. *See also* Abelian function.

algebraic algebra   An algebra A over a field K such that every $a \in A$ is algebraic over K. *See* algebra.

algebraically closed field   A field k, in which every polynomial in one variable, with coefficients in k, has a root.

algebraic closure   The smallest algebraically closed extension field of a given field F. The algebraic closure exists and is unique up to isomorphism.

algebraic correspondence   Let C be a nonsingular algebraic curve. By an *algebraic correspondence* is meant a divisor in the product variety $C \times C$. More generally, an *algebraic correspondence* means a Zariski closed subset T of the product $V_1 \times V_2$ of two irreducible varieties. Points $P_1 \in V_1$ and $P_2 \in V_2$ are said to *correspond* if $(P_1, P_2) \in T$. *See also* correspondence ring.

algebraic curve   An algebraic variety of dimension one. *See also* algebraic variety.

algebraic cycle   By an *algebraic cycle* of dimension m on an algebraic variety V is meant a finite formal sum

$$\sum c_i V_i$$

where the c_i are integers and the V_i are irreducible m-dimensional subvarieties of V. The cycle is said to be *effective* or *positive* if all the coefficients c_i are non-negative. The *support* of the cycle is the union of the subvarieties having non-zero coefficients. The set of cycles of dimension m forms an Abelian group under addition, which is denoted $\mathcal{Z}_m(V)$.

algebraic dependence   The property shared by a set of elements in a field, when they satisfy a non-trivial polynomial equation. Such an equation demonstrates that the set of elements is not algebraically independent.

algebraic differential equation   (**1**) An equation of the form

$$F\left(x, y, y', y'', \ldots, y^{(n)}\right) = 0$$

in which F is a polynomial with coefficients that are complex analytic functions of x.

(**2**) An equation obtained by equating to zero a differential polynomial in a set of differential variables in a differential extension field of a differential field. *See also* differential field.

algebraic element   If K is an extension field of the field k, an element $x \in K$ is an algebraic element of K if it satisfies a non-trivial polynomial equation with coefficients in k.

algebraic equation   An equation of the form $P = 0$ where P is a polynomial in one or more variables.

algebraic equivalence   Two cycles X_1 and X_2 in a non-singular algebraic variety V are *algebraically equivalent* if there is a family of cycles $\{X(t) : t \in T\}$ on V, parameterized by $t \in T$, where T is another non-singular algebraic variety, such that there is a cycle Z in $V \times T$ for which each $X(t)$ is the projection to V of the intersection of Z and $V \times \{t\}$, and $X_1 = X(t_1)$, $X_2 = X(t_2)$, for some $t_1, t_2 \in T$. Such a family of cycles $X(t)$ is called an *algebraic family*.

algebraic equivalence of divisors   Two divisors f and g on an irreducible variety X are algebraically equivalent if there exists an algebraic family of divisors, $f_t, t \in T$, and points t_1 and $t_2 \in T$, such that $f = f_{t_1}$, and $g = f_{t_2}$. Thus, algebraic equivalence is an algebraic analog of homotopy, though the analogy is not particularly fruitful.

Algebraic equivalence has the important property of preserving the degree of divisors; that is, two algebraically equivalent divisors have the same degree. It also preserves principal divisors; that is, if one divisor of an algebraically equivalent pair is principal, then so is the other one. (A divisor is *principal* if it is the divisor of a rational function.) Thus, the group D_0/P is a subgroup of the divisor class group $Cl^0(X) = D/P$. Here, D_0 is the group of divisors algebraically equivalent to 0, P is the group of principal divisors, and D is the group of divisors of degree 0. The group D_0/p is exactly the subgroup of the divisor class group realized by the group of points of the Picard variety of X. *See* algebraic family of divisors, divisor. *See also* integral divisor, irreducible variety, Picard variety.

algebraic expression   An expression formed from the elements of a field and one or more variables (variables are also often called indeterminants) using the algebraic operations of addition, subtraction, multiplication, division, and root extraction.

algebraic extension   An extension field K of a field k such that every α in K, but not in k, is algebraic over k, i.e., satisfies a polynomial equation with coefficients in k.

algebraic family   A family of cycles $\{X(t) : t \in T\}$ on a non-singular algebraic variety V, parameterized by $t \in T$, where T is another non-singular algebraic variety, such that there is a cycle Z in $V \times T$ for which each $X(t)$ is the projection to V of the intersection of Z and $V \times \{t\}$.

algebraic family of divisors   A family of divisors f_t, $t \in T$, on an irreducible variety X, where the index set T is also an irreducible variety, and where $f_t = \phi_t^*(D)$ for some fixed divisor D on $X \times T$ and all $t \in T$. Here, for each $t \in T$, ϕ_t^* is the map from divisors on $X \times T$ to divisors on X induced by the embedding $\phi_t : X \to X \times T$, where $\phi(t) = (x, t)$, and $X \times T$ is the Cartesian product of X and T. The variety T is called the *base* for the algebraic family f_t, $t \in T$. *See also* Cartesian product, irreducible variety.

algebraic function   A function $Y = f(X_1, X_2, \ldots, X_N)$ satisfying an equation $R(X_1, X_2, \ldots, X_N, Y) = 0$ where R is a rational function over a field F. *See also* rational function.

algebraic function field   Let F be a field. Any finite extension of the field of rational functions in

$$X_1, X_2, \ldots, X_n$$

over the field F is called an *algebraic function field* over F.

algebraic fundamental group   A generalization of the concept of fundamental group defined for an algebraic variety over a field of characteristic $p > 0$, formed in the context of finite étale coverings.

algebraic geometry   Classically, *algebraic geometry* has meant the study of geometric properties of solutions of algebraic equations. In modern times, algebraic geometry has become synonymous with the study of geometric objects associated with commutative rings.

algebraic group   An algebraic variety, together with group operations that are regular functions. *See* regular function.

algebraic homotopy group   A generalization of the concept of homotopy group, defined for an algebraic variety over a field of characteristic $p > 0$, formed in the context of finite étale coverings.

algebraic identity   An algebraic equation involving a variable or variables that reduces to an arithmetical identity for all substitutions of numerical values for the variable or variables.

algebraic independence   Let k be a subfield of the field K. The elements $a_1, a_2, \ldots, a_n$ of K are said to be *algebraically independent over k* if, for any polynomial $p(X_1, X_2, \ldots, X_n)$ with coefficients in k, $p(a_1, a_2, \ldots, a_n) = 0$ implies $p \equiv 0$. When a set of complex numbers is said to be *algebraically independent,* the field k is understood to be the rational numbers.

algebraic integer   A complex number that satisfies some monic polynomial equation with integer coefficients.

algebraic Lie algebra   Let k be a field. An algebraic group G, realized as a closed subgroup of the general linear group $GL(n, k)$, is called a linear algebraic group, and its tangent space at the identity, when given the natural Lie algebra structure, is called an *algebraic Lie algebra.*

algebraic multiplication   In elementary algebra, the multiplication of algebraic expressions, which extends the operation of multiplication of numbers in arithmetic.

algebraic multiplicity   The multiplicity of an eigenvalue λ of a matrix A as a root of the characteristic polynomial of A. *See also* geometric multiplicity, index.

algebraic number   A complex number z is an *algebraic number* if it satisfies a non-trivial polynomial equation $P(z) = 0$, for which the coefficients of the polynomial are rational numbers.

algebraic number field   A field $F \subset \mathbf{C}$, which is a finite degree extension of the field of rational numbers.

algebraic operation   In elementary algebra, the operations of addition, subtraction, multiplication, division, and root extraction. In a general algebraic system A, an algebraic operation may be any function from the n-fold cartesian product A^n to A, where $n \in \{1, 2, \dots\}$ (the case $n = 0$ is sometimes also allowed). *See also* algebraic system.

algebraic pencil   A linear system of divisors in a projective variety such that one divisor passes through any point in general position.

algebraic scheme   An algebraic scheme is a scheme of finite type over a field. Schemes are generalizations of varieties, and the algebraic schemes most closely resemble the algebraic varieties. *See* scheme.

algebraic space   A generalization of scheme and of algebraic variety due to Artin and introduced to create a category which would be closed under various constructions. Specifically, an *algebraic space* of finite type is an affine scheme U and a closed subscheme $R \subset U \times U$ that is an equivalence relation and for which both the coordinate projections of R onto U are étale. *See also* étale morphism.

algebraic subgroup   A Zariski closed subgroup of an affine algebraic group.

algebraic surface   A two-dimensional algebraic variety. *See also* algebraic variety.

algebraic system   A set A, together with various operations and relations, where by an operation we mean a function from the n-fold cartesian product A^n to A, for some $n \in \{0, 1, 2, \dots\}$.

algebraic system in the wider sense   While an *algebraic system* is a set A, together with various operations and relations on A, an algebraic system *in the wider sense* may also include higher level structures constructed by the power set operation.

algebraic torus   An algebraic group, isomorphic to a direct product of the multiplicative group of a universal domain. A *universal domain* is an algebraically closed field of infinite transcendence degree over the prime field it contains.

algebraic variety   Classically, the term "algebraic variety" has meant either an affine algebraic set or a projective algebraic set, but in the second half of the twentieth century, various more general definitions have been introduced. One such more general definition, in terms of sheaf theory, considers an algebraic variety V to be a pair $(T, \mathcal{O})$, in which T is a topological space and $\mathcal{O}$ is a sheaf of germs of mappings from V into a given field k, for which the topological space has a finite open cover $\{U_i\}_{i=1}^{N}$ such that each $(U_i, \mathcal{O}|U_i)$ is isomorphic to an affine variety and for which the image of V under the diagonal map is Zariski closed. *See also* abstract algebraic variety.

algebra isomorphism   An algebra homomorphism that is also a one-to-one and onto mapping between the algebras. *See* algebra homomorphism.

algebra of matrices   The $n \times n$ matrices with entries taken from a given field together with the operations of matrix addition and matrix multiplication. Also any nonempty set of such matrices, closed under those operations and containing additive inverses, and thus forming an algebra.

algebra of vectors   The vectors in three-dimensional space, together with the operations of vector addition, scalar multiplication, the scalar product (also called the inner product or the dot product), the vector product (also called the cross product), and the vector triple product.

algebroidal function   An analytic function f satisfying a non-trivial algebraic equation

$$a_0(z) f^n + a_1(z) f^{n-1} + \cdots + a_n(z) = 0,$$

in which the coefficients $a_j(z)$ are meromorphic functions in a domain in the complex z-plane.

all-integer algorithm   An algorithm for which the entire calculation will be carried out in integers, provided the given data is all given in integers. Such algorithms are of interest for linear programming problems that involve additional integrality conditions. A notable example of such an algorithm was given in the early 1960s by Gomory.

allowed submodule   In a module M with operator domain A, an *allowed submodule* is a submodule $N \subset M$ such that $a \in A$ and $x \in N$ implies $ax \in N$. Also called an *A-submodule*.

almost integral   Let R be a subring of the ring R'. An element $a \in R'$ is said to be *almost integral* over R if there exists an element $b \in R$ which is not a zero divisor and for which $a^n b \in R$ holds for every positive integer n.

alternating group   For fixed n, the subgroup of the group of permutations of $\{1, 2, \ldots, n\}$, consisting of the even permutations. More specifically, the set of permutations $\sigma : \{1, 2, \ldots, n\} \to \{1, 2, \ldots, n\}$ such that

$$\prod_{1 \leq i < j \leq n} (\sigma(j) - \sigma(i)) > 0 \,.$$

Usually denoted by A_n.

alternating law   Any binary operation $R(\cdot, \cdot)$ on a set S is said to satisfy an *alternating law* if

$$R(a, b) = -R(b, a)$$

holds for all $a, b \in S$. The term is particularly used for exterior products and for the bracket operation in Lie algebras.

alternating polynomial   Any polynomial $P(X_1, X_2, \ldots, X_n)$ that is transformed into $-P$ by every odd permutation of the indeterminants $X_1, X_2, \ldots, X_n$.

alternative algebra   A distributive algebra, in which the equations $a \cdot (b \cdot b) = (a \cdot b) \cdot b$ and $(a \cdot a) \cdot b = a \cdot (a \cdot b)$ hold for all a and b in the algebra.

alternative field   An alternative ring with unit in which, given any choices of $a \neq 0$ and b, the two equations

$$ax_1 = b \text{ and } x_2 a = b$$

are uniquely solvable for x_1 and x_2. Also called *alternative skew-field*.

amalgamated product   Given a family of groups $\{G_\alpha\}_{\alpha \in A}$ and embeddings $\{h_\alpha\}_{\alpha \in A}$ of a fixed group H into the G_α, the *amalgamated product* is the group G, unique up to isomorphism, having the universal properties that (i.) there exist homomorphisms $\{g_\alpha\}_{\alpha \in A}$ such that $g_\alpha \circ h_\alpha = g_\beta \circ h_\beta$ for all $\alpha, \beta \in A$ and (ii.) for any family $\{\ell_\alpha\}_{\alpha \in A}$ of homomorphisms of the groups G_α to a fixed group L satisfying $\ell_\alpha \circ h_\alpha = \ell_\beta \circ h_\beta$ for all $\alpha, \beta \in A$, there exists a unique homomorphism $\ell : G \to L$ such that $\ell_\alpha = \ell \circ g_\alpha$.

For the case of two groups G_1 and G_2 with isomorphic subgroups $H_1 \subset G_1$ and $H_2 \subset G_2$, the amalgamated product of the groups can be identified with the set of finite sequences of elements of the union of the two groups with the equivalence relation generated by identifying a sequence with the sequence formed when adjacent elements are replaced by their product if they are in the same G_i or with the sequence formed when an element of an H_1 is replaced by its isomorphic image in H_2 and *vice-versa*. Multiplication is then defined by concatenation of sequences.

The amalgamated product is also called the *free product with amalgamation*.

ambig ideal   Let k be a quadratic field, i.e., $k = \mathbf{Q}(\sqrt{m})$ where m is a non-zero integer with no factor that is a perfect square. Conjugation on k is the map sending $\alpha = a + b\sqrt{m}, a, b \in \mathbf{Q}$, to α^c.

ambiguous case   A problem in trigonometry for which there is more than one possible solution, such as finding a plane triangle with two given side lengths and a given non-included angle.

Amitsur cohomology   A cohomology theory defined as follows. Let R be a commutative ring with identity and F a covariant functor from the category $\mathcal{C}_R$ of commutative R-algebras to the category of additive Abelian groups. For

$S \in C_R$ and n a nonnegative integer, let $S^{(n)}$ denote the n-fold tensor product of S over R. For n a nonnegative integer, let $\mathcal{E}_i : S^{(n+1)} \to S^{(n+2)}$ $(i = 0, 1, \ldots, n)$ be the C_R-morphisms defined by

$$\mathcal{E}_i (x_0 \otimes \cdots \otimes x_n) =$$
$$x_0 \otimes \cdots \otimes x_{i-1} \otimes 1 \otimes x_i \otimes \cdots \otimes x_n .$$

Define $d^n : F(S^{(n+1)}) \to F(S^{(n+2)})$ by setting

$$d^n = \sum_{i=0}^{n} (-1)^i F(\mathcal{E}_i) .$$

Then $\{F(S^{(n+1)}), d^n\}$ defines a cochain complex called the Amitsur complex and the cohomology groups are called the Amitsur cohomology groups.

Amitsur cohomology groups   *See* Amitsur cohomology.

Amitsur complex   *See* Amitsur cohomology.

ample   *See* ample vector bundle, ample divisor.

ample divisor   A divisor D such that nD is very ample for some positive integer n. A divisor is *very ample* if it possesses a certain type of canonical projective immersion.

ample vector bundle   A vector bundle E where the line bundle $\mathcal{O}_{E^\vee}(1)$ on $P(E^\vee)$ is ample. That is, there is a morphism f from $P(E^\vee)$ to a projective space $\mathbf{P}^n$ with $\mathcal{O}_{E^\vee}(1)^{\otimes^m} = f^* \mathcal{O}_{P^N}(1)$.

amplification   The process of increasing the magnitude of a quantity.

analytically normal ring   An analytically unramified ring that is also integrally closed. *See* analytically unramified ring.

analytically unramified ring   A local ring such that its completion contains no non-zero nilpotent elements. (An element x of a ring is *nilpotent* if $x \cdot x = 0$.)

analytic function   Same as a holomorphic function, but with emphasis on the fact that such a function has a convergent power series expansion about each point of its domain.

analytic homomorphism   A homomorphism between two Lie groups which is also an analytic function (i.e., expandable in a power series at each point in the Lie group, using a local coordinate system).

analytic isomorphism   An analytic homomorphism between two Lie groups which is one-to-one, onto and has an inverse that is also an analytic homomorphism. *See* analytic homomorphism.

analytic structure   A structure on a differentiable manifold M which occurs when there is an atlas of charts $\{(U_i, \varphi_i) : i \in I\}$ on M, where the transition functions

$$\varphi_j \circ \varphi_i^{-1} : \varphi_i (U_i \cap U_j) \to \varphi_j (U_i \cap U_j)$$

are analytic.

analytic variety   A set that is the simultaneous zero set of a finite collection of analytic functions.

analytic vector   A vector $\mathbf{v}$ in a Hilbert space $\mathcal{H}$ is called an *analytic vector* for a finite set $\{T_j\}_{j=1}^m$ of (unbounded) operators on $\mathcal{H}$ if there exist positive constants C and N such that

$$\|T_{j_1} \cdots T_{j_k} \mathbf{v}\|_\mathcal{H} \leq C N^k k!$$

for all $j_i \in \{1, \ldots, m\}$ and every positive integer k.

anisotropic   A vector space $\mathcal{V}$ with an inner product $(\cdot, \cdot)$ and containing no non-zero isotropic vector. A vector $x \in \mathcal{V}$ is *isotropic* if $(x, x) = 0$.

antiautomorphism   An isomorphism of an algebra A onto its *opposite algebra* A°. *See* opposite.

antiendomorphism   A mapping τ from a ring R to itself, which satisfies

$$\tau(x + y) = \tau(x) + \tau(y), \quad \tau(xy) = \tau(y)\tau(x)$$

for all $x, y \in R$. The mapping τ can also be viewed as an endomorphism (linear mapping) from R to its opposite ring R°. *See* opposite.

antihomomorphism   A mapping σ from a group G into a group H that satisfies $\sigma(xy) = \sigma(y)\sigma(x)$ for all $x, y \in G$. An antihomomorphism can also be viewed as a homomorphism $\sigma : G \to H°$ where $H°$ is the *opposite group* to H. *See* opposite.

anti-isomorphism   A one-to-one, surjective map $f : X \to Y$ that reverses some intrinsic property common to X and Y. If X and Y are groups or rings, then f reverses multiplication, $f(ab) = f(b)f(a)$. If X and Y are lattices, then f reverses the lattice operations, $f(a \cap b) = f(a) \cup f(b)$ and $f(a \cup b) = f(a) \cap f(b)$.

antilogarithm   For a number y and a base b, the number x such that $\log_b x = y$.

antipode   Let S be a sphere in Euclidean space and s a point of S. The line through s and the center of the sphere will intersect the sphere in a uniquely determined second point s' that is called the *antipode* of s. The celebrated Borsuk-Ulam Theorem of algebraic topology considers the antipodal map $P \mapsto -P$. The theory of Hopf algebras contains a notion of antipode which is analogous to the geometric one just described.

antisymmetric decomposition   The decomposition of a compact Hausdorff space X consists of disjoint, closed, maximal sets of antisymmetry with respect to A, where A is a closed subalgebra of $C(X)$, the algebra of all complex-valued continuous functions on X. A is called *antisymmetric* if, from the condition that $f, \bar{f} \in A$, it follows that f is a constant function. A subset $S \beta X$ is called a *set of antisymmetry* with respect to A if any function $f \in A$ that is real on S is constant on this set.

apartment   An element of $\mathcal{A}$, a set of sub-complexes of a complex Δ such that the pair $(\Delta, \mathcal{A})$ is a building. That is, if the following hold:

(i.) Δ is thick;

(ii.) the elements of $\mathcal{A}$ are thin chamber complexes;

(iii.) any two elements of Δ belong to an apartment;

(iv.) if two apartments Σ and Σ' contain two elements $A, A' \in \Delta$, then there exists an isomorphism of Σ onto Σ' which leaves invariant A, A' and all their faces.

approximate functional equations   Equations of the form $f(x) = g(x) + Ev(x)$ where $f(x)$ and $g(x)$ are known functions and the growth of $Ev(x)$ is known.

approximately finite algebra   A C^*-algebra that is the uniform closure of a finite dimensional C^*-algebra.

approximately finite dimensional von Neumann algebra   A von Neumann algebra, $\mathcal{M}$, which contains an increasing sequence of finite dimensional subalgebras, $A_n \subseteq A_{n+1}$, such that $\cup_{n=1}^{\infty} A_n$ is dense in $\mathcal{M}$. (Density is defined in terms of any of a number of equivalent topologies on $\mathcal{M}$, e.g., the weak* topology, or the strong operator topology in any normal representation.)

approximate number   A numerical approximation to the actual value.

approximation theorem   A theorem which states that one class of objects can be approximated by elements from another (usually smaller) class of objects. A famous example is the following.

> **Weierstrass A. T.** *Every continuous function on a closed interval can be uniformly approximated by a polynomial. That is, if $f(x)$ is continuous on the closed interval $[a, b]$ and $\epsilon > 0$, then there exists a polynomial $p_\epsilon(x)$ such that $|f(x) - p_\epsilon(x)| < \epsilon$ for all $x \in [a, b]$.*

Arabic numerals   The numbers $0, 1, 2, 3, 4, 5, 6, 7, 8$, and 9. These numbers can be used to represent all numbers in the decimal system.

arbitrary constant   A constant that can be set to any desired value. For example, in the calculus expression $\int 2x \, dx = x^2 + C$, the symbol C is an arbitrary constant.

arc cosecant   The multiple-valued inverse of the trigonometric function csc θ, e.g., arccsc(2) $= \pi/6 + 2k\pi$ where k is an arbitrary integer ($k = 0$ specifies the *principal value* of arc cosecant). The principal value yields the length of the arc on the unit circle, subtending an angle, whose cosecant equals a given value.

The arc cosecant function is also denoted $\csc^{-1}x$.

arc cosine   The multiple-valued inverse of the trigonometric function cos θ, e.g., arccos(-1) $= \pi + 2k\pi$ where k is an arbitrary integer ($k = 0$ specifies the *principal value* of arc cosine). The principal value yields the length of the arc on the unit circle, subtending an angle, whose cosine equals a given value.

The arc cosine function is also denoted $\cos^{-1}x$.

arc cotangent   The multiple-valued inverse of the trigonometric function cotan θ, e.g., arc-cot ($\sqrt{3}$) $= \pi/6 + 2k\pi$ where k is an arbitrary integer ($k = 0$ specifies the *principal value* of arc cotangent). The principal value yields the length of the arc on the unit circle, subtending an angle, whose cotangent equals a given value.

The arc cotangent function is also denoted $\cot^{-1}x$.

Archimedian ordered field   If K is an ordered field and F a subfield with the property that no element of K is infinitely large over F, then we say that K is *Archimedian*.

Archimedian ordered field   A set which, in addition to satisfying the axioms for a field, also possesses an Archimedian ordering. That is, the field F is *ordered* in that it contains a subset P and the following properties hold:

(i.) F is the disjoint union of P, $\{0\}$, and $-P$. In other words, each $x \in F$ belongs either to P, or equals 0, or $-x$ belongs to P, and these three possibilities are mutually exclusive.

(ii.) If $x, y \in P$, then $x + y \in P$ and $xy \in P$.

The ordered field is also *Archimedian* in that the absolute value function

$$|x| = \begin{cases} x, & \text{if } x \in P \\ 0, & \text{if } x = 0 \\ -x, & \text{if } x \in -P \end{cases}$$

is satisfied.

(iii.) For each $x \in F$ there exists a positive integer n such that $n \cdot 1 > x$.

The rational numbers are an *Archimedian ordered field,* and so are the real numbers. The p-adic numbers are a non-Archimedian ordered field.

Archimedian valuation   A valuation on a ring R, for which $v(x - y) \leq \max(v(x), v(y))$ is false, for some $x, y \in R$. *See* valuation.

arc secant   The multiple-valued inverse of the trigonometric function sec x, sometimes denoted $\sec^{-1}x$.

arc sine   The multiple-valued inverse of the trigonometric function sin θ, e.g., arcsin(1) $= \pi/2 + 2k\pi$ where k is an arbitrary integer ($k = 0$ specifies the *principal value* of arc sine). The principal value yields the length of the arc on the unit circle, subtending an angle, whose sine equals a given value.

The arc sine function is also denoted $\sin^{-1}x$.

arc tangent   The multiple-valued inverse of the trigonometric function tan θ, e.g., arctan ($\sqrt{3}$) $= \pi/3 + 2k\pi$ where k is an arbitrary integer ($k = 0$ specifies the *principal value* of arc tangent). The principal value yields the length of the arc on the unit circle, subtending an angle, whose tangent equals a given value.

The arc tangent function is also denoted $\tan^{-1}x$.

Arens–Royden Theorem   Let $C(M_A)$ denote the continuous functions on the *maximal ideal space M_A* of the Banach algebra A. Suppose that $f \in C(M_A)$ and f does not vanish. Then there exists a $g \in A$, for which $g^{-1} \in A$, and for which $f/\hat{g}$ has a continuous logarithm on M_A. (Here $\hat{g}$ denotes the *Gelfand transform* of g.)

arithmetic   The operations of addition, subtraction, multiplication, and division and their properties for the integers.

arithmetical equivalence   An equivalence relation on the integers which is consistent with the four operations of arithmetic. ($a \sim b$ and $c \sim d$ imply $a \pm c \sim b \pm d$, etc.) An example would be congruence mod n where n is a positive integer. Here, two integers j and k are equivalent if $j - k$ is divisible by n. *See* equivalence relation.

arithmetically effective   Referring to a divisor on a nonsingular algebraic surface, which is numerically semipositive, or numerically effective (nef).

arithmetic crystal class   For an n-dimensional Euclidean space V, an equivalence class of pairs (Γ, G) where Γ is a lattice in V and G is a finite subgroup of $O(V)$. Two pairs (Γ_1, G_1) and (Γ_2, G_2) are equivalent if there is a $g \in GL(V)$ such that $g\Gamma_1 = \Gamma_2$, and $gG_1g^{-1} = G_2$.

arithmetic genus   An integer, defined in terms of the characteristic polynomial of a homogeneous ideal $\mathcal{U}$ in the ring of polynomials, $k[x_1, \ldots, x_n]$, in the variables $x_1, \ldots, x_n$ over a commutative ring k. If $\bar{\chi}(\mathcal{U}; q)$ denotes this characteristic polynomial, then

$$\bar{\chi}(\mathcal{U}; q) = a_0 \binom{q}{r} + a_1 \binom{q}{r-1} + \cdots + a_{r-1} \binom{q}{1} + a_r$$

where $a_0, \ldots, a_r \in k$ and $\{\binom{q}{j}\}$ are the binomial coefficients. The integer $(-1)^r(a_r - 1)$ is the *arithmetic genus* of $\mathcal{U}$.

arithmetic mean   For a positive integer n, the *arithmetic mean* of the n real numbers $a_1, \ldots, a_n$ is $(a_1 + \cdots + a_n)/n$.

arithmetic of associative algebras   An area of mathematics devoted to the study of simple algebras over local fields, number fields, or function fields.

arithmetic progression   A sequence $\{s_n\}$ of real numbers such that

$$s_n = s_{n-1} + r, \quad \text{for} \quad n > 1.$$

The number s_1 is the *initial term,* the number r is the *difference term.* The general term s_n satisfies $s_n = s_1 + (n - 1)r$.

arithmetic series   A series of the form $\sum_{n=1}^{\infty} a_n$ where for all $n \geq 1$, $a_{n+1} = a_n + d$.

arithmetic subgroup   For a real algebraic group $G \subset GL(n, \mathbf{R})$, a subgroup Γ of G, commensurable with $G_{\mathbf{Z}} = G \cap GL(n, \mathbf{R})$. That is,

$$[\Gamma : \Gamma \cap G_{\mathbf{Z}}] < \infty \text{ and } [G_{\mathbf{Z}} : \Gamma \cap G_{\mathbf{Z}}] < \infty.$$

Arrow-Hurewicz-Uzawa gradient method
A technique used in solving convex or concave programming problems. Suppose $\psi(x, u)$ is concave or convex in $x \in A \subset \mathbf{R}^n$ and convex in $u \in 0 \subset \mathbf{R}^m$. Usually $\varphi(x, u) = \psi(x) + u \cdot g(x)$ where φ is the function we wish to minimize or maximize and our constraints are given by the functions $g_j(x) \leq 0$ $1 \leq j \leq m$. The method devised by Arrow-Hurewicz and Uzawa consists of solving the system of equations

$$\frac{dx_i}{dt} = \begin{cases} 0 & \text{if } x_i = 0 \\ & \text{and } \frac{\partial \psi}{\partial x_i} < 0, \\ & i = 1, \ldots, n \\ \frac{\partial \psi}{\partial x_i} & \text{otherwise} \end{cases}$$

$$\frac{du_j}{dt} = \begin{cases} 0 & \text{if } u_j = 0 \\ & \text{and } \frac{\partial \psi}{\partial u_j} > 0, \\ & j = 1, \ldots, m \\ \frac{-\partial \psi}{\partial u_j} & \text{otherwise} \end{cases}$$

If $(x(t), u(t))$ is a solution of this system, under certain conditions, $\lim_{t \to \infty} x(t) = \bar{x}$ solves the programming problem.

artificial variable   A variable that is introduced into a linear programming problem, in order to transform a constraint that is an inequality into an equality. For example, the problem of minimizing

$$C = 3x_1 + 2x_2$$

subject to the constraints

$$\begin{aligned} 4x_1 - 5x_2 &\leq 7 \\ x_1 + x_2 &= 9 \end{aligned}$$

with $x_1 \geq 0, x_2 \geq 0$, is transformed into

$$C = 3x_1 + 2x_2 + 0A_1$$

subject to the constraints

$$\begin{aligned} 4x_1 - 5x_2 + A_1 &= 7 \\ x_1 + x_2 + 0A_1 &= 9 \end{aligned}$$

with $x_1 \geq 0$, $x_2 \geq 0$, $A_1 \geq 0$, by introducing the artificial variable A_1. This latter version is in the standard form for a linear programming problem.

Artin-Hasse function   For k a p-adic number field with k_0 a maximal subfield of k unramified over $\mathbf{Q}_p$, a an arbitrary integer in k_0 and $x \in k$, the function $E(a, x) = \exp -L(a, x)$ where $L(a, x) = \sum_{i=0}^{\infty}((a^\sigma)^i/p^i)x^{p^i}$ and σ is the Frobenius automorphism of $k_o/\mathbf{Q}_p$.

Artinian module   A (left) module for which every descending sequence of (left) submodules

$$M_1 \supset M_2 \supset \cdots \supset M_n \supset M_{n+1} \supset \cdots$$

is finite, i.e., there exists an N such that $M_n = M_{n+1}$ for all $n \geq N$.

Artinian ring   A ring for which every descending sequence of left ideals

$$I_1 \supset I_2 \supset \cdots \supset I_n \supset I_{n+1} \supset \cdots$$

is finite. That is, there exists an N such that $I_n = I_{n+1}$ for all $n \geq N$.

Artin L-function   The function $L(s, \varphi)$, defined as follows. Let K be a finite Galois extension of a number field k with $G = \text{Gal}(K/k)$. Let $\varphi : G \to \text{GL}(V)$ be a finite dimensional representation (characteristic 0). For each prime $\wp$ of k, set $L_\wp(s, \varphi) = \det(I - \varphi_\wp N(\wp)^{-s})^{-1}$, where $\varphi_\wp = \frac{1}{e}\sum_{\tau \in T} \varphi(\sigma\tau)$, T is the inertia group of $\wp$, $|T| = e$ and σ is the Frobenius automorphism of $\wp$. Then

$$L(s, \varphi) = \prod_\wp L_\wp(s, \varphi), \text{ for } \Re s > 1 .$$

Artin-Rees Lemma   Let R be a Noetherian ring, I an ideal of R, F a finitely generated submodule over R, and E a submodule of F. Then, there exists an integer $m \geq 1$ such that, for all integers $n \geq m$, it follows that $I^n F \cap E = I^{n-m}(I^m F \cap E)$.

Artin-Schreier extension   For K a field of characteristic $p \neq 0$, an extension of the form $L = K(Pa_1, \ldots, Pa_N)$ where $a_1, \ldots, a_N \in$

K, Pa_i is a root of $x^p - x - a_i = 0$, L/K is Galois, and the Galois group is an Abelian group of exponent p.

Artin's conjecture   A conjecture of E. Artin that the Artin L-function $L(s, \varphi)$ is entire in s, whenever φ is irreducible and $s \neq 1$. *See* Artin L-function.

Artin's general law of reciprocity   If K/k is an Abelian field extension with conductor $\mathcal{F}$ and $A_\mathcal{F}$ is the group of ideals prime to the conductor, then the Artin map $\mathcal{A} \mapsto \left(\frac{K/k}{\mathcal{A}}\right)$ is a homomorphism $A_\mathcal{F} \to \text{Gal}(K/k)$. The reciprocity law states that this homomorphism is an isomorphism precisely when $\mathcal{A}$ lies in the subgroup $H_\mathcal{F}$ of $A_\mathcal{F}$ consisting of those ideals whose prime divisors split completely. That is, $A_\mathcal{F}/H_\mathcal{F} \cong \text{Gal}(K/k)$.

Artin's symbol   The symbol $\left(\frac{K/k}{\wp}\right)$ defined as follows. Let K be a finite Abelian Galois extension of a number field k with σ the principal order of k and $\mathcal{D}$ the principal order of K. For each prime $\wp$ of K there is a $\sigma = \left(\frac{K/k}{\wp}\right)$ $\in G = \text{Gal}(K/k)$ such that

$$A^\sigma \equiv A^{N(\wp)} \pmod{\wp}, \ A \in \mathcal{D} ;$$

$\left(\frac{K/k}{\wp}\right)$ is called the *Artin symbol* of $\wp$ for the Abelian extension K/k. For an ideal $a = \Pi\wp^e$ of k relatively prime to the relative discriminant of K/k, define $\left(\frac{K/k}{a}\right) = \Pi\left(\frac{K/k}{\wp}\right)^e$.

ascending central series   A sequence of subgroups

$$\{1\} = H_0 < H_1 < H_2 < \cdots < G$$

of a group G with identity 1, where H_{n+1} is the unique normal subgroup of H_n for which the quotient group H_{n+1}/H_n is the center of G/H_n.

ascending chain of subgroups   A sequence of subgroups

$$H_1 < \cdots < H_n < H_{n+1} < \cdots < G$$

of a group G.

associate   A relation between two elements a and b of a ring R with identity. It occurs when $a = bu$ for a unit u.

associated factor sets   Related by a certain equivalence relation between factor sets belonging to a group. Suppose N and F are groups and G is a group containing a normal subgroup $\overline{N}$ isomorphic to N with $G/\overline{N} \cong F$. If $s : F \to G$ is a splitting map of the sequence $1 \to \overline{N} \to G \to F \to 1$ and $c : F \times F \to \overline{N}$ is the map, $c(\sigma, \tau) = s(\sigma)s(\tau)s(\sigma\tau)^{-1}$ (s,c) is called a factor set. More generally, a pair of maps (s, c) where $s : F \to \mathrm{Aut}N$ and $c : F \times F \to N$ is called a factor set if

(i.) $s(\sigma)s(\tau)(a) = c(\sigma, \tau)s(\sigma\tau)(a)c(\sigma, \tau)^{-1}(a \in N)$,

(ii.) $c(\sigma, \tau)c(\sigma\tau, \rho) = s(\sigma)(c(\tau, \rho))c(\sigma, \tau\rho)$.

Two factor sets (s, c) and (t, d) are said to be *associated* if there is a map $\varphi : F \to N$ such that $t(\sigma)(a) = s(\sigma)(\varphi(\sigma)(a)\varphi(\sigma)^{-1})$ and $d(\sigma, \tau) = \varphi(\sigma)(s(\sigma)(\varphi(\tau)))c(\sigma, \tau)\varphi(\sigma\tau)^{-1}$.

associated form   Of a projective variety X in $\mathbf{P}^n$, the form whose zero set defines a particular projective hypersurface associated to X in the Chow construction of the parameter space for X. The construction begins with the irreducible algebraic correspondence $\{(x, H_0, \ldots, H_d) \in X \times \mathbf{P}^n \times \cdots \times \mathbf{P}^n : x \in X \cap (H_0 \cap \cdots \cap H_d)\}$ between points $x \in X$ and projective hyperplanes H_i in $\mathbf{P}^n$, $d = \dim X$. The projection of this correspondence onto $\mathbf{P}^n \times \cdots \times \mathbf{P}^n$ is a hypersurface which is the zero set of a single multidimensional form, the *associated form*.

associative algebra   An algebra A whose multiplication satisfies the associative law; i.e., for all $x, y, z \in A$, $x(yz) = (xy)z$.

associative law   The requirement that a binary operation $(x, y) \mapsto xy$ on a set S satisfy $x(yz) = (xy)z$ for all $x, y, z \in S$.

asymmetric relation   A relation $\sim$, on a set S, which does **not** satisfy $x \sim y \Rightarrow y \sim x$ for some $x, y \in S$.

asymptotic ratio set   In a von Neumann algebra M, the set

$$r_\infty(M) = \{\lambda \in]0, 1[: M \otimes R_\lambda$$
$$\text{is isomorphic to } M\}.$$

augmentation   An *augmentation* (over the integers $\mathbf{Z}$) of a chain complex $\mathcal{C}$ is a surjective homomorphism $C_0 \overset{\alpha}{\to} \mathbf{Z}$ such that $C_1 \overset{\partial_1}{\to} C_0 \overset{\epsilon}{\to} \mathbf{Z}$ equals the trivial homomorphism $C_1 \overset{0}{\to} \mathbf{Z}$ (the trivial homomorphism maps every element of C_1 to 0).

augmented algebra   *See* supplemented algebra.

augmented chain complex   A non-negative chain complex $\mathcal{C}$ with augmentation $\mathcal{C} \overset{\epsilon}{\to} \mathbf{Z}$. A chain complex $\mathcal{C}$ is *non-negative* if each $C_n \in \mathcal{C}$ with $n < 0$ satisfies $C_n = 0$. *See* augmentation.

automorphic form   Let D be an open connected domain in $\mathbf{C}^n$ with Γ a discontinuous subgroup of $\mathrm{Hol}(D)$. For $g \in \mathrm{Hol}(D)$ and $z \in D$ let $j(g, z)$ be the determinant of the Jacobian transformation of g evaluated at z. A meromorphic function f on D is an automorphic form of weight ℓ (an integer) for Γ if $f(\gamma z) = f(z)j(\gamma, z)^{-\ell}, \gamma \in \Gamma, z \in D$.

automorphism   An isomorphism of a group, or algebra, onto itself. *See* isomorphism.

automorphism group   The set of all automorphisms of a group (vector space, algebra, etc.) onto itself. This set forms a group with binary operation consisting of composition of mappings (the automorphisms). *See* automorphism.

average   Often synonymous with *arithmetic mean*. Can also mean *integral average*, i.e.,

$$\frac{1}{b-a} \int_a^b f(x)\,dx,$$

the integral average of a function $f(x)$ over a closed interval $[a, b]$, or

$$\frac{1}{\mu(X)} \int_X f\,d\mu,$$

the integral average of an integrable function f over a measure space X having finite measure $\mu(A)$.

axiom   A statement that is assumed as true, without proof, and which is used as a basis for proving other statements (theorems).

axiom system A collection (usually finite) of axioms which are used to prove all other statements (theorems) in a given field of study. For example, the axiom system of Euclidean geometry, or the Zermelo-Frankel axioms for set theory.

Azumaya algebra A central separable algebra A over a commutative ring R. That is, an algebra A with the center of A equal to R and with A a projective left-module over $A \otimes_R A^\circ$ (where A° is the opposite algebra of A). *See* opposite.

B

back substitution   A technique connected with the Gaussian elimination method for solving simultaneous linear equations. After the system

$$a_{11}x_1 + a_{12}x_2 + \cdots + a_{1n}x_n = b_1$$
$$a_{21}x_1 + a_{22}x_2 + \cdots + a_{2n}x_n = b_2$$
$$\cdots\cdots\cdots$$
$$a_{n1}x_1 + a_{n2}x_2 + \cdots + a_{nn}x_n = b_n$$

is converted to *triangular form*

$$t_{11}x_1 + t_{12}x_2 + \cdots + t_{1n}x_n = c_1$$
$$t_{22}x_2 + \cdots + t_{2n}x_n = c_2$$
$$\cdots\cdots\cdots$$
$$t_{nn}x_n = c_n .$$

One then solves for x_n and then *back substitutes* this value for x_n into the equation

$$t_{n-1\,n-1}x_{n-1} + t_{nn}x_n = c_{n-1}$$

and solves for x_{n-1}. Continuing in this way, all of the variables $x_1, x_2, \ldots, x_n$ can be solved for.

backward error analysis   A technique for estimating the error in evaluating $f(x_1, \ldots, x_n)$, assuming one knows $f(a_1, \ldots, a_n) = b$ and has control of $|x_i - a_i|$ for $1 \leq i \leq n$.

Baer's sum   For given R-modules A and C, the sum of two elements of the Abelian group $\text{Ext}_R(C, A)$.

Bairstow method of solving algebraic equations   An iterative method for finding quadratic factors of a polynomial. The goal being to obtain complex roots that are conjugate pairs.

Banach algebra   An algebra over the complex numbers with a norm $\| \cdot \|$, under which it is a Banach space and such that

$$\|xy\| \leq \|x\|\|y\|$$

for all $x, y \in B$. If B is an algebra over the real numbers, then B is called a *real Banach algebra*.

base   *See* base of logarithm. *See also* base of number system, basis.

base of logarithm   The number that forms the base of the exponential to which the logarithm is inverse. That is, a logarithm, base b, is the inverse of the exponential, base b. The logarithm is usually denoted by $\log_b$ (unless the base is Euler's constant e, when ln or log is used, log is also used for base 10 logarithm). A conversion formula, from one base to another, is

$$\log_a x = \log_b x \log_a b .$$

base of number system   The number which is used as a base for successive powers, combinations of which are used to express all positive integers and rational numbers. For example, 2543 in the base 7 system stands for the number

$$2\left(7^3\right) + 5\left(7^2\right) + 4\left(7^1\right) + 3 .$$

Or, -524.37 in the base 8 system stands for the number

$$-\left[5\left(8^2\right) + 2\left(8^1\right) + 4 + 3\left(8^{-1}\right) + 7\left(8^{-2}\right)\right] .$$

The *base 10* number system is called the *decimal* system. For base n, the term n-*ary* is used; for example, *ternary,* in base 3.

base point   The point in a set to which a bundle of (algebraic) objects is attached. For example, a vector bundle $\mathcal{V}$ defined over a manifold M will have to each point $b \in M$ an associated vector space V_b. The point b is the *base point* for the vectors in V_b.

base term   For a spectral sequence $E = \{E^r, d^r\}$ where $d^r : E^r_{p,q} \to E^r_{p-r,q+r-1}, r = 2, 3, \ldots$ and $E^r_{p,q} = 0$ whenever $p < 0$ or $q < 0$, a term of the form $E^r_{p,o}$.

basic feasible solution   A type of solution of the linear equation $Ax = b$. If A is an $m \times n$ matrix with $m \leq n$ and $b \in \mathbf{R}^m$, suppose $A = A_1 + A_2$ when A_1 is an $m \times m$ nonsingular matrix. It is a solution of the form $A(x_1, 0) = A_1x_1 + A_20 = b$ where $x_1 \in \mathbf{R}^m$ and $x_1 \geq 0$.

basic form of linear programming problem   The following form of a linear programming problem: Find a vector $(x_1, x_2, \ldots, x_n)$ which minimizes the linear function

$$C = c_1 x_1 + c_2 x_2 + \cdots + c_n x_n ,$$

subject to the constraints

$$
\begin{aligned}
a_{11}x_1 + a_{12}x_2 + \cdots + a_{1n}x_n &= b_1 \\
a_{21}x_1 + a_{22}x_2 + \cdots + a_{2n}x_n &= b_2 \\
&\cdots \\
a_{m1}x_1 + a_{m2}x_2 + \cdots + a_{mn}x_n &= b_m
\end{aligned}
$$

and $x_1 \geq 0, x_2 \geq 0, \ldots, x_n \geq 0$.

Here a_{ij}, b_i, and c_j are all real constants and $m < n$.

basic invariants   For a commutative ring K with identity and a ring R containing K and G a subgroup of $\text{Aut}_K(R)$, a minimal set of generators of the ring R^G.

basic optimal solution   A solution of a linear programming problem that minimizes the objective function (cost function) and is *basic* in the sense that, in the linear constraints

$$
\begin{aligned}
a_{11}x_1 + a_{12}x_2 + \cdots + a_{1n}x_n &= b_1 \\
a_{21}x_1 + a_{22}x_2 + \cdots + a_{2n}x_n &= b_2 \\
&\cdots\cdots\cdots \\
a_{m1}x_1 + a_{m2}x_2 + \cdots + a_{mn}x_n &= b_m
\end{aligned}
$$

for the problem, $n - m$ values of the n variables $x_1, x_2, \ldots, x_n$ are zero.

basic variable   A variable that has value zero in a linear programming problem. The basic variables lie on the boundaries of the convex regions determined by the constraints in the problem.

basis   A subset B of a vector space V which has the property that every vector $v \in V$ can be expressed uniquely as a finite linear combination of elements of B. That is, if V is a vector space over the field F, then for a given $v \in V$, there exists a unique, finite, collection of vectors $x_1, x_2, \ldots, x_n \in B$ and scalars $\alpha_1, \alpha_2, \ldots, \alpha_n \in F$ such that $x = \alpha_1 x_1 + \alpha_2 x_2 + \cdots + \alpha_n x_n$.

By definition, V is *finite dimensional* if it has a finite basis. In an infinite dimensional vector space, if there is a topology on V, the sum representing a vector x may be allowed to be infinite (and convergent). If only finite sums are permitted, a basis is referred to as a *Hamel basis*.

Bernoulli method for finding roots   An iterative method for finding a root of a polynomial equation. If $p(x) = a_0 x^n + a_{n-1} x^{n-1} + \cdots + a_n$ is a polynomial, then this method, applied to $p(x) = 0$, consists of the following steps. First, choose some set of initial-values $x_0, x_{-1}, \ldots, x_{-n+1}$. Second, define subsequent values x_m by the recurrence relation

$$x_m = -\frac{a_1 x_{m-1} + a_2 x_{m-2} + \cdots + a_n x_{m-n}}{a_0}$$

for $m \geq 1$. Third, form the sequence of quotients $r_m = x_{m+1}/x_m$ for $m \geq 1$. If the polynomial has a single root, r, of largest magnitude, then the sequence $\{r_m\}$ will converge to r.

Bernoulli number   Consider numbers B_n^* defined by the functional equation

$$
\frac{x}{e^x - 1} + \frac{x}{2} - 1 = \sum_{n=1}^{\infty} \frac{(-1)^{n-1} B_n^* x^{2n}}{(2n)!}
$$

$$
= \frac{B_1^* x^2}{2!} - \frac{B_2^* x^4}{4!} + \frac{B_3^* x^6}{6!} - + \cdots
$$

or, alternatively, by the equation

$$
1 - \frac{x}{2}\cot\left(\frac{x}{2}\right) = \sum_{n=1}^{\infty} \frac{B_n^* x^{2n}}{(2n)!}
$$

$$
= \frac{B_1^* x^2}{2!} + \frac{B_2^* x^4}{4!} + \frac{B_3^* x^6}{6!} + \cdots .
$$

The numbers B_n^* are called the Bernoulli numbers. The definition given here is the classical one. There are several alternative, and more modern, definitions. Bernoulli numbers arise in the theory of special functions, in the study of hypergeometric functions, and as the coefficients of the Taylor expansions of many classical transcendental functions.

Betti numbers   The nth Betti number B_n, of a manifold M, is the dimension of the nth cohomology group, $H^n(M, \mathbf{R})$. [The group $H^n(M, \mathbf{R})$ is the quotient group consisting of equivalence classes of the closed n forms modulo the differentials of $(n - 1)$ forms.]

Bezout's Theorem   If $p_1(x)$ and $p_2(x)$ are two polynomials of degrees n_1 and n_2, respectively, having no common zeros, then there are two unique polynomials $q_1(x)$ and $q_2(x)$ of degrees $n_1 - 1$ and $n_2 - 1$, respectively, such that

$$p_1(x)q_1(x) + p_2(x)q_2(x) = 1 .$$

biadditive mapping   For A-modules M, N and L, the mapping $f : M \times N \to L$ such that

$$f\left(x + x', y\right) = f(x, y) + f\left(x', y\right), \quad \text{and}$$

$$f\left(x, y + y'\right) = f(x, y) + f\left(x, y'\right) ,$$

$x, x' \in M$, $y, y' \in L$.

bialgebra   A vector space A over a field k that is both an algebra and a coalgebra over k. That is, $(A, \mu, \eta, \Delta, \varepsilon)$ is a bialgebra if (A, μ, η) is an algebra over k and (A, Δ, ε) is a coalgebra over k, $\mu : A \otimes_k A \to A$ (multiplication). $\eta : k \to A$ (unit), $\Delta : A \to A \otimes_k A$ (comultiplication). $\varepsilon : A \to k$ (counit) and these maps satisfy

$$\mu \circ (\mu \otimes I_A) = \mu \circ (I_A \otimes \mu) ,$$

$$\mu \circ (\eta \otimes I_A) = \mu \circ (I_A \otimes \eta) ,$$

$$(\Delta \otimes I_A) \circ \Delta = (I_A \otimes \Delta) \circ \Delta ,$$

$$(\varepsilon \otimes I_A) \circ \Delta = (I_A \otimes \varepsilon) \circ \Delta.$$

bialgebra-homomorphism   For $(A, \mu, \eta, \Delta, \varepsilon)$ and $(A', \mu', \eta', \Delta', \varepsilon')$ bialgebras over a field k, a linear mapping $f : A \to A'$ where $f \circ \eta = \eta'$, $f \circ \mu = \mu' \circ (f \otimes f)$, $(f \otimes f) \circ \Delta = \Delta' \circ f$, $\varepsilon = \varepsilon' \circ f$. See bialgebra.

biideal   A linear subspace I of A, where $(A, \mu, \eta, \Delta, \varepsilon)$ is a bialgebra over k, such that $\mu(A \otimes_k I) = I$ and $\Delta(I) \subset A \otimes_k I + I \otimes_k A$.

bilinear form   A mapping $b : V \times V \to F$, where V is a vector space over the field F, which satisfies

$$b(\alpha x + \beta y, z) = \alpha b(x, z) + \beta(y, z)$$

and

$$b(x, \alpha y + \beta z) = \alpha b(x, y) + \beta b(x, z)$$

for all $x, y, z \in V$ and $\alpha, \beta \in F$.   See also quadratic form.

bilinear function   See multilinear function.

bilinear mapping   A mapping $b : V \times V \to W$, where V and W are vector spaces over the field F, which satisfies

$$b(\alpha x + \beta y, z) = \alpha b(x, z) + \beta b(y, z)$$

and

$$b(x, \alpha y + \beta z) = \alpha b(x, y) + \beta b(x, z)$$

for all $x, y, z \in V$ and $\alpha, \beta \in F$.

bilinear programming   The area dealing with finding the extrema of functions

$$f(x_1, x_2) = C_1^t x_1 + C_2^t x_2 + x_1^t Q x_2$$

over

$$X = \{(x_1, x_2) \in \mathbf{R}^{n_1} \times \mathbf{R}^{n_2} : A_1 x_1 \le b_1 ,$$

$$A_2 x_2 \le b_2, x_1 \ge 0, x_2 \ge 0\}$$

where Q is an $n_1 \times n_2$ real matrix, A_1 is an $n_1 \times n_1$, real matrix and A_2 is an $n_2 \times n_2$ real matrix.

binary Diophantine equation   A Diophantine equation in two unknowns.   See Diophantine equation.

binary operation   A mapping from the Cartesian product of a set with itself into the set. That is, if the set is denoted by S, a mapping $b : S \times S \to S$. A notation, such as $\star$, is usually adopted for the operation, so that $b(x, y) = x \star y$.

binomial   A sum of two monomials. For example, if x and y are variables and α and β are constants, then $\alpha x^p y^q + \beta x^r y^s$, where p, q, r, s are integers, is a *binomial* expression.

binomial coefficients   The numbers, often denoted by $\binom{n}{k}$, where n and k are nonnegative integers, with $n \ge k$, given by

$$\binom{n}{k} = \frac{n!}{k!(n - k)!}$$

where $m! = m(m-1)\cdots(2)(1)$ and $0! = 1$ and $1! = 1$. The binomial coefficients appear in the Binomial Theorem expansion of $(x+y)^n$ where n is a positive integer. *See* Binomial Theorem.

binomial equation　An equation of the form $x^n - a = 0$.

binomial series　The series $(1+x)^\alpha = \sum_{n=0}^{\infty} \binom{\alpha}{n} x^n$. It converges for all $|x| < 1$.

Binomial Theorem　For any nonnegative integers b and n, $(a+b)^n = \sum_{j=0}^{n} \binom{n}{j} a^j b^{n-j}$.

birational isomorphism　A k-morphism $\varphi : G \to G'$, where G and G' are algebraic groups defined over k, that is a group isomorphism, whose inverse is a k-morphism.

birational mapping　For V and W irreducible algebraic varieties defined over k, a closed irreducible subset T of $V \times W$ where the closure of the projection $T \to V$ is V, the closure of the projection $T \to W$ is W, and $k(V) = k(T) = k(W)$. Also called *birational transformation*.

birational transformation　*See* birational mapping.

Birch-Swinnerton-Dyer conjecture　The rank of the group of rational points of an elliptic curve E is equal to the order of the 0 of $L(s, E)$ at $s = 1$. Consider the elliptic curve $E : y^2 = x^3 - ax - b$ where a and b are integers. If $E(Q) = E \cap (Q \times Q)$, by Mordell's Theorem $E(Q)$ is a finitely generated Abelian group. Let N be the conductor of E, and if $p \nmid N$, let $a_p + p$ be the number of solutions of $y^2 = (x^3 - ax - b) \pmod{p}$. The L-function of E,

$$L(s, E) = \prod_{p|N} \left(1 - \varepsilon_p p^{-s}\right)$$
$$\prod_{p \nmid N} \left(1 - a_p p^{-s} + p^{1-2s}\right)$$

where $\varepsilon_p = 0$ or ± 1.

block　A term used in reference to vector bundles, permutation groups, and representations.

blowing up　A process in algebraic geometry whereby a point in a variety is replaced by the set of lines through that point. This idea of Zariski turns a singular point of a given manifold into a smooth point. It is used decisively in Hironaka's celebrated "resolution of singularities" theorem.

blowing up　Let N be an n-dimensional compact, complex manifold ($n \geq 2$), and $p \in N$. Let $\{z = (z_i)\}$ be a local coordinate system, in a neighborhood U, centered at p and define

$$\tilde{U} = \left\{ (z, l) \in U \times P^{n-1} : z \in l \right\},$$

where P^{n-1} is regarded as a set of lines l in $\mathbf{C}^n$. Let $\pi : \tilde{U} \to U$ denote the projection $\pi(z, l) = z$. Identify $\pi^{-1}(p)$ with P^{n-1} and $\tilde{U} \setminus \pi^{-1}(p)$ with $U \setminus \{p\}$, via the map π and set

$$\tilde{N} = (N \setminus \{p\}) \cup \tilde{U}, \quad B_p(N) = \tilde{N}/\sim,$$

where $z \sim w$ if $z \in N \setminus \{p\}$ and $w = (z, l) \in \tilde{U}$. The *blowing up* of N at p is $\pi : B_p(N) \to N$.

BN-pair　A pair of subgroups (B, N) of a group G such that:
(i.) B and N generate G;
(ii.) $B \cap N = H \triangle N$; and
(iii.) the group $W = N/H$ has a set of generators R such that for any $r \in R$ and any $w \in W$
(a) $rBw \subset BwB \cap BrwB$,
(b) $rBr \neq B$.

Bochner's Theorem　A function, defined on $\mathbf{R}$, is a Fourier-Stieltjes transform if and only if it is continuous and positive definite. [A function f, defined on $\mathbf{R}$, is defined to be *positive definite* if

$$\int_{\mathbf{R}} f(y) f(x-y) dy > 0$$

for all x-values.]

Borel subalgebra　A maximal solvable subalgebra of a reductive Lie algebra defined over an algebraically closed field of characteristic 0.

Borel subgroup　A maximal solvable subgroup of a complex, connected, reductive Lie group.

Borel-Weil Theorem　If G_c is the complexification of a compact connected group G, any irreducible holomorphic representation of G_c is holomorphically induced from a one-dimensional holomorphic representation of a Borel subgroup of G_c.

boundary (1) (Topology.) The intersection of the complements of the interior and exterior of a set is called the *boundary* of the set. Or, equivalently, a set's boundary is the intersection of its closure and the closure of its complement.

(2) (Algebraic Topology.) A boundary in a differential group C (an Abelian group with homomorphism $\partial : C \to C$ satisfying $\partial\partial = 0$) is an element in the range of ∂.

boundary group The group Im∂, which is a subgroup of a differential group C consisting of the image of the *boundary operator* $\partial : C \to C$.

boundary operator A homomorphism $\partial : C \to C$ of an Abelian group C that satisfies $\partial\partial = 0$. Used in the field of algebraic topology. *See also* boundary, boundary group.

bounded homogeneous domain A bounded domain with a transitive group of automorphisms. In more detail, a *domain* is a connected open subset of complex N space $\mathbf{C}^N$. A domain is *homogeneous* if it has a transitive group of analytic (holomorphic) automorphisms. This means that any pair of points z and w can be interchanged, i.e., $\phi(z) = w$, by an invertible analytic map ϕ carrying the domain onto itself. For example, the unit ball in complex N space, $\{z = (z_1, \ldots, z_N) : |z_1|^2 + \cdots + |z_N|^2 < 1\}$, is homogeneous. A domain is *bounded* if it is contained in a ball of finite radius. A *bounded homogeneous domain* is a bounded domain which is also homogeneous. Thus, the unit ball in $\mathbf{C}^N$ is a bounded homogeneous domain. There are many others. *See also* Siegel domain, Siegel domain of the second kind.

bounded matrix A continuous linear map $K : \ell^2(\mathbf{N}) \otimes \ell^2(\mathbf{N}) \to \ell^1(\mathbf{N})$ where $\mathbf{N}$ is the set of natural numbers.

bounded torsion group A torsion group T where there is an integer $n \geq 0$ such that $t^n = 1$ for all $t \in T$.

bounded variation Let $I = [a, b] \subseteq \mathbf{R}$ be a closed interval and $f : I \to \mathbf{R}$ a function. Suppose there is a constant $C > 0$ such that, for any partition $a = a_0 \leq a_1 \leq \cdots \leq a_k = b$ it holds that

$$\sum_{j=1}^{k} |f(a_j) - f(a_{j-1})| \leq C.$$

Then f is said to be of *bounded variation* on the interval I.

bracket product If a and b are elements of a ring R, then the *bracket product* is defined as $[a, b] = ab - ba$. The bracket product satisfies the distributive law.

branch and bound integers programming At step j of *branch and bound integers programming* for a problem list P a subproblem P_j is selected and a lower bound is estimated for its optimal objective function. If the lower bound is worse than that calculated at the previous step, then P_j is discarded; otherwise P_j is separated into two subproblems (the branch step) and the process is repeated until P is empty.

branch divisor The divisor $\sum i_X X$, where i_X is the differential index at a point X on a nonsingular curve.

Brauer group The Abelian group formed by the tensor multiplication of algebras on the set of equivalence classes of finite dimensional central simple algebras.

Brauer's Theorem Let G be a finite group and let χ be any character of G. Then χ can be written as $\sum n_k \chi_{\psi_k}$, where n_k is an integer and each χ_{ψ_k} is an induced character from a certain linear character ψ_k of an elementary subgroup of G.

Bravais class An arithmetic crystal class determined by $(L, B(L))$, where L is a lattice and $B(L)$ is the Bravis group of L. *See* Bravais group.

Bravais group The group of all orthogonal transformations that leave invariant a given lattice L.

Bravais lattice   A representative of a Bravais type. *See* Bravais type.

Bravais type   An equivalence class of arithmetically equivalent lattices. *See* arithmetical equivalence.

Brill-Noether number   The quantity $g - (k + 1)(g - k + m)$, where g is the genus of a nonsingular curve C and k and m are positive integers with $k \leq g$. This quantity acts as a lower bound for the dimension of the subscheme $\{\varphi(D) : l(D) > m, \deg D = k\}$ of the Jacobian variety of C, where φ is the canonical function from C to this variety.

Bruhat decomposition   A decomposition of a connected semisimple algebraic group G, as a union of double cosets of a Borel subgroup B, with respect to representatives chosen from the classes that comprise the Weyl group W of G. For each $w \in W$, let g_w be a representative in the normalizer $N(B \cap B^-)$ in G of the maximal torus $B \cap B^-$ formed from B and its opposite subgroup B^-. Then G is the disjoint union of the double cosets $B g_w B$ as w ranges over W.

building   A thick chamber complex C with a system S of Coxeter subcomplexes (called the apartments of C) such that every two simplices of C belong to an apartment and if A, B are in S, then there exists an isomorphism of A onto B that fixes $A \cap B$ elementwise.

building of Euclidian type   A building is of Euclidean type if it could be used like a simplical decomposition of a Euclidean space. *See* building.

building of spherical type   A building that has finitely many chambers. *See* building.

Burnside Conjecture   A finite group of odd order is solvable.

Burnside problem   (1) The original Burnside problem can be stated as follows: If every element of a group G is of finite order and G is finitely generated, then is G a finite group? Golod has shown that the answer for p-groups is negative.

(2) Another form of the Burnside problem is: If a group G is finitely generated and the orders of the elements of G divide an integer n, then is G finite?

C

C_i-field Let F be a field and let i, j be integers such that $i \geq 0$ and $j \geq 1$. Also, let P be a homogeneous polynomial of m variables of degree j with coefficients in F. If the equation $P = 0$ has a solution $(s_1, s_2, \ldots, s_m) \neq (0, 0, \ldots, 0)$ in F for any P such that $m > j^i$, then F is called a $C_i(j)$ field. If, for any $j \geq 1$, F is a $C_i(j)$ field, then F is called a C_i-field.

Calkin algebra Let H be a separable infinite dimensional Hilbert space, $B(H)$ the algebra of bounded linear operators on H, and $I(H)$ be the ideal of H consisting of all compact operators. Then, the quotient C^*-algebra $B(H)/I(H)$ is called the *Calkin algebra.*

Campbell-Hausdorff formula A long formula for computation of $z = \ln(e^x e^y)$ in the algebra of formal power series in x and y with the assumption that x and y are associative but not commutative. It was first studied by Campbell. Then Hausdorff showed that z can be written in terms of the commutators of x and y.

cancellation Let x, y, and z be elements of a set S, with a binary operation $*$. The acts of eliminating z in $x * z = y * z$ or $z * x = z * y$ to obtain $x = y$ is called *cancellation.*

cancellation law An axiom that allows cancellation.

canonical class A specified divisor class of an algebraic curve.

canonical cohomology class The 2-cocycle in

$$H^2\left(\mathrm{Gal}(K/k), I_K\right) \cong \mathbf{Z}/n\mathbf{Z}$$

in the Galois cohomology of the Galois extension K/k of degree n with respect to the idéle class group I_K that corresponds to 1 in $\mathbf{Z}/n\mathbf{Z}$ under the above isomorphism.

canonical coordinates of the first kind For each basis $B_1, \ldots, B_n$ of a Lie algebra L of the Lie group G, there exists a positive real number r with the property that $\{\exp(\sum b_i B_i) : |b_i| < r \ (i = 1, \ldots, n)\}$ is an open neighborhood of the identity element in G such that $\exp(\sum b_i B_i) \to (b_1, \ldots, b_n)(|b_i| < r, i = 1, \ldots, n)$ is a local coordinate system. These local coordinates are called the canonical coordinates of the first kind associated with the basis (B_i) of this Lie algebra L.

canonical coordinates of the second kind For each basis $B_1, \ldots, B_n$ of a Lie algebra L of the Lie group G, we have a local coordinate system $\prod \exp(b_i B_i) \to (b_1, \ldots, b_n)$ $(i = 1, 2, \ldots, n)$ in a neighborhood of the identity element in G. These $b_1, \ldots, b_n$ are called the canonical coordinates of the second kind associated with the basis (B_i) of this Lie algebra L.

canonical divisor Any one of the linearly equivalent divisors in the sheaf of relative differentials of a (nonsingular) curve.

canonical function A rational mapping $\phi : X \to J$, from a nonsingular curve X to its Jacobian variety J, defined by $\phi(P) = \Phi(P - P_0)$, where P is a generic point of X and P_0 is a fixed rational point, $\Phi : G_0(X)/G_l(X) \to J$ is the associated isomorphism, $G(X)$ is the group of divisors, $G_0(X)$ is the subgroup of divisors of degree 0 and $G_l(X)$ the subgroup of divisors of functions. Such a ϕ is determined uniquely by Φ up to translation of J.

canonical homology basis A one-dimensional homology basis $\{\beta_i, \beta_{k+i}\}_{i=1}^{k}$ such that $(\beta_i, \beta_j) = (\beta_{k+i}, \beta_{k+j}) = 0$, $(\beta_i, \beta_{k+i}) = 1$, and $(\beta_i, \beta_{k+j}) = 0 \, (i \neq j), (i, j = 1, 2, \ldots, k)$.

canonical homomorphism **(1)** Let R be a commutative ring with identity and let L, M be algebras over R. Then, the tensor product $L \otimes_R M$ of R-modules is an algebra over R. The mappings $l \to l \otimes 1 \ (l \in L)$ and $m \to m \otimes 1 \ (m \in M)$ give algebra homomorphisms $L \to L \otimes_R M$ and $M \to L \otimes_R M$. Each one of these homomorphisms is called a *canonical homomorphism* (on tensor products of algebras).

(2) Let the ring $R = \prod_{i \in I} R_i$ be the direct product of rings R_i. The mapping $\phi_i : R \to R_i$ that assigns to each element r of R its ith component r_i is called a *canonical homomorphism* (of direct product of rings).

canonical injection   For a subgroup H of a group G, the injective homomorphism $\theta : H \to G$, defined by $\theta(h) = h$ for all $h \in H$. (θ is also called the *natural injection*.)

canonically bounded complex   Let $F^0(C)$ and $F^{m+1}(C)$ (m an integer) be subcomplexes of a complex C such that $F^0(C) = C$, and $F^{m+1}(C^m) = 0$, then the complex C is called a *canonically bounded complex.*

canonically polarized Jacobian variety   A pair, (J, P), where J is a Jacobian variety whose polarization P is determined by a theta divisor.

canonical projection   Let $S/\sim$ denote the set of equivalence classes of a set S, with respect to an equivalence relation $\sim$. The mapping $\mu : S \to S/\sim$ that carries $s \in S$ to the equivalence class of s is called the *canonical projection* (or *quotient map*).

canonical surjection   **(1)** Let H be a normal subgroup of a group G. For the factor group G/H, the surjective homomorphism $\theta : G \to G/H$ such that $g \in \theta(g)$, for all $g \in G$, is called the *canonical surjection* (or *natural surjection*) to the factor group.

(2) Let $G = G_1 \times G_2 \times \ldots \times G_n$ be the direct product of the groups $G_1, G_2, \ldots, G_n$. The mapping $(g_1, g_2, \ldots, g_n) \to g_i$ ($i = 1, 2, \ldots, n$) from G to G_i is a surjective homomorphism, called the *canonical surjection* on the direct product of groups.

capacity of prime ideal   Let A be a separable algebra of finite degree over the field of quotients of a Dedekind domain. Let P be a prime ideal of A and let M be a fixed maximal order of A. Then, M/P is the matrix algebra of degree d over a division algebra. This d is called the *capacity* of the prime ideal P.

cap product   **(1)** In a lattice or Boolean algebra, the fundamental operation $a \wedge b$, also called the meet or product, of elements a and b.

(2) In cohomology theory, where $H_r(X, Y; G)$ and $H^s(X, Y; G)$ are the homology and cohomology groups of the pair (X, Y) with coefficients in the group G, the operation that associates to the pair (f, g), $f \in H_{r+s}(X, Y \cup Z; G_1)$, $g \in H^r(X, Y; G_2)$ the element $f \cup g \in H_s(X, Y \cup Z; G_3)$ determined by the composition

$$H_{r+s}(X, Y \cup Z; G_1)$$
$$\longrightarrow H_{r+s}((X, Y) \times (X, Z); G_1)$$
$$\longrightarrow \mathrm{Hom}\left(H^r(X, Y; G_2), H_s(X, Z; G_3)\right)$$

where the first map is induced by the diagonal map $\Delta : (X, Y \cup Z) \to (X, Y) \times (X, Z)$.

Cardano's formula   A formula for the roots of the general cubic equation over the complex numbers. Given the cubic equation $ax^3 + bx^2 + cx + d = 0$, let $A = 9abc - 2b^3 - 27a^2d$ and $B = b^2 - 3ac$. Also, let y_1 and y_2 be solutions of the quadratic equation $Y^2 - AY + B^3 = 0$. If ω is any cube root of 1, then $(-b + \omega \sqrt[3]{y_1} + \omega^2 \sqrt[3]{y_2})/3a$ is a root of the original cubic equation.

cardinality   A measure of the size of, or number of elements in, a set. Two sets S and T are said to have the same *cardinality* if there is a function $f : S \to T$ that is one-to-one and onto. *See also* countable, uncountable.

Cartan integer   Let R be the root system of a Lie algebra L and let $F = \{x_1, x_2, \ldots, x_n\}$ be a fundamental root system of R. Each of the n^2 integers $x_{ij} = -2(x_i, x_j)/(x_j, x_j)$ ($1 \le i$, $j \le n$) is called a *Cartan integer* of L, relative to the fundamental root system F.

Cartan invariants   Let G be a finite group and let n be the number of p-regular classes of G. Then, there are exactly n nonsimilar, absolutely irreducible, modular representations, $M_1, M_2, \ldots, M_n$, of G. Also, there are n nonsimilar, indecomposable components, denoted by $R_1, R_2, \ldots, R_n$, of the regular representation R of G. These can be numbered in a such a way that M_n appears in R_n as both its top and bottom component. If the degree of M_n is m_n and the degree of R_n is r_n, then R_n appears m_n times in R and M_n appears r_n times in R. The multiplicities μ_{nt} of M_t in R_n are called the *Cartan invariants* of G.

Cartan involution   Let G be a connected semisimple Lie group with finite center and let M be a maximal compact subgroup of G. Then there exists a unique involutive automorphism of G whose fixed point set coincides with M. This automorphism is called a *Cartan involution* of the Lie group G.

Cartan-Mal'tsev-Iwasawa Theorem   Let M be a maximal compact subgroup of a connected Lie group G. Then M is also connected and G is homeomorphic to the direct product of M with a Euclidean space $\mathbf{R}^n$.

Cartan's criterion of semisimplicity   A Lie algebra L is semisimple if and only if the Killing form K of L is nondegenerate.

Cartan's criterion of solvability   Let $gl(n, K)$ be the general linear Lie algebra of degree n over a field K and let L be a subalgebra of $gl(n, K)$. Then L is solvable if and only if $tr(AB) = 0$ ($tr(AB) = $ trace of AB), for every $A \in L$ and $B \in [L, L]$.

Cartan's Theorem   (1) *E. Cartan's Theorem.* Let W_1 and W_2 be the highest weights of irreducible representations w_1, w_2 of the Lie algebra L, respectively. Then w_1 is equivalent to w_2 if and only if $W_1 = W_2$.

(2) *H. Cartan's Theorem.* The sheaf of ideals defined by an analytic subset of a complex manifold is coherent.

Cartan subalgebra   A subalgebra A of a Lie algebra L over a field K, such that A is nilpotent and the normalizer of A in L is A itself.

Cartan subgroup   A subgroup H of a group G such that H is a maximal nilpotent subgroup of G and, for every subgroup K of H of finite index in H, the normalizer of K in G is also of finite index in K.

Cartan-Weyl Theorem   A theorem that assists in the characterization of irreducible representations of complex semisimple Lie algebras. Let G be a complex semisimple Lie algebra, $\mathcal{H}$ a Cartan subalgebra, Σ the root system of G relative to $\mathcal{H}$, $\alpha = \sum_{\sigma \in \Sigma} r_\sigma \sigma, r_\sigma \in \mathbf{R}$, a complex-valued linear functional on $\mathcal{H}$, and $\rho : G \to GL_n(\mathbf{C})$ a representation of G. The functional α is a weight of the representation if the space of vectors $v \in \mathbf{C}^n$ that satisfy $\rho(h)v = \alpha(h)v$ for all $h \in \mathcal{H}$ is nontrivial; $\mathbf{C}^n$ decomposes as a direct sum of such spaces associated with weights $\alpha_1, \ldots, \alpha_k$. If we place a lexicographic linear order $\leq$ on the set of functionals α, the Cartan-Weyl Theorem asserts that there exists an irreducible representation ρ of G having α as its highest weight (with respect to the order $\leq$) if and only if $\frac{2[\alpha,\sigma]}{[\sigma,\sigma]}$ is an integer for every $\sigma \in \Sigma$, and $w(\alpha) \leq \alpha$ for every permutation w in the Weyl group of G relative to $\mathcal{H}$.

Carter subgroup   Any finite solvable group contains a self-normalizing, nilpotent subgroup, called a *Carter subgroup*.

Cartesian product   If X and Y are sets, then the *Cartesian product* of X and Y, denoted $X \times Y$, is the set of all ordered pairs (x, y) with $x \in X$ and $y \in Y$.

Cartier divisor   A divisor which is linearly equivalent to the divisor 0 on a neighborhood of each point of an irreducible variety V.

Casimir element   Let $\beta_1, \ldots, \beta_n$ be a basis of the semisimple Lie algebra L. Using the Killing form K of L, let $m_{ij} = K(\beta_i, \beta_j)$. Also, let m^{ij} represent the inverse of the matrix (m_{ij}) and let c be an element of the quotient associative algebra $Q(L)$, defined by $c = \sum m^{ij} \beta_i \beta_j$. This element c is called the *Casimir element* of the semisimple Lie algebra L.

Casorati's determinant   The $n \times n$ determinant

$$D(c_1(x), \ldots, c_n(x)) =$$

$$\begin{vmatrix} c_1(x) & c_2(x) & \cdots & c_n(x) \\ c_1(x+1) & c_2(x+1) & \cdots & c_n(x+1) \\ & & \cdots & \\ c_1(x+n-1) & c_2(x+n-1) & \cdots & c_n(x+n-1) \end{vmatrix},$$

where $c_1(x), \ldots, c_n(x)$ are n solutions of the homogeneous linear difference equation

$$\sum_{k=0}^{n} p_k(x) y(x+k) = 0 .$$

casting out nines   A method of checking base-ten multiplications and divisions. *See* excess of nines.

casus irreducibilis   If the cubic equation $ax^3 + bx^2 + cx + d = 0$ is irreducible over the extension $Q(a, b, c, d)$ of the rational number field Q, and if all the roots are real, then it is still impossible to find the roots of this cubic equation, by only rational operations with real radicals, even if the roots of the cubic equation are real.

category   A graph equipped with a notion of identity and of composition satisfying certain standard domain and range properties.

Cauchy inequality   The inequality

$$\left(\sum_{i=1}^{n} a_i b_i \right)^2 \leq \sum_{i=1}^{n} a_i^2 \sum_{i=1}^{n} b_i^2 ,$$

for real numbers $a_1, \ldots, a_n, b_1, \ldots, b_n$. Equality holds if and only if $a_i = c b_i$, where c is a constant.

Cauchy problem   Given an nth order partial differential equation (PDE) in z with two independent variables, x and y, and a curve Γ in the xy-plane, a *Cauchy problem* for the PDE consists of finding a solution $z = \phi(x, y)$ which meets prescribed conditions

$$\frac{\partial^{j+k} z}{\partial x^j \partial y^k} = f_{jk}$$

$j + k \leq n - 1$, $j, k = 0, 1, \ldots, n - 1$ on Γ. Cauchy problems can be defined for systems of partial differential equations and for ordinary differential equations (then they are called *initial value problems*).

Cauchy product   The *Cauchy product* of two series $\sum_{n=1}^{\infty} a_n$ and $\sum_{n=1}^{\infty} b_n$ is $\sum_{n=1}^{\infty} c_n$ where

$$c_n = a_1 b_n + a_2 b_{n-1} + \cdots + a_n b_1 .$$

If $A = \sum a_n$ and $B = \sum b_n$, then $\sum c_n = AB$ (if all three series converge). The Cauchy product series converges if $\sum a_n$ and $\sum b_n$ converge and at least one of them converges absolutely (*Merten's Theorem*).

Cauchy sequence   (**1**) A sequence of real numbers, $\{r_n\}$, satisfying the following condition. For any $\epsilon > 0$ there exists a positive integer N such that $|r_m - r_n| < \epsilon$, for all $m, n > N$.

(**2**) A sequence $\{p_n\}$ of points in a metric space (X, ρ), satisfying the following condition:

$$\rho(p_n, p_m) \to 0 \text{ as } n, m \to \infty .$$

Cauchy sequences are also called *fundamental sequences*.

Cauchy transform   The *Cauchy transform*, $\hat{\mu}$, of a measure μ, is defined by $\hat{\mu}(\zeta) = \int (z - \zeta)^{-1} d\mu(z)$. If K is a compact planar set with connected complement and $A(K)$ is the algebra of complex functions analytic on the interior of K, then the *Cauchy transform* is used to show that every element of $A(K)$ can be uniformly approximated on K by polynomials.

Cayley algebra   Let F be a field of characteristic zero and let Q be a quaternion algebra over F. A *general Cayley algebra* is a two-dimensional Q-module $Q + Qe$ with the multiplication $(x + ye)(z + ue) = (xz + vu'y) + (xu + yz')e$, where $x, y, z, u \in Q$, $v \in F$ and z', u' are the conjugate quarternions of z and u, respectively. A *Cayley algebra* is the special case of a general Cayley algebra where Q is the quaternian field, F is the real number field, and $v = -1$.

Cayley-Hamilton Theorem   *See* Hamilton-Cayley Theorem.

Cayley number   The elements of a general Cayley algebra. *See* Cayley algebra.

Cayley projective plane   Let H be the set of all 3×3 Hermitian matrices M over the Cayley algebra such that $M^2 = M$ and $\text{tr } M = 1$. The set H, with the structure of a projective plane, is called the Cayley projective plane. *See* Cayley algebra.

Cayley's Theorem   Every group is isomorphic to a group of permutations.

Cayley transformation   The mapping between $n \times n$ matrices N and M, given by $M = (I - N)(I + N)^{-1}$, which acts as its own inverse. The *Cayley transformation* demonstrates a one-to-one correspondence between the real alternating matrices N and proper orthogonal matrices M with eigenvalues different from -1.

CCR algebra   A C*-algebra A, which is mapped to the algebra of compact operators under any irreducible $*$-representation. Also called *liminal C*-algebra.*

center   (1) *Center of symmetry in Euclidean geometry.* The midpoint of a line, center of a triangle, circle, ellipse, regular polygon, sphere, ellipsoid, etc.
     (2) *Center of a group, ring, or Lie algebra X.* The set of all elements of X that commute with every element of X.
     (3) *Center of a lattice L.* The set of all central elements of L.

central extension   Let G, H, and K be groups such that G is an extension of K by H. If H is contained in the center of G, then G is called a *central extension* of H.

centralizer   Let X be a group (or a ring) and let $S \subset X$. The set of all elements of X that commute with every element of S is called the *centralizer* of S.

central separable algebra   An R-algebra which is central and separable. Here a central R-algebra A which is projective as a two-sided A-module, where R is a commutative ring.

central simple algebra   A simple algebra A over a field F, such that the center of A coincides with F. (Also called *normal simple algebra.*)

chain complex   Let R be a ring with identity and let C be a unitary R-module. By a *chain complex* (C, α) over R we mean a graded R-module $C = \sum_n C_n$ together with an R-homomorphism $\alpha: C \to C$ of degree -1, where $\alpha \circ \alpha = 0$.

chain equivalent   Let C_1 and C_2 be chain complexes. If there are chain mappings $\alpha: C_1 \to C_2$ and $\beta: C_2 \to C_1$ such that $\alpha \circ \beta = 1_{C_2}$ and $\beta \circ \alpha = 1_{C_1}$, then we say that C_1 is *chain equivalent* to C_2. *See* chain complex, chain mapping.

chain homotopy   Let C_1 and C_2 be chain complexes. Let $\alpha, \beta: C_1 \to C_2$ be two chain mappings, and let R be a ring with identity. If there is an R-homomorphism $\gamma : C_1 \to C_2$ of degree 1, such that $\alpha - \beta = \gamma \circ \alpha' + \beta' \circ \gamma$, where (C_1, α') and (C_2, β') are chain complexes over R. Then γ is called a *chain homotopy* of α to β. *See* chain complex, chain mapping.

chain mapping   Let (C_1, α) and (C_2, β) be chain complexes over a ring R with identity. An R-homomorphism $\gamma : C_1 \to C_2$ of degree 0 that satisfies $\beta \circ \gamma = \gamma \circ \alpha$ is called a *chain mapping* of C_1 to C_2. *See* chain complex.

chain subcomplex   Let R be a ring with identity and let (C, α) be a chain complex over R. If $H = \sum_n H_n$ is a homogeneous R-submodule of C such that $\alpha(H) \subset H$, then H is called a *chain subcomplex* of C. *See* chain complex.

Chain Theorem   Let A, B, and C be algebraic number fields such that $C \subset B \subset A$ and let $\Delta_{A/C}$, $\Delta_{A/B}$, and $\Delta_{B/C}$ denote the relative difference of A over C, A over B, and B over C, respectively. Then $\Delta_{A/C} = \Delta_{A/B}\Delta_{B/C}$. *See* different.

chamber   In a finite dimensional real affine space A, any connected component of the complement of a locally finite union of hyperplanes. *See* locally finite.

chamber complex   A complex with the property that every element is contained in a chamber and, for two given chambers C, C', there exists a finite sequence of chambers $C = C_0, C_1, \ldots, C_r = C'$ in such a way that $\operatorname{codim}_{C_{k-1}}(C_k \cap C_{k-1}) = \operatorname{codim}_{C_k}(C_k \cap C_{k-1}) \leq 1$, for $k = 1, 2, \ldots, r$. *See* chamber.

character   A *character* $\mathcal{X}$ of an Abelian group G is a function that assigns to each element x of G a complex number $\mathcal{X}(x)$ of absolute value 1 such that $\mathcal{X}(xy) = \mathcal{X}(x)\mathcal{X}(y)$ for all x and y in

G. If *G* is a topological Abelian group, then $\mathcal{X}$ must be continuous.

character group   The set of all characters of a group *G*, with addition defined by $(\mathcal{X}_1 + \mathcal{X}_2)(x) = \mathcal{X}_1(x) \cdot \mathcal{X}_2(x)$. The character group is Abelian and is sometimes called the *dual group* of *G*. *See* character.

characteristic   Let *F* be a field with identity 1. If there is a natural number *c* such that $c1 = 1 + \cdots + 1$ (*c* 1s) $= 0$, then the smallest such *c* is a prime number *p*, called the characteristic of the field *F*. If there is no natural number *c* such that $c1 = 0$, then we say that the characteristic of the field *F* is 0.

characteristic class   (1) Of an *R*-module extension $0 \rightarrow N \rightarrow X \rightarrow M \rightarrow 0$, the element $\Delta^0(\mathrm{id}_N)$ in the extension module $\mathrm{Ext}^1_R(M, N)$, where id_N is the identity map on *N* in $\mathrm{Hom}_R(N, N) \cong \mathrm{Ext}^0_R(N, N)$ and Δ^0 is the connecting homomorphism $\mathrm{Ext}^0_R(N, N) \rightarrow \mathrm{Ext}^1_R(M, N)$ obtained from the extension sequence. *See* connecting homomorphism.

(2) Of a vector bundle over base space *X*, any of a number of constructions of a particular cohomology class of *X*, chosen so that the bundle induced by a map $f : Y \rightarrow X$ is the image of the characteristic class of the bundle over *X* under the associated cohomological map $f^* : H^*(X) \rightarrow H^*(Y)$. *See* Chern class, Euler class, Pontrjagin class, Stiefel-Whitney class, Thom class.

characteristic equation   (1) If we substitute $y = e^{\lambda x}$ in the general *n*th order linear differential equation

$$y^{(n)}(x) + a_{n-1}y^{(n-1)}(x) + \ldots$$
$$+ a_1 y'(x) + a_0 y(x) = 0$$

with constant coefficients a_i ($i = n - 1, \ldots, 0$) and then divide by $e^{\lambda x}$, we obtain

$$\lambda^n + a_{n-1}\lambda^{n-1} + \cdots + a_1\lambda + a_0 = 0,$$

which is called the characteristic equation associated with the given differential equation.

(2) If we substitute $y_n = \lambda^n$ in the general *k*th order difference equation

$$y_n + a_{n-1}y_{n-1} + \cdots + a_{n-k}y_{n-k} = 0$$

with constant coefficients a_i ($i = n - 1, \ldots, n - k$) and then divide by λ^{n-k}, we obtain

$$\lambda^k + a_{n-1}\lambda^{k-1} + \cdots + a_{n-k+1}\lambda + a_{n-k} = 0,$$

which is again called the characteristic equation associated with this given difference equation.

(3) The above two definitions can be extended for a system of linear differential (difference) equations.

(4) Moreover, if $M = (m_{ij})$ is a square matrix of degree *n* over a field *F*, then the algebraic equation $|\lambda I - M| = 0$ is also called the characteristic equation of *M*.

characteristic linear system   Let *S* be a nonsingular surface and let *A* be an irreducible algebraic family of positive divisors of dimension *d* on *S* such that a generic member M of *A* is an irreducible non-singular curve. Then, the characteristic set forms a $(d - 1)$-dimensional linear system and contains $\mathrm{Tr}_M |M|$ (the trace of $|M|$ on *M*) as a subfamily. This linear system is called the characteristic linear system of *A*.

characteristic multiplier   Let *Y(t)* be a fundamental matrix for the differential equation

$$y' = A(t)y. \qquad (*)$$

Let ω be a period for the matrix $A(t)$. Suppose that *H* is a constant matrix that satisfies

$$Y(t + \omega) = Y(t)H, \quad t \in (-\infty, \infty).$$

Then an eigenvalue μ for *H* of index *k* and multiplicity *m* is called a characteristic multiplier for (*), or for the periodic matrix $A(t)$, of index *k* and multiplicity *m*.

characteristic multiplier   Let *Y(t)* be a fundamental matrix for the differential equation

$$y' = A(t)y. \qquad (*)$$

Let ω be a period for the matrix $A(t)$. Suppose that *H* is a constant matrix that satisfies

$$Y(t + \omega) = Y(t)H, \quad t \in (-\infty, \infty).$$

Then an eigenvalue μ for *H* of index *k* and multiplicity *m* is called a *characteristic multiplier*, of index *k* and multiplicity *m*, for (*), or for the periodic matrix $A(t)$.

characteristic of logarithm   The integral part of the common logarithm.

characteristic polynomial   The polynomial on the left side of a characteristic equation. *See* characteristic equation.

characteristic series   Let G be a group. If we take the group Aut(G) (the group of automorphisms of G) as an operator domain of G, then a composition series is called a *characteristic series*. *See* composition series.

characteristic set   A one-dimensional set of positive divisors D of a nonsingular curve of dimension n so that, with respect to one such generic divisor D_0 of the curve, the degree of the specialization of the intersection $D \cdot D_0$ over the specialization of D over D_0 is a divisor of degree equal to that of $D \cdot D_0$.

character module   Let G be an algebraic group, with the sum of two characters $\mathcal{X}_1$ and $\mathcal{X}_2$ of G defined as $(\mathcal{X}_1 + \mathcal{X}_2)(x) = \mathcal{X}_1(x) \cdot \mathcal{X}_2(x)$, for all $x \in G$. The set of all characters of G forms an additive group, called the character module of G. *See* character of group, algebraic group.

character of a linear representation   For the representation $\rho : A \rightarrow \mathrm{GL}_n(k)$ of the algebra A over a field k, the function χ_ρ on A given by $\chi_\rho(a) = \mathrm{tr}(\rho(a))$.

character of group   A rational homomorphism α of an algebraic group G into $GL(1)$, where $GL(1)$ is a one-dimensional connected algebraic group over the prime field. *See* algebraic group.

character system   For the quadratic field k, with discriminant d and ideal class group $I \cong F/H$ (F the group of fractional ideals and H the subgroup of principal ideals generated by positive elements), a collection $\left\{ \chi_p(N(\mathcal{A})) \right\}_{(p|d)}$ of numbers, indexed by the prime factors of d, in which χ_p is the Legendre symbol mod p and $\mathcal{A}$ is any representative ideal in its ideal class mod H. The character system is independent of the choice of representative and uniquely determines each class in I.

Chebotarev Density Theorem   Let F be an algebraic number field with a subfield f, F/f be a Galois extension, C be a conjugate class of the Galois group G of F/f, and $I(C)$ be the set of all prime ideals P of k such that the Frobenius automorphism of each prime factor F_i of P in F is in C. Then the density of $I(C)$ is $|C|/|G|$.

Chern class   The ith *Chern class* is an element of $H^{2i}(M; \mathbf{R})$, where M is a complex manifold. The Chern class measures certain properties of vector bundles over M. It is used in the Riemann-Roch Theorem.

Chevalley complexification   Let G be a compact Lie group, $r(G)$ the representative ring of G, A the group of all automorphisms of $r(G)$, and G' the centralizer of a subgroup of A in A. If G' is the closure of G relative to the Zariski topology of G', then G' is called the *Chevalley complexification* of G.

Chevalley decomposition   Let G be an algebraic group, defined over a field F and R_u the unipotent radical of G. If F is of characteristic zero, then there exists a reductive, closed subgroup C of G such that G can be written as a semidirect product of C and R_u. *See* algebraic group.

Chevalley group   Let F be a field, f an element of F, L_F a Lie algebra over F, B a basis of L_F over F and $t_\theta(f)$ the linear transformation of L_F with respect to B, where θ ranges over the root system of L_F. Then, the group generated by the $t_\theta(f)$, for each root θ and each element f, is called the *Chevalley group* of type over F.

Chevalley's canonical basis   Of a complex, semisimple Lie algebra $\mathcal{G}$ with Cartan subalgebra $\mathcal{H}$ and root system Σ, a basis for $\mathcal{G}$ consisting of a basis $\{H_1, \ldots, H_s\}$ of $\mathcal{H}$ and, for each root $\sigma \in \Sigma$, a basis $\{X_\sigma\}$ of its root subspace $\mathcal{G}_\sigma$ that satisfy: (i.) $\sigma(H_i)$ is an integer for every $\sigma \in \Sigma$ and each H_i; (ii.) $\beta(X_\sigma, X_{-\sigma}) = \frac{2}{(\sigma,\sigma)}$ for every $\sigma \in \Sigma$, where (,) represents the inner product on the roots induced by the Killing form β on $\mathcal{G}$; (iii.) if σ, τ, and $\sigma + \tau$ are all roots and $[X_\sigma, X_\tau] = n_{\sigma,\tau} X_{\sigma+\tau}$, then the numbers $n_{\sigma,\tau}$ are integers that satisfy $n_{-\sigma,-\tau} = -n_{\sigma,\tau}$.

Chevalley's Theorem   Let G be a connected algebraic group, defined over a field F, and let N be a (F-closed) largest, linear, connected, closed, normal subgroup of G. If C is a closed, normal subgroup of G, then the factor group G/C is complete if and only if $N \subset C$.

Cholesky method of factorization   A method of factoring a positive definite matrix A as a product $A = LL^T$ where L is a lower triangular matrix. Then the solution x of $Ax = b$ is found by solving $Ly = b$, $L^T x = y$.

Choquet boundary   Let X be a compact Hausdorff space and let A be a function algebra on X. The *Choquet boundary* is $c(A) = \{x \in X : \text{the evaluation at } x \text{ has a unique representing measure}\}$.

Chow coordinates   Of a projective variety X, the coefficients of the associated form of the variety, viewed as homogeneous coordinates of points on X. *See* associated form.

Chow ring   Of a nonsingular, irreducible, projective variety X, the graded ring whose objects are rational equivalence classes of cycles on X, with addition given by addition of cycles and multiplication induced by the diagonal map $\Delta : X \to X \times X$. The ring is graded by codimension of cycles.

Chow variety   Let V be a projective variety. The set of Chow coordinates of positive cycles that are contained in V is a projective variety called a *Chow variety*.

circulant   *See* cyclic determinant.

circular units   The collection of units of the form $\frac{1-\zeta^s}{1-\zeta^t}$, where ζ is a p^nth root of unity, p is a prime, and $s \not\equiv t \pmod{p}$ (and $p \nmid s, t$).

class   (**1**) (*Algebra.*) A synonym of set that is used when the members are closely related, like an equivalence class or the class of residues modulo m.

(**2**) (*Logic.*) A generalization of set, including objects that are "too big" to be sets. Consideration of classes allows one to avoid such difficulties as Russell's paradox, concerning the set of all sets that do not belong to themselves.

class field   Let F be an algebraic number field and E be a Galois extension of F. Then, E is said to be a *class field* over F, for the ideal group $I(G)$, if the following condition is met: a prime ideal P of F of absolute degree 1 which is relatively prime to G is decomposed in E as the product of prime ideals of E of absolute degree 1 if and only if P is in $I(G)$.

class field theory   A theory created by E. Artin and others to determine whether certain primes are represented by the principal form.

class field tower problem   Let F be a given algebraic number field, and let $F = F_0 \subset F_1 \subset F_2 \subset \cdots$ be a sequence of fields such that F_n is the absolute class field over F_{n-1}, and F_∞ is the union of all F_n. Now we ask, is F_∞ a finite extension of F? The answer is positive if and only if F_k is of class number 1 for some k. *See* absolute class field.

class formation   An axiomatic structure for class field theory, developed by Artin and Tate. A *class formation* consists of

(**1**) a group G, the *Galois group of the formation,* together with a family $\{G_K : K \in \mathbf{Z}\}$ of subgroups of G indexed by a collection Σ of fields K so that

(i.) each G_K has finite index in G;

(ii.) if H is a subgroup of G containing some G_K, then $H = G_{K'}$ for some K';

(iii.) the family $\{G_K\}$ is closed under intersection and conjugation;

(iv.) $\bigcap_\Sigma G_K$ is the trivial subgroup of G;

(**2**) a G-module A, the *formation module,* such that A is the union of its submodules $A^{(G_K)}$ that are fixed by G_K;

(**3**) cohomology groups $H^r(L/K)$, defined as $H^r(G_K/G_L, A^{(G_K)})$, for which $H^1(L/K) = 0$ whenever G_L is normal in G_K;

(**4**) for each field K, there is an isomorphism $A \mapsto \mathrm{inv}_K A$ of the *Brauer group* $H^2(* /K)$ into $\mathbf{Q}/\mathbf{Z}$ such that (i.) if G_L is normal in G_K of index n,

$$\operatorname{inv}_K H^2(L/K) = \left(\frac{1}{n}\mathbf{Z}\right)/\mathbf{Z}$$

and

(ii.) even when G_L is not normal in G_K,

$$\operatorname{inv}_L \circ \operatorname{res}_{K,L} = n \operatorname{inv}_K$$

where $\operatorname{res}_{K,L}$ is the natural restriction map $H^2(*/K) \to H^2(*/E)$.

classical compact real simple Lie algebra
A compact real simple Lie algebra of the type A_n, B_n, C_n, or D_n, where A_n, B_n, C_n, and D_n are the Lie algebras of the compact Lie groups $SU(n + 1)$, $SO(2n + 1)$, $Sp(n)$, and $SO(2n)$, respectively.

classical compact simple Lie group   Any of the connected compact Lie groups $SU(n + 1)$, $SO(nl + 1)$, $Sp(n)$, or $SO(2n)$, with corresponding compact real simple Lie algebra A_n ($n \geq 1$), B_n ($n \geq 2$), C_n ($n \geq 3$), or D_n ($n \geq 4$) as its Lie algebra.

classical complex simple Lie algebra   Let A_n, B_n, C_n, and D_n be the Lie algebras of the complex Lie groups $SL(n + 1, C)$, $SO(2n + 1, C)$, $Sp(n, C)$, and $SO(2n, C)$. Then A_n ($n \geq 1$), B_n ($n \geq 2$), C_n ($n \geq 3$), and D_n ($n \geq 4$) are called *classical complex simple Lie algebras*.

classical group   Groups such as the general linear groups, orthogonal groups, symplectic groups, and unitary groups.

classification   Let R be an equivalence relation on a set S. The partition of S into disjoint union of equivalence classes is called the *classification* of S with respect to R.

class number   The order of the ideal class group of an algebraic number field F. Similarly, the order of the ideal class group of a Dedekind domain D is called the *class number* of D.

class of curve   The degree of the tangential equation of a curve.

clearing of fractions   An equation is *cleared of fractions* if both sides are multiplied by a common denominator of all fractions appearing in the equation.

Clebsch-Gordon coefficient   One of the coefficients, denoted

$$\left(j_1 m_1 j_2 m_2 \middle| j_1 j_2 j m\right)$$

in the formula

$$\psi(jm) = \sum_{-j \leq m_1, m_2 \leq j} \left(j_1 m_1 j_2 m_2 \middle| j_1 j_2 j m\right)$$
$$\times \psi(j_1 m_1)\psi(j_2 m_2)$$

which relates the basis elements of the representation space $\mathbf{C}^2 \otimes \cdots \otimes \mathbf{C}^2$ of $2j$ copies of $\mathbf{C}^2$ for a representation of $SO(3) \cong SU(2)/\{\pm I\}$. The coefficients are determined by the formula

$$\left(j_1 m_1 j_2 m_2 \middle| j_1 j_2 j m\right) = \delta_{m_1 + m_2, m}$$
$$\times \sqrt{\frac{(2j + 1)(j_1 + j_2 - j)!(j + j_1 - j_2)!(j + j_2 - j_1)!}{(j_1 + j_2 + j + 1)!}}$$
$$\times \sum_{\nu}(-1)^\nu \sqrt{\frac{(j_1 + m_1)!(j_1 - m_1)!(j_2 + m_2)!}{\nu!(j_1 + j_2 - j - \nu)!(j_1 - m_1 - \nu)!}}$$
$$\times \sqrt{\frac{(j_2 - m_2)!(j + m)!(j - m)!}{(j_2 + m_2 - \nu)!(j - j_2 + m_1 + \nu)!(j - j_1 - m_2 + \nu)!}}.$$

Clifford algebra   Let L be an n-dimensional linear space over a field F, Q a quadratic form on L, $A(L)$ the tensor algebra over L, $I(Q)$ the two-sided ideal on $A(L)$ generated by the elements $l \otimes l - Q(l) \cdot 1$ ($l \in L$), where $\otimes$ denotes tensor multiplication. The quotient associative algebra $A(L)/I(Q)$ is called the *Clifford algebra* over Q.

Clifford group   Let L be an n-dimensional linear space over a field F, Q a quadratic form on L, $C(Q)$ the Clifford algebra over Q, G the set of all invertible elements g in $C(Q)$ such that $gLg^{-1} = L$. Then, G is a group with respect to the multiplication of $C(Q)$ and is called the *Clifford group* of the Quadratic form Q. *See* Clifford algebra.

Clifford numbers   The elements of the Clifford algebra. *See* Clifford algebra.

closed boundary   If X is a compact Hausdorff space, then a *closed boundary* is a boundary closed in X.

closed image   Let $\mu : V \to V'$ be a morphism of varieties. If V is not complete, then $\mu(V)$ may not be closed. The closure of $\mu(V)$ which is in V', is called the *closed image* of V.

closed subalgebra   A subalgebra B_1 of a Banach algebra B that is closed in the norm topology. B_1 is then a Banach algebra, with respect to the original algebraic operations and norm of B.

closed subgroup   A subgroup H of a group G such that $x \in H$, whenever some nontrivial power of x lies in H.

closed subsystem   Let M be a character module. Let s be a subset of a root system r, and let s' be the submodule of M generated by s. If $s' \cap r = s$, then s is called a *closed subsystem*. *See* character module.

closed under operation   A set S, with a binary operation $*$ such that $a * b \in S$ for all $a, b \in S$.

closure property   The property of a set of being closed under a binary operation. *See* closed under operation.

coalgebra   Let ρ and ρ' be linear mappings defined as $\rho : V \to V \otimes_F V$, and $\rho' : V \to F$, where F is a field, V is vector space over F, and $\otimes$ denotes tensor product. Then, the triple (V, ρ, ρ') is said to be a *coalgebra* over F, provided $(1_V \otimes \rho) \circ \rho = (\rho \otimes 1_V) \circ \rho$ and $(1_V \otimes \rho') \circ \rho = (\rho' \otimes 1_V) \circ \rho = 1_V$.

coalgebra homomorphism   A k-linear map $f : C \to C'$, where (C, Δ, ε) and $(C', \Delta', \varepsilon')$ are coalgebras over a field k, such that $(f \otimes f) \circ \Delta = \Delta' \circ f$ and $\varepsilon = \varepsilon' \circ f$.

coarse moduli scheme   A scheme M and a natural transformation $\varphi : \mathcal{M} \to \text{Hom}(-, M)$ where $\mathcal{M}$ is a contravariant functor from schemes to sets such that
(i.) $\varphi(\text{Spec}(k)) : \mathcal{M}(\text{Spec}(k)) \to \text{Hom}(\text{Spec}$

$(k), M)$ is bijective for any algebraically closed field k, and
(ii.) for any scheme N and any natural transformation $\psi : \mathcal{M} \to \text{Hom}(-, N)$ there is a unique natural transformation $\lambda : \text{Hom}(-, M) \to \text{Hom}(-, N)$ such that $\psi = \lambda \circ \varphi$.

coarser classification   For $R, S \subset X \times X$ two equivalence relations on X, S is coarser than R if $R \subset S$.

coboundary   *See* cochain complex.

coboundary operator   *See* cochain complex.

cochain   *See* cochain complex.

cochain complex   A *cochain complex* $\mathbf{C} = \{C^i, \delta^i : i \in \mathbf{Z}\}$ is a sequence of modules $\{C^i : i \in \mathbf{Z}\}$, together with, for each i, a module homomorphism $\delta^i : C^i \to C^{i+1}$ such that $\delta^{i-1} \circ \delta^i = 0$. Diagramatically, we have

$$\cdots \xrightarrow{\delta^{i-2}} C^{i-1} \xrightarrow{\delta^{i-1}} C^i \xrightarrow{\delta^i} C^{i+1} \xrightarrow{\delta^{i+1}} \cdots ,$$

where the composition of any two successive δ^j is zero. In this context, the elements of C^i are called *i-cochains*, the elements of the kernel of δ^i, *i-cocycles*, the elements of the image of δ^{i-1}, *i-coboundaries*, the mapping δ^i, the ith *coboundary operator*, the factor module $H^i(\mathbf{C}) = \ker \delta^i / \text{im} \, \delta^{i-1}$, the ith *cohomology module* and the set $H(\mathbf{C}) = \{H^i(\mathbf{C}) : i \in \mathbf{Z}\}$, the *cohomology module* of $\mathbf{C}$. If c is an i-cocycle (i.e., an element of $\ker \delta^i$), the corresponding element $c + \text{im} \, \delta^{i-1}$ of $H^i(\mathbf{C})$ is called the *cohomology class of c*. Two i-cocycles belong to the same cohomology class if and only if they differ by an i-coboundary; such cocycles are called *cohomologous*.

cochain equivalence   Two cochain complexes $\mathbf{C}$ and $\tilde{\mathbf{C}}$ are said to be *equivalent* if there exist cochain mappings $\phi : \mathbf{C} \to \tilde{\mathbf{C}}$ and $\psi : \tilde{\mathbf{C}} \to \mathbf{C}$ such that $\phi\psi$ and $\psi\phi$ are homotopic to the identity mappings on $\mathbf{C}$ and $\tilde{\mathbf{C}}$, respectively.

cochain homotopy   Let $\mathbf{C} = \{C^i, \delta^i : i \in \mathbf{Z}\}$ and $\tilde{\mathbf{C}} = \{\tilde{C}^i, \epsilon^i : i \in \mathbf{Z}\}$ be two cochain complexes and let $\phi, \psi : \mathbf{C} \to \tilde{\mathbf{C}}$ be two cochain mappings. A *homotopy* $\zeta : \phi \to \psi$ is a sequence of mappings $\{\zeta^i : i \in \mathbf{Z}\}$ such that, for each i, ζ^i is a homomorphism from C^i to $\tilde{C}^i$ and

$$\phi^i - \psi^i = \delta^i \zeta^{i+1} + \zeta^i \epsilon^{i-1} .$$

When such a homotopy exists, ϕ and ψ are said to be *homotopic* and $H(\phi) = H(\psi)$, where $H(\phi)$ and $H(\psi)$ are the morphisms from the cohomology module $H(\mathbf{C})$ to the cohomology module $H(\tilde{\mathbf{C}})$ induced by ϕ and ψ.

cochain mapping Let $\mathbf{C} = \{C^i, \delta^i : i \in \mathbf{Z}\}$ and $\tilde{\mathbf{C}} = \{\tilde{C}^i, \epsilon^i : i \in \mathbf{Z}\}$ be two cochain complexes. A *cochain mapping* $\phi : \mathbf{C} \to \tilde{\mathbf{C}}$ is a sequence of mappings $\{\phi^i : i \in \mathbf{Z}\}$ such that, for each i, ϕ^i is a homomorphism from C^i to $\tilde{C}^i$ and

$$\phi^i \epsilon^i = \delta^i \phi^{i+1} .$$

The mapping ϕ^i induces a homomorphism from the ith cohomology module $H^i(\mathbf{C})$ to the ith cohomology module $H^i(\tilde{\mathbf{C}})$. Hence ϕ can be regarded as inducing a morphism ϕ^* from the cohomology module $H(\mathbf{C})$ to the cohomology module $H(\tilde{\mathbf{C}})$. The morphism ϕ^* is also denoted $H(\phi)$. This enables $H(\cdot)$ to be regarded as a functor from the category of chain complexes to a category of graded modules; it is called the *cohomological functor.*

cocommutative algebra A coalgebra C, in which the comultiplication Δ is *cocommutative,* i.e., has the property that $\delta\tau = \delta$, where τ is the *flip mapping,* i.e., the mapping from $C \otimes C$ to itself that interchanges the two C factors.

cocycle *See* cochain complex.

codimension Complementary to the dimension. For example, if X is a subspace of a vector space V and V is the direct sum of subspaces X and X', then the dimension of X' is the codimension of X.

coefficient A number or constant appearing in an algebraic expression. (For example, in $3 + 4x + 5x^2$, the coefficients are 3, 4, and 5.)

coefficient field *See* coefficient ring.

coefficient module *See* coefficient ring.

coefficient of equation A number or constant appearing in an equation. (For example, in the equation $2\tan x = 3x + 4$, the coefficients are 2, 3, and 4.)

coefficient of linear representation When a representation χ (of a group or ring) is isomorphic to a direct sum of linear representations and other irreducible representations, the number of times that a particular linear or irreducible representation ϕ occurs in the direct sum is called the *coefficient* of ϕ in χ.

coefficient of polynomial term In a polynomial

$$a_0 + a_1 x + a_2 x^2 + \ldots + a_n x^n ,$$

the constants $a_0, a_1, \ldots, a_n$ are called the *coefficients* of the polynomial. More specifically, a_0 is called the *constant term,* a_1 the *coefficient of* x, a_2 the *coefficient of* $x^2, \ldots, a_n$ the *coefficient of* x^n; a_n is also called the *leading coefficient.*

coefficient ring Consider the set of numbers or constants that are being used as coefficients in some algebraic expressions. If that set happens to be a ring (such as the set of integers), it is called the *coefficient ring.* Likewise, if that set happens to be a field (such as the set of real numbers) or a module, it is called the *coefficient field* or *coefficient module,* respectively.

cofactor Let A be an $n \times n$ matrix. The (k, ℓ) *cofactor* of A written $A_{k\ell}$ is $(-1)^{k+\ell}$ times the determinant of the $n - 1 \times n - 1$ matrix obtained by deleting the kth row and jth column of A.

cofunction The trigonometric function that is the function of the complementary angle. For example, cotan is the *cofunction* of tan.

cogenerator An element A of category $\mathcal{C}$ such that the functor $\text{Hom}(-, A) : \mathcal{C} \to \mathcal{A}$ is faithful where $\mathcal{A}$ is the category of sets. Also called a *coseparator.*

Cohen's Theorem A ring is Noetherian if and only if every prime ideal has a finite basis. There are several other theorems which may be called Cohen's Theorem.

coherent algebraic sheaf   A sheaf $\mathcal{F}$ of $\mathcal{O}_V$-modules on an algebraic variety V, such that if, for every $x \in V$, there is an open neighborhood U of x and positive integers p and q such that the $\mathcal{O}_V|_U$-sequence

$$\mathcal{O}_V^p|_U \to \mathcal{O}_V^q|_U \to \mathcal{F}|_U \to 0$$

is exact.

coherent sheaf of rings   A sheaf of rings $\mathcal{A}$ on a topological space X that is coherent as a sheaf of $\mathcal{A}$-modules.

cohomological dimension   For a group G, the number m, such that the cohomology group $H^i(G, A)$ is zero for every $i > m$ and every G-module A, and $H^m(G, A)$ is non-zero for some G-module A.

cohomology   The name given to the subject area that comprises cohomology modules, co-homology groups, and related topics.

cohomology class   *See* cochain complex.

cohomology functor   *See* cochain mapping.

cohomology group   (1) Because any **Z**-module is an Abelian group, a cohomology module $H(\mathbf{C})$, where **C** is a cochain complex, is called a *cohomology group* in the case where all the modules under consideration are **Z**-modules.
    (2) Let G be a group and A a G-module. Then $H^i(G, A)$, the ith *cohomology group of G with coefficients in A*, is defined by

$$H^i(G, A) = \text{Ext}_G^i(\mathbf{Z}, A) .$$

(When interpreting the Ext functor in this context, **Z** should be interpreted as a trivial G-module.)

cohomology module   *See* cochain complex.

cohomology set   The set of cohomology classes, when these classes do not possess a group structure.

cohomology spectral sequence   *See* spectral sequence.

coideal   Let C be a coalgebra with comultiplication Δ and counit ϵ. A subspace I of C is called a *coideal* if $\Delta(I)$ is contained in

$$I \otimes C + C \otimes I$$

and $\epsilon(I)$ is zero.

coimage   Let $\phi : M \to N$ be a homomorphism between two modules M and N. Then the factor module $M/\ker\phi$ (where $\ker\phi$ denotes the kernel of ϕ) is called the *coimage* of ϕ, denoted $\text{coim}\,\phi$. By the First Isomorphism Theorem, the coimage of ϕ is isomorphic to the image of ϕ, as a consequence of which the term coimage is not often used.

cokernel   Let $\phi : M \to N$ be a homomorphism between two modules M and N. Then the factor module $N/\text{im}\,\phi$ (where $\text{im}\,\phi$ denotes the image of ϕ) is called the *cokernel* of ϕ, denoted $\text{coker}\,\phi$.

collecting terms   The name given to the process of rearranging an expression so as to combine or group together terms of a similar nature. For example, if we rewrite

$$x^2 + 2x + 1 + 3x^2 + 5x \quad \text{as} \quad 4x^2 + 7x + 1$$

we have collected together the x^2 terms and the x terms; if we rewrite

$$x^2 + y^2 + 2x - 2y \quad \text{as} \quad \left(x^2 + 2x\right) + \left(y^2 - 2y\right)$$

we have collected together all the terms involving x, and all the terms involving y.

color point group   A pair of groups (K, K_1) such that $K = G/T$, G is a space group, T is the group of translations, and for some positive integers r, K_1 is the group of conjugacy classes of all subgroups G_1 of G with $T \subset G_1$ and $[G : G_1] = r$.

color symmetry group   A pair of groups (G, G') where G is a space group and $[G : G'] < \infty$.

column finite matrix   An infinite matrix with an infinite number of rows and columns, such that no column has any non-zero entry beyond the nth entry, for some finite n.

column in matrix *See* matrix.

column nullity For an $m \times n$ matrix A, the number $m - r(A)$, where $r(A)$ is the rank of A.

column vector A matrix with only one column, i.e., a matrix of the form

$$
\begin{bmatrix} a_1 \\ a_2 \\ \vdots \\ a_m \end{bmatrix}.
$$

Also called *column matrix*.

combination of things When r objects are selected from a collection of n objects, the r selected objects are called a *combination* of r objects from the collection of n objects, *provided that* the selected objects are all regarded as having equal status and not as being in any particular order. (If the r objects are put into a particular order, they are then called a *permutation,* not a combination, of r objects from n.) The number of different combinations of r objects from n is denoted by ^nC_r or $\begin{pmatrix} n \\ r \end{pmatrix}$ and equals $\dfrac{n!}{r!(n-r)!}$.

commensurable Two non-zero numbers a and b such that $a = nb$, for some rational number n. For example, $2\sqrt{2}$ and $3\sqrt{2}$ are commensurable because $2\sqrt{2} = \frac{2}{3}(3\sqrt{2})$. All rational numbers are commensurable with each other. No irrational number is commensurable with any rational number.

common denominator When two or more fractions are about to be added, it is helpful to re-express them first, so that they have the same denominator, called a *common denominator.* This process is also described as *putting the fractions over a common denominator.* For example, to simplify $\frac{3}{x-1} - \frac{2}{x+1}$ we might write

$$
\frac{3}{x-1} - \frac{2}{x+1}
$$
$$
= \frac{3(x+1)}{(x-1)(x+1)} - \frac{2(x-1)}{(x-1)(x+1)}
$$
$$
= \frac{x+5}{(x-1)(x+1)}.
$$

Here $(x-1)(x+1)$ is the common denominator.

common divisor A number that divides all the numbers in a list. For example, 3 is a common divisor of 6, 9, and 12. *See also* greatest common divisor.

common fraction A quotient of the form a/b where a, b are integers and $b \neq 0$.

common logarithm Logarithm to the base 10 (the logarithm that was most often used in arithmetic calculations before electronic calculators were invented). *See also* logarithm.

common multiple A number that is a multiple of all the numbers in a list. For example, 12 is a common multiple of 3, 4, and 6. *See also* least common multiple.

commutant If S is a subset of a ring R, the *commutant of S* is the set $S' = \{a \in R : ax = xa$ for all $x \in S\}$. Also called: *commutor.*

commutative algebra (1) The name given to the subject area that considers rings and modules in which multiplication obeys the commutative law, i.e., $xy = yx$ for all elements x, y.

(2) An algebra in which multiplication obeys the commutative law. *See* algebra.

commutative field A field in which multiplication is commutative, i.e., $xy = yx$, for all elements x, y. (The axioms for a field require that addition is always commutative. Some versions of the axioms insist that multiplication has to be commutative too. When these versions of the axioms are in use, a *commutative field* becomes simply a *field,* and the term *division ring* is used for structures that obey all the field axioms except commutativity of multiplication.) *See also* field.

commutative group *See* Abelian group.

commutative law The requirement that a binary operation $*$, on a set X, satisfy $x * y = y * x$, for all $x, y \in X$. Addition and multiplication of real numbers both obey the commutative law; matrix multiplication does not.

commutative ring   A ring in which multiplication is commutative, i.e., $xy = yx$, for all elements x, y. (The axioms for a ring ensure that addition is always commutative.) The real numbers are a commutative ring. Rings of matrices are generally not commutative.

commutator   (1) An element of the form $x^{-1} y^{-1}xy$ or $xyx^{-1}y^{-1}$ in a group. Such an element is usually denoted $[x, y]$ and it has the property that it equals the identity element if and only if $xy = yx$, i.e., if and only if x and y commute.

(2) An element of the form $xy - yx$ in a ring. Such an element is denoted $[x.y]$ and is also called the *Lie product* of x and y. It equals zero if and only if $xy = yx$.

In a ring with an involution *, the element $x^*x - xx^*$ is often called the *self-commutator* of x.

commutator group   *See* commutator subgroup.

commutator subgroup   The subgroup G' of a given group G, generated by all the commutators of G, i.e., by all the elements of the form $x^{-1}y^{-1}xy$ (where $x, y \in G$). G' consists precisely of those elements of G that are expressible as a product of a finite number of commutators. So G' may contain elements that are not themselves commutators. G' is a characteristic subgroup of G. G/G' is Abelian and in fact G' is the unique smallest normal subgroup of G with the property that the factor group of G by it is Abelian. G' is also called the *commutator group* or the *derived group* of G.

commutor   *See* commutator.

compact group   A topological group that is compact, as a topological space. A topological group is a group G, with the structure of a topological space, such that the map

$$(x, y) \mapsto xy^{-1} \quad \text{from} \quad G \times G \text{ to } G$$

is continuous.

compact real Lie algebra   A real Lie algebra whose Lie group is a compact group. *See* compact group.

compact simple Lie group   A Lie group that is compact as a topological space and whose Lie algebra is *simple* (i.e., not Abelian and having no proper invariant subalgebra).

compact topological space   A topological space X with the following property: whenever $\mathcal{O} = \{O_\alpha\}_{\alpha \in \mathcal{A}}$ is an open covering of X, then there exists a finite subcovering $O_1, O_2, \ldots, O_k$.

companion matrix   Given the monic polynomial over the complex field

$$p(t) = t^n + a_{n-1}t^{n-1} + \cdots + a_1 t + a_0 ,$$

the $n \times n$ matrix

$$A = \begin{pmatrix} 0 & 0 & \cdots & 0 & -a_0 \\ 1 & 0 & \cdots & 0 & -a_1 \\ 0 & 1 & \ddots & \vdots & \vdots \\ \vdots & \ddots & \ddots & 0 & -a_{n-2} \\ 0 & \cdots & 0 & 1 & -a_{n-1} \end{pmatrix}$$

is called the *companion matrix* of $p(t)$. The *characteristic polynomial* and the *minimal polynomial* of A are known to coincide with $p(t)$. In fact, a matrix is *similar* to the companion matrix of its characteristic polynomial if and only if the minimal and the characteristic polynomials coincide.

complementary degree   Let F be a filtration of a differential $\mathbf{Z}$-graded module A. Then $\{F_p A_n\}$ is a $\mathbf{Z}$-bigraded module and the module $F_p A_n$ has *complementary degree* $q = n - p$.

complementary law of reciprocity   A reciprocity law due to Hasse and superseded by Artin's general law of reciprocity. Let p be a prime number. If $\alpha \in k$ is such that $\alpha \equiv 1 \mod p(1 - \tau_p)$ where $\tau_p = \exp 2\pi i/p \in k$, then

$$\left(\tfrac{p}{d} \right)_p = \tau_p^c ,$$

$$c = \operatorname{Tr}_{k/\mathbf{Q}} \left(\tfrac{\alpha}{p(1-\tau_p)} \right) ;$$

$$\left(\frac{1 - \tau_p}{\alpha} \right)_p = \tau_p^d ,$$

$$d = -\operatorname{Tr}_{k/\mathbf{Q}} \left(\frac{\alpha - 1}{p(1 - \tau_p)} \right) .$$

complementary series   The irreducible, unitarizable, non-unitary, principal series representations of a reductive group G.

complementary slackness   This refers to *Tucker's Theorem on Complementary Slackness*, which asserts that, for any real matrix A, the inequalities $Ax = 0, x \geq 0$ and $^t uA \geq 0$ have solutions x, u satisfying $A^t u + x > 0$.

complementary submodule   Let N be a submodule of a module M. Then a *complementary submodule* to N in M is a submodule N' of M such that $M = N \oplus N'$.

complementary trigonometric functions The functions cosine, cosecant, and cotangent ($\cos\theta$, $\csc\theta$ and $\cot\theta$), so called because the cosine, cosecant, and cotangent of an acute angle θ equal, respectively, the sine, secant, and tangent of the *complementary angle* to θ, i.e., the acute angle ϕ such that θ and ϕ form two of the angles in a right-angled triangle. ($\phi = \frac{\pi}{2} - \theta$ in radians or $90 - \theta$ in degrees.)

complete cohomology theory   A cohomology theory of the following form. Let π be a finite multiplicative group and $B = B(\mathbf{Z}(\pi))$ the bar resolution. If A is a π-module define $H^*(\pi, A) = H^*(B, A)$. *See* cohomology.

complete field   A field F with the following property: whenever $p(x) = a_0 + a_1 x + \cdots a_k x^k$ is a polynomial with coefficients in F then p has a root in F.

complete group   A group G whose center is trivial and all its automorphisms are inner. Thus, G and the automorphism group of G are canonically isomorphic.

complete integral closure   Let $\mathcal{O}$ be an integral domain in a field K and M an $\mathcal{O}$-module contained in K. Let S be the set of all valuations on K that are nonnegative on $\mathcal{O}$. If $\upsilon \in S$, let R_υ be the valuation ring of υ. Then $M' = \bigcap_{\upsilon \in S} R_n M$ is the completion of M. If $\overline{\mathcal{O}}$ is the integral closure of $\mathcal{O}$, $\overline{M'}$ is the *complete integral closure* of the $\overline{\mathcal{O}}$-module $\overline{M'} = \overline{\mathcal{O}} M$.

complete intersection   A variety V in $P^n(k)$ of dimension r, where $I(V)$ is generated by $n - r$ homogeneous polynomials. *See* variety.

complete linear system   The set of all effective divisors linearly equivalent to a divisor.

complete local ring   *See* local ring.

completely positive mapping   For von Neumann algebras A and B, a linear mapping $T : A \rightarrow B$ such that for all $n \geq 1$ the induced mappings

$$I_n \otimes T : M_n(A) = M_n(\mathbf{C}) \otimes A \rightarrow M_n(B)$$

are positive.

completely reduced module   An R module (R a ring) which is the direct sum of irreducible R modules. An R module is *irreducible* if it has no sub R-modules.

completely reducible   Let k be a commutative ring and E a module over k. Let R be a k-algebra and let $\varphi : R \rightarrow \text{End}_k(E)$ be a representation of R in E. We say that φ is *completely reducible* (or *semi-simple*) if E is an R-direct sum of R-submodules E_i,

$$E = E_1 \oplus \cdots \oplus E_m ,$$

with each E_i irreducible.

completely reducible representation   A representation σ such that the relevant R-module E is an R-direct sum of R-submodules E_i,

$$E = E_1 \oplus E_2 \oplus \cdots \oplus E_m ,$$

such that each E_i is irreducible. *See* irreducible representation.
    Also called *semi-simple*.

completely solvable group   A group that is the direct product of simple groups.

complete pivoting   A process of solving an $n \times n$ linear system of equations $Ax = b$. By a succession of row and column operations, one may solve this equation once A has been transformed to an upper triangular matrix. Suppose

$A^{(0)} = A, A^{(1)}, \cdots, A^{(k-1)}$ have been determined so that $A^{(k-1)} = (a_{ij}^{(k-1)})$. Let $a_{pq}^{(k-1)}$ be the entry so that

$$\left| a_{pq}^{(k-1)} \right| = \max \left\{ \left| a_{ij}^{(k-1)} \right| : i, j \geq k \right\} .$$

Interchange the kth row and pth row and the kth column and qth column to obtain a matrix $B^{(k-1)}$. Now $A^{(k)}$ is obtained from $B^{(k-1)}$ by subtracting $b_{ik}^{(k-1)}/b_{kk}^{(k-1)}$ times the kth row of $B^{(k-1)}$ from the ith row of $B^{(k-1)}$.

complete resolution   Let π be a finite multiplicative group. The bar resolution $B\mathbf{Z}((\pi))$ of B is also called the complete free resolution. Here B_n is the free Π-module with generators $[x_1|\cdots|x_n]$ for all $x_1 \neq 1, \ldots, x_n \neq 1$ in π. So, B_n is the free Abelian group generated by all $x[x_1|\ldots|x_n]$ with $x \in \pi$ and no $x_i \neq 1$. Define $\partial : B_n \to B_{n-1}$ for $n > 0$ by setting

$$\partial [x_1|\ldots|x_n] = x_1 [x_2|\ldots|x_n]$$

$$+ \sum_{j=1}^{n-1} (-1)^j [x_1|\ldots|x_j x_{j-1}|\ldots|x_n]$$

$$+ (-1)^n [x_1|\ldots|x_{n-1}]$$

where $[y_1|\ldots|y_n] = 0$ if some $y_i = 1$.

complete scheme   A scheme X over a field K together with a morphism from X to Spec K that is proper and of finite type. *See* scheme.

complete valuation ring   Let k be a field. A subring R of k is called a *valuation ring* if, for any $x \in k$, we have either $x \in R$ or $x^{-1} \in R$. A valuation ring gives rise to a valuation, or norm, on K. The valuation ring is *complete* if every Cauchy sequence in this valuation converges.

complete Zariski ring   *See* Zariski ring.

completing the square   When a quadratic $ax^2 + bx + c$ is rewritten as

$$ax^2 + bx + c = a \left(x^2 + \frac{b}{a}x \right) + c$$

$$= a \left[\left(x + \frac{b}{2a} \right)^2 - \frac{b^2}{4a^2} \right] + c$$

$$= a \left(x + \frac{b}{2a} \right)^2 - \left(\frac{b^2}{4a} - c \right)$$

$$= aX^2 - d$$

$$\left(\text{where } X = x + \frac{b}{2a} \text{ and } d = \frac{b^2 - 4ac}{4a} \right)$$

the process is called *completing the square in x*. It is often used when solving equations or determining the sign of an expression because the absence of an X term in the final form $aX^2 - d$ makes it easy to determine whether the expression is positive, negative, or zero.

completion   The act of enlarging a set (minimally) to a complete space. This occurs in ring theory, measure theory, and metric space theory.

complex   (1) Involving *complex numbers*. *See* complex number.

(2) A set of elements from a group (not necessarily forming a group in their own right).

(3) A sequence of modules $\{ C^i : i \in \mathbf{Z} \}$, together with, for each i, a module homomorphism $\delta^i : C^i \to C^{i+1}$ such that $\delta^{i-1} \circ \delta^i = 0$. Diagrammatically, we have

$$\cdots \xrightarrow{\delta^{i-2}} C^{i-1} \xrightarrow{\delta^{i-1}} C^i \xrightarrow{\delta^i} C^{i+1} \xrightarrow{\delta^{i+1}} \cdots$$

where the composition of any two successive δ^j is zero.

complex algebraic variety   *See* algebraic variety.

complex analytic geometry   (1) Analytic geometry, i.e., the study of geometric shapes through the use of coordinate systems, but within a complex vector space rather than the more usual real vector space so that the coordinates are complex numbers rather than real numbers.

(2) The study of analytic varieties (the sets of common zeros of systems of analytic functions),

as opposed to *algebraic* geometry, the study of algebraic varieties.

complex conjugate representation   Let G be a group and let ϕ be a complex representation of G (so that ϕ is a homomorphism from G to the group GL(n, **C**) of all $n \times n$ invertible matrices with complex entries, under matrix multiplication). If we define a new mapping $\psi : G \to$ GL(n, **C**) by setting $\psi(g) = \overline{\phi(g)}$ for all $g \in G$, where $\overline{\phi(g)}$ is the matrix obtained from the matrix $\phi(g)$ by replacing all its entries by their complex conjugates, then ψ is also a complex representation of G, called the *complex conjugate* of ϕ.

complex fraction   An expression of the form $\frac{z}{w}$ where z and w are expressions involving complex numbers or complex variables. A complex fraction $\frac{z}{w}$ is frequently simplified by observing that it equals $\frac{z\bar{w}}{w\bar{w}}$ (where $\bar{w}$ is the complex conjugate of w), which is simpler because the denominator $w\bar{w}$ is real.

complex Lie algebra   *See* Lie algebra.

complex Lie group   *See* Lie group.

complex multiplication   (**1**) The multiplication of two complex numbers $a + ib$ and $c + id$ (where a, b, c, d are real) using the rule

$$(a + ib)(c + id) = (ac - bd) + i(ad + bc),$$

which is simply the usual rule for multiplying binomials, coupled with the property that $i^2 = -1$.

(**2**) The multiplication of two complexes of a group, which is defined as follows. Let G be a group and let A, B be complexes, i.e., subsets of G. Then $AB = \{ab : a \in A, b \in B\}$. This definition obeys the associative law, i.e., $A(BC) = (AB)C$ for all complexes A, B, C of G.

complex number   A number of the form $z = x + iy$ where x and y are real and $i^2 = -1$. The set of all complex numbers is usually denoted **C** or $\mathbb{C}$.

complex orthogonal group   *See* complex orthogonal matrix.

complex orthogonal matrix   A square matrix A such that its entries are complex numbers and it is *orthogonal,* i.e., has the property that $AA^T = I$, or equivalently that $A^T = A^{-1}$ (where A^T denotes the transpose of A). Such a matrix has determinant ± 1. The set of all $n \times n$ complex orthogonal matrices forms a group under matrix multiplication, called the *complex orthogonal group* O(n, **C**). The set of all $n \times n$ complex orthogonal matrices of determinant 1 is a normal subgroup of O(n, **C**), called the *complex special orthogonal group* SO(n, **C**).

complex plane   The set **C** of complex numbers can be represented geometrically as the points of a plane by identifying each complex number $a + ib$ (where a and b are real) with the point with coordinates (a, b). This geometrical representation of **C** is called the *complex plane* or *Argand diagram.* In this representation, points on the x-axis correspond to real numbers and points on the y-axis correspond to numbers of the form ib (where b is real). So the x- and y-axes are often called the *real* and *imaginary axes,* respectively. Note also that the points representing a complex number z and its complex conjugate $\bar{z}$ are reflections of each other in the real axis.

complex quadratic field   A field of the form $\mathbf{Q}[\sqrt{m}]$ (the smallest field containing the rational numbers and $\sqrt{m}$) where m is a negative integer.

complex quadratic form   An expression of the form

$$\sum_{i=1}^{n} a_i z_i^2 + \sum_{i \neq j} a_{ij} z_i z_j$$

where, for all i and j, z_i is a complex variable, and a_i and a_{ij} are complex constants.

complex representation   A homomorphism ϕ from a group G to the general linear group GL(n, **C**), for some n. (GL(n, **C**) is the group of all $n \times n$ invertible matrices with complex entries, the group operation being matrix multiplication.) n is called the *degree* of ϕ. Complex representations of degree 1 are called *linear.*

complex root   A root of an equation that is a complex number but not a real number. For example, the equation $x^3 = 1$ has roots 1, $e^{2\pi i/3}$, $e^{4\pi i/3}$. Of these, 1 is called the *real cube root* of unity, while $e^{2\pi i/3}$ and $e^{4\pi i/3}$ are called the *complex cube roots* of unity.

complex semisimple Lie algebra   A complex Lie algebra that is *semisimple*, i.e., does not have an Abelian invariant subalgebra.

complex semisimple Lie group   A complex Lie group whose Lie algebra is *semisimple*, i.e., does not have an Abelian invariant subalgebra.

complex simple Lie algebra   A complex Lie algebra that is *simple*, i.e., not Abelian and having no proper invariant subalgebra.

complex simple Lie group   A complex Lie group whose Lie algebra is *simple*, i.e., not Abelian and having no proper invariant subalgebra.

complex special orthogonal group   *See* complex orthogonal matrix.

complex sphere   A *complex n-sphere* with *center z_0* and *radius r* is the set of all points at distance r from z_0 in the n-dimensional complex metric space $\mathbf{C}^n$. ($\mathbf{C}^n$ is the set of all n-tuples $(a_1, a_2, \ldots, a_n)$, where each a_i is a complex number.)

complex spinor group   The universal cover of $SO(n, \mathbf{C}) = \{A \in \mathrm{GL}(n, \mathbf{C}) : A^t A = I$ and $\det A = 1\}$ it is denoted $\mathrm{Spin}(n, \mathbf{C})$.

complex structure   A complex analytic structure on a differentiable manifold. Complex structures may also be put on real vector spaces and on pseudo groups. *See* analytic structure.

complex symplectic group   The set of all $n \times n$ matrices A with complex numbers as entries and having the property that $A^t J A = J$ (where A^T is the transpose of A, J is the matrix

$$\begin{bmatrix} 0 & I \\ -I & 0 \end{bmatrix}$$

and I denotes the $\frac{n}{2} \times \frac{n}{2}$ identity matrix) forms a group under matrix multiplication, called the *complex symplectic group* $\mathrm{Sp}(\frac{n}{2}, \mathbf{C})$.

complex torus   A torus of the form $\mathbf{C}^n / \Gamma$ where Γ is a lattice in $\mathbf{C}^n$.

complex variable   A variable whose values are complex numbers.

component   (**1**) When a vector or force is expressed as an ordered pair (a_1, a_2) in two dimensions, as a triple (a_1, a_2, a_3) in three dimensions, or as an n-tuple $(a_1, a_2, \ldots, a_n)$ in n dimensions, the numbers $a_1, a_2, \ldots, a_n$ are called its *components*.
(**2**) When a vector $\mathbf{v}$ in a vector space V is expressed in the form $a_1\mathbf{e}_1 + a_2\mathbf{e}_2 + \cdots + a_n\mathbf{e}_n$ where $a_1, a_2, \ldots, a_n$ are scalars and $\mathbf{e}_1, \mathbf{e}_2, \ldots, \mathbf{e}_n$ is a basis of V, the scalars $a_1, a_2, \ldots, a_n$ are called the *components* of $\mathbf{v}$ with respect to the basis $\mathbf{e}_1, \mathbf{e}_2, \ldots, \mathbf{e}_n$.
(**3**) When a vector or force $\mathbf{v}$ is expressed as $\mathbf{w} + \mathbf{x}$, where $\mathbf{w}$ is parallel to a given direction and $\mathbf{x}$ is perpendicular to that direction, $\mathbf{w}$ is called the *component* of $\mathbf{v}$ in the given direction.
(**4**) The word *component* is also used loosely to mean simply a part of a mathematical expression.

composite field   The smallest subfield of a given field K containing a given collection $\{k_\alpha : \alpha \in A\}$ of subfields of K.

composite number   An integer that is not zero, not 1, not -1, and not prime.

composition algebra   An alternative algebra A over a field F (characteristic $\neq 2$), with identity 1 and a quadratic norm $n : A \to F$ such that $n(x, y) = n(x)n(y)$.

composition factor   *See* composition series.

composition factor series   For a group G with composition series $G = G_0 \supset G_1 \supset \cdots \supset G_r = \{e\}$, the sequence $G_0/G_1, \ldots, G_{r-1}/G_r$.

composition series   A series of subgroups $G_0, G_1, \ldots, G_n$ of a group G such that $G_0 = 1$,

$G_n = G$ and, for each i, G_i is a proper normal subgroup of G_{i+1} such that G_{i+1}/G_i is simple.

The Jordan-Hölder Theorem states that if H_0, $H_1, \ldots, H_m$ is another composition series for G, then $m = n$ and there is a one-one correspondence between the two sets of factor groups $\{G_{i+1}/G_i : i = 0, \ldots, n-1\}$ and $\{H_{i+1}/H_i : i = 0, \ldots, n-1\}$ such that corresponding factor groups are isomorphic. The factor groups $G_{i+1}/G_i : i = 0, \ldots, n-1$ are therefore called *the composition factors* of G. There are similar definitions for *composition series* and *composition factors* of other algebraic structures such as rings. These are obtained by making obvious changes to the definitions for groups. For example, in the case of rings, take the above definitions, replace $G_0 = 1$ by $G_0 = 0$, and replace the words group, subgroup, and normal subgroup by ring, subring, and ideal throughout.

Composition Theorem (class field theory)
Let K_1 and K_2 be class fields over k for the respective ideal groups H_1 and H_2. Then the composite field $K_1 K_2$ is the class field over k for $H_1 \cap H_2$.

compound matrix   Given positive integers n, ℓ ($\ell \leq n$), denote by $Q_{\ell,n}$ the ℓ-tuples of $\{1, 2, \ldots, n\}$ with elements in increasing order. $Q_{\ell,n}$ has $\binom{n}{\ell}$ members ordered lexicographically. For any $m \times n$ matrix A and $\emptyset \neq \alpha \subseteq \{1, 2, \ldots, m\}$, $\emptyset \neq \beta \subseteq \{1, 2, \ldots, n\}$, let $A[\alpha \mid \beta]$ denote the submatrix of A containing the rows and columns indexed by α and β, respectively.

Given an integer ℓ, $0 < \ell \leq \min(m, n)$, the ℓth compound matrix of A is defined as the $\binom{m}{\ell} \times \binom{n}{\ell}$ matrix

$$A^{(\ell)} = (\det A[\alpha \mid \beta])_{\alpha \in Q_{1,m}, \beta \in Q_{1,n}} .$$

To illustrate this definition, if $A = (a_{ij})$ is a 3×4 matrix, then $A^{(2)}$ is a 3×6 matrix; its $(1, 1)$, $(1, 2)$, and $(2, 1)$ entries are, respectively,

$$\det \begin{pmatrix} a_{11} & a_{12} \\ a_{21} & a_{22} \end{pmatrix}, \det \begin{pmatrix} a_{11} & a_{13} \\ a_{21} & a_{23} \end{pmatrix},$$

and

$$\det \begin{pmatrix} a_{11} & a_{12} \\ a_{31} & a_{32} \end{pmatrix} .$$

computation by logarithms   This is the name given to a method of solving an equation $A = B$ where A and B are complicated expressions involving products and powers. The method is to take logarithms of both sides, i.e., to say that $\log A = \log B$, and then to use the laws of logarithms to simplify and rearrange that equation so as to obtain the logarithm of the unknown variable and hence obtain that variable itself.

comultiplication   *See* coalgebra.

concave programming   The subject dealing with problems of the following type. Suppose X is a closed convex subset of $\mathbf{R}^n$ and $g_1, \ldots, g_m$ are convex functions on X. Let f be a concave function on X. Determine $\bar{x} \in X$ such that $f(\bar{x}) = \min \{f(x) : x \in X \text{ and } g_1(x) \leq 0 \text{ for } i = 1, \ldots, m\}$.

conditional equation   An equation involving variable quantities which fails to hold for some values of the variables.

conditional inequality   An inequality involving variable quantities which fails to hold for some values of the variables.

condition number   The quantity $\|A\| \cdot \|A^{-1}\|$, where $\|A\|$ is the norm of the matrix A.

conductor   (1) Of an Abelian extension K/k, the product $\mathcal{F} = \prod_\wp \mathcal{F}_\wp$ (over all prime ideals) of the conductors of the local fields $K_\wp/k_\wp$; $\mathcal{F}_\wp = \wp^n$, where n is minimal with the property that the norm of every nonzero element of K satisfies $N_{K/k}(x) \equiv 1 \pmod{\wp^n}$. (If $\wp$ is infinite, $\mathcal{F}_\wp = \wp$ when $K_\wp \neq k_\wp$ and $\mathcal{F}_\wp = 1$ otherwise.)

(2) Of a character χ of some representation of the Galois group G of a local field K/k, the function

$$f(x) = \frac{1}{|G_0|} \sum_{i=0}^\infty \sum_{g \in G_i} \chi(1) - \chi(g),$$

where G_i are the ramification subgroups

$$\{\sigma \in G : v(\sigma(x) - x) \geq i + 1 \text{ for all } x \in K^+\},$$

v the discrete valuation on K (with respect to which K is complete) and $K^+ = \{x \in K :$

$v(x) \geq 0$}. The ideal $\wp^{f(\chi)}$, where $\wp$ is the maximal ideal of the ring of integers in K, is also called the (Artin) *conductor* of χ.

Conductor Ramification Theorem (class field theory) If $\mathcal{F}$ is the conductor of the class field K/k, then it is prime to all unramified prime divisors for K/k, and $\mathcal{F}$ factors as $\mathcal{F} = \prod_{\wp} \mathcal{F}_{\wp}$, where each $\mathcal{F}_{\wp}$ is the $\wp$-conductor of the local field $K_{\wp}/k_{\wp}$ at some ramified prime $\wp$. *See* conductor, class field.

conformal transformation A mapping of Riemannian manifolds that preserves angles in the respective tangent spaces. In classical complex analysis, the same as a holomorphic or analytic function with nonzero derivative.

congruence A form of equivalence relation of two sets or collections of objects. The term will have different specific meanings in different contexts.

congruence zeta function The (complex valued) function $\zeta_K(s) = \sum \frac{1}{N(\mathcal{A})^s}$ where the sum is over all integral divisors $\mathcal{A}$ of the algebraic function field K over $k(x)$ where k is a finite field.

congruent integers With respect to a positive integer (modulus) m, two integers a and b are *congruent modulo* m, written $a \equiv b \pmod{m}$, when $a - b$ is divisible by m.

conjugacy class Assume S is a set. A binary relation R on S is a subset of $S \times S$. The *conjugacy class* determined by an element $a \in S$ is the set of elements $b \in S$ so that $(a, b) \in R$. In case R is reflexive ($(a, a) \in R$ for all $a \in S$), symmetric (if $(a, b) \in R$ then $(b, a) \in R$) and transitive (if $(a, b), (b, c) \in R$ then $(a, c) \in R$), then distinct conjugacy classes are disjoint and the union of all the conjugacy classes is S.

conjugate (1) Of a complex number $z = a + bi$ (a, b real), the related complex number $\bar{z} = a - bi$.

(2) Of a group element h, the group element ghg^{-1}, where g is another element of the group. *See also* conjugate radicals.

conjugate complex number For $z = x + iy$ a complex number, the complex conjugate of z is written as $\bar{z}$ or z^* and is given by $\bar{z} = x - iy$. This operation preserves multiplication and addition in the sense that $\overline{z + w} = \bar{z} + \bar{w}$ and $\overline{zw} = \bar{z}\bar{w}$. A consequence is that, if P is a polynomial with real coefficients, then $\overline{P(z)} = P(\bar{z})$, so that roots of P occur in complex conjugate pairs.

conjugate field To the field extension F over a base field k (within some algebraic closure $\bar{k}$), any subfield F' of $\bar{k}$ isomorphic to F. In one of the fundamental theorems of Galois theory, it is found that if F is a subfield of a normal extension E of k, then the conjugate fields of F inside E are precisely those fields F' for which the Galois groups $\mathrm{Gal}(E : F)$ and $\mathrm{Gal}(E : F')$ are conjugate subgroups of $\mathrm{Gal}(E : k)$.

conjugate ideal To a fractional ideal $\mathcal{A}$ of a number field K/k, the image ideal $\varphi(\mathcal{A})$ of the conjugate field $\varphi(K)$ under some k-isomorphism $\varphi : \bar{k} \to \bar{k}$.

conjugate radicals Expressions of the form $\sqrt{a} + \sqrt{b}$ and $\sqrt{a} - \sqrt{b}$. More generally, the expressions $\sqrt[n]{a} - \zeta^i \sqrt[n]{b}$, $i = 0, 1, \ldots, n - 1$, where ζ is an nth root of unity, are *conjugate radicals*.

conjugate subgroup For a subgroup G' of a group G, any of the subgroups $gG'g^{-1} = \{ghg^{-1} : h \in G'\}$.

conjugation mapping An automorphism of a group G of the form $a \mapsto gag^{-1}$ for some $g \in G$.

conjugation operator Given a uniform algebra (function algebra) A on some compact Hausdorff space X with φ in the maximal ideal space of A, and μ a representing measure for φ, the operator that assigns to each continuous real-valued function $u \in \Re A$, the continuous real-valued function $*u$ so that $u + i * u \in A$ and $\int *u d\mu = 0$. *See* function algebra, maximal ideal space, representing measure.

connected graded module A graded module $M = \sum_{n=0}^{\infty} M_n$, over a field k, for which M_0 is isomorphic to k. *See* graded module.

connected group   An algebraic group which is irreducible as a variety. *See* algebraic group, variety.

connected Lie subgroup   A Lie subgroup which is connected, as a differentiable manifold.

connected sequence of functors   A sequence $F^i : C \to C'$ of functors between Abelian categories for which there exist connecting morphisms

$$\partial_* : F^i(C) \to F^{i-1}(A)$$

(or

$$\partial^* : F^i(C) \to F^{i+1}(A)),$$

for every exact sequence $0 \to A \to B \to C \to 0$ of objects in C that turns

$$\cdots \to F^{i+1}(C) \xrightarrow{\partial_*} F^i(A) \longrightarrow F^i(B)$$

$$\longrightarrow F^i(C) \xrightarrow{\partial_*} F^{i-1}(A) \longrightarrow \cdots$$

(respectively,

$$\cdots \to F^{i-1}(C) \xrightarrow{\partial^*} F^i(A) \longrightarrow F^i(B)$$

$$\longrightarrow F^i(C) \xrightarrow{\partial^*} F^{i+1}(A) \longrightarrow \cdots)$$

into a chain complex, and, whenever

$$
\begin{array}{ccccccccc}
0 & \longrightarrow & A & \longrightarrow & B & \longrightarrow & C & \longrightarrow & 0 \\
 & & \downarrow f & & \downarrow g & & \downarrow h & & \\
0 & \longrightarrow & A' & \longrightarrow & B' & \longrightarrow & C' & \longrightarrow & 0
\end{array}
$$

is a morphism of exact sequences, then

$$\partial_* \circ F^i(h) = F^{i-1}(f) \circ \partial_*$$

(respectively,

$$\partial^* \circ F^i(h) = F^{i+1}(f) \circ \partial^*).$$

See chain complex.

connecting homomorphism   The boundary homomorphism

$$\partial_* : H_r(K, L; G) \to H_{r-1}(L; G),$$

connecting the homology groups of the simplicial pair (K, L) with coefficients in the group G and, dually, the coboundary homomorphism

$$\partial^* : H^r(L; G) \to H^{r+1}(K, L; G),$$

connecting the corresponding cohomology groups.

connecting morphism   *See* connected sequence of functors.

consistent equations   A set of equations which have some simultaneous solution.

constant   A function $F : A \to B$ such that there is a $c \in B$ with $F(x) = c$, for all $x \in A$.

constant of proportionality   The constant k relating one quantity to others in a relation of direct, inverse, or joint proportionality. For example, quantity x is *directly proportional* to quantity y if there is a constant $k \neq 0$ such that $x = ky$. In this case, the constant k is the *constant of proportionality*. *See also* direct proportion, inverse proportion, joint proportion.

constant term   Given an equation in a variable x, any part of the equation that is independent of x is a constant term. If $g(x) = \sin(x) + x^2 + 3\pi$, then the constant term is 3π.

constituent   A **Z**-representation of a finite group G, **Z** the rational integers.

constituent   Let $\mathcal{Z}$ denote the rational integers and $\mathcal{Q}$ the rational field. Let T be a $\mathcal{Z}$-representation of a finite group G. Then T is called a constituent of the group G.

constructible sheaf   A sheaf $\mathcal{F}$ on a scheme X, decomposable into locally closed subschemes so that the restriction of $\mathcal{F}$ to each subscheme is locally constant.

continuation method of finding roots   A method for approximating roots of the equation $f(x) = 0$ on the closed interval $[a, b]$ by introducing a parameter t, so that $f(x) \equiv g(x, t)|_{t=b}$ and so that $g(x, a) = 0$ is easily solved to obtain $x = x_0$. We partition the interval to give $a = t_0 < t_1 < \cdots < t_n = b$, then successively solve $g(x, t_i) = 0$ to obtain $x = x_i$ by some iterative method that begins with the previous solution $x = x_{i-1}$.

continued fraction   A number of the form

$$a_0 + \cfrac{1}{a_1 + \cfrac{1}{a_2 + \cfrac{1}{a_3 + \cdots}}}$$

where the a_i are real numbers and $a_1, a_2, \ldots$ are all positive. A continued fraction is *simple* if all the a_i are integers and *finite*, as opposed to *infinite*, if the sequence of a_i is finite. For typographical convenience, the continued fraction is often written as $[a_0; a_1, a_2, \ldots]$.

continuous analytic capacity   Of a subset A of $\mathbf{C}$, the measure $\sup|f'(\infty)|$ over all continuous functions on the Riemann sphere S^2 that vanish at ∞, are analytic outside some compact subset of A, and have sup norm $\|f\| \leq 1$ on S^2.

continuous cocycle   A matrix that arises as the derivative of a group action on a manifold.

contragredient representation   Of a representation ρ of a group G on some vector space V, the representation $\rho*$ of G on the dual space $V*$ defined by $\rho * (g) = \rho(g^{-1})*$. *See* representation, dual space.

contravariant   *See* functor.

convergent matrix   An $n \times n$ matrix A such that every entry of A^m converges to 0 as m approaches ∞. Convergent matrices arise in numerical analysis; for example, in the study of iterative methods for the solution of linear systems of equations. It is known that A is convergent if and only if its *spectral radius* is strictly less than 1; namely, all eigenvalues of A have modulus strictly less than 1.

convex hull   The smallest convex set containing a given set. (A set S is *convex* if, whenever $x, y \in S$, then the straight line joining x and y also lies in S.) If E is a vector space and $e_1, \ldots, e_k$ are elements of E, then the convex hull of the set $\{e_1, \ldots, e_k\}$ is the set $\{\sum_1^k c_j e_j : c_j \geq 0 \text{ and } \sum c_j = 1\}$.

convex programming   The theory that deals with the problem of minimizing a convex func-

tion on a convex set obtained as the solution set to a family of inequalities.

coordinate ring   Of an *affine algebraic set X* (over a field k), the quotient ring

$$k[X] = k[x_1, x_2, \ldots, x_n]/I(X)$$

of the ring of polynomials over k by the ideal of polynomials that vanish on X.

coproduct   Of objects A_i $(i \in I)$ in some category, the universal object $\coprod A_i$, with morphisms $\varphi_i : A_i \to \coprod A_i$ satisfying the condition that if X is any other object and $\psi_i : A_i \to X$ are morphisms, there will exist a unique morphism $\psi : \coprod A_i \to X$ so that $\psi_i = \psi \circ \varphi_i$. If the category is Abelian, the coproduct coincides with the direct sum $\sum A_i$.

coradical   Of a coalgebra, the sum of its simple subcoalgebras. *See* coalgebra, simple coalgebra.

coregular representation   The contragredient representation of the right regular representation reciprocal to a given left regular representation. *See* contragredient representation.

Corona Theorem   If $f_1, \ldots, f_n$ belong to H^∞, the set of bounded analytic functions in the unit disk D in the complex plane, and if $|f_1(z)| + \cdots + |f_n(z)| \geq \delta > 0$ in D, then there exist $g_1, \ldots, g_n \in H^\infty$ with $\sum f_j g_j = 1$ in D.

In functional analytic terms, the theorem says this. H^∞ is a vector space under pointwise addition and scalar multiplication and a Banach space with the norm $\|f\| = \sup |f(z)|$, for $z \in D$. Further, since the pointwise product of two bounded functions is bounded it is an algebra with pointwise multiplication. As a Banach space, H^∞ has a dual space $H^{\infty'}$, which is the set of all continuous linear mappings from H^∞ into the complex numbers $\mathbf{C}$. A functional $\lambda \in H^{\infty'}$ may have the added property of being multiplicative, $\lambda(fg) = \lambda(f)\lambda(g)$. We denote the set of all such multiplicative linear functionals by M. For example, if $a \in D$, then the linear functional $\lambda(f) = f(a)$ is in M and is called a point evaluation. The set $H^{\infty'}$ has a topology whereby λ_a converges to λ if, for each $f \in H^\infty$, the numbers $\lambda_a(f)$ converge to $\lambda(f)$. M inherits this method of convergence from $H^{\infty\prime}$ and the Corona Theorem states that the "point evaluations" are dense in M in this topology. This was proved by Carleson.

correspondence ring   Of a nonsingular curve X, the ring $\mathcal{C}(X)$ whose objects, called *correspondences,* are linear equivalence classes of divisors of the product variety $X \times X$, modulo the relation that identifies divisors if they are linearly equivalent to a degenerate divisor. The addition in this ring is addition of divisors, and the multiplication is defined by composition, that is, if C_1, C_2 are correspondences of X and $x \in X$, then $C_1 \circ C_2$ is the correspondence $C_1(C_2(x))$ where $C(x)$ is the projection on to the second component of $C(x, X)$.

cosecant function   The reciprocal of the sine function, denoted $\csc\theta$. Hence, $\csc\theta = \frac{1}{\sin\theta}$. *See* sine function.

cosecant of angle   The reciprocal of the sine of the angle. Hence, $\csc\theta = \frac{1}{\sin\theta}$. *See* sine of angle.

cosemisimple coalgebra   A coalgebra which is equal to its coradical. *See* coalgebra, coradical.

cosine function   One of the fundamental trigonometric functions, denoted $\cos x$. It is (1) periodic, satisfying $\cos(x + 2\pi) = \cos x$; (2) bounded, satisfying $-1 \le \cos x \le 1$ for all real x; and (3) intimately related with the sine function, $\sin x$, satisfying the important identities $\cos x = \sin(\frac{\pi}{2} - x)$, $\sin^2 x + \cos^2 x = 1$, and many others. It is related to the exponential function via the identity $\cos x = \frac{e^{ix}+e^{-ix}}{2}$ ($i = \sqrt{-1}$), and has series expansion

$$\cos x = 1 - \frac{x^2}{2!} + \frac{x^4}{4!} - \frac{x^6}{6!} + \cdots$$

valid for all real values of x.
    See also cosine of angle.

cosine of angle   Written $\cos\alpha$, the x-coordinate of the point where the terminal ray of the angle α whose initial ray lies along the positive x-axis intersects the unit circle. If $0 < \alpha < \frac{\pi}{2}$ (α

in radians) so that the angle is one of the angles in a right triangle with adjacent side a, opposite side b, and hypotenuse c, then $\cos\alpha = \frac{a}{c}$.

cospecialization   Let A and B be sets and $\Phi : A \to B$ a function. *Cospecialization* is a process of selecting a subset of B with reference to subsets of A, using the mapping Φ as a referencing operator.

cotangent function   The quotient of the cosine and the sine functions. Also the reciprocal of the tangent function. Hence, $\cot\theta = \frac{\cos\theta}{\sin\theta} = \frac{1}{\tan\theta}$. *See* sine function, cosine function.

cotangent of angle   Written $\cot\alpha$, the x-coordinate of the point where the terminal ray of the angle α whose initial ray lies along the positive x-axis intersects the line ℓ with equation $y = 1$. If α measures more than π radians ($= 180°$), the terminal is taken to extend back to intersect the line ℓ. If $0 < \alpha < \frac{\pi}{2}$ (α in radians), so that the angle is one of the angles in a right triangle with adjacent side a, opposite side b, and hypotenuse c, then $\cot\alpha = \frac{a}{b}$.

coterminal angles   Directed angles whose terminal sides agree. (Their initial sides need not agree.)

cotriple   A functor $T : \mathcal{C} \to \mathcal{C}$ on a category $\mathcal{C}$ for which there exist natural transformations $\varepsilon : \mathrm{Id}_{\mathcal{C}} \to T$, $\delta : T \to T^2$, for which the following diagrams commute:

$$T(X) \xleftarrow{l_{T(X)}} T(X) \xrightarrow{l_{T(X)}} T(X)$$
$$\varepsilon_{T(X)} \searrow \quad \delta_X \downarrow \quad \swarrow T(\varepsilon_X)$$
$$T^2(X)$$

$$\begin{array}{ccc} T(X) & \xrightarrow{\delta_X} & T^2(X) \\ \delta_X \downarrow & & \downarrow \delta_{T(X)} \\ T^2(X) & \xrightarrow{T(\delta_X)} & T^3(X) \end{array}$$

countable set   A set S such that there is a one-to-one mapping $f : S \to \mathbf{N}$ from S onto the set of natural numbers. *See also* cardinality.

counting numbers   The positive integers 1, 2, 3, . . .

Courant-Fischer (min-max) Theorem   Let A be an $n \times n$ Hermitian matrix with eigenvalues

$$\lambda_1 \geq \lambda_2 \geq \cdots \geq \lambda_n .$$

Then

$$\lambda_i = \max_{\dim X = i} \min_{x \in X, \, x^*x=1} x^*Ax$$

and

$$\lambda_i = \min_{\dim X = n-i+1} \max_{x \in X, \, x^*x=1} x^*Ax .$$

This theorem was first proved by Fischer for matrices (1905) and later (1920) it was extended by Courant to differential operators.

covariant   A term describing a type of functor, in constrast with a contravariant functor. A covariant functor $F : C \to D$ assigns to every object c of the domain category C an object $F(c)$ of the codomain category D and to every arrow $\alpha : c \to c'$ of C an arrow $F(\alpha) : F(c) \to F(c')$ of D in such a way that $F(\mathrm{id}_C) = \mathrm{id}_D$, where id_C and id_D are the identity arrows of the respective categories, and that $T(\beta \circ \alpha) = T(\beta) \circ T(\alpha)$ ($\circ$ represents composition of arrows in both categories) whenever $\beta \circ \alpha$ is defined in C. *See also* contravariant.

covering   Of a (nonsingular) algebraic curve X over the field k, a curve Y for which there exists a k-rational map $Y \to X$ which induces an inclusion of function fields $k(x) \hookrightarrow k(Y)$ making $k(Y)$ separably algebraic over $k(X)$.

covering family   A family of morphisms in a category C which define the Grothendieck topology on C.

coversine function   The function

$$\mathrm{covers}\,\alpha = 1 - \sin\alpha .$$

Coxeter complex   A thin chamber complex in which, to every pair of adjacent chambers, there exists a root containing exactly one of the chambers. When $C \subset C'$ we say that C is a *face* of C'. Two chambers are adjacent if their intersection has *codimension* 1 in each: the codimension of C in C' is the number of minimal nonzero faces

of C' lying in the *star complex* $\mathrm{St}(C)$ of objects containing C. A *root* is the image of the complex under some idempotent endomorphism, called a *folding*.

Coxeter diagram   A labeled graph whose nodes are indexed by the generators of a Coxeter group which has (P_i, P_j) as an edge labeled by M_{ij} whenever $M_{ij} > 2$. Here M_{ij} are elements of the Coxeter matrix. Also called Coxeter-Dynkin diagrams. These are used to visualize Coxeter groups.

Coxeter group   A group with generators r_i, $i \in I$, and relations of the form $(r_i r_j)^{a_{ij}} = 1$, where all $a_{ii} = 1$ and whenever $i \neq j$, $a_{ij} > 1$ (a_{ij} may be infinite, implying no relation between r_i and r_j in such a case).

Cramer's rule   Assume we are given n linear equations in n unknowns. That is, we are given the linear equations $L_i(X) = \sum_{j=1}^{n} C_{ij} X_j$, where the C_{ij} are given numbers, the X_j are unknowns and $i = 1, \ldots, n$. We are asked to solve the n equations $L_i(X) = b_j$, where $j = 1, \ldots, n$. One can write these equations in matricial form $AX = B$, where A is a square $n \times n$ matrix and $X = (X_1, \ldots, X_n)$ and $B = (b_1, \ldots, b_n)$ are vectors in $\mathbf{R}^n$. Cramer gave a formula for solving these equations providing the matrix A has an inverse. Let $|\cdot|$ denote the determinant of a matrix, which maps a square matrix to a number. (*See* determinant.) The matrix A will have an inverse provided the number $|A|$ is not zero. So, assuming A has an inverse (this is also referred to by saying that A is *non-singular*), the solution values of X_j are given as follows. In the matrix A, replace the ith column of A (that is, the column made up of $C_{i1}, \ldots, C_{in}$) by B. We again obtain a square matrix A_i and the solution numbers $X_i = \frac{|A_i|}{|A|}$. If X is the n vector made up of these numbers, it will solve the system and this is the only solution.

Cremona transformation   A birational map from a projective space over some field to itself. *See* birational mapping.

cremona transformation   A birational map of the projective plane to itself.

criterion of ruled surfaces   A theorem of Nakai, characterizing ample divisors D on a ruled surface X as those for which the intersection numbers of D with itself and with every irreducible curve in X are all positive. *See* ample, ruled surface, intersection number.

crossed product   Of a commutative ring R with a commutative monoid G (usually a group) with respect to a factor set $\{a_{g,h} \in R : g, h \in G\}$. If G acts on R in such a way that the mapping $r \mapsto r^g$ is an automorphism of R, then the crossed product of R by G with respect to the factor set is the R-algebra with canonical basis elements b^g for $g \in G$ and multiplication law

$$\left(\sum_{g \in G} r^g b^g \right) \left(\sum_{h \in G} s^h b^h \right)$$

$$= \sum_{g,h \in G} a_{g,h} r^g (s^{hg} b^{gh}) \ .$$

To ensure that this multiplication is associative, the elements of the factor set must satisfy the relations $a_{g,h} a_{gh,k} = a_{h,k}^g a_{g,hk}$ for all $g, h, k \in G$. Further, if $e \in G$ is the identity element, the unit element of the algebra is b^e, which also requires that, in the factor set, $a_{e,g} = a_{g,e}$ equals the unit element in R for all $g \in G$.

Crout method of factorization   A type of LU-decomposition of a matrix in which L is lower triangular, U is upper triangular, and U has 1s on the diagonal.

crystal family   A collection of crystallographic groups whose *point groups*, Γ, are all conjugate in $GL_3(\mathbf{R})$ and whose *lattice groups* are minimal in their crystal class with respect to the ordering $(\Lambda, \Gamma) \leq (\Lambda', \Gamma')$, defined as: there exists a $g \in GL_3(\mathbf{R})$ so that $(\Lambda') = g(\Lambda)$, $\Gamma' = g\Gamma g^{-1}$ and $B(\Lambda) \subseteq B(\Lambda')$ where $B(\Lambda)$ is the Bravais group $\{g \in O(3) : (\Lambda) = (\Lambda)\}$. *See* crystallographic group, Bravais group.

crystallographic group   A discrete group of motions in $\mathbf{R}^n$ containing n linearly independent translations. *See also* crystal system.

crystallographic restriction   The proper subgroup H_0 of the point group H of a crystallographic group Γ is cyclic of order q, where the only possible values of q are 1, 2, 3, 4, or 6.

The terms used are defined as follows. The group Γ is a discrete subgroup of the Euclidean group $E(2) = \mathbf{R}^2 \times O(2)$. Let $j : E(2) \to O(2)$ be the homomorphism $(x, A) \mapsto A$. Then $H = j(\Gamma)$ is a subgroup of $O(2)$, called the point group of Γ. Let L be the kernel of j, a discrete subgroup of $\mathbf{R}^2$. If the rank of L is 2, then Γ is called a crystallographic group. Then H_0 is $H \cap SO(2)$.

crystallographic space group   A crystallographic group of motions in Euclidean 3-space. *See* crystallographic group.

crystal system   A classification of 3-dimensional crystallographic lattice groups. Where the lattice constants a, b, c represent the lengths of the three linearly independent generators and α, β, γ the angles between them, the seven crystal systems are (where x, y are distinct and not 1, and $\theta \neq 90°$):

Name	$a : b : c$	(α, β, γ)
cubic	$1 : 1 : 1$	$(90°, 90°, 90°)$
tetragonal	$1 : 1 : x$	$(90°, 90°, 90°)$
rhombic	$1 : x : y$	$(90°, 90°, 90°)$
monoclinic	$1 : x : y$	$(90°, \theta, 90°)$
hexagonal	$1 : 1 : x$	$(90°, 90°, 120°)$
rhombohedral	$1 : 1 : 1$	(θ, θ, θ)
triclinic	other	than	above

See crystallographic group.

cube   (1) In geometry, a three-dimensional solid bounded by six square faces which meet orthogonally in a total of 12 edges and, three faces at a time, at a total of eight vertices. One of the five platonic solids.

(2) In arithmetic, referring to the third power $x^3 = x \cdot x \cdot x$ of the number x.

cube root   Of a number x, a number t whose cube is x: $t^3 = x$. If x is real, it has exactly one real cube root, which is denoted $t = \sqrt[3]{x}$.

cubic   (1) A polynomial of degree 3: $p(X) = a_0 + a_1 X + a_2 X^2 + a_3 X^3$, where $a_3 \neq 0$.

(2) A curve whose analytic representation in some coordinate system is a polynomial of degree 3 in the coordinate variable.

cubic equation   An equation of the form $ax^3 + bx^2 + cx + d = 0$, where a, b, c, d are constants (and $a \neq 0$).

cup product   (1) In a lattice or Boolean algebra, the fundamental operation $a \vee b$, also called the *join* or *sum* of the elements a and b.

(2) In cohomology theory, where $H^r(X, Y; G)$ is the cohomology of the pair (X, Y) with coefficients in the group G, the operation that sends the pair (f, g), $f \in H^r(X, Y; G_1)$, $g \in H^s(X, Z; G_2)$, to the image

$$f \cup g \in H^{r+s}(X, Y \cup Z; G_3)$$

of the element $f \otimes g$ under the map

$$H^r(X, Y; G_1) \otimes H^s(X, Z; G_2)$$
$$\to H^{r+s}(X, Y \cup Z; G_3),$$

induced by the diagonal map

$$\Delta : (X, Y \cup Z) \to (X, Y) \times (X, Z).$$

Cup Product Reduction Theorem   A theorem of Eilenberg-MacLane in the theory of cohomology of groups. Suppose the group G can be presented as the quotient of the free group F with relation subgroup H, so that $1 \to H \to G \to F \to 1$ is exact. If A is a G-module, then the induced extension of G given by $1 \to H/[H, H] \to F/[H, H] \to G \to 1$ is a 2-cocycle ζ in $H^2(G, A)$. Let $K = H/[H, H]$. If $\varphi \in H^r(G, \text{Hom}(K, A))$, the cup product $\varphi \cup \zeta \in H^{r+2}(G, A)$ (which makes use of the natural map $\text{Hom}(K, A) \otimes K \to A$) determines an isomorphism

$$H^r(G, \text{Hom}(K, A)) \cong H^{r+2}(G, A).$$

cuspidal parabolic subgroup   A closed subgroup $G_1 G_2 G_3$ of a connected algebraic group G which is the product of (i.) a reductive Lie subgroup G_1, stable under the Cartan involution, (ii.) a vector subgroup G_2 whose centralizer in G is $G_2 G_1$, and (iii.) a group $G_3 = \exp(\sum \mathcal{G}_\alpha)$ where $\mathcal{G}$ is the Lie algebra of $G_1 \mathcal{G}_\alpha = \{X \in \mathcal{G} : [H, X] = \alpha(H)X \text{ for all } H \in \mathcal{H}\}$, where $\mathcal{H}$ is the Lie algebra of G_2, α is some functional on $\mathcal{H}$, and the sum that defines G_3 is over all positive α for which $\mathcal{G}_\alpha$ is nonzero.

cycle   (1) In graph theory, a graph C_n on vertices $v_1, v_2, \ldots, v_n$ whose only edges are between v_1 and v_2, v_2 and v_3, $\ldots$, v_n and v_1.

(2) In a permutation group S_n, a permutation with at most one orbit containing more than one object.

(3) In homology theory, any element in the kernel of a homomorphism in a homology complex.

cyclic algebra   A crossed product algebra of a cyclic extension field F (over a base field k) with its (cyclic) Galois group $G = \langle g \rangle$ with respect to a factor set that is determined by a single nonzero element a of k. The elements of this algebra are uniquely of the form $\alpha_0 + \alpha_1 b + \cdots + \alpha_{n-1} b^{n-1}$, where the α_i come from F, $\{1, b, b^2, \ldots, b^{n-1}\}$ are the formal basis elements of the algebra, and n is the order of G. The multiplication is determined by the relations $b\alpha = \alpha^g b$, $b^i b^j = b^{i+j}$ and $b^n = a$.

cyclic determinant   A determinant of the form

$$\begin{vmatrix} a_0 & a_1 & \cdots & a_{n-1} \\ a_{n-1} & a_0 & \cdots & a_{n-2} \\ & & \cdots & \\ a_1 & a_2 & \cdots & a_0 \end{vmatrix},$$

in which the entries of successive rows are shifted to the right one position (modulo n). Also called a *circulant*.

cyclic group   A group, all of whose elements are powers a^n ($n = 0, 1, \ldots$) of a single generating element a. Such a group is finite if, for some d, a^d equals the identity element. The group is often denoted $\langle a \rangle$.

cyclic representation   A unitary representation ρ of a topological group G, having a cyclic vector x, i.e., an element x of the representation space for which the span of vectors $\rho(g)x$, as g runs through G, is dense. *See* representation.

cyclic subgroup   Any subgroup H of a group G in which all elements of H are powers a^n ($n = 0, 1, \ldots$) of a single element $a \in H$, the generator of the subgroup.

cyclotomic field   Any extension of a field obtained by adjoining roots of unity.

cyclotomic polynomial   Any of the polynomials $\Phi_n(x)$ whose roots are precisely those roots of unity of degree exactly n. That is, $\Phi_1(x) = x - 1$ and, for $n > 1$,

$$\Phi_n(x) = \frac{x^n - 1}{\prod_{d|n} (\Phi_d(x))},$$

where the product is over the proper divisors d of n.

D

Danilevski method of matrix transformation
A method for computing eigenvalues of a matrix M, involving application of row and column operations that produce the *companion matrix* of M.

decimal number system   The base 10 positional system for representing real numbers. Every real number x has a representation of the form

$$d_{n-1} d_{n-2} \cdots d_1 d_0 \cdot d_{-1} d_{-2} \cdots ,$$

in which the d_i are the digits of x; d_0 and d_{-1} are separated by the decimal point (.). The digits of x are determined recursively by the formulas $d_{n-1} = \lfloor \frac{x}{10^{n-1}} \rfloor$ and for $k > 1$,

$$d_{n-k} = \left\lfloor \frac{x - \sum_{j=1}^{k-1}(d_{n-j} 10^{n-j})}{10^{n-k}} \right\rfloor ,$$

where n is the unique integer that satisfies $10^{n-1} \le x < 10^n$ and $\lfloor t \rfloor$ is the *floor* function (the greatest integer $\le t$). For example, if $x = 238\frac{3}{4}$, then $n = 3, d_2 = 2, d_1 = 3, d_0 = 8, d_{-1} = 7, d_{-2} = 5$. The only possible digits are 0, 1, 2, 3, 4, 5, 6, 7, 8, 9. The digit d_0 is called the *ones* digit, d_1 the *tens* digit, d_2 the *hundreds* digit, d_3 the *thousands* digit, etc.; also, d_{-1} is the *tenths* digit, d_{-2} the *hundredths* digit, etc. It can be shown that, precisely when x is rational, the sequence of digits of x eventually repeats. That is, there is a smallest integer p for which $d_{i-p} = d_i$ for every i less than some fixed index. Here we call the string of digits $d_{i-1} d_{i-2} \cdots d_{i-p}$ a *repeating block* and p the *period of the representation* of x. In the special case that the repeating block is the single digit 0, the convention is to drop all the trailing zeros from the representation and say that x has a *finite* or *terminating* decimal expansion. Further, it is possible in this case to give a second, distinct expansion of x: if x has finite decimal expansion with final nonzero digit

d_i, then another representation of x can be obtained by replacing d_i with $d_i - 1$ and defining $d_{i-1} = d_{i-2} = \cdots = 9$. For example, 238.75 = 238.74999⋯.

decomposable operator   A bounded linear operator T, on the separable Hilbert space L^2 $(\Gamma, \mu; \mathcal{H})$ of square-integrable, measurable, $\mathcal{H}$-valued functions on some measure space (Γ, μ) where $\mathcal{H}$ is also a Hilbert space, so that for each measurable $\xi(\gamma)$, the function $\gamma \mapsto T(\gamma)\xi(\gamma)$ is measurable, and so that, for each $\xi \in L^2(\Gamma, \mu; \mathcal{H})$, T can be represented as the *direct integral*

$$T\xi = \int_\Gamma \oplus T(\gamma)\xi(\gamma) d\mu(\gamma) .$$

decomposition field   Let the ring A be closed in its quotient field K. Suppose that B is its integral closure in a finite Galois extension L, with group G. Then B is preserved by elements of G. Let $\wp$ be a maximal ideal of A and $\mathcal{B}$ a maximal ideal of B that lies above $\wp$. Now $G_\mathcal{B}$ is the subgroup of G consisting of those elements that preserve $\mathcal{B}$. Observe that $G_\mathcal{B}$ acts in a natural way on the residue class field $B/\mathcal{B}$ and it leaves $A/\wp$ fixed. To any $\sigma \in G_\mathcal{B}$, we can associate an element $\overline{\sigma} \in B/\mathcal{B}$ over $A/\wp$; the map

$$\sigma \longmapsto \overline{\sigma}$$

thereby induces a homomorphism of $G_\mathcal{B}$ into the group of automorphisms of $B/\mathcal{B}$ over $A/\wp$.

The fixed field in $G_\mathcal{B}$ is called the *decomposition field* of $\mathcal{B}$, and is denoted L^{dec}.

decomposition field   Let the ring A be closed in its quotient field K. Suppose that B is its integral closure in a finite Galois extension L, with group G. Then B is preserved by elements of G. Let $\wp$ be a maximal ideal of A and $\mathbf{B}$ a maximal ideal of B that lies above $\wp$. Now $G_\mathbf{B}$ is the subgroup of G consisting of those elements that preserve $\mathbf{B}$. Observe that $G_\mathbf{B}$ acts in a natural way on the residue class field $B/\mathbf{B}$ and it leaves $A/\wp$ fixed. To any $\sigma \in G_\mathbf{B}$ we can associate an element $\overline{\sigma} \in B/\mathbf{B}$ over $A/\wp$; the map

$$\sigma \longmapsto \overline{\sigma}$$

thereby induces a homomorphism of $G_\mathbf{B}$ into the group of automorphisms of $B/\mathbf{B}$ over $A/\wp$.

The fixed field in $G_\mathbf{B}$ is called the *decomposition field* of **B**, and is denoted L^{dec}

decomposition group   Of a prime ideal $\wp$ for a Galois extension K/k, the stabilizer of $\wp$ in $\text{Gal}(K/k)$. *See* stabilizer, Galois group.

decomposition number   The multiplicity of an absolutely irreducible modular representation of a group G, in the splitting field K (for which it is the Galois group) as it appears in a decomposition of one of the irreducible representations of G in some number field for which K is the residue class field. *See* absolutely irreducible representation, modular representation.

Dedekind cut   One of the original ways of defining irrational numbers from rational ones. A *Dedekind cut* (L, R) is a decomposition of the rational numbers into two sets L and R such that (i.) L and R are nonempty and disjoint; (ii.) if $x \in L$ and $y \in R$ then $x < y$; (iii.) L has no largest element. Thus, a rational number can be identified with a *Dedekind cut* (L, R) for which R has a least member and an irrational number can be identified with a *Dedekind cut* (L, R) for which R has no least element.

Dedekind domain   An integral domain which is Noetherian, integrally closed, and whose nonzero prime ideals are all maximal. *See* integral domain, Noetherian ring, integrally closed.

Dedekind's Discriminant Theorem   Let $F = Q(\sqrt[3]{d})$ be a pure cubic field and let $d = ab^2$ with ab square-free. If F is of the first kind, then the discriminant of F is given by $d(F) = -27(ab)^2$ and if F is of the second kind, then $d(F) = -3(ab)^2$.

Dedekind zeta function   Let k be a number field and O_k be the ring of all algebraic integers in k. If I is a nonzero O_k-ideal, we write $N(I)$ for the finite index $[O_k : I]$. The *Dedekind zeta function* is then defined by

$$\zeta_k(x) = \sum_I N(I)^{-x} = \prod_J \left(1 - N(J)^{-x}\right)^{-1} ,$$

where $x > 1$ and the sum extends over all nonzero ideals of O_k, while the product runs over all nonzero prime ideals of O_k.

defect   If k is a field which is complete under an arbitrary valuation, and if E is a finite extension of degree n, with ramification e and residue class degree f, then ef divides n: $n = ef\delta$, and δ is called the *defect* of the extension.

defective equation   An equation, derived from another equation, which has fewer roots than the original equation. For example, if $x^2 + x = 0$ is divided by x, the resulting equation $x + 1 = 0$ is defective because the root 0 was lost in the process of division by x.

defective number   A positive integer which is greater than the sum of all its factors (except itself). For example, the number 10 is defective, since the sum of its factors (except itself) is $1 + 2 + 5 = 8$.

If a positive integer is equal to the sum of all its factors (except itself), then it is called a perfect number. If the number is less than the sum of its factors (except itself), then it is called abundant.

deficiency   Let $R \mapsto F \mapsto G$ be a free presentation of a group G. Let $\{x_k\}$ be a set of generators of F and $\{r_k\}$ be a set of elements of F generating R as a normal subgroup. Then the data $P = (\{x_k\}, \{r_k\})$ is called a group presentation of G, x_k are called generators, and r_k are called relators. The group presentation P is called finite if both $\{x_k\}$ and $\{r_k\}$ are finite. A group G is finitely presentable if there exists a finite group presentation for G.

The deficiency of a finite group presentation P is the integer

$$\text{def}(P) = \#\{\text{generators}\} - \#\{\text{relators}\} .$$

The deficiency of a finitely presentable group G, denoted $\text{def}(G)$, is the maximum deficiency of finite group presentations for G.

defining relation   A defining relation for a quantity or property τ is an equation or property that uniquely determines τ.

definite Hermitian form   Let H be a (complex) Hilbert space. A Hermitian form is a function $f : H \times H \to \mathbf{C}$, such that $f(x, y)$ is linear in x, and conjugate linear in y, and $f(x, x)$

is real. A Hermitian form f is called *positive definite* [resp., *nonnegative definite, negative definite, nonpositive definite*] if, for $x \neq 0$, $f(x, x) > 0$ [resp., $\geq 0, < 0, \leq 0$].

definite quadratic form   Let E be a finite dimensional vector space over the complex numbers **C** . Let L be a symmetric bilinear form on E. (*See* bilinear form, symmetric form.) The quadratic form associated with L is the function $Q(x) = L(x, x)$. If $Q(x) > 0$, [resp., $Q(x) \geq 0, < 0, \leq 0$] the form is called *positive definite* [resp., *nonnegative definite, negative definite, nonpositive definite*]. As an important application, assume $F(x, y)$ is a smooth function of two real variables x and y. Let f be the quadratic part of the Taylor expansion of F

$$f(x, y) = ax^2 + 2bxy + cy^2 ,$$

where a, b, and c are determined as the appropriate second partial derivatives of F, evaluated at a given point. Questions involving the minimum points of F can be solved by considering the two by two matrix

$$A = \begin{bmatrix} a & b \\ c & d \end{bmatrix} ,$$

with its determinant and the upper left one by one determinant a both positive. In this case we are considering

$$(x, y)A \begin{bmatrix} x \\ y \end{bmatrix} = f(x, y) .$$

deflation   A process of finding other eigenvalues of a matrix when one eigenvalue and eigenvector are known. More specifically, if A is an $n \times n$ matrix with eigenvalues $\lambda_1, \ldots, \lambda_n$, and $A\mathbf{v} = \lambda_1 \mathbf{v}$, with $\mathbf{v}$ a nonzero (column) vector, then, for any other vector $\mathbf{u}$, the eigenvalues of the matrix $B = A - \mathbf{v}\mathbf{u}^T$ are

$$\lambda_1 - \mathbf{u}^T \mathbf{v}, \lambda_2, \ldots, \lambda_n .$$

By choosing $\mathbf{u}^T$ to be a multiple of the first row of A and scaled so that $\mathbf{u}^T v = \lambda_1$, the first row of B becomes identically zero.

deformation   A deformation is a transformation which shrinks, twists, expands, etc. in

any way without tearing. Additional conditions, such as continuity, are usually attached to a deformation. Thus, one can talk about a continuous deformation, or smooth deformation, etc.

degenerate   A term found in numerous subjects in mathematics. For example, in algebraic geometry, when considering the homogeneous bar resolutions of Abelian groups, one uses subgroups generated by $(n + 1)$-tuples $(y_0, y_1, \ldots, y_n)$ with $y_i = y_{i+1}$ for at least one value of i; such an $(n + 1)$-tuple is called *degenerate*.

degree of divisor   If a polynomial $p(x)$ is factored as follows:

$$p(x) = a_0 (x - x_1)^{n_1} \cdots (x - x_k)^{n_k} ,$$

where $x_1, \ldots, x_k$ are distinct. Then each $x - x_i$ is called a divisor and the corresponding n_i is called the degree of the divisor.

degree of equation   In a polynomial equation, the highest power is called the *degree of equation*. In a differential equation, the highest order of differentiation is called the *degree of equation*.

degree of polynomial   A polynomial is an expression of the form

$$p(x) = a_0 + a_1 x + \cdots + a_n x^n, \qquad a_n \neq 0 .$$

The integer n is called the *degree of polynomial* $p(x)$.

degree of polynomial term   *See* degree of polynomial.

De Moivre's formula   *See* De Moivre's Theorem.

De Moivre's Theorem   For any integer n and any angle θ the complex equation (De Moivre's formula)

$$(\cos \theta + i \sin \theta)^n = \cos(n\theta) + i \sin(n\theta)$$

holds.

denominator   The quantity B, in the fraction $\frac{A}{B}$ (A is called the *numerator*).

density　Weight per unit (volume, area, length, etc.).

dependent variable　In a function $y = f(x)$, x is called the *independent variable* and y the *dependent variable*.

derivation　A map $D : A \to M$, from a commutative ring A to an A-module M such that

$$D(a + b) = D(a) + D(b)$$

and

$$D(ab) = aD(b) + bD(a)$$

for all a and b in A.

derivation of equation　A proof of an equation, by modifying a known identity, using certain rules.

derivative　If a function $y = f(x)$ is defined on a real interval (a, b), containing the point x_0, then the limit

$$\lim_{x \to x_0} \frac{f(x) - f(x_0)}{x - x_0},$$

if it exists, is called the *derivative* of f at x_0 and may be denoted by

$$f'(x_0), \frac{df}{dx}(x_0), D_x f(x_0), f_x(x_0), \text{ etc}.$$

If the function $y = f(x)$ has a derivative at every point of (a, b), then the derivative function $f'(x)$ is sometimes simply called the *derivative* of f.

derived equation　*See* derivation of equation.

derived functor　If T is a functor then its left-derived functors are defined inductively as follows: Let $T_0 = T$. If S_n is any connected sequence of (additive) functors, then each natural transformation $S_0 \to T_0$ extends to a unique morphism $\{S_n : n \geq 0\} \to \{T_n : n \geq 0\}$ of connected sequences of functors. The right derived functors are defined similarly.

derived series　Given a Lie algebra G, we define its derived series $G_0, G_1, \ldots,$ inductively by $G_0 = G$, $G_{n+1} = [G_n, G_n]$, $n \geq 0$, where, for any subsets S and T of G, $[S, T]$ denotes the

Lie subalgebra generated by all $[s, t]$ for $s \in S$ and $t \in T$.

Descartes's Rule of Signs　A rule setting an upper bound to the number of positive or negative zeros of a function. For example, the positive zeros of the function $f(x)$ cannot exceed the number of changes of sign in $f(x)$.

descending central series　The series of normal subgroups

$$G = N_0 \supseteq N_1 \supseteq N_2 \supseteq \cdots,$$

of a group G, defined recursively by: $N_0 = G$, $N_{i+1} = [G, N_i]$, where

$$[G, N_i] = \left\{ x^{-1}y^{-1}xy : x \in G, \ y \in N_i \right\}$$

is a commutator subgroup. *See also* commutator subgroup.

descending chain of subgroups　A (finite or infinite) sequence $\{G_i\}$ of subgroups of a group G, such that each G_{i+1} is a subgroup of G_i. *See also* subgroup.

determinant　A number, defined for every square matrix, which encapsulates information about the matrix. Common notation for the *determinant* of A is det A and $|A|$. For a 1×1 matrix, the *determinant* is the unique entry in the matrix. For a 2×2 matrix,

$$\begin{pmatrix} a_{11} & a_{12} \\ a_{21} & a_{22} \end{pmatrix}$$

the *determinant* is $a_{11}a_{22} - a_{12}a_{21}$. For a larger $n \times n$ matrix A, the *determinant* is defined recursively, as follows: Let $A_{i,j}$ be the $(n-1) \times (n-1)$ matrix created from A by removing the ith row and the jth column. Then

$$\det A = \sum_{i=1}^{n} (-1)^{i+j} a_{ij} \det A_{i,j}$$

This sum can be computed, and is the same, for any choice of j between 1 and n.

determinant factor　If A is a matrix with elements in a principal ideal ring (for example, the integers or a polynomial ring over a field), then the *determinant factors* of A are the numbers $d_1, \ldots, d_r$ where d_i is the greatest common divisor of all minors of A of degree i and r is the rank of A.

determinant of coefficients   The *determinant of coefficients* of a set of n linear equations

$$a_{11}x_1 + \cdots + a_{1n}x_n = b_1$$
$$\vdots \qquad\qquad \vdots$$
$$a_{n1}x_1 + \cdots + a_{nn}x_n = b_n$$

in n unknowns over a commutative ring R is denoted by

$$\det \begin{bmatrix} a_{11} & \cdots & a_{1n} \\ \vdots & \ddots & \vdots \\ a_{n1} & \cdots & a_{nn} \end{bmatrix},$$

or by

$$\begin{vmatrix} a_{11} & \cdots & a_{1n} \\ \vdots & \ddots & \vdots \\ a_{n1} & \cdots & a_{nn} \end{vmatrix}.$$

Its value is $\sum_P (\operatorname{sgn} P) a_{1p_1} \cdots a_{np_n}$, where the sum is over all permutations $P = (p_1, \ldots, p_n)$ of $1, 2, \ldots, n$. Usually, R is the real or the complex numbers. The set of equations is uniquely solvable if and only if the determinant of the coefficients is nonzero.

diagonalizable linear transformation   A linear transformation from a vector space V into another vector space W which can be represented by a diagonal matrix (with respect to some choice of bases for V and W). *See* diagonal matrix. *See also* diagonalizable operator.

diagonalizable operator   A linear transformation of a vector space V into itself which can be represented by a diagonal matrix with respect to some basis of V. An operator is diagonalizable if and only if there is a basis for V made up entirely of eigenvectors of the operator. *See also* Jordan normal form.

diagonally dominant matrix   An $n \times n$ matrix $A = (a_{ij})$ with entries from the complex field is called *row diagonally dominant* if

$$|a_{ii}| \geq \sum_{k \neq i} |a_{ik}|$$

for each $i \in \{1, 2, \ldots, n\}$. When the inequality above holds strictly for every $i \in \{1, 2, \ldots, n\}$, we say that A is *strictly row diagonally dominant*. Similarly, we can define diagonal dominance with respect to the sums of the moduli of the off-diagonal entries in each column.

diagonal matrix   An $n \times n$ matrix (a_{ij}) where $a_{ij} = 0$ if $i \neq j$.

diagonal sum   The sum of the diagonal entries of a square matrix A, which also equals the sum of the eigenvalues of A. If the entries in A are complex and the diagonal sum is positive (negative), then at least one of the eigenvalues of A has a positive (negative) real part. Also called *spur* or *trace*.

difference   The (set theoretical) difference of two sets A and B is defined by:

$$A \backslash B = \{x : x \in A \quad \text{and} \quad x \notin B\}.$$

The (algebraic) difference of subsets A and B of a group G is defined by:

$$A - B = \{x \in G : x = a - b, \ a \in A, \ b \in B\}.$$

difference equation   An equation of the form

$$x_{n+1} = F(x_n, x_{n-1}, \ldots, x_0),$$

which defines a sequence of numbers, provided that initial values (e.g., x_0) are given. Difference equations are the discrete analog of differential equations. The difference equation known as the logistic equation, $x_{n+1} = ax_n(1 - x_n)$, a constant, is one of the original examples of a system that exhibits chaotic behavior.

difference group   The quotient group of an additive group G by a subgroup H (written as $G - H$). For example, the additive group of integers has the even integers as a subgroup, and the *difference group* is the mod2 group $\{0, 1\}$. *See* quotient group.

difference of like powers   The factorization:

$$a^n - b^n = (a - b)\left(a^{n-1} + a^{n-2}b \right.$$
$$\left. + a^{n-3}b^2 + \cdots + ab^{n-2} + b^{n-1}\right).$$

difference of the nth order   If $y(x)$ is a function of a real variable x and Δx is a fixed number, then the first order difference $\Delta y(x)$ is defined by $\Delta y(x) = y(x + \Delta x) - y(x)$. Scaling Δx to 1, the *difference of the nth order* is defined by

$$\Delta^n y(x) = \Delta \left(\Delta^{n-1} y(x) \right)$$

$$= \sum_{k=0}^{n} (-1)^{n-k} \binom{n}{k} y(x + k) .$$

difference of two squares   The factorization: $a^2 - b^2 = (a - b)(a + b)$. *See also* difference of like powers.

difference product   The polynomial defined over an integral domain by $p(x_1, \ldots, x_n) = \prod_{i<j}(x_i - x_j)$. Also called *simplest alternating polynomial* in $x_1, \ldots, x_n$. The *difference product* p is invariant with respect to even permutations of $x_1, \ldots, x_n$ and becomes $-p$ with respect to an odd permutation. If the characteristic of the integral domain is different from 2 and q is any alternating polynomial in $x_1, \ldots, x_n$, then $q = ps$, where s is symmetric.

different   Suppose that K is an algebraic number field and $\mathbf{Q}$ denotes the rational subfield of the complex number field. Let $M = \{ A \in K : \text{trace}(A\Theta) \subset \theta$, where Θ and θ are the principal orders of K and $\mathbf{Q}$, respectively$\}$. Then the *different* $D_{K/Q}$ of K is M^{-1}. *See* algebraic number field, principal order.

differential automorphism   An automorphism A of a differential field F such that A commutes with each derivation of F and leaves the ground field fixed. Kolchin has determined the structure of the group of *differential automorphisms. See* differential field.

differential extension ring   If R is a differential ring and S is a differential subring of R, then R is a *differential extension ring* of S. For example, the differential ring of all real-valued differentiable functions on the real line is an extension of the subring of polynomials. *See* differential ring, differential subring.

differential field   A differential ring which also forms a field. *See* differential ring.

differential form of the first kind   Suppose that Γ is a nonsingular curve and that ω is a differential form on Γ. If (ω) is a positive divisor of the free Abelian group generated by points of Γ, then ω is a *differential form of the first kind* (or a *regular 1-form*). If, for any point P of Γ, there is a rational function f_P such that $\omega - df_P$ is a regular 1-form, then ω is a *differential form of the second kind*. If ω has nonzero residues, then it is a *differential form of the third kind*.

differential form of the second kind   *See* differential form of the first kind.

differential form of the third kind   *See* differential form of the first kind.

differential ideal   Suppose that R is a differential ring with derivations $D_1, \ldots, D_k$. Then an ideal a of R is a *differential ideal* if $D_i a \subset a$ for all i. *See* differential ring.

differential index   Suppose that Γ_1 and Γ_2 are nonsingular curves, that $\pi : \Gamma_1 \to \Gamma_2$ is singular, and that t_1 and t_2 are local parameters at P on Γ_1 and $Q = \pi(P)$ on Γ_2, respectively. The *differential index* at P is the nonnegative integer $v_P(ds/dt)$, where v_P is the valuation at P.

differential polynomial   Suppose that $X_1, \ldots, X_r$ belong to a field which is a differential extension (ring) of a differential field K, with derivations $D_1, \ldots, D_k$. Then $X_1, \ldots, X_r$ are *differential variables.* If $s_1, \ldots, s_k$ are nonnegative integers and $D_1^{(s_1)} \cdots D_k^{(s_k)} X_i$ are algebraically independent over K, then a polynomial in these elements is a *differential polynomial.*

differential representation   Suppose that U is a unitary representation of a Lie group G with Lie algebra g, that $X \in g$ and that x is an analytic vector with respect to U. Suppose that $V(X)x$ is the derivative at $t = 0$ of $U_{\exp tX}(x)$. Then the linear mapping $V : X \to V(X)$ is the *differential representation* of U.

differential ring   A commutative ring R, with unit, together with a finite number of commuting derivations on R. (A *derivation* on R is a mapping $D : R \to R$ such that $D(x+y) = Dx + Dy$

and $D(xy) = Dx \cdot y + x \cdot Dy$, for $x, y \in R$.) The ring of all real-valued differentiable functions of a real variable is a *differential ring*.

differential subring   If R is a differential ring with derivations $D_1, \ldots, D_k$, then a subring of S of R is a *differential subring* if $D_i S \subset S$ for all i. *See* differential ring.

differential variable   *See* differential polynomial.

differentiation   (1) In a chain complex, a map, usually denoted d_n or δ_n, from one module to the next. An example of a differentiation is the boundary operator encountered in the study of simplicial complexes. *See also* chain complex, boundary operator.

(2) Of a function $f(x)$ of a real variable, at a real number $x = a$, the limit

$$f'(a) = \lim_{h \to 0} \frac{f(a+h) - f(a)}{h}.$$

Many generalizations to other topological spaces exist.

digit   A symbol in a number system. For example, in the binary number system, the only digits that are used are 0 and 1. *See also* duodecimal number system.

dihedral group   An algebraic group D_n, generated by two elements: a, which is a rotation of the Euclidean plane about the origin through angle of $\frac{2\pi}{n}$, and b, which is reflection through the y-axis. D_n is the group of symmetries of the regular n-gon, and has order $2n$.

dimension   The number of vectors in the basis of a vector space V. If the basis is finite, then V is called finite dimensional. In this case, if V is a vector space of *dimension* n over the real numbers $\mathbf{R}$, then it is isomorphic to the Euclidean space $\mathbf{R}^n$. If the basis is infinite, then V is called infinite dimensional. *See also* basis.

Diophantine equation   An equation in which solutions are restricted to the integers.

direct decomposition   A group G has a *direct decomposition* $G = H_1 \times H_2 \times \cdots \times H_n$

if each H_i is a normal subgroup of G, $G = H_1 H_2 \ldots H_n$, and $H_1 \ldots H_{i-1} \cap H_i = \{e\}$, $i = 2, \ldots, n$.

directed set   A set X equipped with a partial ordering and such that if $x, y \in X$ then there exists $z \in X$ such that $x \leq z$ and $y \leq z$.

direct factor   Either H or K, in the direct product $H \times K$ or $H \otimes K$. *See* direct product.

direct integral   *See* integral direct sum.

direct limit   Suppose $\{G_\mu\}_{\mu \in I}$ is an indexed family of Abelian groups, where I is a directed set. Suppose that there is also a family of homomorphisms $\varphi_{\mu\nu} : G_\mu \to G_\nu$, defined for all $\mu < \nu$, such that: $\varphi_{\mu\mu} : G_\mu \to G_\mu$ is the identity, and if $\mu < \nu < \kappa$ then $\varphi_{\nu\kappa} \circ \varphi_{\mu\nu} = \varphi_{\mu\kappa}$. Consider the disjoint union of the groups G_μ and form an equivalence relation by $x_\mu \sim x_\nu$, $x_\mu \in G_\mu$ and $x_\nu \in G_\nu$, if for some upper bound κ of μ and ν we have $\varphi_{\mu\kappa}(x_\mu) = \varphi_{\nu\kappa}(x_\nu)$. Then the direct limit is defined to be the set of equivalence classes and is denoted by

$$\lim_{\to \ \mu \in I} G_\mu.$$

See also directed set.

direct product   (1) A group G is called the *internal direct product* of subgroups H and K if the following three conditions hold: H and K are normal subgroups of G, $H \cap K$ contains only the identity element, and $G = HK$. This internal direct product is denoted $H \times K$.

(2) If H and K are any two groups, then the *external direct product* of H and K, denoted $H \otimes K$, is the Cartesian product: $\{(h, k) : h \in H, k \in K\}$. $H \otimes K$ is a group, defining multiplication componentwise, i.e., $(h_1, k_1) \cdot (h_2, k_2) = (h_1 \cdot h_2, k_1 \cdot k_2)$. *See also* internal product, external product.

direct proportion   Quantity x is *directly proportional* to quantity y, or *varies directly as y*, if there is a constant $k \neq 0$ such that $x = ky$. (k is called the *constant of proportionality*.) *See also* inverse proportion, joint proportion.

direct sum   (1) In the case where $V_1, V_2, \ldots,$ V_n are all vector spaces over the same field F, one can define the direct sum V to be a vector space made up of n-tuples of the form $(v_1, v_2, \ldots, v_n)$, where $v_i \in V_i$. The common notation for this direct sum is:

$$V = V_1 \oplus V_2 \oplus \cdots \oplus V_n .$$

(2) In the case where $V_1, V_2, \ldots, V_n$ are all vector subspaces of the same vector space W, and $V_i \perp V_j$ if $i \neq j$, we can define the direct sum V to be a vector space made up of sums of the form: $v_1 + v_2 + \cdots + v_n$ where $v_i \in V_i$. The same notation is used as above. *See also* orthogonal subset.

(3) In the case where $H_1, H_2, \ldots, H_n$ are all subgroups of the same Abelian group G, we can define the direct sum H to be a subgroup of G made up of sums of the form: $h_1 + h_2 + \cdots + h_n$ where $h_i \in H_i$, provided that each $x \in H$ has a unique representation as the sum of elements from the subgroups $\{H_i\}$. The same notation is used as above.

(4) If A is a $k \times l$ matrix and B is an $m \times n$ matrix, then the direct sum of A and B is the $(k + m) \times (l + n)$ partitioned matrix

$$\begin{pmatrix} A & 0 \\ 0 & B \end{pmatrix} .$$

direct trigonometric functions   The usual trigonometric functions (sine, cosine, tangent, etc.) as opposed to the inverse trigonometric functions.

direct variation   *See* direct proportion.

Dirichlet algebra   A closed subalgebra A of the continuous complex-valued functions $C(X)$ on a compact Hausdorff space X such that (i.) A contains the constant functions, (ii.) A separates points of X, and (iii.) $\{\Re(f), f \in A\}$ is dense in $C_{\mathbf{R}}(X)$, the space of all real-valued and continuous functions on X.

Dirichlet L-function   The function $L(s)$, defined by

$$L(s) = \sum_{n=1}^{\infty} \chi(n)/n^s ,$$

where χ is a character of the group of classes coprime to some positive integer m and

$$\chi(n) = \begin{cases} \chi((n)) \text{ if } (n,m)=1 \\ 0 \text{ if } (n,m)\neq 1 , \end{cases}$$

where (n) is the residue class of $n \pmod m$. The function $L(s)$ converges absolutely for $\Re(s) > 1$, and is used widely in the study of rational number fields and of quadratic and cyclotomic number fields. *See also* L-function.

Dirichlet Unit Theorem   Suppose that k is an extension field (of first degree) of the rational subfield $\mathbf{Q}$ of the complex number field. Then the group of units of k is the direct product of a cyclic group and a free Abelian multiplicative group.

discrete filtration   A finite collection $\{F^1, \ldots, F^n\}$ of submodules of a module A such that $F^i \supset F^{i+1}$ and $F^n = 0$.

discrete series   Suppose that G is a connected, semisimple Lie group, with a square integrable representation. The set of all square integrable representations of G is the discrete series of the irreducible unitary representations of G.

discrete valuation   A non-Archimedean valuation v is *discrete* if the valuation ideal of v is a nonzero principal ideal. In this case the valuation ring for v is also said to be *discrete*.

discrete valuation ring   *See* discrete valuation.

discriminant   (1) For the quadratic equation $ax^2 + bx + c = 0$, the number $\Delta = b^2 - 4ac$. If $\Delta > 0$, then the equation has two real-valued solutions. If $\Delta < 0$, then the equation has two complex-valued solutions which are complex conjugates. If $\Delta = 0$, then the equation has a double root which is real valued.

(2) For the conic section $Ax^2 + Bxy + Cy^2 + Dx + Ey + F = 0$, the number $\Delta = B^2 - 4AC$. If $\Delta > 0$, then the conic section is a hyperbola. If $\Delta < 0$, then the conic section is an ellipse. If $\Delta = 0$, then the conic section is a parabola. The discriminant is invariant under rotation of the axes.

discriminant of equation   *See* discriminant.

disjoint unitary representations   A pair, U_1 and U_2, of unitary representations of a group such that no subrepresentation of one is equivalent to a subrepresentation of the other.

disjunctive programming   In mathematical programming, the task is to find an extreme value of a given function f, which maps a set A into an ordered set R. Usually A is a closed subset of a Euclidean space and (usually) A is defined by a collection of inequalities or equalities. In *disjunctive programming,* the set A is not connected.

distributive algebra   A linear space A, over a field K, such that there is a bilinear mapping (or multiplication) $A \times A \to A$. If the multiplication does not satisfy the associative law, the algebra is *nonassociative.*

distributive law   A law from algebra that states that if a, b, and c belong to a set with two binary operations $+$ and $\cdot$, then it is true that $a \cdot (b+c) = a \cdot b + a \cdot c$ and $(b+c) \cdot a = b \cdot a + c \cdot a$.

dividend   The quantity a in the division algorithm. It is a quantity which is to be divided by another quantity; that is, the number a in $\frac{a}{b}$. *See* division algorithm.

divisibility relation   Suppose that a, b, and c lie in a ring R and that $a = bc$. Then we say that b divides a (written $b|a$), and call this a *divisibility relation* where b is a *divisor* or *factor* of a.

divisible   An integer a is *divisible* by an integer b if there exists another integer k such that $a = bk$.

divisible group   An Abelian group G (under the operation of addition) such that, for every $g \in G$ and every $n \in \mathbf{N}$, there exists $x \in G$ such that $g = nx$. In other words, each element in G is divisible by every natural number. An example of a divisible group is the factor group G/T, where T is the torsion subgroup of G. *See* torsion group.

division   (1) Finding the quotient q and the remainder r in the division algorithm. *See* division algorithm.

(2) A binary operation which is the inverse of the multiplication operation. *See also* quotient.

division algebra   An algebra such that every nonzero element has a multiplicative inverse.

division algorithm   Given two integers a and b, not both equal to zero, there exist unique integers q and r such that $a = qb + r$ and $0 \le r < |q|$. q is called the quotient and r is called the remainder. Also called *Euclidian Algorithm.*

The *division algorithm* is used in the development of a number of ideas in elementary number theory, including greatest common divisor and congruence. There are other situations in which a division algorithm holds. *See* greatest common divisor, congruence, division of polynomials, Gaussian integer.

division by logarithms   To compute $\frac{x}{y}$, first compute $c = \log_b x - \log_b y$. Then $\frac{x}{y} = b^c$. This can be an aid to computation, when a table of logarithms to the base b is available. *See also* logarithm.

division by zero   For any real number x, $\frac{x}{0}$ is not defined.

division of complex numbers   For real numbers a, b, c, and d

$$\frac{a+bi}{c+di} = \frac{ac+bd}{c^2+d^2} + \frac{bc-ad}{c^2+d^2}i$$

This formula comes from multiplying the numerator and denominator of the original expression by $c - di$.

division of polynomials   If polynomials $f(x)$ and $g(x)$ belong to the polynomial ring $F[x]$, and the degree of $g(x)$ is at least 1, then there exist unique polynomials $q(x)$ and $r(x)$ in $F[x]$ such that

$$f(x) = q(x)g(x) + r(x)$$

where $r(x) \equiv 0$ or the degree of $r(x)$ is less than the degree of $q(x)$. The process is sometimes called *synthetic division. See also* division algorithm, degree of polynomial.

division of whole numbers *See* division algorithm. *See also* divisible.

divisor (**1**) The quantity b in the division algorithm. *See* division algorithm.

(**2**) For an integer a, the integer d is called a divisor of a if there is another integer b so that $a = bd$. Colloquially, d is a divisor of a if d "evenly divides" a.

divisor class An element of the factor group of the group of divisors on a Riemann surface by the subgroup of meromorphic functions. The factor group is called the *divisor class group*.

divisor class group If X is a smooth algebraic variety, then the *divisor class group* is the free Abelian group on the irreducible codimension one subvarieties of X (these are called "divisors"), modulo divisors of the form $(f) = (f)_0 - (f)_\infty$, for all rational (or meromorphic) functions f on X; here $(f)_0$ is the divisor of zeros and $(f)_\infty$ is the divisor of poles.

If X is not smooth, "divisor" in "divisor class group" means Cartier divisor, i.e., a divisor which is locally given by one equation; that is to say, one which is locally of the form (f) for a rational function f on X.

Divisor class groups also exist in commutative algebra, and in geometry for singular germs.

See divisor class.

divisor of set Let X be a non-singular variety defined over an algebraically closed field k. A closed irreducible subvariety $Y \subseteq X$ having codimension 1 is called a *prime divisor*. An element of the free Abelian group generated by the set of prime divisors is called a *divisor on X*.

domain The domain of a function is the set on which the function is defined. For example, the function $f(x) = \sin x$ has domain $\mathbf{R}$ since the sine of every real number is defined, while the domain of the function $f(x) = \sqrt{x}$ is the non-negative real numbers. *See also* integral domain, unique factorization domain, range.

Doolittle method of factorization An "LU-factorization" method for a square matrix A. The method concerns factoring A into the product of two square matrices: L is a lower triangular matrix and U is an upper triangular matrix. All of the diagonal elements of L are required to be 1. Explicit formulas can then be created for the rest of the entries in L and U. *See also* Gaussian elimination.

double chain complex A double complex of chains $\mathbf{B}$ over Γ is an object in $M_\Gamma^{\mathbf{Z} \times \mathbf{Z}}$, together with two endomorphisms $\partial' : \mathbf{B} \to \mathbf{B}$ and $\partial'' : \mathbf{B} \to \mathbf{B}$ of degree $(-1, 0)$ and $(0, -1)$, respectively, called the differentials, such that

$$\partial'\partial' = 0, \quad \partial''\partial'' = 0, \quad \partial''\partial' + \partial'\partial'' = 0 .$$

In other words, we are given a bigraded family of Γ-modules $\{B_{pq}\}$, $p, q \in \mathbf{Z}$, and two families of Γ-module homomorphisms

$$\left\{ \partial'_{pq} : B_{pq} \to B_{p-1q} \right\}, \quad \left\{ \partial''_{pq} : B_{pq} \to B_{pq-1} \right\} ,$$

such that the earlier three equations involving the operators ∂' and ∂'' hold.

double invariance Let Γ be a dense subgroup of the additive group $\mathbf{R}$ of real numbers, with the discrete topology. Let G be the character group of Γ. Any element $a \in \Gamma$, as a character of G, defines a continuous function χ_a on G. Let σ be the Haar measure of G. A closed subspace M of $L^2(\sigma)$ is called *invariant* if $\chi_a M \subseteq M$, for all $a \in \Gamma$ with $a \geq 0$. M is called *doubly invariant* if $\chi_a M \subseteq M$ for all $a \in \Gamma$. Such invariance is called *double invariance*.

Douglas algebra Let L^∞ be the space of bounded functions on the unit circle and H^∞ be the subspace of L^∞ consisting of functions whose harmonic extension to the unit disk is bounded and analytic. An inner function is a function in H^∞ whose modulus is 1 almost everywhere. A Douglas algebra is a subalgebra of L^∞ generated by H^∞ and the conjugates of finitely many inner functions (in the uniform topology).

A theorem of Chang and Marshall asserts that every uniform algebra A, with $H^\infty \subset A \subset L^\infty$ is a Douglas algebra.

downhill method of finding roots To find the real roots of

$$f(x, y) = 0, \quad g(x, y) = 0,$$

find points (a, b) where $\phi = f + g$ has its extreme values. In the *downhill method* the coordinates of (a, b) are approximated by choosing an estimate (x_0, y_0), selecting a value for h, evaluating ϕ at the nine points $x_j = x_0 + \epsilon_1 h$, $y_k = y_0 + \epsilon_2 h$ where $\epsilon_i = -1, 0, 1$, and constructing the quadratic surface, $\phi_1 = a_0 + a_1 x + a_2 y + a_3(3x^2 - 2) + a_4(3y^2 - 2) + a_5 xy$, where the values of the coefficients a_j are calculated by applying the *method of least squares* and using the values of ϕ_1 at the nine points (x_j, y_k). Then replace the 0th approximation (x_0, y_0) to (a, b) by the center (x_1, y_1) of the quadratic surface defined by ϕ_1. The process is repeated with smaller and smaller values of h.

Drazin inverse *See* generalized inverse.

dual algebra Suppose that (A, μ, η) is an algebra over a field k (with multiplication μ and unit mapping η), and (C, Δ, ϵ) is a *coalgebra* of (A, μ, η). Then (C^*, μ, η) is the *dual algebra* of (C, Δ, ϵ) if C^* is the dual space of C and μ, η and Δ, ϵ are correspondingly dual.

dual coalgebra Suppose that (A, μ, η) is an algebra over a field k and that A° is the collection of elements of the dual space A^* whose kernels each contain an ideal I where A/I is finite dimensional. Then $(A^\circ, \Delta, \epsilon)$, where Δ and ϵ are induced dually by μ and η, respectively, is the *dual coalgebra*.

dual curve Suppose that Γ is an irreducible plane curve of degree $m \geq 1$ in a projective plane. The *dual curve* $\hat{\Gamma}$ in the dual projective plane is the closure of the set of tangent lines to Γ at its nonsingular points. The dual of $\hat{\Gamma}$ is Γ.

dual graded module Suppose $A = \sum_{n \leq 0} A_n$ is a graded module over a field k. If A_n^* is the dual of the module A_n, then $A^* = \sum A_n^*$ is the *dual graded module* of A.

dual homomorphism A mapping $\phi : L_1 \to L_2$, of one lattice to another, such that $\phi(x \cap y) = \phi(x) \cup \phi(y)$ and $\phi(x \cup y) = \phi(x) \cap \phi(y)$.

dual Hopf algebra Suppose that (A, ϕ, ψ) is a graded Hopf algebra. Then (A^*, ϕ^*, ψ^*) is also a graded Hopf algebra which is called the *dual Hopf algebra. See* graded Hopf algebra.

duality If X is a normed complex vector space, then the set of all bounded linear functionals on X is called the dual of X and is usually denoted X^*. The dual space can be defined for many other classes of spaces, including topological vector spaces, Banach spaces, and Hilbert spaces. An identification of the dual space is usually referred to as *duality.* For example, the *duality* of L^p spaces, where $L^{p*} = L^q$, $p^{-1} + q^{-1} = 1$, $1 \leq p < \infty$.

duality principle in projective geometry Suppose that P^n is a finite dimensional projective geometry of dimension n. Suppose that T is a proposition in P^n and P^{n-r-1} ($0 \leq r \leq n$) and that "contains" and "contained in" are reversed, in the statement of T, obtaining in this way a new statement $\hat{T}$, called the *dual* of T. Then T is true if and only if $\hat{T}$ is true.

Duality Theorem A theorem in the study of linear programming. The *Duality Theorem* states that the minimum value of $c_1 x_1 + c_2 x_2 + \cdots + c_n x_n$ in the original problem is equal to the maximum value of $b_1 y_1 + b_2 y_2 + \cdots + b_m y_m$ in the dual problem, provided an optimal solution exists. If an optimal solution does not exist, then there are two possibilities: either both feasible sets are empty, or else one is empty and the other is unbounded. *See* dual linear programming problem.

dual linear programming problem (**1**) Another linear programming problem which is intimately related to a given one. Consider the linear programming problem: minimize $c_1 x_1 + c_2 x_2 + \cdots + c_n x_n$, where all $x_i \geq 0$ and subject to the system of constraints:

$$
\begin{aligned}
a_{11} x_1 + a_{12} x_2 + \cdots + a_{1n} x_n &\geq b_1 \\
a_{21} x_1 + a_{22} x_2 + \cdots + a_{2n} x_n &\geq b_2 \\
&\vdots \\
a_{m1} x_1 + a_{m2} x_2 + \cdots + a_{mn} x_n &\geq b_m .
\end{aligned}
$$

The following linear programming problem is known as the dual problem: maximize $b_1 y_1 +$

$b_2 y_2 + \cdots + b_m y_m$ where all $y_i \geq 0$ and subject to the system of constraints:

$$
\begin{aligned}
a_{11} y_1 + a_{21} y_2 + \cdots + a_{m1} y_m &\leq c_1 \\
a_{12} y_1 + a_{22} y_2 + \cdots + a_{m2} y_m &\leq c_2 \\
&\vdots \\
a_{1n} y_1 + a_{2n} y_2 + \cdots + a_{mn} y_m &\leq c_n .
\end{aligned}
$$

The Duality Theorem describes the relation between the optimal solution of the two problems. *See* Duality Theorem.

(2) The definition above may also be stated in matrix notation. If $\vec{c}$ is an $1 \times n$ vector, $\vec{b}$ is a $m \times 1$ vector, and A is an $m \times n$ matrix, then the original linear programming problem above can be stated as follows: Find the $n \times 1$ vector $\vec{x}$ which minimizes $\vec{c}\vec{x}$, subject to $\vec{x} \geq 0$ and $A\vec{x} \geq \vec{b}$. The dual problem is to maximize $\vec{y}\vec{b}$, subject to $\vec{y} \geq 0$ and $\vec{y}A \leq c$.

(3) The dual of a dual problem is the original problem.

dual module   For a module M over an arbitrary ring R, the module denoted M^*, equal to the set of all module homomorphisms (also called linear functionals) from M to R. This dual module is also denoted $\mathrm{Hom}_R(M, R)$ and is, itself, a module. *See also* homomorphism, linear function.

dual quadratic programming problem
Suppose that P is the quadratic programming problem:

$$\text{maximize} \quad z = c^T x - (1/2)x^T D x ,$$

subject to the constraints $Ax \leq b, x \geq 0, x$ in $\mathbf{R}^n$.

Then the *dual quadratic programming problem* PD is:

$$\text{minimize} \quad w = b^T y + (1/2)x^T D x ,$$

subject to the constraints $A^T y + Dx \geq c$, all components $x_i \geq 0, y_i \geq 0$.

It can be shown that if x^* solves P, then PD has solution x^*, y^*, and max $z = $ max w.

dual representation   Let $\pi : G \to \mathrm{GL}(V)$ be a representation of the group G in the linear space V. Then the dual representation $\pi^\vee :$ $G \to \mathrm{GL}(V_\pi^*)$ is given by $\pi^\vee(g) = \pi(g^{-1})^*$.

dual space   (1) For a vector space V over a field F, the vector space V^*, equal to the set of all linear functions from V to F. V^* is called the *algebraic dual space*.

(2) In the case where H is a normed vector space over a field F, the *continuous dual space* H^* is the set of all bounded linear functions from H to F. By bounded, we mean that each linear function f satisfies

$$\sup_{\substack{x \neq 0 \\ x \in H}} \frac{|f(x)|}{\|x\|} < \infty .$$

This type of dual space is the focus of theorems in functional analysis such as the Riesz Representation Theorem.

duodecimal number system   A number system using a base of 12 rather than 10. The Arabic numerals 0 through 9 are utilized, along with two other symbols X and Σ, which are used to represent 10 (base 10) and 11 (base 10). The number 12 (base 10) is then represented in the duodecimal system as 10.

Durand-Kerner method of solving algebraic equations   A method of solving an algebraic equation $f(z) = z^n + a_1 z^{n-1} + \cdots + a_n = 0$ (with complex coefficients and $a_n \neq 0$) using an iteration formula for approximating the n roots $z_1, \ldots, z_n$ of f_i:

$$z_{i,k+1} = z_{i,k} - f\left(z_{i,k}\right) / \prod_{j=1}^{n} \left(z_{i,k} - z_{j,k}\right) ,$$

$j \neq i, i = 1, \ldots, n, k = 0, 1, 2, \ldots.$ The method approximates all n roots of $f(z)$ simultaneously. Speed of convergence is second order.

dynamic programming   An approach to a multistep decision process, in which an outcome is calculated for each stage. In Richard Bellman's approach to dynamic programming, an *optimal policy* has the property that, for each initial state and decision, the subsequent decisions must generate an optimal policy with respect to the outcome of the initial state and initial decision. This is now the most widely used approach to *dynamic programming*.

E

e   One of the most important constants in mathematics. The following are two common ways of approximating this number:

$$e = \lim_{n \to \infty} \left(1 + \frac{1}{n}\right)^n$$

and

$$e = \sum_{n=0}^{\infty} \frac{1}{n!},$$

where $n! = 1 \cdot 2 \cdot 3 \cdots n$ is the factorial.

The number e is transcendental and its value is approximately 2.7182818 .... The letter e stands for Euler.

effective divisor   *See* integral divisor.

effective genus   Suppose that Γ is an irreducible algebraic curve and $\tilde{\Gamma}$ is a nonsingular curve that is birationally equivalent to Γ. Then the genus of $\tilde{\Gamma}$ is the *effective genus* of Γ. An algebraic curve Γ is *rational* if its effective genus is 0 and *elliptic* if its effective genus is 1.

eigenfunction   A function f which satisfies the equation $T(f) = \lambda f$, for some number λ, where T is a linear transformation from a space of functions into itself. For example, if T is the linear transformation on the space of twice differentiable functions of one variable:

$$T(f) = \frac{d^2 f}{dx^2}$$

then $cos(x)$ is an eigenfunction of T, with $\lambda = -1$. Eigenfunctions arise in the analysis of partial differential equations such as the heat equation. *See also* eigenvalue.

eigenspace   For an eigenvalue λ of a matrix (or linear operator) A, the set of all eigenvectors associated with λ, along with the zero vector. The eigenspace is the space of all possible solutions of the vector equation

$$(\lambda I - A)\vec{x} = \vec{0}.$$

See also eigenvalue, eigenvector.

eigenspace in the weaker sense   Suppose that L is a linear space, T is a linear transformation on L, and λ is an eigenvalue of T. Then the collection of all elements v in L such that $(T - \lambda I)^k v = 0$, for some integer $k > 0$, is the *eigenspace in the weaker sense*, corresponding to the eigenvalue λ. Such an element v is sometimes called a *root vector* of T.

eigenvalue   For an $n \times n$ matrix (or a linear operator) A, a number λ, such that there exists a non-zero $n \times 1$ vector v satisfying the equation $Av = \lambda v$. The product of the eigenvalues of a matrix equals its determinant. *See also* eigenvector.

eigenvector   A non-zero $n \times 1$ vector $\vec{v}$ that satisfies the equation $Av = \lambda \vec{v}$ for some number λ, for a given $n \times n$ matrix A. *See also* eigenvalue.

Eisenstein series   One of the simplest examples of a modular form, defined as a sum over a lattice. In detail: Let Γ be a discontinuous group of finite type operating on the upper half plane H, and let $\kappa_1, \ldots, \kappa_h$ be a maximal set of cusps of Γ which are not equivalent with respect to Γ. Let Γ_i be the stabilizer in Γ of κ_i, and fix an element $\sigma_i \in G = \mathrm{SL}(2, \mathbf{R})$ such that $\sigma_i \infty = \kappa_i$ and such that $\sigma_i^{-1} \Gamma_i \sigma_i$ is equal to the group Γ_0 of all matrices of the form

$$\begin{pmatrix} 1 & b \\ & 1 \end{pmatrix}$$

with $b \in \mathbf{Z}$. Denote by $y(z)$ the imaginary part of $z \in H$. The Eisenstein series $E_i(z, s)$ for the cusp κ_i is then defined by

$$E_i(z, s) = \sum y(\sigma_i^{-1} \sigma z)^s, \quad \sigma \in \Gamma_1 \setminus \Gamma,$$

where s is a complex variable.

Eisenstein's Theorem   If $f(x) = a_0 + a_1 x + a_2 x^2 + \cdots + a_n x^n$ is a polynomial of positive degree with integral coefficients, and if there exists a prime number p such that p divides all of the coefficients of $f(x)$ except a_n, and if p^2 does not divide a_0, then $f(x)$ is irreducible (prime) over the field of rational numbers; that is, it cannot be factored into the product of two polynomials with rational coefficients and positive degrees.

elementary divisor (of square matrix)   For a matrix A with entries in a field F, one of the finitely many monic polynomials h_i, $1 \leq i \leq s$, over F, such that $h_1 | h_2 | \cdots | h_s$, and A is similar to the block diagonal sum of the companion matrices of the h_i.

elementary divisor of a finitely generated module   For a module M over a principal ideal domain R, a generator of one of the finitely many ideals I_i of the polynomial ring $R[X]$, $1 \leq i \leq s$, such that $I_1 \subseteq I_2 \subseteq \cdots \subseteq I_s$ and M is isomorphic to the direct sum of the modules $R[X]/I_i$.

elementary Jordan matrix   An $n \times n$ square matrix of the form

$$\begin{pmatrix} \lambda & 0 & 0 & \ldots & 0 \\ 1 & \lambda & 0 & \ldots & 0 \\ 0 & 1 & \ddots & & \\ \vdots & & \ddots & & \\ 0 & 0 & \ldots & 1 & \lambda \end{pmatrix}.$$

In the special case where $n = 1$, every matrix (λ) is an *elementary Jordan matrix*. (In other words, one can forget about the 1s below the diagonal if there is no room for them.)

It is easier to understand this if we phrase it in the language of linear operators rather than matrices. An *elementary Jordan matrix* is a square matrix representing a linear operator T with respect to a basis $e_1, \ldots, e_n$, such that $Te_1 = \lambda e_1 + e_2, Te_2 = \lambda e_2 + e_3, \ldots, Te_{n-1} = \lambda e_{n-1} + e_n$, and $Te_n = \lambda e_n$. Thus $(T - \lambda I)^n = 0$, that is $T - \lambda I$ must be nilpotent, but n is the smallest positive integer for which this is true. Here, I is the identity operator, and the scalar λ is called the *generalized eigenvalue* for T.

It is a theorem of linear algebra that every matrix with entries in an algebraically complete field, such as the complex numbers, is similar to a direct sum of elementary Jordan matrices. *See* Jordan canonical form. The infinite dimensional analog of this theorem is false. However, *shift* operators, that is operators S such that $Se_i = e_{i+1}$, for $i = 1, 2, 3, \ldots$, play a prominent role in operator theory on infinite dimensional Hilbert and Banach spaces. The connection between shift operators and elementary Jordan matrices is that the nilpotent operator $T - \lambda I$ may be thought of as the finite dimensional analog of the shift S.

elementary symmetric polynomial   A polynomial of several variables that is invariant under permutation of its variables, and which cannot be expressed in terms of similar such polynomials of lower degree. For polynomials of two variables, $x + y$ and xy are all the elementary symmetric polynomials.

elementary symmetric polynomials   For n variables $x_1, \ldots, x_n$, the elementary symmetric polynomials are $\sigma_1, \cdots, \sigma_n$, where σ_k is the sum of all products of k of the variables $x_1, \ldots, x_n$. For example, if $n = 3$, then $\sigma_1 = x_1 + x_2 + x_3$, $\sigma_2 = x_1 x_2 + x_1 x_3 + x_2 x_3$ and $\sigma_3 = x_1 x_2 x_3$.

elimination of variable   A method used to solve a system of equations in more than one variable. First, in one of the equations, we solve for one of the variables. We then substitute that solution into the rest of the equations. For example, when solving the system of equations:

$$\begin{aligned} 2x + 4y &= 10 \\ 8x + 9y &= 47 \end{aligned}$$

we can solve the first equation for x, yielding: $x = 5 - 2y$. Then, substituting this solution into the second equation, we have $8(5 - 2y) + 9y = 47$, which we can then solve for y. Once we have a value for y, we can then determine the value for x. *See also* Gaussian elimination.

elliptic curve   A curve given by the equation

$$y^2 + a_1 xy + a_2 y = x^3 + a_3 x^2 + a_4 x + a_5$$

where each a_i is an integer. Elliptic curves were important in the recent proof of Fermat's Last Theorem. *See* Fermat's Last Theorem.

elliptic function field   The field of functions of an elliptic curve; a field of the form $k(x, \sqrt{f(x)})$, where k is a field, x is an indeterminant over k, and $f(x)$ is a separable, cubic polynomial, with coefficients in k.

elliptic integral   One of the following three types of integral.

Type I:

$$\int_0^x \frac{dt}{\sqrt{(1-t^2)(1-k^2t^2)}} = \int_0^\phi \frac{dt}{\sqrt{1-k^2\sin^2 t}} .$$

Type II:

$$\int_0^x \sqrt{\frac{1-k^2t^2}{1-t^2}}\, dt = \int_0^\phi \sqrt{1-k^2\sin^2 t}\, dt .$$

Type III:

$$\int_0^x \frac{dt}{(t^2-a)\sqrt{(1-t^2)(1-k^2t^2)}}$$

$$= \int_0^\phi \frac{dt}{(\sin^2 t - a)\sqrt{1-k^2\sin^2 t}} .$$

Here $0 < k^2 < 1$ and a is an arbitrary constant. The constant k is called the modulus. If $x = 1$, or equivalently, if $\phi = \pi/2$, then the elliptic integral is called complete; otherwise, it is incomplete.

elliptic transformation   A linear fractional transformation $w = Tz$ of the complex plane which can be written as

$$\frac{w-a}{w-b} = k\frac{z-a}{z-b}$$

for some unimodular constant k.

endomorphism   (1) A function from a space into itself, satisfying additional conditions depending on the nature of the space. For example, when one studies groups, an endomorphism is a function F from a group G into itself such that $F(x+y) = F(x) + F(y)$ for all x and y in G.

(2) A morphism in a category, with the property that its domain and range coincide.

endomorphism ring   The ring consisting of all endomorphisms (from a space A with an additive structure to itself) with addition given by the addition in A and multiplication coming from composition. *See* endomorphism.

entire algebroidal function   An algebroidal function $f(z)$ that has no pole in $|z| < \infty$. *See* algebroidal function.

entire linear transformation   Consider a linear fractional transformation (also known as a Möbius transformation) in the complex plane; that is, a function given by

$$S(z) = \frac{az+b}{cz+d} ,$$

where a, b, c, and d are complex numbers. If $c = 0$ and $d \neq 0$, $S(z)$ becomes a typical linear transformation, which is also an entire function.

enveloping algebra   Let $\mathcal{B}$ be a subset of an associative algebra $\mathcal{A}$ with multiplicative identity 1. The subalgebra $\mathcal{B}^\dagger$ of $\mathcal{A}$ containing 1 and generated by $\mathcal{B}$ is called the enveloping algebra of $\mathcal{B}$ in $\mathcal{A}$. *See* associative algebra.

enveloping von Neumann algebra   Let A denote a C^*-algebra and S denote the state space of A. For each state $\phi \in S$, let $(\pi_\phi, H_\phi, \psi_\phi)$ denote the cyclic representation of ϕ. Here π_ϕ is a representation of A on the Hilbert space H_ϕ. The linear span of vectors of the form $\pi_\phi(a)\psi_\phi$, for $a \in A$, is dense in H_ϕ. Let H_S be the direct sum Hilbert space of the spaces H_ϕ, as ϕ varies through S, and let π_S be the representation of A on H_S obtained from the direct sum of the representations π_ϕ, for $\phi \in S$. The *enveloping von Neumann algebra* of A is the closure A'' of $\pi_S(A)$ with respect to the weak operator topology. Because π_S is a faithful representation, A can now be viewed as a C^*-subalgebra of the von Neumann algebra A''.

The enveloping von Neumann algebra A'' of a C^*-algebra A is isomorphic, as a Banach space, to the second dual of A. If m is a von Neumann algebra, G is a locally compact group and if $\alpha : G \to \mathrm{Aut}(m)$ is a continuous group homomorphism, where the automorphism group $\mathrm{Aut}(m)$ has the topology of weak convergence, then the W^*-crossed product algebra $m \times_\alpha G$ arises as the enveloping von Neumann algebra of a certain C^*-crossed product affiliated with the W^*-dynamical system (m, G, α).

epimorphism   A morphism e in a category, such that the equation $f \circ e = g \circ e$ for morphisms f and g in the category implies that $f = g$. In most familiar categories, such as the category of sets and functions, an epimorphism is simply a *surjective* or *onto* function in the category. A function e from set X to set Y is *surjective* or *onto* if $e(X) = Y$, that is if every element $y \in Y$ is the image $e(x)$ of an element $x \in X$. *See also* morphism in a category, monomorphism, surjection.

Epstein zeta function A function, $\zeta_Q(s, M)$, defined by Epstein in 1903. Let V be a vector space over $\mathbf{R}$ having dimension n. Let M be a lattice in V and $Q(X)$ a positive definite quadratic form defined on V. Then, $\zeta_Q(s, M)$ is defined for complex numbers s by

$$\zeta_Q(s, M) = \sum_{x \in M \backslash \{0\}} \frac{1}{Q(x)^s} \, .$$

The Epstein zeta function is absolutely convergent when $\Re s > \frac{m}{2}$. If $Q(x)$ is a positive integer for all $x \in M \backslash \{0\}$, then

$$\zeta_Q(s, M) = \sum_{k=1}^{\infty} \frac{a(k)}{k^s} \, ,$$

where $a(k)$ denotes the number of distinct $x \in M$ with $Q(x) = k$. The Epstein zeta function, in general, has no Euler product expansion. *See also* Riemann zeta function.

equal fractions Two fractions (of positive integers), $\frac{m}{n}$ and $\frac{k}{\ell}$, are equal if $m\ell = nk$. If the greatest common divisor of m and n is 1 and the greatest common divisor of k and ℓ is also 1, then the two fractions are equal if and only if $m = k$ and $n = \ell$.

equality (1) The property that two mathematical objects are identical. For example, two sets A, B are equal if they have the same elements; we write $A = B$. Two vectors in a finite dimensional vector space are equal if their coefficients with respect to a fixed basis of the vector space are the same.

(2) An equation. *See* equation.

equation An assertion of equality, usually between two mathematical expressions f, g involving numbers, parameters, and variables. We write $f = g$. When the equation involves one or more variables, the equality asserted may be true for some or all values of the variables. A natural question then arises: For which values

of the variables is the equality true? The task of answering this question is referred to as *solving the equation*.

equivalence class Given an equivalence relation R on a set S and an $x \in S$, the *equivalence class* of x, usually denoted by $[x]$, consists of all $y \in S$ such that $(x, y) \in R$. Clearly, $x \in [x]$, and $(x, y) \in R$ if and only if $[x] = [y]$. As a consequence, the equivalence classes of R induce a partition of the set S into non-overlapping subsets.

equivalence properties The defining properties of an equivalence relation $R \subseteq S \times S$; namely, that R is reflexive $((x, x) \in R$ for all $x \in S)$, symmetric $((x, y) \in R$ whenever $(y, x) \in R)$, and transitive $((x, z) \in R$ whenever $(x, y) \in R$ and $(y, z) \in R)$. *See also* relation, equivalence class.

equivalence relation A relation R on a set S (that is, a subset of $S \times S$), which is reflexive, symmetric, and transitive. (*See* equivalence properties.) For example, let S be the set of all integers and R the subset of $S \times S$ defined by $(x, y) \in R$ if $x - y$ is a multiple of 2. Then R is an equivalence relation because for all $x, y, z \in S$, $(x, x) \in R$ (reflexive), $(x, y) \in R$ whenever $(y, x) \in R$ (symmetric), and $(x, z) \in R$ whenever $(x, y) \in R$ and $(y, z) \in R$ (transitive).

equivalent divisors *See* algebraic equivalence of divisors.

equivalent equations Equations that are satisfied by the same set of values of their respective variables. For example, the equation $x^2 = 3y - 1$ is equivalent to the equation $6w - 2z^2 = 2$ because their solutions coincide.

equivalent valuations Let $\phi : F \to \mathbf{R}^+$ be a valuation on a field F (here $\mathbf{R}^+$ denotes the set of all nonnegative real numbers). The valuation ϕ gives rise to a metric on F, where the open neighborhoods are the open spheres centered at $a \in F$ defined by

$$\{b \in F \mid \phi(b - a) < \epsilon\}, \ \epsilon \in \mathbf{R}^+ \, .$$

Two valuations are called equivalent if they induce the same topology on F.

error   In the context of numerical analysis, an *error* occurs when a real number x is being approximated by another real number $\hat{x}$. A typical example arises in the implementation of numerical operations on computing machines, where an error occurs whenever a real number x is made machine representable either by rounding off or by truncating at a certain digit. For example, if $x = 4.567$, truncation at the third digit yields $\hat{x} = 4.56$ while rounding off at the third digit yields $\hat{x} = 4.57$.

There are two common ways to measure the error in approximating x by $\hat{x}$: The quantity

$$|x - \hat{x}|$$

is referred to as the absolute error, and the quantity

$$\frac{|x - \hat{x}|}{|x|} \ (x \neq 0)$$

is referred to as the relative error. The concept of error is also used in the approximation of a vector $x \in \mathbf{R}^n$ by $\hat{x} \in \mathbf{R}^n$; the absolute value in the formulae above are then replaced by some vector norm. In particular, if we use the *infinity norm,* and if the relative error is approximately 10^{-k}, then we can deduce that the largest in absolute value entry of $\hat{x}$ has approximately k correct significant digits.

étale   Let X and Y be schemes of finite type over k. A morphism $f : X \to Y$ is called *étale* if it is smooth of relative dimension 0.

étale morphism   A morphism $f : X \to Y$ is said to be *étale* at a point $x \in X$ if $d_x f$ induces an isomorphism $C_x X \to C_{f(x)} Y$ of the corresponding tangent cones (viewed as schemes).

étale site   Let X be a scheme, let S/X be the category of schemes over X and let C/X be a full subcategory of S/X that is closed under fiber products and is such that, for any morphism $Y \to X$ in C/X and any étale morphism $U \to Y$, the composite $U \to X$ is in C/X. The category C/X, together with the étale topology of C/X, is the *étale site* over X. *See* étale topology.

étale topology   Let Y be a connected, closed subscheme of a normal variety X and assume that Y is $G2$ in X. Let X' be the normalization of X in $K(\hat{X})$, and let $f : X' \to X$ be the natural map. Then there is a subvariety Y' of X' which is $G3$ in X' and such that $f\big|_{Y'}$ is an isomorphism of Y' onto Y, and such that f is étale at points of X' in a suitable neighborhood of Y'. Then (X', Y') is an *étale neighborhood* of (X, Y). The étale neighborhoods form a subbasis for the *étale topology. See also* étale morphism.

Euclidean domain   Let R be a ring. Then R is an integral domain if $x, y \in R$ and $xy = 0$ implies either $x = 0$ or $y = 0$. An integral domain is Euclidean if there is a function d from the non-zero elements of R to the non-negative integers such that
(i.) For $x \neq 0$, $y \neq 0$, both elements of R, we have $d(x) \leq d(xy)$;
(ii.) Given non-zero elements $x, y \in R$, there exists $s, t \in R$ such that $y = sx + t$, where either $t = 0$ or $d(t) < d(x)$.

Euclidian Algorithm   An algorithm for finding the greatest common divisor g.c.d. (m, n), of two positive integers m, n satisfying $m > n$. It can be described as follows:

1. Divide n into m, i.e., find a positive integer p and a real number r so that
   $m = pn + r, \ 0 \leq r < n$.
2. If $r = 0$, then g.c.d. $(m, n) = n$.
3. If $r \neq 0$, then g.c.d. $(m, n) =$ g.c.d. (n, r); replace m by n and n by r and repeat step 1.

After possibly several iterations, the process always terminates by detecting a zero remainder r. Then g.c.d. (m, n) equals the value of the last nonzero remainder detected. The following example illustrates this algorithm:

$954 = 29 \times 32 + 26,$
$32 = 1 \times 26 + 6,$
$26 = 4 \times 6 + 2,$
$6 = 3 \times 2 + 0.$

Hence g.c.d. $(954, 32) = 2$. The algorithm can be modified so that the remainder in step 1 is the smallest number r (in absolute value) satisfying $0 \leq |r| \leq n$. This may lead to fewer required iterations. For example, the modified algorithm for the above numbers yields

$$954 = 30 \times 32 - 6,$$
$$32 = 5 \times 6 + 2,$$
$$6 = 3 \times 2 + 0.$$

Hence g.c.d. $(954, 32) = 2$ in three iterations.

Euclid ring   An integral domain R such that there exists a function $d(\cdot)$ from the nonzero elements of R into the nonnegative integers which satisfies the following properties:
(i.) For all nonzero $a, b \in R$, $d(a) \le d(ab)$.
(ii.) For any nonzero $a, b \in R$, there exist $p, r \in R$ such that $a = pb + r$, where either $r = 0$ or $d(r) < d(b)$.

The integers with ordinary absolute value as the function $d(\cdot)$ is an example of a Euclid ring. Also the Gaussian integers, consisting of all the complex numbers $x + iy$ where x, y are integers, form a Euclid ring; $d(x + iy) = x^2 + y^2$ serves as the required function.

Also called Euclidian ring.

Euler class   The *Euler class* of a compact, oriented manifold X, denoted by $\chi \in H^n(X, \dot{X})$, is defined by

$$\chi = (U_1 \cup U_2)/z .$$

Euler-Poincaré characteristic   Let K be a simplicial complex of dimension n, and let α_r denote the number of r-simplexes of K. The Euler-Poincaré characteristic is

$$\chi(K) = \sum_{r=0}^{n} (-1)^r \alpha_r .$$

$\chi(K)$ is a generalization of the Euler characteristic, $V - E + F$, where V, E, F are, respectively, the numbers of vertices, edges, and faces of a simple closed polyhedron (namely, a polyhedron that is topologically equivalent to a sphere). Euler's Theorem in combinatorial topology states that $V - E + F = 2$. *See also* Lefschetz number.

Euler product   Consider the Riemann zeta function

$$\zeta(s) = 1 + \frac{1}{2^s} + \cdots + \frac{1}{k^s} + \cdots ,$$

which converges for all real numbers $s > 1$. Euler observed that

$$\zeta(s) = \prod \frac{1}{1 - p^{-s}} ,$$

where p runs over all prime numbers. This infinite product is called the *Euler product*. One of the main questions regarding zeta functions is whether they have similar infinite product expansions, usually referred to as Euler product expansions.

Euler's formula   The expression

$$e^{iz} = \cos z + i \sin z ,$$

where $i = \sqrt{-1}$. This is proved by considering the infinite series expansions of e^z, $\sin z$ and $\cos z$ for $z \in \mathbf{C}$.

See also polar form of a complex number.

even element   Let Q be a quadratic form with nonzero discriminant on an n-dimensional vector space V, over a field F, of characteristic not equal to 2. Let $C(Q)$ denote the Clifford algebra of Q. Then $C(Q)$ is the direct sum $C^+(Q) + C^-(Q)$, where

$$C^+(Q) = Fi + V^2 + V^4 + \cdots ,$$
$$C^-(Q) = V + V^3 + V^5 + \cdots .$$

The elements of $C^+(Q)$ and $C^-(Q)$ are called the even and odd elements, respectively, of the Clifford algebra.

even number   An integer that is divisible by 2. An even number is typically represented by $2n$, where n is an integer.

evolution   Another term for the process of extracting a root. *See* extraction of root.

exact sequence   Consider a sequence of modules $\{M_i\}$ and a sequence of homomorphisms $\{h_i\}$ with

$$h_i : M_{i-1} \longrightarrow M_i$$

for all $i = 1, 2, \ldots$ The sequence of homomorphisms is called *exact at* M_i if $\mathrm{Im} h_i = \mathrm{Ker} h_{i+1}$. The sequence is called *exact* if it is exact at every M_i. For a sequence

$$0 \longrightarrow M_1 \xrightarrow{h_2} M_2 ,$$

it is understood that the first arrow (from 0 to M_1) represents the 0 map, so the sequence is exact if and only if h_2 is injective.

exact sequence of cohomology   (**1**) Suppose G is an Abelian group and A is a subspace of X. Then there is a long exact sequence of cohomology:

$$\cdots \to H^n(X, A, G) \to H^n(X, G)$$

$$\to H^n(A, G) \to H^{n+1}(X, A, G) \to \cdots .$$

Here H^n are the cohomology groups.

(**2**) If $0 \to A \to B \to C \to 0$ is a short exact sequence of complexes, then there are natural maps $\delta^i : H^i(C) \to H^{i+1}(A)$, giving rise to an exact sequence

$$\cdots \to H^i(A) \to H^i(B) \to H^i(C)$$

$$\to H^{i+1}(A) \to \cdots$$

of cohomology.

exact sequence of ext (contravariant)   If

$$0 \longrightarrow A \longrightarrow B \longrightarrow C \longrightarrow 0$$

is a short exact sequence of modules over a ring R and M is an R-module (on the same side as A), then the associated *contravariant exact sequence of ext* is a certain long exact sequence of the form

$$0 \to \operatorname{Hom}(C, M) \to \operatorname{Hom}(B, M)$$

$$\to \operatorname{Hom}(A, M)$$

$$\to \operatorname{Ext}_R^1(C, M) \to \operatorname{Ext}_R^1(B, M)$$

$$\to \operatorname{Ext}_R^1(A, M)$$

$$\to \cdots \to \operatorname{Ext}_R^n(C, M) \to \operatorname{Ext}_R^n(B, M)$$

$$\to \operatorname{Ext}_R^n(A, M) \to \cdots .$$

exact sequence of ext (covariant)   If

$$0 \longrightarrow A \longrightarrow B \longrightarrow C \longrightarrow 0$$

is a short exact sequence of modules over a ring R and M is an R-module (on the same side as A),

then the associated *covariant exact sequence of ext* is a certain long exact sequence of the form

$$0 \to \operatorname{Hom}(M, A) \to \operatorname{Hom}(M, B)$$

$$\to \operatorname{Hom}(M, C)$$

$$\to \operatorname{Ext}_R^1(M, A) \to \operatorname{Ext}_R^1(M, B)$$

$$\to \operatorname{Ext}_R^1(M, C)$$

$$\to \cdots \to \operatorname{Ext}_R^n(M, A) \to \operatorname{Ext}_R^n(M, B)$$

$$\to \operatorname{Ext}_R^n(M, C) \to \cdots .$$

exact sequence of homology   (**1**) Suppose A is a subspace of X. Then there exists a long exact sequence

$$\cdots \to H_n(A) \to H_n(X)s$$

$$\to H_n(X, A) \to H_{n-1}(A) \to \cdots ,$$

called the *exact sequence of homology,* where $H_n(X)$ denotes the nth homology group of X and $H_n(X, A)$ is the nth relative homology group.

(**2**) If $0 \to A \to B \to C \to 0$ is a short exact sequence of complexes, then there are natural maps $\delta_i : H_i(C) \to H_{i-1}(A)$, giving rise to an exact sequence

$$\cdots \to H_i(A) \to H_i(B) \to H_i(C)$$

$$\to H_{i-1}(A) \to \cdots$$

of homology.

exact sequence of Tor   Let A be a right Γ-module and let $B' \mapsto B \mapsto B''$ be an exact sequence of left Γ-modules. Then there exists an exact sequence

$$\operatorname{Tor}_n^\Gamma(A, B') \to \operatorname{Tor}_n^\Gamma(A, B) \to \operatorname{Tor}_n^\Gamma(A, B'')$$

$$\to \cdots \to \operatorname{Tor}_1^\Gamma(A, B') \to \operatorname{Tor}_1^\Gamma(A, B)$$

$$\to \operatorname{Tor}_1^\Gamma(A, B'') \to A \otimes_\Gamma B' \to A \otimes_\Gamma B$$

$$\to A \otimes_\Gamma B'' \to 0 .$$

exceptional compact real simple Lie algebra
Since compact real semisimple Lie algebras are in one-to-one correspondence with complex semisimple Lie algebras (via complexification), the classification of compact real simple Lie algebras reduces to the classification of complex simple Lie algebras. Thus, the compact real

simple Lie algebras corresponding to the exceptional complex simple Lie algebras E_l ($l = 6, 7, 8$), F_4 and G_2 are called *exceptional compact real simple Lie algebras*.

exceptional complex simple Lie algebra   In the classification of complex simple Lie algebras α, there are seven categories: A,B,C,D,E,F,G, resulting from all possible Dynkin diagrams. The notation in each category includes a subscript l (e.g., E_l), which denotes the rank of α. The algebras in categories E_l ($l = 6, 7, 8$), F_4 and G_2 are called *exceptional complex simple Lie algebras* (in contrast to the classical complex simple Lie algebras).

exceptional complex simple Lie group   A complex connected Lie group associated with one of the exceptional complex simple Lie algebras E_l ($l = 6, 7, 8$), F_4 or G_2.

exceptional curve of the first kind   Given two mutually nonsingular surfaces F, F' and a birational transformation $T : F \to F'$, the total transform E of a simple point in F' by T^{-1} is called an *exceptional curve*. It is called an *exceptional curve of the first kind* if, in addition, T is regular along E. Otherwise, it is called an *exceptional curve of the second kind*. *See also* algebraic surface.

exceptional curve of the second kind   *See* exceptional curve of the first kind.

exceptional Jordan algebra   A Jordan algebra that is not special. *See* special Jordan algebra.

excess of nines   A method for verifying the accuracy of operations among integers, also known as the method of *casting out nines*. It uses the sums of the digits of the integers involved, in modulo 9 arithmetic. We illustrate the method for addition of integers. Consider the sum

$$683 + 256 = 939 .$$

The sum of digits of 683, 256, and 939 are $17 \equiv 8 \bmod 9$, $13 \equiv 4 \bmod 9$ and $21 \equiv 3 \bmod 9$, respectively. Indeed, $8 + 4 = 12 \equiv 3 \bmod 9$.

exhaustive filtration   A filtration $\{M_k : k \in \mathbf{Z}\}$ of a module M is called *exhaustive* (or *convergent from above*) if

$$\cup_k M_k = M .$$

See filtration. *See also* discrete filtration.

Existence Theorem (class field theory)   For any ideal group there exists a unique class field. *See* ideal group, class field.

expansion of determinant   Given an $n \times n$ matrix $A = (a_{ij})$, the determinant of A is formally defined by

$$\det A = \sum_{\sigma \in S_n} \operatorname{sgn}(\sigma) a_{1\sigma(1)} a_{2\sigma(2)} \cdots a_{n\sigma(n)} ,$$

where S_n denotes the symmetric group of degree n (the group of all $n!$ permutations of the set $\{1, 2, \ldots, n\}$), and where $\operatorname{sgn}(\sigma)$ denotes the sign of the permutation σ. ($\operatorname{sgn}(\sigma) = 1$ if σ is an even permutation and $\operatorname{sgn}(\sigma) = -1$ if σ is an odd permutation.) The formula in the equation above is referred to as the *expansion of the determinant* of A.

exponent   Given an element a of a multiplicative algebraic structure, the product of $a \cdot a \cdot \ldots \cdot a$, in which a appears k times, is written as a^k and k is referred to as the exponent of a^k. The simplest example is when a is a real number. In this case, we can also give meaning to negative exponents by $a^{-k} = 1/a^k$ (assuming that k is a positive integer and that $a \neq 0$). We define $a^0 = 1$. The basic laws of exponents of real numbers are the following:
(1) $a^k a^m = a^{k+m}$,
(2) $a^k / a^m = a^{k-m}$ ($a \neq 0$),
(3) $(a^k)^m = a^{km}$,
where k and m are nonnegative integers. We can extend the definition to rational exponents m/n, where m is any integer and n is a positive integer, by defining $a^{m/n} = \sqrt[n]{a^m}$. Synonyms for the exponent are the words index and power. *See also* exponential mapping.

exponential function of a matrix   Let A be an $n \times n$ matrix over the complex numbers $\mathbf{C}$. The exponential function $f(A) = e^A$ is defined by the infinite series

$$e^A = I + A + \frac{1}{2!}A^2 + \frac{1}{3!}A^3 + \cdots$$

$$= \sum_{k=0}^{\infty} \frac{1}{k!} A^k .$$

Letting $\| \cdot \|$ denote the Euclidean vector norm, as well as the induced matrix norm, we can show that the exponential function of a matrix is well defined (i.e., the series is convergent) by showing that it is absolutely convergent. Indeed, invoking well-known inequalities, and the definition of the real exponential function e^x, we have that $\|e^A\|$ equals

$$\left\| \sum_{k=0}^{\infty} \frac{1}{k!} A^k \right\| \leq \sum_{k=0}^{\infty} \frac{1}{k!} \|A\|^k = e^{\|A\|} < \infty .$$

One basic property of e^A is that its eigenvalues are of the form e^λ, where λ is an eigenvalue of A. It follows that e^A is always a nonsingular matrix. Also, the exponential property $e^{A+B} = e^A e^B$ holds if and only if A and B commute, that is, $AB = BA$.

The exponential function of a matrix arises in the solution of systems of linear differential equations in the vector form $dx(t)/dt = Ax(t)$, $t \geq 0$, where $x(t) \in \mathbf{C}^n$. If the initial condition $x(0) = x_0$ is specified, then the unique solution to this differential problem is given by $x(t) = e^{tA}x_0$.

exponential mapping   The mapping (function) $f(x) = e^x$, where $x \in \mathbf{R}$ and e is the base of the natural logarithm. (*See* e.) The exponential mapping is the inverse mapping of the natural logarithm function: $y = e^x$ if and only if $x = \ln y$. Mclaurin's Theorem yields

$$e^x = \sum_{k=0}^{\infty} \frac{x^k}{k!} .$$

More generally, the exponential mapping to base a ($a \neq 1$) is defined by $f(x) = a^x$.

exponentiation   The process of evaluating a^k, that is, evaluating the product $a \cdot a \cdot \ldots \cdot a$, in which a appears k times. *See also* exponent.

expression   A mathematical statement, using mathematical quantities such as scalars, variables, parameters, functions, and sets, as well as relational and logical operators such as equality, conjunction, existence, union, etc.

exsecant function   A trigonometric function defined via the secant of an angle as $\operatorname{exsec}\theta = \sec\theta - 1$. Similarly, we define the excosecant function as $\operatorname{excosec}\theta = \csc\theta - 1$.

extension   Given a subfield E of a field F, namely, a subset of F that is a field with respect to the operations defined in F, we call F an extension field of E. The field F can be regarded as a vector space over E. The dimension of F over E is called the degree of the extension field F over E. If $f_1, \ldots, f_p \in F$, then by $E(f_1, \ldots, f_p)$ we denote the smallest subfield of F containing E and $f_1, \ldots, f_p$. $E(f_1)$ is called a simple extension of E.

If every element of E is algebraic over F, we call E an algebraic extension. Otherwise, we call E a transcendental extension.

The notion of extension also applies to rings. *See also* number field.

extension of coefficient ring   Let $R[t]$ be the ring of polynomials over the (coefficient) ring R in the indeterminate t. As the notion of extension can also concern a ring, an extension of R is usually referred to as the extension of the coefficient ring of $R[t]$. *See* extension, ring of polynomials.

extension of valuation   If v is a valuation on a field F and if K is an extension of F, then an extension of v to K is a valuation w on K such that $w(x) = v(x)$, for $x \in F$. *See* valuation.

exterior algebra   Let V be a vector space over a field F. Let also $T_0(V)$ denote the direct sum of the tensor products $V \otimes \ldots \otimes V$. $T_0(V)$ is called the contravariant tensor algebra over V and is equipped with the product $\otimes$ as well as addition and scalar multiplication. Let S be formed by all elements of $T_0(V)$ of the type $v \otimes v$, as well as their sums, scalar multiples and their products with arbitrary elements in $T_0(V)$. Then S is a subgroup of (the Abelian group) $T_0(V)$ and the quotient group $T_0(V)/S$ can be considered. This quotient group can be made into an algebra by defining the operations $\cdot$ and $\wedge$ as follows:

$$c \cdot (t + S) = ct + S \quad (c \in F),$$

$$(t_1 + S) \bigwedge (t_2 + S) = (t_1 \otimes t_2) + S.$$

The operation $\wedge$ is called the exterior product of the exterior algebra $T_0(V)/S$ of V, which is denoted by $\wedge V$. $\wedge V$ is also known as the Grassmann algebra of V. The image of $v_1 \otimes \ldots \otimes v_p$ under the natural mapping $T_0(V) \to \wedge V$ is denoted by $v_1 \wedge \ldots \wedge v_p$ and is called the exterior product of $v_1, \ldots, v_p \in V$. In general, the image of $V \otimes \ldots \otimes V$ with p factors under the above natural mapping is called the p-fold exterior power of V and is denoted by $\wedge^p V$.

The exterior product satisfies some important rules and properties. For example, it is multilinear, $v_1 \wedge \ldots \wedge v_p = 0$ if and only if $v_1, \ldots, v_p$ are linearly dependent, and $u \wedge v = (-1)^{pq} v \wedge u$ whenever $u \in \wedge^p V$ and $v \in \wedge^q V$.

See also algebra, tensor product.

external product   Given two groups G_1, G_2, the *external (direct) product* of G_1, G_2 is the group $G = G_1 \times G_2$ formed by the set of all pairs (g_1, g_2) with $g_1 \in G_1$ and $g_2 \in G_2$; the operation in G is defined to be

$$(g_1, g_2)(g_1', g_2') = (g_1 g_1', g_2 g_2').$$

This definition can be extended in the obvious way to any collection of groups $G_1, G_2, \ldots$. The operation in each component is carried out in the corresponding group. The external product coincides with the *external sum* of groups if the number of groups is finite. *See also* internal product.

Ext group   The Ext group is defined in several subjects. For example, if A is an Abelian group and if

$$0 \to R \to F \to A \to 0$$

is any free resolution of A, then for every Abelian group B, there exists a group $\text{Ext}(A, B)$ such that

$$0 \to \text{Hom}(A, B) \to \text{Hom}(F, B)$$
$$\to \text{Hom}(R, B) \to \text{Ext}(A, B) \to 0$$

is exact; moreover, the group is independent of the choice of the free resolution of A. The elements of $\text{Ext}(A, B)$ are equivalence classes of short exact sequences $0 \to B \to M \to A \to 0$ and addition is induced by Baer sum.

The Ext group is also defined in homological algebra, topology, and operator algebras. The group $\text{Ext}(A, B)$ is called the group of extensions of B by A.

extraction of root   The process of finding a root of a number (e.g., that the fifth root of 32 is 2) or the process of finding the roots of an equation.

extraneous root   A root, obtained by solving an equation, which does not satisfy the original equation. Such roots are usually introduced when exponentiation or clearing of fractions is performed.

extreme terms of proportion   Given a proportion, namely, an equality of two ratios $\frac{a}{b} = \frac{c}{d}$, the numbers a and d are called the *extreme* (or *outer*) *terms of the proportion*. *See* proportion.

F

factor (1) An integer n is a *factor* (or *divisor*) of an integer m if $m = nk$ for some integer k. Thus, $\pm 1, \pm 2$, and ± 4 are all the factors of 4. More generally, given a commutative monoid M that satisfies the cancellation law, $b \in M$ is a factor of $a \in M$ if $a = bc$ for some $c \in M$. We then usually write $b|a$. *See also* prime, factor of polynomial.

(2) A von Neumann algebra whose center is the set of scalar multiples of the identity operator. The study of von Neumann algebras is carried out by studying the factors which are type I, II, or III with subtypes for each. *See also* type-I factor, type-II factor, type-III factor, Krieger's factor.

factorable polynomial A polynomial that has factors other than itself or a constant polynomial. *See also* factor of polynomial.

factor group Let G be a group and let H denote a normal subgroup of G. The set of left (or right) cosets of G, denoted by

$$G/H = \{aH : a \in G\},$$

forms a group under the operation $(aH)(bH) = (ab)H$ and is called the *factor group* or the *quotient group* of G relative to H.

The factor groups of a group G can be useful in revealing important information about G. For example, letting $C(G)$ represent the center of G, if $G/C(G)$ is cyclic it follows that G is an Abelian group.

factorial The factorial of a positive integer n (read n *factorial*) is denoted by $n!$ and is defined by

$$n! = n\,(n-1)\,(n-2)\ldots 3\,2\,1\,.$$

For example, $4! = 4\,3\,2\,1 = 24$. By definition, $0! = 1$. An approximation of $n!$ for large values of n is given by Stirling's formula:

$$n! \approx \left(\frac{n}{e}\right)^n \sqrt{2\pi n}\,.$$

factorial series The infinite series, involving factorials,

$$\sum_{k=0}^{\infty} \frac{1}{k!} = 1 + \frac{1}{1!} + \frac{1}{2!} + \cdots$$

This series converges to the number e. *See* factorial.

factoring The process of finding factors; of an integer or of a polynomial, for example. Given an integer $n > 1$, the Fundamental Theorem of Arithmetic states that n can be expressed as a product of positive prime numbers, uniquely, apart from the order of the factors. The process of finding these prime factors is referred to as the *prime factoring* or the *prime factorization* of n. *See also* division algorithm, factoring of polynomials.

factoring of polynomials The process of finding factors of a polynomial $f(x) \in F[x]$ over a field F. If $f(x)$ has positive degree, then $f(x)$ can be expressed as a product

$$f(x) = cg_1(x)g_2(x)\ldots g_r(x)\,,$$

where $c \in F$ and $g_1, g_2, \ldots, g_r$ are irreducible and monic polynomials in $F[x]$. This is referred to as the prime factoring or the prime factorization of $f(x)$ and is unique apart from the order of the factors.

If $f(x) \in F[x]$ is monic and has positive degree, then there is an extension field E of F, so that $f(x)$ can be factored into

$$f(x) = (x - r_1)\,(x - r_2)\ldots(x - r_k)$$

in $E[x]$. The field E is the splitting field of $f(x)$ and it satisfies $E = F(r_1, r_2, \ldots, r_k)$, where $r_1, r_2, \ldots, r_k$ are the roots of $f(x)$ in E. If the field F is algebraically closed (for example the complex numbers), then $E = F$. *See also* Factor Theorem, division algorithm.

factor of polynomial A polynomial $g(x) \in F[x]$ over a field F is a *factor* of $f(x) \in F[x]$ if $f(x) = g(x)h(x)$ for some $h(x) \in F[x]$. For example, $g(x) = x - 1$ is a factor of $f(x) = x^2 - 1$. *See also* factoring of polynomials.

factor representation Consider a nontrivial Hilbert space H and a topological group G

(i.e., a group with the structure of a topological space so that the mappings $(x, y) \rightarrow xy$ and $x \rightarrow x^{-1}$ are continuous). Let U be a unitary representation of G, namely, a homomorphism of G into the group of unitary operators on H that is strongly continuous. If the von Neumann algebra M generated by $\{U_g : g \in G\}$ and its commutant M' satisfy

$$M \cap M' = \{z1 : z \in \mathbf{C}\},$$

then U is called a *factor representation* of G.

factor set   Suppose A is an Abelian group and G is an operator group acting on A. Then every element $\sigma \in G$ defines an automorphism $a \rightarrow a^\sigma$ of A such that $(a^t a u)^\sigma = a^{\sigma\tau}$. A *factor set* is a collection of elements $\{a_{\sigma,\tau} : \sigma, \tau \in G\}$ in A such that

$$a_{\sigma,\tau} a_{\sigma\tau,\rho} = a_{\tau,\rho}^\sigma a_{\sigma,\tau\rho}.$$

Factor Theorem   The linear term $(x - a)$ is a factor of the polynomial $f(x) \in F[x]$ over a field F if and only if $f(a) = 0$.

It is an immediate corollary of the Remainder Theorem. (*See* Remainder Theorem.) The *Factor Theorem* can be useful in finding factors of polynomials: If $f(x) = x^4 + x^3 + x^2 + 3x - 6$, then one knows that if a is an integer, then $(x-a)$ is a factor of $f(x)$ only if a divides 6. Hence it makes sense to search for integer roots of $f(x)$ among the factors of 6.

faithful function   Let V and W be partially ordered vector spaces with positive proper cones V^+ and W^+, respectively. A function $f : V \rightarrow W$ is said to be *faithful* if $f(x) = 0$, with $x \in V^+$, occurs only in the case where $x = 0$.

As an example, let V be the complex vector space of $n \times n$ matrices with positive cone V^+ consisting of all positive semidefinite matrices. Let W be the complex field and $W^+ = [0, \infty)$, and let $f : V \rightarrow W$ be the trace function: $f(x) = \text{trace}(x) = \sum_{i=1}^n x_{ii}$, for all $x = (x_{ij}) \in V$. Then f is a faithful function.

Faithful functions arise in the theory of C^*-algebras, as follows: Let A be a C^*-algebra. By a theorem of Gelfand, Naimark, and Segal, there is a faithful C^*-algebra homomorphism $\rho : A \rightarrow \mathcal{B}(H)$, where $\mathcal{B}(H)$ is the C^*-algebra

of bounded linear operators acting on a Hilbert space H. *See* partially ordered space.

faithful R-module   A module M over a ring R such that, whenever $r \in R$ satisfies $rM = 0$, then $r = 0$. Also called faithfully flat.

false position   The *method of false position* (or *regular falsi method*) is a numerical method for approximating a root r of a function $f(x)$, given initial approximations r_0 and r_1 that satisfy $f(r_0)f(r_1) < 0$. The next approximation r_2 is chosen to be the x-intercept of the line through the points $(r_0, f(r_0))$ and $(r_1, f(r_1))$. Then r_3 is chosen as follows: If $f(r_1)f(r_2) < 0$ we choose r_3 to be the x-intercept of the line through the points $(r_1, f(r_1))$ and $(r_2, f(r_2))$. Otherwise, we choose r_3 as the x-intercept of the line through the points $(r_0, f(r_0))$ and $(r_2, f(r_2))$, and swap the indices of p_0 and p_1 to continue. This process ensures that successive approximations enclose the root r. *See also* secant method.

feasible region   The set of all feasible solutions of a linear programming problem. *See also* feasible solution.

feasible solution   In linear programming, the objective is to minimize or maximize a linear function of several variables, subject to one or more constraints that are expressed as linear equations or inequalities. A solution (choice) of the variables that satisfies these constraints is called a *feasible solution*. *See also* feasible region.

Feit-Thompson Theorem   Every non-Abelian simple group must have even order. This result was conjectured by Burnside and proved by Feit and Thompson in 1963. It was an important step and the driving force behind the effort to classify the finite simple groups.

Fermat numbers   Integers of the form $2^{2^n} + 1$, where n is a nonnegative integer. For $n = 1, 2, 3, 4$, the Fermat numbers are prime integers. Euler proved, contrary to a conjecture by Fermat, that the Fermat number for $n = 5$ is not a prime.

Fermat's Last Theorem   There are no positive integers x, y, z, and n, with $n > 2$, satisfying $x^n + y^n = z^n$.

Pierre Fermat wrote a version of this theorem in the margin of his copy of Diophantus' *Arithmetica*. He commented that he knew of a marvelous proof but that there was not enough space in the margin to present it.

This assertion is known as Fermat's Last Theorem, because it was the last unresolved piece of Fermat's work. A proof eluded mathematicians for over 300 years. In 1993, A. J. Wiles announced a proof of the conjecture. Some gaps and errors that were found in the original proof were corrected and published in the *Annals of Mathematics* in 1995.

fiber   (**1**) Preimage, as of an element or a set; inverse image.

(**2**) In homological algebra, if $f : X \to Y$ is a morphism of schemes, $y \in Y, k(y)$ is the residue field of y, and Spec $k(y) \to Y$ is the natural morphism, then the *fiber* of the morphism f over the point y is the scheme

$$X_y = X \times_Y \text{ Spec } k(y) \ .$$

Also spelled *fibre*.

Fibonacci numbers   The sequence of numbers f_n given by the recursive formula $f_n = f_{n-1} + f_{n-2}$ for $n = 2, 3, \ldots$, where $f_1 = f_2 = 1$ (or sometimes $f_1 = 0$, $f_2 = 1$). It can be shown that every positive integer is the sum of distinct Fibonacci numbers. Any two consecutive Fibonacci numbers are relatively prime. The ratios $\frac{f_{n+1}}{f_n}$ form a convergent sequence whose limit as $n \to \infty$ is the golden ratio, $\frac{\sqrt{5}+1}{2}$.

fibre   *See* fiber.

field   A commutative ring F with multiplication identity 1, all of whose nonzero elements are invertible with respect to multiplication: for any nonzero $a \in F$ there exists $c \in F$ such that $ac = 1$. We usually write $c = a^{-1}$. It follows that if $a, b \in F$ are nonzero elements of a field, then so is ab, namely, every field is an integral domain. Well-known examples of fields are the rational numbers, the real numbers, and the complex numbers with the familiar operations

of multiplication and addition. Also the residue ring modulo p, Z_p, is a field when p is a prime integer. *See* ring.

field of quotients   Let $D \neq 0$ be a commutative integral domain and let D^* denote its nonzero elements. Consider the relation $\equiv$ in $D \times D^*$ defined by $(a, b) \equiv (c, d)$ if $ad = bc$. It can be shown that $\equiv$ is an equivalence relation. (*See* equivalence relation.) Denote the equivalence class determined by (a, b) as a/b (called a quotient or a fraction) and let $F = \{a/b\}$ be the quotient set determined by $\equiv$. We can now equip F with addition, multiplication, an element 0, and an element 1 so that it becomes a field as follows:

$$a/b + c/d = (ad + bc)/bd \ ,$$
$$(a/b) \cdot (c/d) = ac/bd \ ,$$
$$0 = 0/1, \text{ and } 1 = 1/1 \ .$$

It can be shown that the above operations $+, \cdot$ define single-valued compositions in F and that F with the above 0 and 1 is a commutative ring. Moreover, if $a/b \neq 0$, then $a \neq 0$ and b/a is the inverse of b/a. This shows that F is a field; it is called the *field of quotients* or the *field of fractions* (or *rational expressions*) of D.

field of rational expressions   *See* field of quotients.

field of values   *See* numerical range.

field theory   In algebra, the theory and research area associated with fields. *See* field.

figure   (**1**) A symbol used to denote an integer.

(**2**) In topology, a set of points in the space under consideration.

filtration   A *filtration* of a module M is a collection $\{M_k : k \in \mathbf{Z}\}$, of submodules of M, such that $M_{k+1} \subset M_k$ for all $k \in \mathbf{Z}$. *See also* exhaustive filtration, filtration degree.

filtration degree   Suppose that M is a graded module with differentiation d of degree 1, i.e., M is a complex, and that the filtration $\{F^k M\}_{k \in \mathbf{Z}}$ of M is homogeneous. Define the graded module $E_0(M)$ to be the direct sum $\sum_{k \in \mathbf{Z}} E_0^k(M)$

where $E_0^k(M) = F^k M/F^{k+1} M$. Define

$$F^{k,l} M = M^{k+l} \cap F^k M = F^k M^{k+l}$$

and $E_0^{k,l}(M) = F^{k,l} M/F^{k+1,l-1} M$, where $k, l \in \mathbf{Z}$. Then $E_0(M)$ is doubly graded as the direct sum $\sum_{k,l \in \mathbf{Z}} E_0^{k,l}(M)$. In the same way, the module $E_0(H(M))$ is doubly graded by the modules $E_0^{k,l}(H(M)) = F^k H^{k+l}(M)/F^{k+1} H^{k+l}(M)$, where $H(M)$ is the homology module of M and $F^k H^{k+l}(M) = F^{k,l} H(M)$.

For $1 \le r \le \infty$ define

$$Z_r^{k,l}(M) = \mathrm{Im}\Big(H^{k+l}\Big(F^k M/F^{k+r} M\Big)$$

$$\to H^{k+l}\Big(F^k M/F^{k+1} M\Big)\Big),$$

$$B_r^{k,l}(M) = \mathrm{Im}\Big(H^{k+l-1}\Big(F^{k-r+1} M/F^k M\Big)$$

$$\to H^{k+l}\Big(F^k M/F^{k+1} M\Big)\Big),$$

$$E_r^{k,l}(M) = Z_r^{k,l}(M)/B_r^{k,l}(M).$$

Then $E_r(M)$ is doubly graded by identifying $E_r^k(M)$ with the direct sum

$$\sum_{l \in \mathbf{Z}} E_r^{k,l}(M).$$

Finally, the differentiation operator $d_r : E_r \to E_r$ is composed of homomorphisms $d_r^{k,l} : E_r^{k,l} \to E_r^{k+r,l-r+1}$, in other words, d_r has bi-degree $(r, 1-r)$.

In all of these doubly graded modules the first degree k is called the *filtration degree,* the second degree l is called the *complementary degree,* and $k + l$ is called the *total degree.*

fine moduli scheme   A moduli scheme M_g of curves of genus g such that there exists a flat family $F \to M_g$ of curves of genus g such that, for any other flat family $X \to Y$ of curves of genus g, there is a unique map $Y \to M_g$ via which X is the pullback of F.

finer   If $\mathcal{T}$ and $\mathcal{U}$ are two topologies on a space X, and if $\mathcal{T} \subseteq \mathcal{U}$, then $\mathcal{U}$ is said to be *finer* than $\mathcal{T}$.

finite   A term associated with the number of elements in a set. A set S is finite if there exists a natural number n and a one-to-one correspondence between the elements of S and the

elements of the set $\{1, 2, \ldots, n\}$. We may then write $|S| = n$. The fact that a set S is finite is denoted by $|S| < \infty$ and we say that S consists of a finite number of elements. *See also* cardinality, finite function.

finite Abelian group   A group G that is finite, as a set, and commutative; namely for any $a, b \in G$, $ab = ba$. *See also* finite, Abelian group.

finite basis   A basis of a vector space V over a field F, comprising a finite number of vectors. More precisely, a finite basis of V is a finite set of vectors $\mathcal{B} = \{v_1, v_2, \ldots, v_n\} \subset V$ with two properties:

(i.) $\mathcal{B}$ is a spanning set of V, that is, every vector of V is a linear combination of the vectors in $\mathcal{B}$.

(ii.) $\mathcal{B}$ consists of linearly independent vectors (over the specified field F).

A (finite) basis of a vector space is not, in general, unique. However, all the bases of V have the same number of vectors. This number is called the dimension of V. Here the dimension of V is n and thus V is referred to as a finite dimensional vector space.

The set of vectors $\{e_i\}_{i=1}^n$, consisting of the vectors $e_i \in \mathbf{R}^n$ whose ith entry is one and all other entries equal zero, is a finite basis for the vector space $\mathbf{R}^n$. It is usually called the *standard basis* of $\mathbf{R}^n$.

finite continued fraction   Let q be a continued fraction defined by $q = p_1 + 1/q_1$, where $q_1 = p_2 + 1/q_2$, $q_2 = p_3 + 1/q_3, \ldots$, and where p_i, q_i are numbers or functions of a variable. *See* continued fraction. If the expression q terminates after a finite number of terms, then q is a finite continued fraction. For example,

$$q = 1 + \cfrac{1}{2 + \frac{1}{3}}$$

is a finite continued fraction usually denoted by $1 + \frac{1}{2+} \frac{1}{3}$.

finite field   A field F which is a finite set. In such a case, the prime field of F can be identified with Z_p, the field of residue classes, modulo p, for some prime integer p. (*See* prime field.) It follows that $|F| = p^n$ for some integer n. We thus have the fundamental fact that the number of elements (cardinality) of a finite field is a power of a prime integer. Moreover, all finite fields of the same cardinality are isomorphic.

finite function (1) A function $f : X \to Y$ which is finite, when thought of as a subset of the Cartesian product $X \times Y$. It follows that f is finite if and only if its domain X is a finite set.

(2) A function from a set X to the extended real or complex numbers ($\mathbf{R} \cup \{\pm\infty\}$ or $\mathbf{C} \cup \{\pm\infty\}$), never taking the values $\pm\infty$. *See also* semifinite function.

finite graded module *See* graded module.

finite group A group, which is finite as a set. An example of a finite group is the symmetric group, S_n, of degree n, having $n!$ elements. In fact, any finite group with n elements is isomorphic to some subgroup of S_n. *See* symmetric group.

finitely generated group A group G that consists of all possible finite products of the elements of a finite set S. We usually write

$$G = \langle S \rangle = \{s_1 s_2 \ldots s_k : s_i \in S\}$$

and we call S a *set of generators* of S. The simplest example is a cyclic group, namely, a group G generated by one element a. We then write $G = \langle \{a\} \rangle = \{a^k : k \in \mathbf{Z}\}$.

finitely generated ideal An ideal I in a ring R such that I contains elements $i_1, \ldots, i_k$ which, under sums and products by elements of R, generate all of I.

finitely generated module Let A be a ring and M an A-module. We say that M is *finitely generated* if there is a finite set $\{x_1, x_2, \ldots, x_k\}$ of elements of M such that, for each element $x \in M$, there exist scalars $a_i \in A$, so that

$$x = \sum_{i=1}^{k} a_i x_i \, .$$

We refer to $\{x_1, x_2, \ldots, x_k\}$ as a *system of generators* of M.

finitely presented group Let F be the free group generated by $a_1, a_2, \ldots, a_n$ and G be a group generated by $b_1, b_2, \ldots, b_n$. Then there is a homomorphism h of F onto G. If the kernel of h is the minimal normal subgroup of F containing the classes of words

$$w_1 (a_1 \ldots, a_n), \ldots, w_m (a_1 \ldots, a_n) \, ,$$

then

$$w_1 (b_1 \ldots, b_n) = 1, \ldots, w_m (b_1 \ldots, b_n) = 1$$

are the *defining relations* of G. If m and n are both finite, we call G a finitely presented group.

Finiteness Theorem (1) (*Finiteness theorem of Hilbert concerning first syzygies*) There are only finitely many first syzygies. *See* first syzygy.

(2) (*Completeness of predicate calculus*) If a proposition H is provable (derivable) from a set of statements X, then there exists a finite subset $X^* \subset X$ from which H can also be derived.

finite nilpotent group A group G, which is both finite and nilpotent. *See* nilpotent group. The following conditions are equivalent to being nilpotent for a finite group:

(i.) G has at least one central series.

(ii.) The upper central series of subgroups Z_i of G,

$$\{e\} \equiv Z_0 \subset Z_1 \subset Z_2 \subset \ldots$$

ends with $Z_n \equiv G$ for some finite $n \in \mathbf{N}$.

(iii.) The lower central series of G

$$G \equiv \tilde{H}_0 \supset \tilde{H}_1 \supset \ldots$$

ends with $\tilde{H}_m = \{e\}$ for some finite $m \in \mathbf{N}$.

(iv.) Every maximal (proper) subgroup is normal.

(v.) Every subgroup differs from its normalizer.

(vi.) G is a direct product of Sylow p-subgroups of G.

finite prime divisor An equivalence class of non-Archimedean valuations of an algebraic number field K.

finite simple group A group G of finite order $|G|$ ($|G| > 1$) that contains no proper normal subgroup.

finite solvable group   A group G that has a composition series

$$G \equiv G_0 \supset G_1 \supset \cdots \supset G_n \equiv \{e\},$$

whose factor groups G_i/G_{i+1}, $i = 0, \dots, n - 1$, are of prime order. Equivalently, the composition factors G_i/G_{i+1} are simple Abelian groups (i.e., cyclic groups of prime order).

first factor of class number   The class number h of the p-th cyclotomic field K_p, where p is a prime, is a product $h = h_1 h_2$ of two factors h_1 and h_2, called, respectively, the first and the second factor of the class number h.

If $K_m = \mathbf{Q}(\xi_m)$, $\xi_m = e^{2\pi i/m}$, then h_2 is the class number of the real subfield $K'_m = \mathbf{Q}(\xi + \xi^{-1})$. Explicitly

$$h_1 = \frac{(-1)^{r+1}}{(2p)^r} \prod_{i=1}^{r+1} \left(\sum_{j=1}^{p-1} j \chi_i(j) \right),$$

$$h_2 = \frac{|E|}{R_0},$$

where $r = (p - 3)/2$ gives the number of multiplicatively independent units, χ_i, $i = 1, \dots,$ $p - 1$, are the multiplicative characters of the reduced residue classes of $\mathbf{Z}$ modulo p, and χ_i, $i = 1, \dots, r - 1$, are those characters for which $\chi_i(-1) = -1$. Further, $0 \neq E = R[\varepsilon_0, \dots, \varepsilon_{r-1}]$ is the Dedekind regulator of multiplicatively independent units ε_i ($i = 0, \dots, r - 1$), also called circular units and R_0 is the regulator of K'_p.

Remark: $K_m = \mathbf{Q}(\xi_m)$, $\xi_m = e^{2\pi i/m}$ is an algebraic number field obtained by adjoining an m-th primitive root of unity to $\mathbf{Q}$. It is a Galois extension over $\mathbf{Q}$ of degree $\phi(m)$, where ϕ is Euler's function [$\phi(m)$ gives the number of primitive roots of unity].

first syzygy   Let $R = k[x_1, \dots, x_n]$ be a polynomial ring of n variables $x_1, \dots, x_n$ over a field k and, relying on the natural gradation of R [i.e., $\deg(x_i) = 1$, $\deg(c) = 0$ for $c \in k$], let M be a finitely generated graded R-module. Designate by $(f_1, \dots, f_m)$ a minimal basis of M over R consisting of homogeneous elements. Introduce m indeterminates g_i ($i = 1, \dots, m$) and the free R-module F, $F = \sum_{i=1}^{m} R g_i$ generated by

them. Requiring that $\deg(g_j) = \deg(f_j)$, $j = 1, \dots, m$, we supply F with the structure of a graded R-module. The kernel of a graded R-homomorphism $\phi : F \to M$, $M = \phi(F)$, defined by $\phi(g_j) = f_j$ is referred to as the *first syzygy*. It is uniquely determined by M up to a graded R-module isomorphism.

(*More generally:* Let R be a Noetherian ring, M a finitely generated R-module. Then one can find a finitely generated free R-module F and an R-homomorphism $\phi : F \to M$ (onto), whose kernel defines the first syzygy of M.)

fixed component   The maximal positive divisor Σ_0 that is contained in all divisors of a linear system Σ is called the fixed component of Σ.

Let V be a complete irreducible variety, f_0, $f_1, \dots, f_n$ the elements of the function field $k(V)$ of V, and D a divisor ring on V such that $(f_i) + D \geq 0$, for all i. Then the set Σ of the divisors of the form $(\Sigma a_i f_i) + D$, with $a_i \in k$ not all zero, is called *a linear system*.

flat dimension   A left R module B, where R is a ring with unit, has *flat dimension n* if there is a flat resolution

$$0 \longrightarrow E_n \longrightarrow \cdots \longrightarrow E_0 \longrightarrow B \longrightarrow 0,$$

but no shorter flat resolution of B. The definition of the *flat dimension* of a right R module is entirely similar. *See* flat resolution. Flat dimensions have little relation to more elementary notions of dimension, such as the dimension of a vector space, but they are related to the idea of *injective dimension*. *See also* injective dimension, projective dimension.

flat module   Let R be a ring and M a right R-module. If for any exact sequence

$$0 \to N' \to N \to N'' \to 0,$$

the induced sequence

$$0 \to M \otimes_R N' \to M \otimes_R N \to M \otimes_R N'' \to 0$$

is exact, then R is a flat R-module.

Here $M \otimes_R N$ is a tensor product of a right R-module M and a left R-module N.

Remark: In view of the isomorphism between $M \otimes_R N$ and $N \otimes_R M$, we could drop the qualifiers left and right.

flat morphism of schemes   A morphism of schemes $f : X \to Y$ such that, for each point $x \in X$, a stalk $\mathcal{O}_{X,x}$ at that point is a flat $\mathcal{O}_{Y,f(x)}$-module. If f is surjective, f is *faithfully* flat.

flat resolution   Let B be a left R module, where R is a ring with unit. A *flat resolution* of B is an exact sequence,

$$\cdots \xrightarrow{\phi_2} E_1 \xrightarrow{\phi_1} E_0 \xrightarrow{\phi_0} B \longrightarrow 0 ,$$

where every E_i is a flat left R module. (We shall define *exact sequence* shortly.) There is a companion notion for right R modules. Flat resolutions are important in homological algebra and enter into the dimension theory of rings and modules. *See also* flat dimension, flat module, injective resolution, projective resolution.

An *exact sequence* is a sequence of left R modules, such as the one above, where every ϕ_i is a left R module homomorphism (the ϕ_i are called *connecting homomorphisms*), such that $\mathrm{Im}(\phi_{i+1}) = \mathrm{Ker}(\phi_i)$. Here $\mathrm{Im}(\phi_{i+1})$ is the image of ϕ_{i+1}, and $\mathrm{Ker}(\phi_i)$ is the kernel of ϕ_i. In the particular case above, because the sequence ends with 0, it is understood that the image of ϕ_0 is B, that is ϕ_0 is onto. There is a companion notion for right R modules.

formal group   Formal groups are analogs of the local Lie groups, for the case of algebraic k-groups with a nonzero prime characteristic.

formal scheme   A topological local ringed space that is locally isomorphic to a formal spectrum $\mathrm{Spf}(A)$ of a Noetharian ring A.

formal spectrum   Let R be a Noetherian ring which is complete with respect to its ideal I (in the I-adic topology), so that its completion along I can be identified with R. The formal spectrum $\mathrm{Spf}(R)$, of R, is a pair $(\mathcal{X}, \mathcal{O}_{\mathcal{X}})$ consisting of a formal scheme $\mathcal{X} = V(I) \subset \mathrm{Spec}(R)$ and a sheaf of topological rings $\mathcal{O}_{\mathcal{X}}$ defined as follows:

$$\Gamma(D(f) \cap \mathcal{X}, \mathcal{O}_{\mathcal{X}}) = \varprojlim_{n>0} R_f / I^n R_f, \ f \in R .$$

Here $V(I)$ is a set of primitive ideals of R containing I, $D(f) = \mathrm{Spec}(R) - V(f)$, $f \in R$ are elementary open sets forming a base of the Zariski topology, and $\Gamma(Q, \mathcal{O})$ designates the set of sections over Q of a sheaf space $\mathcal{O}$ over

$\mathcal{X}$. A section of $\mathcal{O}$ over Q is a continuous map $\sigma : A \subset \mathcal{X} \to \mathcal{O}$ such that $\pi \circ \sigma = 1_A$, where $\pi : \mathcal{O} \to \mathcal{X}$ is such that $\pi(\mathcal{O}_x) = x$, $\mathcal{O}_x$ being a stalk over x.

form ring   Let (R, P) be a local ring and Q its P-primary ideal. Set

$$F_i = Q^i / Q^{i+1}, \quad i = 0, 1, \ldots; \ Q^0 = R ,$$

and for $A = A' \pmod{Q^{i+1}} \in F_i$ $B = B' \pmod{Q^{j+1}} \in F_j$, require

$$AB = A'B' \left(\mathrm{mod}\, Q^{i+j+1} \right) \in F_{i+j} .$$

Then the form ring of R with respect to Q is defined as a graded ring F generated by F_1 over F_0 and equal to the direct sum of modules

$$F = \bigoplus_{i=0}^{\infty} F_i ,$$

where F_i is a module of homogeneous elements of degree i.

formula   (1) A formal expression of a proposition in terms of local symbols.

(2) A formal expression of some rule or other results (e.g., Frenet formula, Stirling formulas, etc.).

(3) Any sequence of symbols of a formal calculus.

forward elimination   A step in the Gauss elimination method of solving a system of linear equations

$$\sum_{j=1}^{n} a_{ij} x_j = b_i, \quad (i = 1, \ldots, n)$$

consisting of the following steps

$$a_{ij}^{(m+1)} = a_{ij}^{(m)} - a_{im}^{(m)} a_{mj}^{(m)} / a_{mm}^{(m)} ,$$

$$b_i^{(m+1)} = b_i^{(m)} - a_{im}^{(m)} b_m^{(m)} / a_{mm}^{(m)} ,$$

$i, j = m + 1, \ldots, n$, with

$$a_{ij}^{(1)} \equiv a_{ij}, \ b_i^{(1)} = b_i ,$$

for all i, j. After the forward elimination, we get a system with a triangular coefficient matrix

$$\sum_{j=m}^{n} a_{mj}^{(m)} x_j = b_m^{(m)}, \quad m = 1, \ldots, n$$

and apply backward elimination to solve the system.

Also called *forward step*.

four group   The simplest non-cyclic group (of order 4). It may be realized by matrices

$$\begin{pmatrix} \pm I & 0 \\ 0 & \pm I \end{pmatrix}$$

(any two elements different from the identity generate V), or as a non-cyclic subgroup of the alternating group A_4, involving the permutations

$$(1), \ (12)(34), \ (13)(24), \ (14)(23) \ .$$

It is Abelian and, as a transitive permutation group, it is imprimitive with the imprimitivity system {12}, {34}.

Also called *Klein's four group* or *Vierergruppe V*.

Fourier series   An infinite trigonometric series of the form

$$\frac{1}{2} a_0 + \sum_{n=1}^{\infty} [a_n \cos nx + b_n \sin nx]$$

with $a_0, a_1, \ldots, b_1, \cdots \in \mathbf{R}$, referred to as (real) *Fourier coefficients*. Here

$$a_n = \frac{1}{\pi} \int_{-\pi}^{\pi} f(t) \cos nt \, dt \ ,$$

$$b_n = \frac{1}{\pi} \int_{-\pi}^{\pi} f(t) \sin nt \, dt \ ,$$

$n = 0, 1, 2, \ldots$, with $f(t)$ a periodic function with period 2π. The relationship between $f(x)$ and the series is the subject of the theory of Fourier series. The complex form is

$$\sum_{k=-\infty}^{\infty} c_k e^{ikx} \ ,$$

so that

$$c_k = \frac{1}{2} (a_k - ib_k) = \bar{c}_{-k} \ .$$

Fourier's Theorem   Consider an nth degree polynomial in x,

$$f(x) = a_0 x^n + a_1 x^{n-1} + \cdots + a_0 \ ,$$

with $a_i \in \mathbf{R}$ and $a_0 \neq 0$, and the algebraic equation

$$f(x) = 0 \ .$$

Designate by $V(c_1, c_2, \ldots, c_p)$ the number of sign changes in the sequence $c_1, c_2, \ldots, c_p$ of real numbers, defined by

$$V(c_1, c_2, \ldots, c_p) = \frac{1}{2} \sum_{j=1}^{q-1} \left(1 - \operatorname{sgn} c_{v_j} c_{v_{j+1}}\right) \ ,$$

where $c_{v_1}, c_{v_2}, \ldots, c_{v_q}$ is obtained from $c_1, c_2, \ldots, c_p$ by deleting the vanishing terms $c_i = 0$. Defining, finally

$$W(x) = V\left(f(x), f'(x), \ldots, f^{(n)}(x)\right)$$

and

$$N \equiv N(a, b) = W(a) - W(b) \ ,$$

the number $m \equiv m(a, b)$ of real roots in the interval (a, b) equals

$$m = N \pmod 2, \quad m \leq N \ ,$$

i.e., $m = N$ or $m = N - 2$ or $m = N - 4, \ldots$, or $m = 0$ or 1.

The precise value of m can be obtained using the theorem of Sturm that exploits the sequence $f(x), f'(x), R_1(x), \ldots, R_{m-1}(x), R_m$, where $-R_i$ is the remainder when dividing R_{i-2} by R_{i-1}, with $R_0 = f'$ and $R_{-1} = f$. Then $m = W(a) - W(b)$.

Fourier's Theorem (on algebraic equations) is also called the Budom-Fourier Theorem.

fraction   A ratio of two integers m/n, where m is not a multiple of n and $n \neq 0, 1$, or any number that can be so expressed. In general, any ratio of one quantity or expression (the numerator) to another nonvanishing quantity or expression (the denominator).

fractional equation   An algebraic equation with rational integral coefficients.

fractional exponent   The number a, in an expression x^a, where a is a rational number.

fractional expression   *See* fraction.

fractional ideal   (**1**) (*Of an algebraic number field k*) Let k be an algebraic number field of finite degree (i.e., an extension field of **Q** of finite degree) and I an integral domain consisting of all algebraic integers. Further, let **p** be the principle order of k (i.e., an integral domain $k \cap I$ whose field of quotients is k) and **a** an integral ideal of k (i.e., an ideal of the principle order **p**). Then a *fractional ideal* of k is a **p**-module that is contained in k (i.e., $\mathbf{pa} \subset \mathbf{a}$) such that $\alpha \mathbf{a} \subset \mathbf{p}$ for some α ($\alpha \in k, \alpha \neq 0$). *See* algebraic number, algebraic integer.

(**2**) (*of a ring R*) An R-submodule **a** of the ring Q of total quotients of R such that there exists a non-zero divisor **q** of R such that $\mathbf{qa} \subset R$. *See* ring of total quotients.

fractional programming   Designating: **x** an n-dimensional vector of decisive variables, **m** an n-dimensional constant vector, **Q** a positive definite, symmetric, constant $n \times n$ matrix and (,) the scalar product, the problem of *fractional programming* is to maximize the expression

$$\frac{(\mathbf{x}, \mathbf{m})}{\sqrt{(\mathbf{x}, \mathbf{Qx})}},$$

subject to nonnegativity of **x** ($\mathbf{x} \geq 0$) and linear constraints $\mathbf{Ax} \leq \mathbf{b}$.

fractional root   A root of a polynomial $p(x)$ with integral coefficients

$$p(x) = a_0 x^n + a_1 x^{n-1} + \cdots + a_n$$

that has the form r/s. One can show that r and s are such that a_n is divisible by r and a_0 is divisible by s. Any rational root of monic polynomial having integral coefficients is thus an integer. Also called *rational root*.

free Abelian group   The direct product (finite or infinite) of infinite cyclic groups. Equivalently, a *free Abelian group* is a free **Z**-module.

An infinite cyclic group is one generated by a simple element x such that all integral powers of x are distinct. *See* cyclic group, direct product.

free additive group   The direct sum of additive groups $A_i, i \in \Xi$, such that each A_i is isomorphic to **Z**. *See* additive group.

free group   A free product of infinite cyclic groups. The number of free factors is called the rank of the group F. Alternatively, (i.) a free group F_n on n generators is a group generated by a set of free generators (i.e., by the generators that satisfy no relations other than those implied by the group axioms); (ii.) a free group is a group with an empty set of defining relations.

To form a *free group* we can start with a *free semigroup*, defined on a set of symbols $S = \{a_1, a_2, \dots\}$, that consists of all *words* (i.e., finite strings of symbols from S, repetitions being allowed), including an empty word representing the unity. Next, we extend S to S' that contains the inverses and the identity e

$$S' = \left\{ e, a_1, a_1^{-1}, a_2, a_2^{-1}, \dots \right\}.$$

Then a set of equivalence classes of words formed from S' with the law of composition defined by juxtaposition (to obtain a product $\alpha\beta$ of two words α and β, we attach β to the end of α; to obtain the inverse of α reverse the order of symbols a_i while replacing a_i by a_i^{-1} and vice versa) is the free group F on the set S.

The equivalence relation (designated by "$\sim$") used to define the equivalence classes is defined via the elementary equivalences $ee \sim e$, $a_i a_i^{-1} \sim e$, $a_i^{-1} a_i \sim e$, $a_i e \sim a_i$, $a_i^{-1} e \sim a_i^{-1}$, $ea_i = a_i$, $ea_i^{-1} = a_i^{-1}$, so that $\alpha \sim \beta$ if α is obtainable from β through a sequence of elementary equivalences.

free module   For a ring R, an R-module that has a basis.
Remarks: (i.) If R is a field, then every R-module is free (i.e., a linear space over R).

(ii.) A finitely generated module V is free if there is an isomorphism $\phi : R^n \to V$, where R is a commutator ring with unity.

(iii.) A free **Z**-module is also called a free Abelian ring.

free product   Consider a family of groups $\{G_i\}_{i\in\Xi}$ and their disjoint union as sets S. A *word* is either empty (or void) or a finite sequence $a_1 a_2 \ldots a_n$, of the elements of S. Designate by W the set of all words and define the following binary relations on W:

(i.) The product of two words w and w' is obtained by juxtaposition of w and w'.

(ii.) $w \succ w'$ if either w contains successive elements $a_k a_{k+1}$ belonging to the same G_i and w' results from w by replacing $a_k a_{k+1}$ by their product $a = a_k a_{k+1}$, or if w contains an identity element and w' results by deleting it.

(iii.) $w \sim w'$ if there exists a finite sequence of words $w = w_0, w_1, \ldots, w_n = w'$ such that for each j ($j = 0, \ldots, n-1$) either $w_j \succ w_{j+1}$ or $w_{j+1} \succ w_j$.

Clearly, the definition of the product immediately implies the associativity and "$\sim$" represents an equivalence relation that is compatible with multiplication. We can thus define the product for the quotient set G of W by the equivalence relation $\sim$, and take as the identity the equivalence class containing the empty word. The resulting group G is called the free product of the system of groups $\{G_i\}_{i\in\Xi}$.

Remarks: (i.) The free product is the dual concept to that of the direct product (and is called the coproduct in the theory of categories and functors).

(ii.) If each G_i is an infinite cyclic group generated by a_i, then the free product of $\{G_i\}_{i\in\Xi}$ is the free group generated by $\{a_i\}_{i\in\Xi}$.

free resolution   (Of **Z**) A certain cohomological functor, defined by Artin and Tate, that can be described as a set of cohomology groups concerning a certain complex in a non-Abelian theory of homological algebras.

free semigroup   All words (i.e., finite strings of symbols from a set of symbols $S = \{a_1, a_2, \ldots\}$) with the product defined by juxtaposition of words [and the identity being the empty (void) word when a semigroup with identity is considered]. *See also* free group.

free special Jordan algebra   Let $A = k[x_1, \ldots, x_n]$ be a noncommutative free ring in the indeterminates $x_1, \ldots x_n$ (i.e., the associative algebra over k). Defining a new product by

$$x * y = (xy + yx)/2 \,,$$

we obtain a Jordan algebra $A^{(J)}$. The subalgebra of $A^{(J)}$ generated by 1 and the x_i is called the *free special Jordan algebra* (of n generators).
Remark: A special Jordan algebra $A^{(J)}$ arises from an associative algebra A by defining a new product $x * y$, as above.

Frobenius algebra   An algebra A over a field k such that its regular and coregular representations are similar.
Remarks: (i.) Any finite dimensional semisimple algebra is a *Frobenius algebra*.

(ii.) An algebra A is a *Frobenius algebra*, if the left A-module A and a dual module A^* of the right A-module A are isomorphic as left A-modules.

Frobenius automorphism   An element of a Galois group of a special kind that plays an important role in algebraic number field theory.

Designate by K/F a relative algebraic number field, with K a Galois extension of F of degree $[K : F] = n$, $G = G(K/F)$ the corresponding Galois group, and Ω the principal order of K. In Hilbert's decomposition theory of prime ideals of F, for a Galois extension K/F in terms of the Galois group G, the subgroup

$$H = \{\sigma \in G : \mathcal{P}^\sigma = \mathcal{P}\} \,,$$

called the *decomposition group* of a prime ideal $\mathcal{P}$ of Ω over F, plays an important role. The normal subgroup $\mathcal{H}$ of H, called the *inertia group* of $\mathcal{P}$ over F, is defined by

$$\mathcal{H} = \{\sigma \in H : a^\sigma \equiv a (\mathrm{mod}\mathcal{P}), a \in \Omega\} \,.$$

The quotient group $H/\mathcal{H}$ is a cyclic group of order k, where k is the relative degree of $\mathcal{P}$. This cyclic group is generated by an element $\sigma_o \in H$ that is uniquely determined $(\mathrm{mod}\mathcal{H})$ by the requirement

$$a^{\sigma_o} \equiv a^{N(\pi)} (\mathrm{mod}\mathcal{P}), \quad a \in \Omega$$

where π is a prime ideal of F that is being decomposed. This element σ_o is referred to as the *Frobenius automorphism* or the *Frobenius substitution* of $\mathcal{P}$ over F.

See also ramification group, ramification numbers, ramification field, Artin's symbol.

Frobenius endomorphism   For a commutative ring R with identity and prime characteristic p, the ring homomorphisn $F : R \to R$, defined by

$$F(a) = a^p, \qquad \text{(for all } a \in R) \,.$$

Clearly, for any $a, b \in R$, we have that

$$(a + b)^p = a^p + b^p \text{ and } (a \cdot b)^p = a^p \cdot b^p \,.$$

For a Galois extension $K/\mathbf{F}_p$ over a prime field $\mathbf{F}_p$ of degree $m := [K : \mathbf{F}_p]$ (with $p^m = q$ distinct elements), referred to as Galois field $\mathbf{F}_q$ of order q, the Frobenius endomorphism

$$F : \mathbf{F}_q \to \mathbf{F}_q, \quad F : x \mapsto x^p$$

is injective and thus an automorphism of $\mathbf{F}_q$. (*See also* Frobenius automorphism.) In fact, F generates the cyclic Galois group $G(\mathbf{F}_q/\mathbf{F}_p)$.

Generally, for a scheme X over a finite field $\mathbf{F}_q$ of $q(= p^n)$ elements, the Frobenius endomorphism is an endomorphism $\psi : X \to X$ such that $\psi = Id$ (the identity mapping) on the set of $\mathbf{F}_q$–points of X (i.e., on the set of points of X defined over $\mathbf{F}_q$) and the mapping of the structure sheaf $\psi^* : \mathcal{O}_X \to \mathcal{O}_X$ raises the elements of $\mathcal{O}_X$ to the qth power.

Thus, for an affine variety $X \subset \mathbf{A}^n$ defined over $\mathbf{F}_q$, we have

$$\psi(x_1, \ldots, x_n) = \left(x_1^q, \ldots, x_n^q\right) \,.$$

Frobenius formula   The characters $\chi_{(\alpha)}^{[\lambda]}$ of the irreducible representation $[\lambda] \equiv (\lambda_1, \lambda_2, \ldots, \lambda_n)$, associated with the class $(\alpha) \equiv (1^{\alpha_1} 2^{\alpha_2} 3^{\alpha_3} \ldots n^{\alpha_n})$ of the symmetric ($\equiv$ permutation) group $\mathcal{S}_n$, where the partitions $[\lambda]$ and (α) of n satisfy the relations

$$\sum_{i=1}^n \lambda_i = n = \sum_{i=1}^n i\alpha_i \,,$$
$$\lambda_1 \geq \lambda_2 \geq \cdots \geq \lambda_n \geq 0 \,,$$

are given by the following Frobenius formula:

$$\sum_\lambda \chi_{(\alpha)}^{[\lambda]} \Delta^{[\lambda]}(x) = F_{(\alpha)}(x) \,,$$

where $F_{(\alpha)}(x) = \Delta(x)s_{(\alpha)}(x)$ is the Frobenius generating function (q.v.) and $\Delta^{[\lambda]}(x)$ is a generalized Vandenmonde determinant

$$\Delta^{[\lambda]}(x) \equiv \Delta^{(\lambda_1\lambda_2\ldots\lambda_n)}(x_1, x_2, \ldots, x_n)$$

$$= \begin{vmatrix} x_1^{\lambda_1+n-1} & x_2^{\lambda_1+n-1} & \cdots & x_n^{\lambda_1+n-1} \\ x_1^{\lambda_2+n-2} & x_2^{\lambda_2+n-2} & \cdots & x_n^{\lambda_2+n-2} \\ \vdots & \vdots & & \vdots \\ x_1^{\lambda_n} & x_2^{\lambda_n} & \cdots & x_n^{\lambda_n} \end{vmatrix} \,.$$

An alternative form is

$$s_{(\alpha)}(x) = \sum_\lambda \chi_{(\alpha)}^{[x]} S_{[\lambda]}(x) \,,$$

where $S_{[\lambda]}(x)$ is Schur polynomial and $s_{(\alpha)}(x) = \prod_{r=1}^n (s_r(x))^{\alpha_i}$ is the product of α_ith powers of Newton polynomials

$$s_r(x) = \sum_{i=1}^n x_i^r \,.$$

See also Frobenius generating function.

Frobenius generating function   For an arbitrary partition (α) of n, written in the form

$$(\alpha) \equiv \left(1^{\alpha_1} 2^{\alpha_2} \ldots n^{\alpha_n}\right), \quad \sum_{i=1}^n i\alpha_i = n \,,$$

the Frobenius generating function $F_{(\alpha)}(x) \equiv F_{(\alpha)}(x_1, \ldots, x_n)$ is given by the product

$$F_{(\alpha)}(x) = \Delta(x)s_{(\alpha)}(x) \,,$$

where $\Delta(x)$ is a polynomial, expressible as a Vandermonde determinant

$$\Delta(x) \equiv \Delta(x_1, x_2, \ldots, x_n) = \prod_{i<j} (x_i - x_j)$$

$$= \begin{vmatrix} x_1^{n-1} & x_2^{n-1} & \cdots & x_n^{n-1} \\ \vdots & \vdots & & \vdots \\ x_1^2 & x_2^2 & \cdots & x_n^2 \\ x_1 & x_2 & \cdots & x_n \\ 1 & 1 & \cdots & 1 \end{vmatrix} \,,$$

and $s_{(\alpha)}(x) \equiv \prod_{i=1}^n (s_r(x))^{\alpha_i}$ is a product of α_ith powers of power sums (or Newton polynomials) s_r,

$$s_r(x) = \sum_{i=1}^n x_i^r \,.$$

See also Frobenius formula.

Frobenius group   A (nonregular) transitive permutation group on a set M, each element of which has at most one fixed point, i.e., the identity of the group is its only element that leaves more than one element of M invariant. *See* regular transitive permutation group.

Frobenius homomorphism   (*For commutative rings.*) Let A be a commutative ring of prime characteristic p, so that $(a + b)^{p^n} = a^{p^n} + b^{p^n}$ for any $a, b \in A$ and any $n \in \mathbf{N}$. Thus, the map $\phi : a \mapsto a^p$ is an endomorphism of the additive group of A. Since further $1^p = 1$ and $(ab)^p = a^p b^p$ for $a, b \in A$, this is also an endomorphism of A, which is referred to as the *Frobenius homomorphism* of A. It is injective when 0 is the only nilpotent element of A, which is the case when A is an integral domain. *See also* Frobenius endomorphism for schemes.

Frobenius inequality   Let $f : X \to Y$, $g : Y \to Z$, and $h : Z \to V$ be linear maps of finite dimensional vector spaces over a division ring (a skew-field). Then

$$\text{rank}\,(hg) + \text{rank}\,(gf) \le \text{rank}\,(g) + \text{rank}\,(hgf)\,.$$

Frobenius morphism   *See* Frobenius automorphism, Frobenius homomorphism, Frobenius endomorphism.

Frobenius norm   For an $n \times n$ matrix A, the length, denoted $\|A\|$, of the n^2-dimensional vector

$(a_{11}, a_{12}, \ldots, a_{1n}, a_{21}, a_{22}, \ldots, a_{2n}, \ldots, a_{nn})$,

i.e.,

$$\|A\| = \left[\sum_{i,j=1}^{n} |a_{ij}|^2 \right]^{1/2}.$$

Frobenius normal form   An $n \times n$ matrix $A = (a_{ij})$ that is block (upper) triangular, with the diagonal blocks being square irreducible matrices that correspond to the strongly connected components of $G(A)$. Here, $G(A)$ is the directed graph (digraph), $G(A) = (V, E)$, consisting of vertices $V = \{1, 2, \ldots, n\}$ and directed edges $E = \{(i, j) : a_{ij} \neq 0\}$. The set V admits a partition into disjoint subsets of vertices so that in each such subset the vertices have access to each other via a directed path (sequence of edges). The subsets of this partition are called the strongly connected components of $G(A)$.

If there is only one strongly connected component, we call A irreducible. Otherwise, A is reducible. If A is reducible, there exists a *permutation matrix P* such that PAP^T, is in Frobenius normal form. For instance, if $G(A)$ has t strongly connected components, then its Frobenius normal form is

$$\begin{pmatrix} A_{11} & A_{12} & \cdots & & \cdots & A_{1t} \\ 0 & A_{22} & A_{23} & & \cdots & A_{2t} \\ 0 & 0 & \ddots & & \ddots & \vdots \\ \vdots & & \ddots & \ddots & A_{t-1,t-1} & A_{t-1,t} \\ 0 & \cdots & & \cdots & 0 & A_{tt} \end{pmatrix},$$

where each A_{ii} for $i = 1, 2, \ldots, t$ is square and irreducible. The Frobenius normal form is not, in general, unique. As for any permutation matrix P, $P^{-1} = P^T$, spectral properties of A can be studied by studying spectral properties of the irreducible blocks A_{ii} in its Frobenius normal form.

Frobenius Reciprocity Theorem   Let G and H be finite groups, Γ and γ some irreducible representations of G and H, respectively, and H a subgroup of G, $H \subset G$. Designate by $(\gamma \uparrow G)$ the representation of G induced by γ and by $(\Gamma \downarrow H)$ the representation of H subduced (restricted) from Γ. Then the multiplicity of Γ in $(\gamma \uparrow G)$ equals the multiplicity of γ in $(\Gamma \downarrow H)$. Equivalently,

$$\langle \chi^{(\Gamma)}, \chi^{(\gamma \uparrow G)} \rangle_G = \langle \chi^{(\Gamma \downarrow H)}, \chi^{(\gamma)} \rangle_H\,,$$

where $\chi^{(\Xi)}$ designates the character of the representation Ξ of X and

$$\langle \chi^{(\Xi)}, X^{(\Xi')} \rangle_X = \frac{1}{|X|} \sum_{x \in X} \overline{\chi^{(\Xi)}(x)} \chi^{(\Xi')}(x)\,,$$

is a normalized Hermitian inner product on the class function space of X, $X = H$ or G.

Frobenius substitution   *See* Frobenius automorphism.

Frobenius Theorem   There are a number of theorems associated with the name of Frobenius. (*See also* Frobenius Theorem on Non-Negative Matrices, Frobenius Reciprocity Theorem.)

(1) (*Frobenius Theorem on Division Algebras.*) The fields **R** (the field of real numbers) and **C** (the field of complex numbers) are the only finite dimensional real associative and commutative algebras without zero divisors (i.e., division algebras), while **H** (the skew-field of quaternions or Hamilton's quaternion algebra) is the only finite dimensional real associative, but noncommutative, division algebra.

For nonassociative algebras, the only alternative algebra without zero divisors is the Cayley algebra. *See* Cayley algebra, alternative algebra.

(2) (*Frobenius Theorem for Finite Groups.*) For a finite group G of order $|G| = g$, the number of solutions of the equation $x^n = c$, where c belongs to a class C having h conjugated elements, is given by g.c.d. (hn, g).

The original, simpler version of this theorem states that the number m of solutions of $x^n = 1$, where $n|g$, is divisible by n, i.e., $n|m$.

(3) (*Frobenius Theorem for Transitive Permutation Groups.*) For a transitive permutation group G of degree n, whose elements, other than the identity, leave at most one of the permuted symbols invariant, the elements of G displacing all the symbols form, together with the identity, a normal subgroup of order n.

(4) (*Also called Zolotarev-Frobenius Theorem.*) Let a be an arbitrary integer and b any odd integer such that a and b are relatively prime, i.e., g.c.d.$(a, b) = 1$. Further, let π_a designate multiplication by a in the additive group $H := \mathbf{Z}/b\mathbf{Z}$. Then π_a (regarded as an automorphism of H) represents a permutation of the set H and as such possesses the parity (or sign) δ given by

$$\delta(\pi_a) = \left(\frac{a}{b}\right),$$

where $\left(\frac{a}{b}\right)$ is the Jacobi (generalized Legendre) symbol.

(5) (*Frobenius Theorem in Finite Group Theory.*) Let H be a selfnormalizing subgroup of a finite group G, $N_G(H) = H$, and

$$H \cap x^{-1}Hx = \{e\},$$

for all $x \in G$. Then the elements of G that do not lie in H, together with the identity element

e, form a normal subgroup N of G,

$$N = (G\backslash H) \cup \{e\},$$

such that $G = NH$, $H \cap N = \{e\}$ and $G/N \cong H$. *See also* Frobenius group.

(6) (*Frobenius Theorem on Abelian Varieties.*) Let G be an additive group of divisors on an Abelian variety A, X a divisor on A, and $\hat{A}$ the Picard variety of A. Denote by ϕ_X a rational homomorphism of A into $\hat{A}$ that maps $a \in A$ into the linear equivalence class of the divisor $X_a - X$, where X_a is the image of X under the translation $A \to A$ defined by $b \mapsto a+b$. There are elements $a_1, \ldots, a_n$ such that the product (intersection) $X_{a_1} \bullet \cdots \bullet X_{a_n}$, $n = \dim A$ is defined. Designating the degree of the zero cycle $X_{a_1} \bullet \cdots \bullet X_{a_n}$ by $(X^{(n)})$, the degree of ϕ_X is given by $(X^{(n)})/n!$

(7) (*Frobenius Theorem on Subduced Representations.*) Let G be a (finite) group having r classes C_ℓ, $(\ell = 1, \ldots, r)$ with r_ℓ elements each, $|C_\ell| = r_\ell$. Further, let $\{\Gamma_i\}$ be the set of irreducible representations (irreps) of G, and χ_i the character of Γ_i. Similarly, let H be a subgroup of G having s classes D_k with s_k elements each, and having the irreps $\{\Delta_j\}$ and characters ϕ_j. Designate, further, the representation of H subduced by Γ_i by $\tilde{\Gamma}_i = \Gamma_i \downarrow H$ and its character by $\tilde{\chi}_i$. Then there exist rs nonnegative integers c_{ij} such that

$$\tilde{\Gamma}_i \cong \bigoplus_{j=1}^{s} c_{ij}\Delta_j, \quad (i = 1, \ldots, r)$$

and

$$\tilde{\chi}_i = \sum_{j=1}^{s} c_{ij}\phi_j, \quad (i = 1, \ldots, r).$$

Clearly,

$$c_{ij} = \frac{1}{|H|} \sum_{k=1}^{s} s_k \tilde{\chi}_i(k)\overline{\phi}_j(k).$$

Finally,

$$\sum_{i=1}^{r} c_{ij}\chi_i(\ell) = \frac{|G|}{r_\ell|H|} \sum_{\ell'} s_{\ell'}\phi_j(\ell'[\ell]),$$

where $\ell'[\ell]$ labels the classes $D_{\ell'}$ of H that are contained in $(C_\ell \cap H)$ and have $s_{\ell'}$ elements.

Frobenius Theorem on Non-Negative Matrices (Also called Peron-Frobenius Theorem.)

Let $\mathbf{A}$ be an indecomposable (or irreducible), non-negative $n \times n$ matrix over $\mathbf{R}$, $\mathbf{A} \equiv \|a_{ij}\|_{n \times n}$ (i.e., all entries a_{ij} of $\mathbf{A}$ are non-negative and there are no invariant coordinate subspaces when we regard $\mathbf{A}$ as an operator on $\mathbf{R}^n$; in other words, $a_{ij} \geq 0$ and there are no permutations of rows and columns that would reduce the matrix to the following block form

$$\begin{pmatrix} \mathbf{A}_{12} & \mathbf{Q} \\ \mathbf{A}_{21} & \mathbf{A}_{22} \end{pmatrix}).$$

Further, let $\lambda_0, \ldots, \lambda_{n-1}$ be eigenvalues of $\mathbf{A}$, labeled in such a way that

$$\rho \equiv |\lambda_0| = |\lambda_1| = \cdots = |\lambda_{h-1}| > |\lambda_h| \geq |\lambda_{h+1}|$$

$$\geq \cdots \geq |\lambda_{n-1}|, \quad (1 < h \leq n).$$

Then

(i.) If $\mathbf{A}$ majorizes a complex matrix $\mathbf{B}$, i.e.,

$$|b_{ij}| \leq a_{ij}, \quad (i, j = 1, \ldots, n)$$

and

$$\rho_{\mathbf{B}} := \max_{0 \leq i < n} |\mu_i| ,$$

with $\mu_0, \mu_1, \ldots, \mu_{n-1}$ the eigenvalues of $\mathbf{B}$, we have that

$$\rho_{\mathbf{B}} \leq \rho .$$

(ii.) $\mathbf{A}$ always has a positive eigenvalue ρ that is a simple root of the characteristic polynomial of $\mathbf{A}$ and ρ majorizes the moduli of all other eigenvalues as the above introduced notation (i) implies. The coordinates c_i ($1 \leq i \leq n$) of an eigenvector $\mathbf{c}$, $\mathbf{c} = (c_1, c_2, \ldots, c_n)^T$ that is associated with this "maximal" eigenvalue ρ are either all positive ($c_i > 0$) or all negative ($c_i < 0$).

(iii.) If $\mathbf{A}$ has h characteristic values $\lambda_0 = \rho, \lambda_1, \ldots, \lambda_{h-1}$ of modulus r, as in (i), then these eigenvalues are all distinct and are given by the roots of the equation $\lambda^h - \rho^h = 0$, i.e.,

$$\lambda_j = \omega^i \rho, \quad (j = 0, 1, \ldots, h - 1)$$

where ω is the hth root of unity, $\omega = e^{2\pi i / h}$. Moreover, any eigenvalue of $\mathbf{A}$, multiplied by ω, is again an eigenvalue of $\mathbf{A}$. Thus, the entire spectrum $\{\lambda_i, 0 \leq i < n\}$ of $\mathbf{A}$ is invariant with respect to a rotation by $2\pi / h$ when represented by points in the complex plane.

(iv.) Finally, if $h > 1$, the matrix $\mathbf{A}$ can be brought to the following cyclic form

$$\mathbf{A} = \begin{Vmatrix} 0 & \mathbf{A}_{12} & 0 & \ldots & 0 \\ 0 & 0 & \mathbf{A}_{23} & \ldots & 0 \\ \vdots & \vdots & \vdots & \vdots & \\ 0 & 0 & 0 & \ldots & \mathbf{A}_{h-1h} \\ \mathbf{A}_{h1} & 0 & 0 & \ldots & 0 \end{Vmatrix},$$

with square blocks along the diagonal, by a suitable permutation of rows and columns.

Fuchsian group A special case of a Kleinian group, i.e., a finitely generated discontinuous group of linear fractional transformations acting on some domain in the complex plane. (*See* Kleinian group, linear fractional function.)

Generally, a *Fuchsian group* is a discrete (or discontinuous) transformation group of an open disc X in $\mathbf{C}$ onto the Riemann sphere. More specifically, one considers transformations of the upper half-plane $X_u = \{z \in \mathbf{C} : \Im z > 0\}$ or of the unit disc $X_d = \{z \in \mathbf{C} : |z| < 1\}$ onto the complex plane. In the former case ($X = X_u$), the elements of a Fuchsian group are Möbius (linear fractional) transformations (or conformal mappings)

$$z \mapsto \frac{az + b}{cz + d}, \quad a, b, c, d \in \mathbf{R}, \quad ad - bc = 1 ,$$

so that the relevant group is a subgroup of PSL(2). In the latter case ($X = X_d$), the group elements are Möbius transformations with pseudo-unitary matrices.

When one considers the disc X as a conformal model of the Lobachevski plane, then a *Fuchsian group* can be regarded as a discrete group of motions in this plane that preserve the orientation. A Fuchsian group is referred to as *elementary* if it preserves a straight line in the Lobachevski plane (or, equivalently, some point in the closure $\overline{X}$ of X). For a *nonelementary* Fuchsian group Γ, one then defines the *limit set of* Γ, designated as $L(\Gamma)$, as the set of limit points of the orbit of a point $x \in \overline{X}$ located on the circle ∂X and independent of x. We then distinguish Fuchsian groups of the first and second kind: for the former kind, we require that

$L(\Gamma) = \partial X$, while in the second case $L(\Gamma)$ is a nowhere dense subset of ∂X.

For any $z \in \mathbf{C} \cup \{\infty\}$ (the extended complex plane) and any sequence $\{\gamma_i\}$ of distinct elements of Γ, we define a *limit point* of Γ as a cluster point of $\{\gamma_i z\}$. If there are at most two limit points, Γ is conjugate to a group of motions of a plane. Otherwise, the set $\mathcal{L}$ of all limit points of Γ is infinite, and Γ is called a *Fuchsoid group*. Such a group is Fuchsian if it is finitely generated.

Fuchsoid group   *See* Fuchsian group. Also called *Fuchsoidal* group.

function   One of the most fundamental concepts in mathematics. Also referred to as a *mapping, correspondence, transformation,* or *morphism,* particularly when dealing with abstract objects. This concept gradually crystalized from its early implicit use into the present day abstract form. The term "function" was first used by Leibniz, and it gradually developed into a general concept through the work of Bernoulli, Euler, Dirichlet, Bolzano, Cauchy, and others. The modern-day definition as a correspondence between two abstract sets is due to Dedekind.

Generally, a function f is a relation between two sets, say X and Y, that associates a unique element $f(x) \in Y$ to an element $x \in X$. (*See* relation.) The sets X and Y need not be distinct. Formally, this many-to-one relation is a set of ordered pairs $f = \{(x, y)\}, x \in X, y \in Y$, that is, a subset of the Cartesian product $X \times Y$, with the property that for any (x', y') and (x'', y'') from f, the inequality $y' \neq y''$ implies that $x' \neq x''$. The first element $x \in X$ of each pair $(x, y) \in f$ is called the *argument* or the *independent variable* of f and the second element $y \in Y$ is referred to as the *abscissa, dependent variable* or the *value of* f for the argument x.

The sets X_f and Y_f of the first and second elements of ordered pairs $(x, y) \in f$ are called *the domain* (or *the set*) *of definition of* f and *the range* (or *the set*) *of values of* f, respectively, while the entire set X is simply called the *domain* and Y the *codomain* of f. For any subset $A \subset X$, the set of values of f, $\{y = f(x) \in Y : x \in A\}$ is called the *image of A under* f and is designated by $f(A)$. In particular, the image of the domain of f is $f(X_f)$, i.e., $Y_f = f(X_f)$ or

$Y_f = f(X)$, and the *image of the element* $x \in X_f$ under f is $y = f(x)$. Often one simply sets $X = X_f$. The set of ordered pairs $f = \{(x, y)\}$, regarded as a subset of $X \times Y$, is referred to as the *graph* of f. (*See* graph.)

The mapping property of a function f is usually expressed by writing $f : X \to Y$, and for $(x, y) \in f$ one often writes $y = f(x)$ or $f : x \mapsto y$, or even $y = fx$ or $y = xf$. In lieu of the symbol $f(x_0)$ one also writes $f(x)|_{x=x_0}$, and often the function itself is denoted by the symbol $f(x)$ rather than $f : x \mapsto y$, since this notation is more convenient for actual computations.

The set of elements of X that are mapped into a given $y_0 \in Y$ is called the pre-image of y_0 and is designated by $f^{-1}(y_0)$, so that

$$f^{-1}(y_0) = \{x \in X : f(x) = y_0\} .$$

For $y_0 \in Y \backslash Y_f$ we have clearly $f^{-1}(y_0) = \emptyset$ (the empty set).

The notation $f : X \to Y$ indicates that the set X is mapped *into* the set Y. When $X = Y$, we say that X is mapped *into itself.* When $Y = Y_f$, we say that f maps X *onto* Y or that f is *surjective* (or a *surjection*). Thus, $f : X \to Y$ is a surjection (or onto) if for each $y \in Y$ there exists at least one $x \in X$ such that $f : x \mapsto y$. If the images of distinct elements of X are distinct, i.e., if $x' \neq x''$ implies that $f(x') \neq f(x'')$ for any $x', x'' \in X$, we say that f is *one-to-one,* or *univalent,* or *injective* (or an *injection*). Thus, f is injective if the preimage of any $y \in Y_f$ contains precisely one element from X, i.e., card $f^{-1}(y) = 1, y \in Y_f$. The mapping $f : X \to Y$ that is simultaneously injective and surjective (or one-to-one and onto) is referred to as *bijective* (or a *bijection*). For a bijective function one defines the *inverse function* by $f^{-1} = \{(y, f^{-1}(y)), y \in Y_f$. See inverse function.

For two functions $f : X \to Y$ and $g : Y \to Z$, with $Y_f \subset Y_g$, the function $h : X \to Z$ that is defined as

$$h(x) = g(f(x)), \quad \text{for all } x \in X_f ,$$

is called the *composite function of f and g* (also the superposition or composition of f and g), and is designated by $h = g \circ f$.

function algebra   For a compact Hausdorff space X, let $C(X)$ [or $C_{\mathbf{R}}(X)$] be the algebra of all complex- [or real-] valued functions on X. Then a closed subalgebra A of $C(X)$ [or $C_{\mathbf{R}}(X)$] is referred to as a *function algebra* on X if it contains the constant functions and separates the points of X [i.e., for any $x, y \in X$, $x \neq y$, there exists an $f \in A$ such that $f(x) \neq f(y)$]. A typical example is the *disk algebra,* that is, the complex-valued functions, analytic in the unit disk $D = \{|z| < 1\}$, which extend continuously to the closure of D, with the supremum norm.

Alternatively, a function algebra is a semi-simple, commutative Banach algebra, realized as an algebra of continuous functions on the space of its maximal ideals (recall that a commutative Banach algebra is semi-simple if its radical reduces to $\{0\}$, the radical being the set of generalized nilpotent elements).

function field   A type of extension of the field of rational functions $\mathbf{C}(x)$ that plays an important role in algebraic geometry and the theory of analytic functions.

For an (irreducible) affine variety $V(\mathcal{P})$, where $\mathcal{P}$ is a prime ideal in $\mathbf{C}[x] \equiv \mathbf{C}[x_1, \ldots, x_n]$, the function field of $V(\mathcal{P})$ is the field of quotients of the affine coordinate ring $\mathbf{C}[x]/\mathcal{P}$. Similarly, for a projective variety $V(\mathcal{P})$, where $\mathcal{P}$ is a homogeneous prime ideal in a projective n-space $\mathbf{P}^n$, the function field of $V(\mathcal{P})$ is the subfield of the quotient field of the homogeneous coordinate ring $\mathbf{C}[x]/\mathcal{P}$ of zero degree [i.e., the ring of rational functions $f(x_0, x_1, \ldots x_n)/g(x_0, x_1, \ldots, x_n)$ (f, g being homogeneous polynomials of the same degree and $g \notin \mathcal{P}$) modulo the ideal of functions (f/g), with $f \in \mathcal{P}$]. *See also* Abelian function field, algebraic function field, rational function field.

function group   A Kleinian group Γ whose region of discontinuity has a nonempty, connected component, invariant under Γ.

functor   A mapping from one category into another that is compatible with their structure. Specifically, a *covariant functor* (or simply a *functor*) $F : \mathcal{C} \to \mathcal{D}$, from a category $\mathcal{C}$ into a category $\mathcal{D}$, represents a pair of mappings (usually designated by the same letter, i.e., F).

$$F : \mathrm{Ob}\,\mathcal{C} \to \mathrm{Ob}\mathcal{D}, \quad F : \mathrm{Mor}\,\mathcal{C} \to \mathrm{Mor}\mathcal{D},$$

associating with each object X of $\mathcal{C}$, $X \in \mathrm{Ob}\,\mathcal{C}$ an object $F(X)$ of $\mathcal{D}$, $F(X) \in \mathrm{Ob}\mathcal{D}$, and with each morphism $\alpha : X \to Y$ in $\mathcal{C}$, $\alpha \in \mathrm{Mor}\,\mathcal{C}$, a morphism $F(\alpha) : F(X) \to F(Y)$ in $\mathcal{D}$, $F(\alpha) \in \mathrm{Mor}\mathcal{D}$, in such a way that the following hold:

(i.) $F(1_X) = 1_{F(X)}$, for all $X \in \mathrm{Ob}\,\mathcal{C}$ and

(ii.) $F(\alpha \circ \beta) = F(\alpha) \circ F(\beta)$ for all morphisms $\alpha \in \mathrm{Hom}_{\mathcal{C}}(X, Y)$ and $\beta \in \mathrm{Hom}_{\mathcal{C}}(Y, Z)$.

Note that a functor $F : \mathcal{C} \to \mathcal{D}$ defines a mapping of each set of morphisms $\mathrm{Hom}_{\mathcal{C}}(X, Y)$ into $\mathrm{Hom}_{\mathcal{D}}(F(X), F(Y))$, associating the morphism $F(\alpha) : F(X) \to F(Y)$ to each morphism $\alpha : X \to Y$. It is called *faithful* if all these maps are injective, and *full* if they are surjective. The identity functor $Id_{\mathcal{C}}$ or $1_{\mathcal{C}}$ of a category $\mathcal{C}$ is the identity mapping of $\mathcal{C}$ into itself.

A *contravariant functor* $F : \mathcal{C} \to \mathcal{D}$ associates with a morphism $\alpha : X \to Y$ in $\mathcal{C}$ a morphism $F(\alpha) : F(Y) \to F(X)$ in $\mathcal{D}$ (or, equivalently, acts as a covariant functor from the dual category $\mathcal{C}^*$ to $\mathcal{D}$), with the second condition (ii.) replaced by (ii.') $F(\alpha \circ \beta) = F(\beta) \circ F(\alpha)$ for all morphisms $\alpha \in \mathrm{Hom}_{\mathcal{C}}(X, Y)$, $\beta \in \mathrm{Hom}_{\mathcal{C}}(Y, Z)$.

A generalization involving a finite number of categories is an *n-place functor* from n categories $\mathcal{C}_1, \ldots \mathcal{C}_n$ into $\mathcal{D}$ that is covariant for the indices $i_1, i_2, \ldots, i_k$ and contravariant in the remaining ones. This is a functor from the Cartesian product $\otimes_{i=1}^{n} \tilde{\mathcal{C}}_i$ into $\mathcal{D}$ where $\tilde{\mathcal{C}}_i = \mathcal{C}_i$ for $i = i_1, \ldots, i_k$ and $\tilde{\mathcal{C}}_i = \mathcal{C}_i^*$ otherwise. A two-place functor that is covariant in both arguments is called a *bifunctor.*

fundamental curve   A concept in the theory of birational mappings (or correspondences) of algebraic varieties. *See* birational mapping.

Consider complete, irreducible varieties V and W and a birational mapping F between them, $F : V \to W$. A subvariety V' of V is called *fundamental* if $\dim F[V'] > \dim V'$.

When V' is a point, it is called a *fundamental point with respect to F* and, likewise, when V' is a curve, it is referred to as a *fundamental curve with respect to F*.

See also Cremona transformation for a birational mapping between projective planes.

Remarks: A variety V is irreducible if it is not the union of two proper subvarieties and any algebraic variety can be embedded in a complete variety.

A projective variety is always complete, while an affine variety over a field F is complete when it is of zero dimension.

fundamental exact sequence   A concept in the theory of homological algebras.

Let $H^i(G, A)$ be the ith cohomology group of G with coefficients in A, where A is a left G-module (that can be identified with a left $\mathbf{Z}[G]$-module). (*See* cohomology group.) Designate, further, the submodule of G-invariant elements in A by A^G and assume that $H^i(H, A) = 0$, $i = 1, \ldots, n$, for some normal subgroup H of G. Then the sequence

$$O \to H^n\left(G/H, A^H\right) \to H^n(G, A)$$
$$\to H^n\left(H, A^G\right)$$
$$\to H^{n+1}\left(G/H, A^H\right)$$
$$\to H^{n+1}(G, A)$$

(composed of inflation, restriction, and transgression mappings) is exact and is referred to as the *fundamental exact sequence*.

fundamental operations of arithmetic   The operations of addition, subtraction, multiplication, and division. *See* operation. Often, extraction of square roots is added to this list.

Starting with the set $\mathbf{N}$ of natural numbers, which is closed with respect to the first three fundamental operations, one carries out an extension to the field of rational numbers $\mathbf{Q}$ representing the smallest domain in which the four fundamental operations can be carried out indiscriminately, excepting division by zero. Adjoining the operation of square root extractions, we arrive at the field of complex numbers $\mathbf{C}$.

These operations can be defined for various algebraic systems, in which case the commutative and associative laws may not hold.

fundamental point   *See* fundamental curve.

fundamental root system   (*of a (complex) semi-simple Lie algebra* $\mathbf{g}$) A similar concept arises in Kac-Moody algebras, in algebraic groups, algebraic geometry, and other fields. Let Π be a subset of the root system Δ of a semi-simple Lie algebra $\mathbf{g}$, relative to a chosen Cartan subalgebra $\mathbf{h}$, $\Pi = \{\alpha_1, \alpha_2, \ldots, \alpha_r\}$. Then Π is called a *fundamental root system* of Δ if

(i.) any root $\alpha \in \Delta$ is a linear combination of the α_i with integral coefficients,

$$\alpha = \sum_{i=1}^{r} m_i \alpha_i, \quad m_i \in \mathbf{Z},$$

and

(ii.) the m_i are either all non-negative (when $\alpha \in \Delta_+$ is a positive root) or all non-positive.

The roots belonging to Π are usually referred to as *simple roots*. They can be defined as positive roots that are not expressible as a sum of two positive roots. They are linearly independent and constitute a basis of the Euclidean vector space spanned by Δ whose (real) dimension equals the rank of $\mathbf{g}$.

fundamental subvariety   *See* fundamental curve.

fundamental system   (*Of solutions of a system of linear homogeneous equations*)

$$\sum_{j=1}^{n} a_{ij} x_j = 0, \quad i = 1, \ldots, m). \quad (1)$$

Let $f_i = \sum_{j=1}^{n} a_{ij} X_j$, $(i = 1, \ldots, m)$ be linear forms over a field F, $f_i : F^n \to F$. Clearly, the solutions of (1) form a (right) linear space V over F, since if $\mathbf{x}_k^i \equiv (x_1^{(k)}, \ldots, x_n^{(k)}) \in F^n$, $(k = 1, \ldots, r)$ are solutions of (1), so is their (right) linear combination $\sum_{j=1}^{r} \mathbf{x}_j c_j$, $c_j \in F$. In fact, V is the kernel of the (left) linear mapping $G : F^n \to F^m$ given by $G : \mathbf{x} \mapsto (f_1(\mathbf{x}), \ldots, f_m(\mathbf{x}))$. Designating the dimension of V by d, $d = \dim V$, we can distinguish the following cases:

(i.) $d = 0$: In this case the system (1) has only the *trivial solution* $\mathbf{x} = \mathbf{0}$.

(ii.) $d > 0$: Choosing a basis $\{\mathbf{x}_1, \ldots, \mathbf{x}_d\}$ for V, we see that any solution of (1) is a (right) linear combination of the $\mathbf{x}_k$, $(k = 1, \ldots, d)$. One then says that $\mathbf{x}_1, \ldots, \mathbf{x}_d$ form a *fundamental system* of solutions of (1). Clearly, a nontrivial solution is found if and only if $r < n$, r being the rank of the matrix $\mathbf{A} = \|a_{ij}\|$ [or, equivalently, the number of linearly independent linear forms f_i, $(i = 1, \ldots, m)$], and the number of linearly independent fundamental solutions is $d = n - r$. Since $m \geq r$, we see that a nontrivial solution always exists if the number of equations is less than the number of unknowns.

fundamental system of irreducible representations (*Of a complex semisimple Lie algebra* **g**.) Consider a complex semisimple Lie algebra **g** and fix its Cartan subalgebra **h**. Let $k = \dim \mathbf{h} = \text{rank } \mathbf{g}$, $\mathbf{h}^*$ the dual of **h**, consisting of complex-valued linear forms on **h**, and $\mathbf{h}_R^*$ the real linear subspace of $\mathbf{h}^*$, spanned by the root system Δ. Let, further, $\Pi = \{\alpha_1, \ldots, \alpha_k\}$ be the system of simple roots (*see* fundamental root system) and define

$$\alpha_i^* = \frac{2\alpha_i}{(\alpha_i, \alpha_i)} ,$$

where (α, β) is a symmetric bilinear form on $\mathbf{h}^*$ defined by the Killing form $K(\cdot, \cdot)$ as follows:

$$(\alpha, \beta) = K\left(t_\alpha, t_\beta\right) ,$$

t_α being a star vector associated to α, $t_\alpha = \nu^{-1}(\alpha)$, where ν designates the bijection $\nu : \mathbf{h} \to \mathbf{h}^*$. Let, further, $\Lambda_1, \ldots, \Lambda_k$ be a basis of $\mathbf{h}_R^*$ that is dual to $\alpha_1^*, \ldots, \alpha_\ell^*$, i.e., $(\Lambda_i, \alpha_j^*) = \delta_{ij}$. Then the set of irreducible representations $\{\Gamma_1, \ldots, \Gamma_k\}$ that have $\Lambda_1, \ldots, \Lambda_k$ as their highest weights is called the *fundamental system of irreducible representations* (irreps) *associated with* Π.

Fundamental Theorem of Algebra Every nonconstant polynomial (i.e., a polynomial with a positive degree) with complex coefficients has a complex root.

Also called *Euler-Gauss Theorem*.

Fundamental Theorem of Arithmetic Every nonzero integer $n \in \mathbf{Z}$ can be expressed as a product of a finite number of positive primes times a unit (± 1), i.e.,

$$n = \mathcal{C} p_1 p_2 \ldots p_k ,$$

where $\mathcal{C} = \pm 1$, $k \geq 0$ and p_i, $i = 1, \ldots, k$ are positive primes. This expression is unique except for the order of the prime factors.

Fundamental Theorem of Galois Theory
Let K be a Galois extension of a field M with a Galois group $G = G(K/M)$. Then there exists a bijection $H \leftrightarrow L$ between the set of subgroups $\{H\}$ of G and the set of intermediate fields $\{L\}$, $K \supset L \supset M$. For a given H, the corresponding subfield $L = L(H)$ is given by a fixed field K^H of H (consisting of all the elements of K that are fixed by all the automorphisms of H). Conversely, to a given L corresponds a subgroup $H = H(L)$ of G that leaves each element of L fixed, i.e., $H(L) = G(K/L)$, so that $[K : L] = |H|$. This bijection has the property that $[L : M] = [G : H]$, where $[L : M]$ is the degree of the extension L/M and $[G : H]$ is the index of H in G.

Also called *Main Theorem of Galois Theory*.

Fundamental Theorem of Proper Mapping
A basic theorem in the theory of formal schemes, also called formal geometry, in algebraic geometry. *See* formal scheme.

Let $f : S \to T$ be a proper morphism of locally Noetherian schemes S and T, T' a closed subscheme of T, S' the inverse image of T' (given by the fiber product $S \times_T T'$) and, finally, $\hat{S}$ and $\hat{T}$ the completions of S and T along S' and T', respectively. Then $\hat{f} : \hat{S} \to \hat{T}$ (the induced proper morphism of formal schemes) defines the canonical isomorphism

$$\left(R^n f_*(X)\right)_{|T'} \cong F^n \hat{f}_*(X_{|S'}), \quad n \geq 0$$

for every coherent $\mathcal{O}_S$-module X on S, i.e., a sheaf of $\mathcal{O}_S$-modules.

For a coherent sheaf X on S, $X_{|S'}$ denotes the completion of X along S'. $R^n f_*$ is the right derived functor of the direct image $f_*(X)$ of X.

Fundamental Theorem of Symmetric Polynomials Any symmetric polynomial in n variables, $p(\mathbf{x}) \equiv p(x_1, x_2, \ldots x_n)$, from a polynomial ring $R[\mathbf{x}] \equiv R[x_1, x_2, \ldots x_n]$, can be uniquely expressed as a polynomial in the elementary symmetric functions (or polynomials) $s_1, s_2, \ldots, s_n$ in the variables x_i. *See* elementary symmetric polynomial.

In other words, for each $p(\mathbf{x}) \in R[\mathbf{x}]$ there exists a unique polynomial $\pi(\mathbf{z}) \in R[\mathbf{z}]$, $\mathbf{z} \equiv (z_1, z_2, \ldots, z_n)$ such that

$$p(x_1, x_2, \ldots, x_n) = \pi(s_1, s_2, \ldots, s_n) ,$$

where s_i are the elementary symmetric polynomials (functions)

$s_1 = \sum_{i=1}^{n} x_i$,
$s_2 = \sum_{i<j} x_i x_j$,
$s_3 = \sum_{i<j<k} x_i x_j x_k$,

$\vdots$

$s_n = x_1 x_2 \cdots x_n$.

Also called *Main Theorem on Symmetric Polynomials*.

Fundamental trigonometric identities   Recall that in a plane $\mathbf{R}^2$ with a Cartesian coordinate system $O - xy$ (i.e., with the origin $O \equiv (0, 0)$ and the x- and y- axes representing the abscissa and the ordinate), any point $P \in \mathbf{R}^2$ is uniquely represented by its coordinates (x, y). Designating the radial distance $\overline{OP}$ by r, $r = \sqrt{x^2 + y^2}$, and the angle POx (i.e., angle between the line PO and the x-axis) by α, we define six ratios as follows:

$$\sin\alpha = \tfrac{y}{r}, \qquad \cos\alpha = \tfrac{x}{r},$$
$$\tan\alpha = \tfrac{y}{x}, \qquad \cot\alpha = \tfrac{x}{y},$$
$$\sec\alpha = \tfrac{r}{x}, \qquad \csc\alpha = \tfrac{r}{y}.$$

These functions of α are called trigonometric or circular functions. They are interrelated as follows

$$\tan\alpha = \tfrac{\sin\alpha}{\cos\alpha}, \qquad \cot\alpha = \tfrac{\cos\alpha}{\sin\alpha} = \tfrac{1}{\tan\alpha},$$
$$\sec\alpha = \tfrac{1}{\cos\alpha}, \qquad \csc\alpha = \tfrac{1}{\sin\alpha},$$
$$1 + \tan^2\alpha = \sec^2\alpha, \qquad 1 + \cot^2\alpha = \csc^2\alpha.$$

Furthermore, we have that

$$\sin^2\alpha + \cos^2\alpha = 1 \, ,$$

implying that the points $P = (x, y)$ lying on the unit circle $x^2 + y^2 = 1$ with its center at the origin $(0, 0)$ can be expressed in terms of the angle α as $P = (\cos\alpha, \sin\alpha)$.

The important addition formulas read

$\sin(\alpha \pm \beta) = \sin\alpha \cos\beta \pm \cos\alpha \sin\beta$,
$\cos(\alpha \pm \beta) = \cos\alpha \cos\beta \mp \sin\alpha \sin\beta$,
$\tan(\alpha \pm \beta) = (\tan\alpha \pm \tan\beta)/(1 \mp \tan\alpha \tan\beta)$.

When $\alpha = \beta$ we have

$\sin 2\alpha = 2\sin\alpha \cos\alpha$,
$\cos 2\alpha = \cos^2\alpha - \sin^2\alpha = 2\cos^2\alpha - 1 = 1 - 2\sin^2\alpha$,

and for a general integral multiple of α

$\sin n\alpha =$

$$\sum_{j=0}^{[(n-1)/2]} \binom{n}{2j+1} (-1)^j \sin^{2j+1}\alpha \cos^{n-(2j+1)}\alpha,$$

$$\cos n\alpha = \sum_{j=0}^{[n/2]} \binom{n}{2j} (-1)^j \sin^{2j}\alpha \cos^{n-2j}\alpha,$$

while the half-angle formulas are

$\sin^2\left(\tfrac{\alpha}{2}\right) = \tfrac{1}{2}(1 - \cos\alpha)$,
$\cos^2\left(\tfrac{\alpha}{2}\right) = \tfrac{1}{2}(1 + \cos\alpha)$,
$\tan^2\left(\tfrac{\alpha}{2}\right) = \tfrac{1-\cos\alpha}{1+\cos\alpha}$.

The addition formulas are

$\sin\alpha + \sin\beta = 2\sin\tfrac{\alpha+\beta}{2}\cos\tfrac{\alpha-\beta}{2}$,
$\sin\alpha - \sin\beta = 2\cos\tfrac{\alpha+\beta}{2}\sin\tfrac{\alpha-\beta}{2}$,
$\cos\alpha + \cos\beta = 2\cos\tfrac{\alpha+\beta}{2}\cos\tfrac{\alpha-\beta}{2}$,
$\cos\alpha - \cos\beta = -2\sin\tfrac{\alpha+\beta}{2}\sin\tfrac{\alpha-\beta}{2}$,

and the product formulas are

$2\sin\alpha \cos\beta = \sin(\alpha + \beta) + \sin(\alpha - \beta)$,
$2\cos\alpha \sin\beta = \sin(\alpha + \beta) - \sin(\alpha - \beta)$,
$2\cos\alpha \cos\beta = \cos(\alpha + \beta) + \cos(\alpha - \beta)$,
$-2\sin\alpha \sin\beta = \cos(\alpha + \beta) - \cos(\alpha - \beta)$.

fundamental unit   By Dirichlet's Unit Theorem, the unit group E_k of an algebraic number field k is the direct product of a cyclic group of a finite order and the free Abelian multiplicative group of rank r (note that $r = r_1 + r_2 - 1$, where r_1 and r_2 designate, respectively, the number of real and complex conjugates $x^{(i)}$, $i = 1, \ldots, n$ of any $x \in k$, so that $r_1 + 2r_2 = n$, n being the degree of k over $\mathbf{Q}$). A basis $(e_1, e_2, \ldots, e_r)$ of this free Abelian group is referred to as a *system of fundamental units* of k. See Dirichlet Unit Theorem, unit group.

fundamental vectors   The vectors $e_1, e_2, \ldots, e_n$ of an n-dimensional vector space V over F, forming a basis of V, are referred to as *fundamental vectors* of V. Clearly, for any $\mathbf{x} \in V$, we have that $\mathbf{x} = \sum_{i=1}^{n} \xi_i e_i$, $\xi_i \in F$, and the ξ_i are called the components of $\mathbf{x}$, with respect to the fundamental vectors $e_1, \ldots, e_n$.

G

Galois cohomology   Let K/k be a finite Galois extension with the Galois group $G(K/k)$. Suppose, further, that $G(K/k)$ acts on some Abelian group A. The *Galois cohomology* groups $H^n(G(K/k), A) \equiv H^n(K/k, A)$, $n \geq 0$ are then the cohomology groups defined by the (cochain) complex (F^n, d), with F^n consisting of all mappings $G(K/k)^n \to A$ and d designating the coboundary operator (*see* cohomology groups). When the extension K/k is of an infinite degree, one also requires that the Galois topological group acts continuously on the discrete group A and the mappings for the cochains in F^n are also continuous.

One also defines Galois cohomology for a non-Abelian group A, in which case one usually restricts oneself to zero- and one-dimensional cohomology groups, H^0 and H^1, respectively. In the first case, $H^0(K/k, A) = A^{G(K/k)}$ represents a set of fixed points in A under the action of the Galois group $G(K/k)$, while in the second case $H^1(K/k, A)$ is the quotient set of the set of 1-dimensional cocycles.

The concept of *Galois cohomology* enables one to define the cohomological dimension of the Galois group G_k of a field k. *See* cohomological dimension. Non-Abelian Galois cohomology enables the classification of principal homogeneous spaces of group schemes and, in particular, to classify types of algebraic varieties (using Galois cohomology groups of algebraic groups).

Also called cohomology of a Galois group. *See also* Tamagawa number, Tate-Shafarevich group.

Galois equation   Let K/k be a finite Galois extension with the Galois group $G(K/k)$. Then K is a minimal splitting field of a separable polynomial $f(X) \in k[X]$, and we call $G(K/k)$ the *Galois group of* $f(X)$ or the *Galois group of the algebraic equation* $f(X) = 0$. (*See* minimal splitting field, separable polynomial.) If $G(K/k)$ is Abelian or cyclic, we call the equation $f(X) = 0$ an *Abelian equation* or a *cyclic equation,* respectively. (*See* Abelian equation.) When the extension field K can be obtained by adjoining a root α of $f(X)$ to k, $K = k[\alpha]$, then the equation $f(X) = 0$ is called a *Galois equation.*

Galois extension   A finite field extension K/k such that the order of the Galois group $G(K/k)$ is equal to the degree $[K : k] = \dim_k K$ of the field extension K, i.e.,

$$|G(K/k)| = [K : k] = \dim_k K .$$

Remarks:

(i.) The degree $[K : k]$ of the field extension K/k, $k \subset K$, equals the dimension of K as a k-vector space, $[K : k] = \dim_k K$. One distinguishes quadratic ($[K : k) = 2$), cubic ($[K : k] = 3$), biquadratic ($[K : k] = 4$), finite ($[K : k] < \infty$), etc., extensions.

(ii.) All quadratic and biquadratic extensions are Galois extensions.

(iii.) For any finite field extension K/k, the order of the Galois group $g \equiv |G(K/k)|$ divides the degree of the extension $[K : k]$, i.e., $g | [K : k]$.

(iv.) For a Galois extension K/k with the Galois group $G(K/k)$, the fixed field K^G is given by k, i.e., $K^G = k$. *See also* Galois theory.

Galois field   Finite fields F_q, $q = p^n$, are also called *Galois fields. See* finite field.

Galois group   For an extension K of a field k, the group of all k-automorphisms of K is called the *Galois group* of the field extension K/k and is denoted by $G(K/k)$.

Remarks:

(i.) A k-automorphism of an extension field K is an automorphism that acts as the identity on the subfield k. It is also referred to as an automorphism of a field extension K.

(ii.) Since every Galois extension is a splitting field of some polynomial $f(x) \in k[x]$, and any two splitting fields K of a polynomial $f(x) \in k[x]$ are isomorphic, the Galois group $G(K/k)$ depends only on f (up to isomorphism). *See* Galois extension. Thus, if K is the splitting field of $f(x) \in k[x]$, the Galois group $G(K/k)$ is also referred to as the *Galois group of the polynomial* over k.

Galois theory   In a broad sense, a theory studying various mathematical objects on the basis of their automorphism groups (e.g., Galois theories of rings, topological spaces, etc.). In a narrower sense, it is the Galois theory of fields that originated in the problem of finding of roots of algebraic equations of higher degrees (e.g., quintic and higher). This problem was solved by Galois in his famous letter that he wrote on the eve of his execution (1832) and laid unread for more than a decade. In today's language, this theory may be summarized as follows.

Consider an arbitrary field k. An *extension* (field) K of k is any field containing k as a subfield, $k \subset K$, and may be regarded as a linear space over k (finite or infinite dimensional; $\dim_k K \equiv [K : k]$ is called the *degree of the extension K/k*). One says that $\alpha \in K$ is *algebraic* over k if it is a root of a non-zero (irreducible) polynomial $p(x) \in k[x]$ from a polynomial ring $k[x]$ (i.e., with coefficients from k). The smallest extension of k containing α is usually denoted by $k(\alpha)$, and the smallest extension of k that contains all the roots of an irreducible polynomial $p(x) \in k[x]$ is called the *splitting field* of $p(x)$. The degree of such an extension is divisible by the degree of $p(x)$ and is equal to this degree if all the roots of $p(x)$ can be expressed as polynomials in one of these roots. A finite extension K/k is *separable* if $K = k(\alpha)$ and the irreducible polynomial $p(x)$ with α as a root has no multiple roots, and *normal* if it is the splitting field of some polynomial in $k[x]$. When the extension is both separable and normal it is called a *Galois extension. See* Galois extension. If char $k = 0$ (e.g., k is a number field), any finite extension is separable.

The group of all automorphisms of a Galois extension K, leaving all elements of k invariant, is the *Galois group $G(K/k)$*. The relationship between its subgroups H, $H \subset G$, and the corresponding intermediate extension fields L, $k \subset L \subset K$, is described by the *Fundamental* (or *Main*) *Theorem of Galois theory. See* Fundamental Theorem of Galois Theory. In this way, the difficult problem of finding all subfields of K is reduced to a much simpler problem of determining the subgroups of $G(K/k)$. Moreover, if H is normal, L is a Galois extension. These results are then exploited in studying the solutions of algebraic equations. If K is the splitting field

of an irreducible polynomial $p(x) \in k[x]$ without multiple roots, the Galois group $G(K/k)$ is referred to as the *Galois group of the equation* (or polynomial) $p(x) = 0$. This group can be computed without actually solving the equation and can be regarded as a subgroup of the group of permutations of the roots of $p(x)$. In the general case, it is simply the permutation group of all the roots, i.e., the symmetric group of degree $n = \deg[p(x)]$. Since such a group is not solvable for $n \geq 5$, there are no solutions in radicals for the quintic and higher degree algebraic equations. This theory can also be employed to decide which problems of geometry are solvable by ruler and compass (by reducing them to an equivalent problem of solving some algebraic equation over the field of rational numbers in terms of quadratic radicals).

Galois theory had an enormous impact on the development of algebra during the nineteenth century. It was extended and generalized in various directions (Galois topological groups, class field theory, inverse problem of Galois theory, etc.), even though many important problems of the classical Galois theory remain unsolved.

Galois theory of differential fields   Let K be a differential field and N a field extension. The corresponding Galois group, $G(N/K)$, is the set of all differential isomorphisms of N over K.

gap value   A concept from the theory of algebraic functions.

Consider a closed Riemann surface $\mathcal{R}$ of genus g. When $\mathcal{R}$ carries no meromorphic function whose only pole of multiplicity m is at a point $p \in \mathcal{R}$, then m is called a *gap value* of $p \in \mathcal{R}$.

The Riemann-Roch Theorem implies that if $g = 0$, then no point has gap values, while if $g \geq 1$, then every $p \in \mathcal{R}$ has exactly g gap values. A point $p \in \mathcal{R}$ is an ordinary point if m at p equals $1, 2, \ldots, g$ and a Weierstrass point otherwise. *See* ordinary point, Weierstrass point. *See also* Riemann-Roch Theorem.

gauge transformation   A concept which arose in Maxwell's formulation of the electromagnetic field theory and was later extended to more general field theories. In mathematics, it is used to designate bundle automorphisms of a principle fiber bundle over a (space-time) manifold that is endowed with a group structure. In general, gauge transformations (in both classical and quantum field theories) change non-observable field properties (potentials) without affecting the physically observable quantities (observables).

In electromagnetic field theory, the *gauge transformation* (also called *gradient transformation* or *gauge transformation of the second kind*) has the form

$$\phi \to \phi' = \phi + \frac{\partial f}{\partial t}, \quad \mathbf{A} \to \mathbf{A}' = \mathbf{A} - \operatorname{grad} f,$$

where ϕ and $\mathbf{A}$ designate, respectively, the scalar and vector potentials of the field and f is an arbitrary (twice differentiable) scalar function of space and time. Equivalently, using the formalism of special relativity, the 4-component electromagnetic vector potential $A_j(x)$, $j = 0, 1, 2, 3$ and $x = (x^0, x^1, x^2, x^3)$, transforms as follows:

$$A_j(x) \to A'_j(x) = A_j(x) + \frac{\partial f}{\partial x^j}.$$

This transformation does not change the fields involved implying the *gauge invariance* of the underlying field theory. It may be used to simplify the relevant field equations through a suitable choice of gauge, i.e., of a function $f(x)$ (Coulomb gauge, Lorentz gauge, etc.).

In Weyl's unified field theory (which originated from the theory of Cartan's connections), one employs a space (time) whose structure is defined by the fundamental tensor g_{ij}, the covariant derivative of which is defined in terms of the electromagnetic potential A_i as follows:

$$\nabla_i g_{jk} = 2A_i g_{jk}.$$

This equation is invariant with respect to the scale transformation $g_{ij} \to g'_{ij} = \rho^2 g_{ij}$ and the gauge transformation

$$A_i \to A'_i = A_i - \frac{\partial \log \rho}{\partial x^i}.$$

For complex valued fields (required in quantum field theories), the theory must also be invariant with respect to gauge transformations (of the first kind) of the wave functions $\Phi(x)$ involved, which have the general form

$$\Phi(x) \to \Phi'(x) = e^{ig(x)}\Phi(x).$$

For example, for the complex valued fields $\Phi(x)$ interacting via the electromagnet field $A_i(x)$ that is generated by the electric charges, the theory (i.e., the field equations and the Lagrangians) should be invariant with respect to gauge transformations of the type

$$\begin{aligned}
\Phi(x) \to \Phi'(x) &= e^{if(x)}\Phi(x), \\
\Phi^*(x) \to \Phi'^*(x) &= e^{-if(x)}\Phi^*(x), \\
A_j(x) \to A'_j(x) &= A_j(x) + \frac{\partial f(x)}{\partial x^j}.
\end{aligned}$$

Such gauge transformations form an Abelian group of transformations [with the binary operation $f(x) = g(x) + h(x)$ for two successive gauge transformations g and h].

The concept of *gauge transformations* has been generalized to various field theories. In mathematics, one employs this concept in exploring principle fiber bundles over a manifold endowed with the group structure. A (gauge) potential is then a connection on this bundle and a gauge transformation is a bundle automorphism that leaves the underlying manifold pointwise invariant. These automorphisms form a group of gauge transformations.

Gaussian elimination A method for successive elimination of unknowns when solving a system of linear algebraic equations (Σ_0),

$$\sum_{j=1}^{n} a_{ij} x_j - a_{i0} = 0, \quad i = 1, \ldots, m, \quad (\Sigma_0)$$

where a_{ij} are elements of some field F. Assuming that $a_{11} \neq 0$ (otherwise, renumber the equations), the key algorithmic step can be described as follows:

Multiply the first equation by (a_{21}/a_{11}) and subtract it (term by term) from the second equation. Next, multiply the first equation by (a_{31}/a_{11}) and subtract it from the third equation, etc., until the first equation is multiplied by (a_{m1}/a_{11}) and subtracted from the last $(i = m)$ equation of the system (Σ_0).

Designate the resulting system of equations with the first equation deleted by (Σ_1), and carry out the same set of operations on (Σ_1) obtaining (Σ_2), etc. Assuming that the rank of the coefficient matrix of (Σ_0) [which is also called the rank of the system of equations (Σ_0)], $r = \operatorname{rank}(\Sigma_0)$, is smaller than m, $r < m$, we obtain, after the rth step, a system (Σ_r) in which all the coefficients of the unknowns vanish. The system (Σ_0) [or (Σ_r)] is called *compatible,* if in (Σ_r) all the absolute terms vanish as well (i.e., when the rank of the coefficient matrix equals the rank of the augmented matrix); otherwise it is *incompatible* and has no solution.

To obtain a solution of a compatible system, we choose some solution $(x_r^{(0)}, \cdots, x_n^{(0)})$ of the system (Σ_{r-1}) and proceed by the back substitution to (Σ_{r-2}), (Σ_{r-3}), etc., until (Σ_0) is reached. *See* back substitution. In general, we can choose a solution for (Σ_{r-1}) by assigning arbitrary values to the $n-r$ variables $x_{r+1}, \ldots, x_n$, say $x_j = c_{j-r}$ $(j = r+1, \ldots, n)$, so that

$$x_r^{(0)} = (a_{r0}^{(r-1)} - \sum_{j=r+1}^{n} a_{rj}^{(r-1)} c_{j-r}) / a_{rr}^{(r-1)}$$

and $x_j^{(0)} = c_{j-r}$ for $j = r+1, \ldots, n$. The (general) solution will then depend on $r - n$ arbitrary parameters $c_j \in F$, $j = 1, \ldots, n - r$.

Once we have a solution of (Σ_{r-1}), the back substitution proceeds by assigning the values $x_r^{(0)}, \ldots, x_n^{(0)}$ to the unknowns $x_r, \ldots, x_n$ in the first equation of (Σ_{r-2}), obtaining $x_{r-1} = x_{r-1}^{(0)}$, and thus a solution $(x_{r-1}^{(0)}, x_r^{(0)}, \ldots, x_n^{(0)})$ of (Σ_{r-2}). These values for $x_{r-1}, \ldots, x_n$ are then substituted into the first equation of (Σ_{r-3}), obtaining $x_{r-2} = x_{r-2}^{(0)}$, etc., until a solution $(x_1^{(0)}, x_2^{(0)}, \ldots, x_n^{(0)})$ of (Σ_0) is obtained. The general solution results when the c_j $(j = 1, \ldots, n - r)$ are regarded as free parameters.

This method can be generalized in various ways. (*See* Gauss-Jordan elimination.) It can also be formulated in terms of a general $m \times n$ (or $m \times n+1$) matrix $\mathbf{A}$ over F [representing the coefficient (or augmented) matrix of a system (Σ_0)], in which case it is normally referred to as the *row reduction* of $\mathbf{A}$. The algorithm can then be conveniently expressed through the so-called *elementary row operations,* which in turn can be represented by the *elementary matrices* of three basic types $[(I + ae_{ij})$, $i \neq j$, replacing the ith row X_i by $X_i + aX_j$, $I + e_{ij} + e_{ji} - e_{ii} - e_{jj}$, interchanging rows i and j, and $I + (c-1)e_{ii}$, $c \neq 0$, multiplying the ith row by c] acting from the left on $\mathbf{A}$. Clearly, the action of the elementary matrices and of their inverses on the augmented matrix $\mathbf{A}$ of a system (Σ_0) produces an equiv-

alent system of linear algebraic equations, and the process of Gauss elimination can thus be represented by a product of corresponding elementary matrices.

In practical applications, when numerical accuracy is at stake, one can also require that the diagonal coefficients (the so-called *pivots*) are not only different from zero, but the largest ones possible: in *partial pivoting* one chooses the absolutely largest a_{ii} (from the ith column), and in *complete pivoting* the absolutely largest element of the entire coefficient matrix (by appropriately renumbering the unknowns).

Also called *Gauss method, Gauss elimination method* or *Gauss algorithm for solving linear systems of algebraic equations. See also* forward elimination.

Gaussian integer   A complex number $(a + bi)$ with integer a and b. The Gaussian integers are thus the points of a square lattice in the complex plane, forming the ring

$$\mathbf{Z}[i] = \{a + bi : a, b \in \mathbf{Z}\}.$$

In fact, $\mathbf{Z}[i]$ is an integral domain (with four units $\pm 1, \pm i$) and, using the absolute value squared as a size function, it is also a Euclidean domain (and, hence, principal ideal domain and thus a unique factorization domain). The prime elements (called *Gauss primes*) are either rational primes that are congruent to 3 modulo 4 (i.e., 3, 7, 11, 19, etc.), or the complex numbers $(a + ib)$ whose norm squared $N = a^2 + b^2$ is either a rational prime congruent to 1 modulo 4 or 2 (i.e., $1 + i$, $1 + 2i$, $3 + 4i$, etc.).

(Also called Gauss integer or Gauss number.)

Gaussian ring   A unique factorization domain. *See* unique factorization domain.

Gaussian sum   Let $\chi(n, k)$ be a numerical character modulo k. Then a trigonometric sum of the form

$$G(a, \chi) = \sum_{n=0}^{k-1} \chi(n, k) e^{\frac{2\pi i (an)}{k}}$$

is referred to as the *Gauss(ian) sum* modulo k. It is thus fully defined by specifying the character $\chi(n, k)$ and the number a. Note that when $a \equiv b \pmod{k}$, then $G(a, \chi) = G(b, \chi)$.

The Gauss sum is exploited in number theory where it enables one to establish a relation between the multiplicative and additive characters.

Gauss-Jordan elimination   A variant of the Gauss elimination method, in which one zeros out the elements in the entire column, rather than only below the diagonal. *See* Gaussian elimination.

The initial step is identical with that of Gauss elimination, while in the subsequent steps one subtracts the ith equation multiplied by (a_{ji}/a_{ii}) from all the other equations. Consequently, the upper left submatrix of the coefficient matrix after the ith iteration is diagonal or, by dividing each equation by the diagonal coefficient, it is a unit matrix. In this way, one obtains the solution of the system directly, without performing the substitution. The coding of this algorithm is simpler than for Gauss elimination, although the required computational effort is larger.

Also called *sweeping-out method*.

Gauss-Manin connection   A way to differentiate cohomology classes with respect to parameters. *See* cohomology class.

The first de Rham cohomology group

$$H^1_{dR}(X/K)$$

of a smooth projective curve X over a field K can be identified with the space of differentials (of the second kind) on X modulo exact differentials. Each derivative θ of K can be lifted in a canonical way to a mapping ∇_θ of $H^1_{dR}(X/K)$ into itself such that

$$\nabla_\theta(f\omega) = f\nabla_\theta(\omega) + \theta(f)\omega \,,$$

where $f \in K$ and $\omega \in H^1_{dR}(X/K)$. This implies an integrable connection

$$\nabla : H^1_{dR}(X/K) \to \Omega^1_K \otimes H^1_{dR}(X/K)$$

called the *Gauss-Manin connection*. This can be generalized to higher dimensions as well as to other algebraic or analytic structures. *See also* Hodge theory.

Gauss-Seidel method for solving linear equations   An iterative numerical method, also called the *single step method,* for approximating the solution to a system of linear equations. In more detail, suppose we wish to approximate the solution to the equation $Ax = b$, where A is an $n \times n$ square matrix, and x and b are n dimensional column vectors. Write $A = L + D + U$, where L is lower triangular, D is diagonal, and U is upper triangular. The matrix $L + D$ is easy to invert, so replace the exact equation $(L + D)x = -Ux + b$ by the relation $(L + D)x_k = -Ux_{k-1} + b$, and solve for x_k in terms of x_{k-1}:

$$x_k = -(L + D)^{-1}Ux_{k-1} + (L + D)^{-1}b \,.$$

This gives us the core of the Gauss-Seidel iteration method. We choose a convenient starting vector x_0 and use the above formula to compute successive approximations $x_1, x_2, x_3, \ldots$ to the actual solution x. Under suitable conditions, the sequence of successive approximations does indeed converge to x. *See also* iteration matrix.

Gauss's Theorem   (1) *See* Fundamental Theorem of Algebra.

(2) Let R be a unique factorization domain. Every polynomial from $R[X]$ or $R[X_1, \ldots, X_n]$ (which are also unique factorization domains) can be uniquely expressed as a product of certain primitive polynomials and an element of R. *See* primitive polynomial. Then a product of primitive polynomials is primitive.

A number of other theorems in analysis are associated with Gauss's name, e.g., the so-called *"Theorema Egregium"* or Gauss curvature for regular surfaces in E^3, Mean Value Theorem for Harmonic Functions, Gauss-Bonnet Theorem, etc.

GCR algebra   A generalization of CCR [completely continuous (= compact) representation] algebras that are also referred to as *liminal* or *liminary* algebras. *See* CCR algebra.

A *GCR algebra* is a C^*-algebra having a (possibly transfinite) composition series whose factor algebras are CCR (i.e., if I_λ is a composition series of our C^*-algebra, then $I_{\lambda+1}/I_\lambda$ is CCR). *See* composition series. This is equivalent to requiring that the trace quotients be continuous.

Equivalently, a C^*-algebra is GCR if the image of every nontrivial representation contains some nonzero compact operator. Thus, starting with a largest CCR ideal I_0 of a given C^*-algebra A that consists of all elements $a \in A$ whose image $\pi(a)$ is compact for all irreducible representations π, we construct the factor algebra A/I_0. Continuing this process, we eventually obtain the largest GCR ideal of A. This will turn out to be the C^*-algebra A itself if it is GCR. Clearly, any CCR algebra is GCR while the converse is false.

Also called *postliminary algebra*.

Gel'fand-Mazur Theorem   A complex Banach algebra is a field if and only if it coincides with the field of complex numbers $\mathbf{C}$.

See Banach algebra.

Gel'fand-Naimark Theorem   Any C^*-algebra admits a faithful (i.e., injective) representation on some Hilbert space.

More precisely, any C^*-algebra A is isometrically $*$-isomorphic to a C^*-subalgebra of some algebra $\mathcal{B}(H)$ of bounded linear operators on a Hilbert space H. Thus, a C^*-algebra is a Banach algebra of (bounded linear) operators on a Hilbert space H (with the usual operations of addition, multiplication by scalars, and product of operators) which is closed under the taking of adjoints.

Moreover, if A is separable, then H can be assumed to be separable as well.

Gel'fand-Pyatetski-Shapiro Reciprocity Law
Let G be a connected semisimple Lie group, Γ a discrete subgroup of G and T the regular representation of G on $\Gamma \backslash G$ [defined by $(T_g f)(x) = f(xg)$, $f \in L^2(\Gamma \backslash G)$]. Then the multiplicity of a unitary, irreducible representation γ in the regular representation T on $\Gamma \backslash G$ equals the dimension of the vector space formed by all automorphic forms of Γ of type γ. *See* automorphic form.

Gel'fand representation   A correspondence between the elements of a commutative Banach algebra R and the continuous functions on the space of regular maximal ideals M of R.

Recall that a maximal ideal M of R is called *regular* if the quotient algebra R/M is a field, in which case R/M is isomorphic to $\mathbf{C}$, the field of complex numbers. Thus, each coset $x(M)$

[i.e., the coset containing $x \in R$] can be regarded as a complex number and the functional $\hat{x} : x \mapsto x(M)$ is multiplicative and linear, i.e., $xy(M) = x(M)y(M)$. Conversely, with each multiplicative linear functional one can associate a regular maximal ideal. Designating by $\mathcal{X}$, the set of all multiplicative functionals on R, endowed with Gel'fand topology, we obtain the *Gel'fand representation,* associating with an element of R a continuous function on $\mathcal{X}$ vanishing at infinity. *See* Gel'fand topology. In fact, $\mathcal{X}$ is a locally compact Hausdorff space and, when R has a unit element, a compact Hausdorff space.

Also called *Gel'fand transform*.

Gel'fand tableau   A triangular pattern

$$[m] =: \begin{bmatrix} (m_n) \\ (m_{n-1}) \\ \vdots \\ (m_2) \\ (m_1) \end{bmatrix} :=$$

$$\begin{bmatrix} m_{1n} & & m_{2n} & \\ & m_{1n-1} & & m_{2n-1} \\ & & & \cdots \\ & & m_{12} & \\ & & & m_{11} \end{bmatrix}$$

$$\begin{array}{cc} \cdots & m_{nn} \\ \cdots & m_{n-1n-1} \\ m_{22} & \end{array}$$

that is employed to label the basis vectors of the carrier spaces for the irreducible representations $\Gamma(m_n)$ of the unitary group U(n) [or SU(n) setting $m_{nn} = 0$] relying on the Gel'fand-Tsetlin group chain. *See* Gel'fand-Tsetlin basis. The irreducible representations $\Gamma(m_n)$ of U(n) are uniquely labeled by their highest weight $(m_n) \equiv (m_{1n}m_{2n} \ldots m_{nn})$, where

$$m_{1n} \geq m_{2n} \geq \cdots \geq m_{nn} \geq 0 .$$

The entries of the lexicographical Gel'fand tableaux satisfy the so-called "betweenness conditions"

$$m_{ij} \geq m_{i,j-1} \geq m_{i+1,j} ,$$

$i = 1, \ldots, n-1$; $j = 2, \ldots, n$, reflecting Weyl's branching law for the subduction of Γ (m_n) from $U(n)$ to $U(n-1)$. The ith row of the Gel'fand tableau thus represents the highest weight of those $U(i)$ irreducible representations that result by a successive subduction of $\Gamma(m_n)$, implied by the Gel'fand-Tsetlin chain.

Arranging the basis vectors in a lexicographical order, we have, for example, for the (210) irreducible representation of $U(3)$ or $SU(3)$,

$$\begin{bmatrix} 210 \\ 21 \\ 2 \end{bmatrix}, \begin{bmatrix} 210 \\ 21 \\ 1 \end{bmatrix}, \begin{bmatrix} 210 \\ 20 \\ 2 \end{bmatrix}, \begin{bmatrix} 210 \\ 20 \\ 1 \end{bmatrix},$$

$$\begin{bmatrix} 210 \\ 20 \\ 0 \end{bmatrix}, \begin{bmatrix} 210 \\ 11 \\ 1 \end{bmatrix}, \begin{bmatrix} 210 \\ 10 \\ 1 \end{bmatrix}, \begin{bmatrix} 210 \\ 10 \\ 0 \end{bmatrix}.$$

Gel'fand topology   The weak topology on the space of multiplicative linear functions on a commutative Banach algebra R. *See* Gel'fand representation.

Gel'fand transform   *See* Gel'fand representation.

Gel'fand-Tsetlin basis   A basis for the carrier space of $U(n)$ or $SU(n)$ irreducible representations exploiting the group chain

$$U(n) \supset U(n-1) \supset \cdots \supset U(2) \supset U(1), \quad (1)$$

where $U(i) \supset U(i-1)$ represents schematically the imbedding $U(n-1) \oplus (1)$ [i.e., $U(n-1)$ represents a subgroup of blocked $n \times n$ unitary matrices (linear operators) consisting of a $(n-1) \times (n-1)$ unitary matrix and the 1×1 matrix (1); clearly this subgroup of $U(n)$ is isomorphic with $U(n-1)$]. According to Weyl's branching law, each irreducible representation of $U(i)$ subduced to $U(i-1)$ is simply reducible. Since, moreover, $U(1)$ is Abelian, the canonical chain (1) provides an unambiguous labeling for mutually orthogonal one-dimensional subspaces spanning the carrier space of a given irreducible representation of $U(n)$ through a set of highest weights characterizing the subduction at each level. Collecting these highest weights we obtain a Gel'fand tableau. *See* Gel'fand tableau. Also called *Gel'fand-Zetlin basis* or *canonical basis*.

general associative law   The associative law for a given binary operation implies that $(a_1 a_2)$

$a_3 = a_1 (a_2 a_3) = a_1 a_2 a_3$ (representing the binary operation involved by a juxtaposition). The general associative law requires that any finite ordered subset of n elements, say $a_1, a_2, \ldots, a_n$, $a_i \in G$, $(i = 1, \ldots, n)$, $n > 2$, uniquely determines their "product" $a_1 a_2 \cdots a_n$, irrespective of the sequence of binary operations employed.

general equation   *See also* Galois equation, Galois group, Galois theory. Recall that the Galois group $G(K/k)$ of the finite field extension K/k is also called the Galois group of the algebraic equation $f(X) = 0$, when K is a minimal splitting field of a separable polynomial $f(X)$ $\in k[X]$. The algebraic equation $f(X) = 0$ is also called a Galois equation when the extension field K can be obtained by adjoining some root of $f(X)$ to k, i.e., when $K = k(\alpha)$ for some root α of $f(x)$. We also note that $G(K/k)$ has a faithful permutation representation based on the permutation group of the roots of $f(X)$. When this representation is primitive, $f(X) = 0$ is called a *primitive equation*. *See* primitive equation. *See also* affect (of $f(X)$).

Now, when $\alpha_1, \alpha_2, \ldots, \alpha_n$ are algebraically independent elements over k, then the equation $F^{(n)}(X) = 0$ for the nth degree polynomial

$$F^{(n)}(X) = X^n - \alpha_1 X^{n-1} + \alpha_2 X^{n-2}$$

$$-\alpha_3 X^{n-3} \pm \cdots + (-1)^n \alpha_n$$

from $k(\alpha_1, \alpha_2, \ldots, \alpha_n)[X]$ is called a general equation of degree n. *See* algebraic independence. The Galois group of this equation is then isomorphic with S_n, the symmetric (or permutation) group of degree n. Moreover, if char $k \neq 2$, then the quadratic subfield L corresponding to the alternating group $\mathcal{A}_n$ of the same degree is obtained by adjoining the square root of the discriminant D of $F^{(n)}(X)$, i.e., $L = k(\sqrt{D})$.

generalization   Motto: *Be wise! Generalize!* "Picayune Sentinel" and M. Artin's *Algebra*.

(1) An extension of a statement (or a concept, a principle, a theorem, etc.) that applies or is valid for some system or structure A to all members of a larger class of systems B containing A as one of its elements (or a proper subsystem).

(2) The process of inferring such a statement (or concept, etc.).

(3) In logic, *generalization* implies the formal derivation of a general statement from a particular one by replacing the subject of the statement with a bounded variable and prefixing a quantifier (in predicate calculus). For example, the statement "the hypothesis $H(i)$ holds for some $i \in \mathbf{Z}$" is the (existential) generalization of "the hypothesis $H(i)$ holds for $i = 1$ and 3." A *universal generalization* applies to all members of a given class while an *existential generalization* applies only to some unspecified members of such a class.

generalized Borel embedding   A generalization of a Borel embedding of a symmetric bounded domain to an embedding of an arbitrary homogeneous bounded domain. Let D be a homogeneous bounded domain and let G_h be the identity component of the full automorphism group $G_h(D)$ of D. Let g_h be the Lie algebra of G_h, then g_h is a j-algebra with a collection of endomorphisms (j). Let g_h^c be the complexification of g_h and let $g_h^- = \{x + ijx \in g_h^c : x \in g_h, \ j \in (j)\}$. If G_h^c is the analytic subgroup of the full linear group corresponding to g_h^c, and G_h^- is the complex closed subgroup of G_h^c generated by g_h^-, then D can be embedded as the G_h orbit of the origin in the complex homogeneous space G_h^c/G_h^-.

generalized decomposition number   Let G be a finite group of order $|G| = g$ and K a splitting field of G of characteristic char $K = p \neq 0$ (i.e., K is a splitting field for the group ring $K[G]$). If p is a divisor of g, $p|g$, we can generate modular representations of G. *See* modular representation.

Let, further, L be an algebraic number field that is a splitting field of G having a prime ideal π dividing p, and Λ_i, $i = 1, \ldots, n$, be the nonsimilar irreducible representations of G in L and χ_i, $i = 1, \ldots, n$, their standard characters.

Now, any $x \in G$ can be uniquely decomposed as follows:

$$x = yz = zy \, ,$$

where y is the so-called p-factor of x (whose order is a power of p) and z is a p-regular element of G (whose order is prime to p). If, further, $\phi_1^{(y)}, \ldots, \phi_{k_y}^{(y)}$ are absolutely irreducible modular characters of the centralizer $C_G(y)$ of y in G, then

$$\chi_i(yz) = \sum_j c_{ij}^{(y)} \phi_j^{(y)}(z) \, ,$$

and the coefficients $c_{ij}^{(y)}$ are referred to as the *generalized decomposition numbers* of G.

When the order of y is a power p^r of p, then the coefficients $c_{ij}^{(y)}$ are algebraic integers of the field of the p^rth roots of unity. *See also* decomposition number, modular representation of finite groups.

generalized eigenvalue problem   The problem of finding scalars $\lambda \in F$ and vectors $\mathbf{x}$ (from a linear space V over F) for linear operators (or matrices) A and B on V satisfying the equation

$$A\mathbf{x} = \lambda B \mathbf{x} \, . \tag{1}$$

Usually, A and B are required to be Hermitian and, moreover, B to be positive definite. When $B = I$, the identity (matrix) on V, the problem reduces to the standard or classical eigenvalue problem.

A further generalization examines the nonlinear problem

$$\left(A_n \lambda^n + A_{n-1} \lambda^{n-1} + \cdots + A_1 \lambda + A_0 \right) \mathbf{x} = 0.$$

Clearly, for $n = 1$ we obtain (1).

generalized Eisenstein series   Let $\Gamma \subset \mathrm{SL}(2, \mathbf{R})$ be a Fuchsian group, acting on the upper half-plane H of the complex plane $\mathbf{C}$, and designate by $\Gamma \backslash H$ the quotient space of H by Γ. *See* Fuchsian group. The Selberg zeta function $Z_p(s, \Lambda)$, defined for $s \in H$ and a matrix representation Λ of Γ, has a number of interesting properties when $\Gamma \backslash H$ is compact and Γ is torsion-free. *See* Selberg zeta function. For a more general case, when $\Gamma \backslash G$ is noncompact (though has a finite volume), the decomposition of $L^2(\Gamma \backslash G)$ into the irreducible representation spaces has a continuous spectrum. Generalized Eisenstein series (defined by Selberg) give the explicit representation for the eigenfunctions of this continuous spectrum. In the special case in which Γ is the elliptic modular group, $\mathrm{SL}(2, \mathbf{Z})$ (or the corresponding group of linear functional transformations) $\Gamma = \mathrm{SL}(2, \mathbf{Z})$, this series is

$$\sum_{(c,d)=1} \frac{y^2}{|c\tau + d|^{2s}}$$

where $y = \Im z$, $z \in H$.

Similar generalized Eisenstein series can be defined for semisimple algebraic groups G and their arithmetic subgroups. *See also* Eisenstein series.

generalized Hardy class A concept arising in the theory of function algebras (or uniform algebras) generalizing that of the Hardy class from the theory of analytic functions.

[Recall that the Hardy class H^p ($p > 0$) consists of analytic functions on the open unit disc $D = \{z : |z| < 1\}$ having the property

$$\sup_{0 < r < 1} \{ \int_{\partial D} |f(r\zeta)|^p d\mu(\zeta) \}^{1/p} < \infty \, ,$$

where $d\mu = |d\zeta|/2\pi$ is normalized Lebesgue measure on the boundary $\partial D = \{\zeta : |\zeta| = 1\}$ of D. This concept is important for harmonic analysis, theory of power series, linear operators, random processes, approximation theory, control theory (in particular the so-called H^∞ control theory; note that H^∞ represents the class of bounded analytic functions in D), etc. Their importance stems from the fact that they are precisely the classes of analytic functions in D with boundary values (of class L^p) from which they can be recovered by means of the Cauchy integral.]

For a function (uniform) algebra A with a positive multiplicative measure μ, one defines the generalized Hardy classes H^p as the closure of A in $L^p(\mu)$ for $1 \leq p < \infty$, while for $p = \infty$, $H^\infty(\mu)$ represents the weak $*$-closure of A in $L^\infty(\mu)$.

Remark: Here one considers a fixed uniform algebra A on a compact Hausdorff space X, the maximal ideal space M_A of A, and a fixed homomorphism $\phi \in M_A$. It is important to recall that there exists a correspondence between non-zero complex-valued homomorphisms ϕ of A and maximal ideals A_ϕ in A that are given by the kernel of ϕ. In fact, M_A is a compact Hausdorff space. Then, designating by μ a measure for ϕ, one defines $H^p(\mu)$ as the closure of A in $L^p(\mu)$

and, similarly, $H^\infty(\mu)$ as the weak-star closure of A in $L^\infty(\mu)$. Note that $L^\infty(\mu)$ is a commutative Banach $*$-algebra (with the pointwise multiplication and the involution given by complex conjugation), which is isometrically isomorphic to the algebra of complex-valued continuous functions on the maximal ideal space of $L^\infty(\mu)$.

generalized inverse For an $m \times n$ matrix A, with entries from the complex field, the unique $n \times m$ matrix A^+ satisfying

(i.) $AA^+A = A$,
(ii.) $A^+AA^+ = A^+$,
(iii.) $(AA^+)^* = AA^+$,
(iv.) $(A^+A)^* = A^+A$.

This definition (by Penrose) is equivalent to the definition (by Moore) that A^+ is the unique matrix so that

(1) AA^+ is the (orthogonal) *projection matrix* onto the range of A,
(2) A^+A is the (orthogonal) projection matrix onto the range of A^+.

The matrix A^+ is also referred to as the Moore-Penrose inverse (or pseudo inverse) of A.

There are other types of generalized inverses. More usually encountered are the group inverse $A^\#$ of a square matrix A ($AA^\#A = A$, $A^\#AA^\# = A^\#$, $AA^\# = A^\#A$), and the Drazin inverse A^D ($A^DAA^D = A^D$, $AA^D = A^DA$, $A^k = A^DA^{k+1}$, $k = 0, 1, \dots$). A^D is unique and $A^\#$ is unique if it exists.

All generalized inverses coincide with A^{-1} when A is invertible.

generalized law of reciprocity The main theorem of class field theory, stating that, for a finite Galois extension L/K, the reciprocity map $r_{L/K}$,

$$r_{L/K} : G(L/K)^{ab} \to A_K/N_{L/K}A_L \, ,$$

defined by

$$r_{L/K}(\sigma) := N_{\Sigma/K}(\pi_\Sigma) \bmod (N_{L/K}A_L) \, ,$$

is an isomorphism. Here, Σ is the fixed field of a Frobenius lift $\tilde{\sigma} \in \phi(\bar{L}/K)$ of $\sigma \in G(L/K)$ and $\pi_\Sigma \in A_\Sigma$ is a prime element. For any field K and a multiplicative G-module A one defines

$$A_K = A^{G_K} = \{a \in A : \sigma a = a, \forall \sigma \in G_K\} \, ,$$

where G_K is a closed subgroup of the profinite group G that is associated with the field K and $A^G/N_G A$ is the norm residue group relative to the norm group $N_G A = \{N_G a = \sum_{\sigma \in G} \sigma a : a \in A\}$. One further defines the Galois group of L/K as

$$G(L/K) = G_K/G_L \,,$$

assuming that G_L is a normal subgroup of G_K. In such a case, the extension L/K is called a *normal* or *Galois extension. See* Galois extension. Note that for a finite extension L/K, when $A_K \subseteq A_L$, there is a normal map

$$N_{L/K} : A_L \to A_K, \quad N_{L/K}(a) = \prod_\sigma \sigma a \,,$$

with σ ranging over the right representatives of G_K/G_L. When L/K is Galois, A_L is a $G(L/K)$-module and $A_K = A_L^{G(L/K)}$.

Recall also that a prime element π_K of A_K, $\pi_K \in A_K$, is an element with $v_K(\pi_K) = 1$, where v_K designates a homomorphism

$$v_K = \frac{1}{f_K} v \circ N_{K/k} : A_K \to \widehat{\mathbf{Z}} \,,$$

with v a henselian valuation $v : A_k \to \mathbf{Z}$ and k the ground field for which $G_k = G$. Further, $\tilde{\sigma}$ is a Frobenius lift of σ if $\sigma = \tilde{\sigma}|_L$,

$$\tilde{\sigma} \in \phi(\tilde{L}/K)$$
$$= \left\{ \tilde{\sigma} \in G(\tilde{L}/K) : \deg_K(\tilde{\sigma}) \in \mathbf{N} \right\} \,,$$

where $\deg_K$ is a surjective homomorphism $\deg_K : G(\tilde{L}/K) \to \widehat{\mathbf{Z}}$, $\widehat{\mathbf{Z}}$ the Prüfer ring

$$\widehat{\mathbf{Z}} = \varprojlim_{n \in \mathbf{N}} \mathbf{Z}/n\mathbf{Z} \,,$$

where $\varprojlim$ denotes projective limit. In fact, $\widehat{\mathbf{Z}} = \prod_p \mathbf{Z}_p$, where p is a prime.

Finally, for every finite extension K/k of the ground field, one defines $\tilde{K} = K \cdot \tilde{k}$,

$$f_K = [K \cap \tilde{k} : k]$$

and

$$\deg_K : \frac{1}{f_K} \deg : G_K \to \widehat{\mathbf{Z}} \,,$$

where

$$\tilde{K} = K\left(\sqrt[n]{K^*}\right)$$

is the maximal Kummer extension of exponent n.

Also called *generalized reciprocity law.*

generalized nilpotent element An element from the kernel of the Gel'fand representation of a commutative Banach algebra R, i.e., an element $x \in R$ such that

$$\lim_{n \to \infty} \left\| x^n \right\|^{1/n} = 0 \,.$$

The kernel of R is referred to as the *radical* of R. When this radical reduces to $\{0\}$, R is said to be *semisimple.*

generalized nilpotent group An extension of the concept of nilpotency to infinite groups, in a nonstandard manner. Thus, a group G is a generalized nilpotent group if any homomorphic image of G that is different from $\{e\}$ has a center that is different from $\{e\}$. This definition reduces to the standard one when G is finite, but not when G is an infinite group.

generalized peak point A point x in a compact Hausdorff space such that $\{x\}$ is a generalized peak set. *See* peak set, generalized peak set.

generalized peak set An intersection of peak sets of a compact Hausdorff space. *See* peak set.

generalized quaternion group A generalization of the quaternion group G, which is the group of order 8 with two generators σ and τ satisfying the relations $\sigma^4 = e$, $\tau^2 = \sigma^2$ and $\tau \sigma \tau^{-1} = \sigma^{-1}$, which is isomorphic to the multiplicative group of quaternion units $\{\pm 1, \pm i, \pm j, \pm k\}$. The *generalized quaternion group* is a group of order 2^n, $(n \geq 3)$, again with two generators σ and τ, which, however, satisfy the following relations:

$$\sigma^{2^{n-1}} = e, \ \tau^2 = \sigma^{2^{n-2}} \quad \text{and} \quad \tau \sigma \tau^{-1} = \sigma^{-1} \,.$$

The standard quaternion group corresponds to $n = 3$.

generalized Siegel domain A domain D in $\mathbf{C}^n \times \mathbf{C}^m$ $(n, m \geq 0)$ with the following properties. (i.) D is holomorphically equivalent to a bounded domain. (ii.) D contains a point of the form $(z, 0)$ where $z \in \mathbf{C}^n$. (iii.) D is invariant under the following holomorphic transformations of $\mathbf{C}^n \times \mathbf{C}^m$ for some $c \in \mathbf{R}$, and for all $s \in \mathbf{R}^n$ and $t \in \mathbf{R} : (z, u) \mapsto (z + s, u)$, $(z, u) \mapsto (z, e^{it}u)$, and $(z, u) \mapsto (e^t z, e^{ct}u)$. In the above situation, D is said to be a *generalized Siegel domain* with exponent c.

generalized solvable group   An extension of the concept of solvability to infinite groups in a nonstandard way. For example, a group G is a generalized solvable group if any homomorphic image of G that is different from $\{e\}$ contains a nontrivial (i.e., different from $\{e\}$) Abelian normal subgroup. As with generalized nilpotency, this definition reduces to a standard one for finite groups but not in the case of infinite groups. *See* generalized nilpotent group.

Generalized Tauberian Theorem   A theorem, due to Wiener, originating in the theory of the Fourier transform. Here we give a version that is pertinent for the exploitation of Banach algebras in the theory of topological groups.

The theorem pertains to the following problem of *spectral synthesis*. *See* spectral synthesis.

Consider an Abelian topological group G and its L^1-algebra (or group algebra) R. Let, further, $\widehat{G}$ designate the character group of G. Then any closed ideal I in R determines a set $Z(I)$ in $\widehat{G}$ consisting of common zeros of the Fourier transforms of the elements of I. [Recall that the Fourier transform of $x \in R$ is given by its Gel'fand transform, in the present case by

$$\hat{x}(\gamma) = \int x(g)\overline{\gamma(g)}d\mu(g) ,$$

where γ is a character of G and μ is a (left-invariant) measure on G (with G regarded as a locally compact Hausdorff group).] The above-mentioned problem then asks whether the converse also holds, namely, whether I can be uniquely characterized by $Z(I)$.

A special case when $Z(I)$ is empty is addressed by the Generalized Tauberian Theorem which states that I must coincide with R when $Z(I) = \emptyset$.

generalized uniserial algebra   Consider a finite dimensional, unitary (associative) algebra A over a field F and designate its system of orthogonal idempotents by $\{e_i^{(s)}\}$, so that

$$\sum_{i=1}^{n}\sum_{s=1}^{d_i} e_i^{(s)} = 1$$

and

$$A = \sum_{i=1}^{n}\sum_{s=1}^{d_i} A e_i^{(s)} = \sum_{i=1}^{n}\sum_{s=1}^{d_i} e_i^{(s)} A ,$$

the latter relationship providing a decomposition of A into a direct sum of indecomposable left (right) ideals. Here n denotes the number of simple algebra components $\overline{A}_i$ in the decomposition of the semisimple quotient algebra A/N, N being the radical of A,

$$A/N \equiv \overline{A} = \sum_{i=1}^{n} \overline{A}_i ,$$

each $\overline{A}_i$ being a full matrix ring of degree d_i, which thus decomposes into the direct sum of d_i minimal left (right) mutually $\overline{A}$-isomorphic ideals $\overline{A}\overline{e}_i^{(s)}$ $(\overline{e}_i^{(s)}\overline{A}_i)$, $i = 1, \ldots, n; s = 1, \ldots, d_i$, where $\overline{e}_i^{(s)}$ designate the orthogonal idempotents of $\overline{A}_i$. The idempotents $e_i^{(s)}$ in the above are representatives from each residue class $\overline{e}_i^{(s)}$, which are chosen in such a way that the relations hold. In the following, we designate $e_i^{(s)}$ simply by e_i.

We call A a *generalized uniserial algebra* if each indecomposable left (right) ideal Ae_i $(e_i A)$ of A has a unique composition series. *See* composition series.

An algebra A is a generalized uniserial algebra if its radical N is a principal left and right ideal, and a uniserial algebra if and only if every two-sided ideal of A is a principal left and right ideal, i.e., if and only if every quotient algebra of A is a Frobenius algebra. *See* uniserial algebra, Frobenius algebra.

An algebra A is an *absolutely uniserial algebra* if the algebra A^K over K, where K is an extension field of a field F, is uniserial for any such extension K/F. This is the case if and only if the radical N of A is a principal ideal generated by an element from the center Z of A and Z decomposes into a direct sum of simple extensions $F[\alpha]$ of F.

generalized valuation   A valuation of a general rank. *See* valuation.

The rank of an additive valuation (or simply of a valuation) $v : F \to G \cup \{\infty\}$ of the field F, with G a (totally) ordered additive group and the element ∞ is defined to be greater than any element of G, is defined to be the Krull dimension of the valuation ring $R_v = \{a \in F : v(a) \geq 0\}$. *See* Krull dimension.

general linear group   On a finite dimensional linear space V, with $\dim V = n$, over a field K, the group of (nonsingular) linear mappings of V onto V, the group operation being defined as composition of mappings, denoted GL(V). More generally, the automorphism group $\text{Aut}_K(V)$ of the free right K-module V with n generators, for K an associative ring with unit.

Equivalently, a *general linear group* of degree n over K, usually designated GL(n, K) or $\text{GL}_n(K)$, is a group of $n \times n$ invertible matrices with entries in K. Clearly, GL(V) and GL(n, K) are isomorphic. When viewed as an affinite variety, GL(n, K) can also be regarded as an algebraic group. *See* algebraic group. In most applications, K is a field. The structure of GL(n, K) over a ring K is studied in algebraic K-theory.

Also called *full linear group*. *See also* special linear group, projective general linear group.

general linear Lie algebra   (Of degree n over a commutative ring with unity [or over a field] K), a Lie algebra, denoted gl(n, K), which results from the total matrix algebra $M(n, K)$ of (all) $n \times n$ matrices with entries from K, when endowed with a Lie (or bracket) product defined by the commutator $[X, Y] := XY - YX$. *See* Lie algebra. A *general linear Lie algebra* gl(n, K) is the Lie algebra of GL(n, K), a general linear group, and may thus also be regarded as the tangent space to GL(n, K) at the identity (represented by the identity matrix). *See also* general linear group.

general position   Let $x_0, \ldots, x_k$ be points in Euclidean space. These points are said to be in *general position* if they are not contained in any plane of dimension less than k. This concept may be generalized to geometric objects of higher dimension.

general solution   As a rule, the *general solution* of a problem is a solution involving a certain number of parameters from which any other solution (except for singular solutions) can be obtained, through a suitable choice of these parameters.

Specifically, in analogy to differential equations, one understands, by a general solution of a nonhomogeneous linear difference equation

$$\sum_{i=0}^{n} p_i(x)\, y(x + i) = q(x) \qquad (1)$$

a solution of the form

$$y(x) = \sum_{i=1}^{n} c_i(x)\phi_i(x) + \psi(x)\,,$$

where $\phi_i(x)$, $i = 1, \ldots, n$ are linearly independent solutions of the corresponding homogeneous equation, $\psi(x)$ is an arbitrary solution of the nonhomogeneous equation (1), and $c_i(x)$ are arbitrary periodic functions of period 1.

general term   (Of a series, of a polynomial, of an equation, of a language, etc.) An expression or an object that forms a separable part of some other expression or object, in particular expressions separated by the plus sign (in a series or a polynomial), comma (in a sequence), inequality or identity sign (in an inequality or a chain of inequalities or equation(s)), etc. Also called *generic term*.

generating representation   Let G be a compact Lie group. Designate, further, the commutative, associative algebra of complex-valued continuous functions on G by $C^{(0)}(G, \mathbf{C}) \equiv C^{(0)}(G)$, and its subalgebra of *representative functions* referred to as the *representative ring* of G by $R(G, \mathbf{C}) \equiv R(G)$. *See* representative function, representative ring. Note that, with the supremum norm $\|f\| = \sup_{g \in G} |f(g)|$, $C^{(0)}(G)$ is a Banach space and the actions of G on this space (given by left and right translations) are continuous. At the same time, $C^{(0)}(G)$ may be completed with respect to the norm arising from the inner product $(u, v) = \int_G u\bar{v}$, yielding the Hilbert space $L^2(G)$ of square integrable functions on G. The actions of G on $L^2(G)$ (again by left and right translations) are unitary.

In view of the Peter-Weyl theorem, $R(G)$ is dense in both $C^{(0)}(G)$ and $L^2(G)$ so that there exists a faithful (i.e., injective) representation $\rho : G \to \mathrm{GL}(n, \mathbf{C})$ of G and $R(G)$ may be regarded as a finitely generated algebra over $\mathbf{C}$. Thus, there exists a faithful representation $g \mapsto \{r_{ij}(g)\}$ such that the functions r_{ij} and $\bar{r}_{ij}$ generate $R(G)$. Such a representation is referred to as a *generating representation*.

generator(s)   (**1**) In *group theory,* the elements of a nonempty subset S of a group G such that G is generated by S, i.e., $G = \langle S \rangle$, where $\langle S \rangle$ consists of all possible products of the elements of S and of their inverses that are possibly subject to a certain number of conditions (called *relations*) of the type

$$s_1^{n_1} s_2^{n_2} \cdots s_m^{n_m} = e, \quad s_i \in S, \, n_i \in \mathbf{Z} .$$

When S is finite, G is said to be *finitely generated* or *finitely presented. See* relation, finitely presented group. For a cyclic group, S consists of a single element.

More precisely, let G be a group generated by some set $S \equiv \{s_1, s_2, \ldots, s_n\}$ and let us designate by F the free group on S. Further, let T be a subset of F, $T \equiv \{t_1, t_2, \ldots, t_m\} \subset F$, and N_T the smallest normal subgroup of F containing T (given by the intersection of all normal subgroups of F containing T). Then, G is said to be given by the *generators* s_i, $(i = 1, \ldots, n)$ subject to the *relations* $t_1 = t_2 = \cdots = t_m = e$, if there is an isomorphism $G \approx F/N_T$ that associates s_i with $s_i N_T$. This is usually expressed by writing

$$G = \langle s_1, \ldots, s_n : t_1 = t_2 = \cdots = t_m = e \rangle , \tag{1}$$

and one also says that G has the *presentation* given by the right-hand side of (1).

Remarks: (i.) Although we have used a finite number of generators and relations in the above definition, this is not necessary.

(ii.) The relation $r_i = e$, e.g., $a^{-1}b^{-2}ab = e$, is often more convenient to write in the form avoiding inverses, i.e., $ab = b^2 a$ in our example.

(iii.) Recall that a free group is "free" of relations, so that we can regard any element in N_T as the identity, as in fact the notation of (1) suggests.

(iv.) Cyclic groups are generated by a single generator (which, clearly, is not unique), e.g., $\mathbf{Z}_8 = \langle 1 \rangle = \langle 3 \rangle = \langle 5 \rangle = \langle 7 \rangle$.

(v.) *See also* finitely generated group.

(**1a**) In the analytic theory of semigroups, one defines the *infinitesimal generator* F of an equicontinuous semigroup T of class (C_0), $T \equiv \{T_t \mid t \geq 0\}$, as a limit

$$Fx = \lim_{t \to 0+} t^{-1} (T_t - e) x .$$

(**2**) In *ring theory,* we say that an ideal I of a (commutative) ring R is generated by a finite subset $S = \{s_1, \ldots, s_n\}$ of R, when

$$I = \left\{ \sum_{i=1}^n r_i s_i : r_i \in R \right\} ,$$

in which case we also write

$$I = (s_1, \ldots, s_n) \quad \text{or} \quad I = \sum_{i=1}^n R s_i ,$$

and refer to the ring elements s_i, $i = 1, \ldots, n$, as *generators* of I. An ideal generated by a single generator, $I = (s)$, is called a *principal ideal.* An extension to noncommutative rings is obvious.

(**3**) An analogous definition also applies to *modules over a ring* (or a *field*).

Consider an A-module M (i.e., a module over a ring A) and a family $X = \{x_i\}_{i \in I}$ of elements of M. The smallest sub-A-module N of M containing all the elements of X consists of all linear combinations of these elements and we write

$$N = \sum_{i \in I} A x_i = \left\{ \sum_{i \in I} a_i x_i : a_i \in A, \, i \in I \right\} .$$

The family X is then referred to as a *system of generators* for N. We also say that N is generated by X. When X is finite, i.e., $\mathrm{Card}(X) < \infty$, N is said to be *finitely generated,* and an A-module Ax, generated by a single element $x \in M$, is called a *monomial.*

When A is a field, the A-module M becomes a linear space over A, and the same terminology is sometimes employed even in this case, although one often defines M to be *spanned* rather than generated by X, and X is referred to as a spanning set (or a basis, if linearly independent).

(4) The method of defining groups by generators and relations can be also applied to *Lie algebras.* Let L be a Lie algebra over a field F generated by a set $X = \{X_i\}_{i \in I}$. If L is free on X (in which case the vector space $V = \mathrm{Span}(x)$ can be given an L-module structure by assigning to each $x \in X$ an element of the general linear Lie algebra $\mathrm{gl}(V)$ and extending canonically to L) and M is the ideal of L generated by a family of elements $A = \{a_j\}_{j \in J}$, J being some index set, then the quotient algebra L/M is referred to as the Lie algebra with *generators* X_i, $(i \in I)$ and *relations* $a_j = 0$, $(j \in J)$, with X_i the image of the element $x_i \in X$ in L/M. This is referred to as a presentation of L.

Remarks: (i.) For semisimple Lie algebras L over an algebraically closed field F, char $F = 0$, one can give a presentation of L in terms of generators and relations that depends only on the root system of L. (*See* Serre's Theorem.)

(ii.) In the physics literature, one often refers to the basis elements of the defining (or standard) representation of matrix Lie groups or algebras as generators. Thus, for example, the matrix units $e_{ij} = \|\delta_{ik}\delta_{j\ell}\|$ of $\mathrm{gl}(n, F)$ are referred to as raising $(i < j)$, lowering $(i > j)$, and weight $(i = j)$ generators according as they raise, lower, or preserve the weight of a representation involved.

(5) In *homological algebra,* an object G of an Abelian category $\mathcal{A}$ (with all functions being additive) is called a *generator* if the natural mapping

$$\mathrm{Hom}(A, B) \to \mathrm{Hom}\,(h_G(A), h_G(B))$$

is one to one. Here Hom designates the functor defining the category $\mathcal{A}$,

$$\mathrm{Hom} : \mathcal{A} \times \mathcal{A} \to (\mathrm{Ab})\,,$$

with (Ab) designating the category of Abelian groups, and with

$$h_G(\cdot) := \mathrm{Hom}(G, \cdot)\,.$$

Similarly, one defines a cogenerator G when

$$\mathrm{Hom}(A, B) \to \mathrm{Hom}\left(h^G(B), h^G(A)\right)$$

is one to one, where now

$$h^G(\cdot) := \mathrm{Hom}(\cdot, G)\,.$$

(6) In *coding theory,* with a *linear code* defined as a subspace W of $V_n = \{0, 1, \ldots, q - 1\}^n$, with q being a prime power $q = p^\alpha$, one associates with each element $a \in V_n$ the polynomial

$$a(X) = a_1 + a_2 X + \cdots + a_n X^{n-1}$$

over a Galois field $GF(q)$ [note that V_n can be regarded as an n-dimensional vector space over $GF(q)$]. Then the *cyclic code W modulo $g(X)$* is defined by

$$W = \{a \in V_n \,:\, a(X) \equiv 0 (\mathrm{mod}\, g(X))\}$$

and $g(X)$ is referred to as the generator of N.

generic point   Consider an irreducible variety V in K^n, where K is the universal domain and designate by k a field of definition for V (which is a subfield of K, $k \subset K$), having a finite transcendence degree over the underlying prime field. *See* irreducible variety, universal domain, transcendence degree, prime field. Then a point $(x) \equiv (x_1, \ldots, x_n)$ of V, $(x) \in V$, with the property that all points of V are specializations of (x) over k, is called a *generic point* of V over k. *See* specialization. Note that in general a generic point of V is not unique.

Remarks:

(i.) In some texts, *generic point* refers to an arbitrary point of a nonempty Zariski open set of a variety V. *See* Zariski open set.

(ii.) A *generic point x* in a topological (sub)-space A is a generic point of A when $A = \overline{\{x\}}$ (where the bar designates the closure in the hull-kernel topology).

genus   **(1)** For an *algebraic variety,* a discrete, numerical invariant representing an important parameter then enables a classification of such varieties. These invariants may be defined in various ways, the most useful ones being based on the concept of differential forms on a variety and on the cohomology of coherent sheaves. One distinguishes the *algebraic genus $p_a(X)$* and the *geometric genus $p_g(X)$* of a variety X. *See* geometric genus.

For a one-dimensional algebraic variety C over a field k, referred to as an *algebraic curve,* the genus g of C is defined as the dimension of the space of regular differential 1-forms on

C, assuming that C is smooth and complete. *See* algebraic curve. We recall here that an algebraic curve can be transformed into a plane curve with only ordinary multiple points by a finite number of plane Cremona transformations (quadratic transformations of a projective plane into itself). *See* Cremona transformation. A plane algebraic curve C of degree n is a point set in an affine 2-space defined by the zero set $f(X, Y) = 0$, of an nth degree polynomial $f(X, Y)$ in X and Y. Setting $F(X_0, X_1, X_2) = X_0^n f(X_1/X_0, X_2/X_0)$, the homogeneous polynomial F defines an algebraic curve C of degree n in a projective plane $\mathbf{P}^2$. We say that C is irreducible if $f(X, Y)$ is irreducible. Clearly, a curve of degree 1 is a line. A point $P = (a, b)$ on C is an *r-ple point* if $f(X + a, Y + b)$ has no terms of degree less than r in X and Y. There are r tangent straight lines (counting multiplicity) at such a point: if these tangents are all distinct, P is referred to as an *ordinary point* and an ordinary double point is called a *node*. An r-ple point with $r > 1$ is called a *multiple* or a *singular point*.

For a nonsingular irreducible curve C, we define a *divisor* $\mathbf{a}$ as an element of the free Abelian group $G(C)$ generated by the points of C that is of the form

$$\mathbf{a} = \sum_i n_i P_i, \quad (n_i \in \mathbf{Z}) \tag{1}$$

and has a *degree* $\deg(\mathbf{a}) = \sum_i n_i$. We say that the representation (1) for $\mathbf{a}$ is *reduced* if $P_i \neq P_j$ for $i \neq j$. A *positive* or *an integral divisor*, written as $\mathbf{a} \succ 0$, involves only positive coefficients n_i, $(n_i > 0)$. We further designate by w a discrete valuation whose value group is the additive group of integers, thus representing a normal (or normalized) valuation w_P of $\mathbf{K}(C)$ defined by a valuation ring R_P given by the subset of the function field $\mathbf{K}(C)$ of C ($\mathbf{K}$ designating the universal domain) consisting of those functions f that are regular at P. The integer $w_P(f)$ is referred to as the order of f at P, and P is said to be a zero of f when $w_P(f) > 0$ and a pole when $w_P(f) < 0$. The linear combination

$$\Sigma w_P(f) P =: (f) \tag{2}$$

is called the *divisor of the function* f, and the set of all positive divisors that are linearly equivalent to a given divisor $\mathbf{a}$ is referred to as a *complete linear system* determined by $\mathbf{a}$ and is designated by $|\mathbf{a}|$. Further, one defines a finite dimensional vector space $\mathcal{V}(\mathbf{a})$ over $\mathbf{K}$ as follows:

$$\mathcal{V}(\mathbf{a}) := \{f \in \mathbf{K}(C) : (f) + \mathbf{a} \succ 0\},$$

whose one-dimensional subspaces are in a bijective correspondence with the elements of $|\mathbf{a}|$. We then define the *dimension* of $|\mathbf{a}|$, $\dim|\mathbf{a}|$, by

$$\dim|\mathbf{a}| := \dim_{\mathbf{K}} \mathcal{V}(\mathbf{a}) - 1,$$

and the *genus* of C, $g \equiv g(C)$, as the supremum of non-negative and bounded integers $\deg(\mathbf{a}) - \dim|\mathbf{a}|$,

$$g := \sup_{\mathbf{a} \in G(C)} (\deg(\mathbf{a}) - \dim|\mathbf{a}|).$$

One can show that for any $g > 0$, $g \in \mathbf{Z}$, there exists an algebraic curve of genus g. The curves with $g = 1$ are the so-called *elliptic curves* and those with $g > 1$ are subdivided into the classes of *hyperelliptic* and *non-hyperelliptic curves*. *See* elliptic curve, hyperelliptic curve. *See also* specialty index, Riemann-Roch Theorem.

For an r-dimensional projective variety Y in $\mathbf{P}^n$ (a projective n-space over an algebraically closed field k), the arithmetic genus of Y can be defined in terms of the constant term of the Hilbert polynomial $P_Y(X)$, $X = (X_0, X_1, \ldots, X_n)$, of Y, as follows:

$$p_a(Y) = (-1)^r (P_Y(0) - 1).$$

It can be shown that $p_a(Y)$ is independent of projective embeddings of Y. When Y is a plane curve of degree d (see above), then

$$p_a(Y) = \frac{1}{2}(d - 1)(d - 2),$$

and when it is a hypersurface of degree d, then

$$p_a(Y) = \binom{d-1}{n}.$$

Equivalently, for a projective scheme Y over a field k, the arithmetic genus $p_a(Y)$ can be defined by

$$p_a(Y) = (-1)^r [\chi(\mathcal{O}_Y) - 1],$$

where $\mathcal{O}_Y$ designates a sheaf of rings of regular functions from Y to k and $\chi(\mathcal{F})$ is the so-called *Euler characteristic* of a coherent sheaf $\mathcal{F}$ on Y that is defined in terms of the cohomology groups $H^i(Y, \mathcal{F})$ of $\mathcal{F}$ as follows:

$$\chi(\mathcal{F}) := \Sigma(-1)^i \dim_k H^i(Y, \mathcal{F}) .$$

See projective scheme. Thus, when Y is a curve, we have that

$$p_a(Y) = \dim_k H^1(Y, \mathcal{O}_Y) ,$$

while for a complete smooth algebraic surface Y, we have

$$p_a(Y) = \chi(\mathcal{O}_Y) - 1$$

$$= \dim_k H^2(Y, \mathcal{O}_Y) - \dim_k H^1(Y, \mathcal{O}_Y) .$$

On the other hand, the geometric genus $p_g(Y)$ of a nonsingular projective variety Y over k is defined as the dimension of the (global) section of the canonical sheaf ω_Y of Y (defined as the nth exterior power of the sheaf of differentials where $n = \dim Y$), $\Gamma(Y, \omega_Y)$, i.e.,

$$p_g(Y) := \dim_k \Gamma(Y, \omega_Y) .$$

For a projective nonsingular curve C, the arithmetic and the geometric genuses coincide, i.e., $p_a(C) = p_g(C) = g$, while for varieties of dimension greater than or equal to 2, they are not necessarily equal, and their difference is referred to as the *irregularity* of Y. *See* irregularity.

(2) For an algebraic *function field* K over k of dimension 1 (or of transcendence degree 1) (i.e., a finite separable extension of a purely transcendental extension $k(x)$ of k such that k is maximally algebraic in K), the *genus* of K/k is defined similarly as for a curve C. Thus, it is achieved by replacing C with K/k, $\mathbf{K}(C)$ by K, and $\mathbf{K}$ by k, while points on C now become prime divisors of K/k.

We can also say that the genus of a function field is given by the genus of its Riemann surface (recall that the Riemann surface S of a function field is homeomorphic to the complement of a finite set of points in a compact oriented two-dimensional manifold $\overline{S}$, while the genus of the latter is defined, loosely speaking, as the number of "holes" in $\overline{S}$, i.e., $g = 0$ if $\overline{S}$ is a sphere, $g = 1$ if $\overline{S}$ is a torus, etc.).

(3) For *quadratic forms* Q over an algebraic number field k of finite degree, one defines equivalence classes of forms having the same *genus* by requiring that Q and Q' be equivalent over the principle order $o_\mathbf{p}$ in $k_\mathbf{p}$ for all non-Achimedean prime divisors $\mathbf{p}$ of k and, at the same time, that they are equivalent over $k_\mathbf{p}$ for all the Archimedian prime divisors $\mathbf{p}$ of k. (Recall that prime divisors of k are equivalence classes of nontrivial multiplicative valuations over k and are referred to as Archimedean or non-Archimedian accordingly if these valuations are Archimedean or not.) *See* valuation.

(4) For a *complex matrix representation* of a finite group G. Consider a finite group G and an algebraic number field K. Recall that there exists a bijective correspondence between linear and matrix representations of G and that every linear representation of G over K is equivalent to some linear representation of the group ring $K[G]$. When K is the ring of rational integers, a linear representation over K is called an integral representation.

Further, for a given algebraic number field K, every $K[G]$-module V contains G-invariant R-lattices (called G-lattices for short), where R is the ring of integers in K, that may be viewed as finitely generated $R[G]$-modules, providing an integral representation of G.

Designating by P a prime ideal of R and by R_P the ring of quotients (or fractions) of the ring R with respect to the prime ideal P, also called the local ring at P, one can also explore the R_P-representations (this approach is closely related with modular representation theory). Thus, with any $R[G]$-module M we can associate a family of $R_P[G]$-modules $M_P = R_P \otimes M$, with P ranging over all prime ideals of R. Then, when, for two G-lattices (or $R[G]$-modules) M and N, belonging to a $K[G]$-module V, we have that $M_P \cong N_P$ for all P, we say that M and N have (or are of) the same *genus*. (*See also* class number.)

(5) Similarly, for a connected *algebraic group* G over an algebraic number field k of finite degree, we consider a ring R of integers in k and designate by L a general R-lattice (also called a transformation space) in the vector space V on which G acts. Defining then an action of G_A, the *adele group* of G, on the set $\{L\}$ of R-lattices

in a natural way, one defines the *genus* of L as the orbit $G_A L$ of L with respect to G_A.

(6) For an *imaginary quadratic field* k. (*See also* principal genus of k.) Each coset of the ideal class group G of k modulo the subgroup H of all ideal classes of k satisfying the condition $(\epsilon_1, \ldots, \epsilon_t) = (1, \ldots, t)$ is referred to as a *genus* of k. *See* ideal class, ideal class group. Here, $\epsilon_1 = \chi_i(N(\mathbf{a}))$, $i = 1, \ldots, t$, where $\mathbf{a}$ is an integral ideal with $(\mathbf{a}, (d)) = 1$, N designates the norm, the χ_i are the Kronecker symbols, and d is the discriminant of k. *See* Kronecker symbol. For each genus, the values of ϵ_i, $(i = 1, \ldots, t)$ are unique and $(\epsilon_1, \ldots, \epsilon_t)$ is called the character system of this genus. *See* character system.

genus of function field *See* genus (2).

geometrical equivalence Let V be an n-dimensional Euclidean space and let $O(V)$ be the orthogonal group of V. If T is an n-dimensional lattice in V and K is a finite subgroup of $O(V)$, let (T, K) denote a faithful linear representation of K on T. Every pair (T, K) corresponds to a set of crystallographic groups. Two pairs (T_1, K_1) and (T_2, K_2) are said to be *geometrically equivalent* if there exists a $g \in GL(V)$ such that $K_2 = g K_1 g^{-1}$. This equivalence relation is denoted by $(T_1, K_1) \sim (T_2, K_2)$, and the equivalence classes are called *geometric crystal classes*.

geometric crystal class *See* geometrical equivalence.

geometric difference equation A difference equation of the form

$$y(px) = f(x, y(x)) ,$$

where $p \in \mathbf{C}$ is an arbitrary complex number. For example, the standard form of a linear difference equation,

$$\sum_{k=0}^{n} a_k(x) \, y(x + k) = b(x) ,$$

can be transformed to this form via the change of variables $z = p^x$, yielding

$$\sum_{k=0}^{n} A_k(z) \, Y\left(z p^k\right) = B(z) .$$

geometric fiber The fiber $X \times_Y \mathrm{Spec}(K)$ of a geometric point $\mathrm{Spec}(K) \to Y$, where X and Y are schemes and K is an algebraically closed field.

Specifically, for a morphism of schemes $\phi : X \to Y$ and a point $y \in Y$, the fiber of the morphism ϕ over the point y is the scheme

$$X_y = X \times_Y \mathrm{Spec}\, k(y) ,$$

where $k(y)$ is the residue field of y. *See* fiber. A point of Y with values in a field K is a morphism $\mathrm{Spec}(K) \to Y$. Also, when K is algebraically closed, such a point is referred to as a geometric point. *See* geometric point.

A suitable embedding of $k(y)$ in K is assumed, in the above definition.

Also called *geometric fiber.*

geometric fibre *See* geometric fiber.

geometric genus (Of an algebraic surface S) a numerical invariant characterizing this surface given by the number of linearly independent holomorphic 2-forms on S.

More generally, for an n-dimensional irreducible variety V, the *geometric genus* is given by the number of linearly independent differential forms of the first kind of degree n.

See also genus.

geometric mean Given any n positive numbers $a_1, \ldots, a_n$, the positive number $G = \sqrt[n]{a_1 a_2 \cdots a_n}$.

geometric multiplicity Given an eigenvalue λ of a matrix A, the geometric multiplicity of λ is the dimension of its *eigenspace*. It coincides with the size of the diagonal block with diagonal element λ in the *Jordan normal form* of A. *See also* algebraic multiplicity, index.

geometric point A morphism $\mathrm{Spec}(K) \to X$, where X is a scheme and K is an algebraically closed field.

geometric programming A special case of nonlinear programming, in which the objective function and the constraint functions are linear combinations of (not necessarily integral) powers of the variables.

geometric progression   A sequence of non-zero numbers, having the form $ar, ar^2, \ldots, ar^n(, \ldots)$. *See also* geometric series.

geometric quotient   Let Z and Y be algebraic schemes over a field. Suppose that G is a reductive algebraic group that operates on Z, and $f : Z \to Y$ is a G-invariant morphism of schemes. Denote by f_* the homological mapping induced by f, and by $\mathcal{O}_Z$ and $\mathcal{O}_Y$ the sheaves of germs of regular functions on Z and Y, respectively. The morphism f is called a *geometric quotient* if

(i.) f is a surjective affine morphism and $f_*(\mathcal{O}_Z)^G = \mathcal{O}_Y$,

(ii.) $f(X)$ is a closed subset of Y whenever X is a G-stable closed subset of Z, and

(iii.) for each z_1 and z_2 in Z, the equality $f(z_1) = f(z_2)$ holds if and only if the G-orbits of z_1 and z_2 are equal.

geometric series   A series of the form

$$\sum_{j=0}^{\infty} ar^j = ar + ar^2 + \cdots + ar^n + \cdots .$$

Sometimes also referred to as a geometric progression.

If $|r| < 1$, the above series converges to the sum $ar/(1 - r)$.

Geršgorin's Theorem   The eigenvalues of an $n \times n$ matrix $A = (a_{ij})$, with entries from the complex field, lie in the union of n closed discs (known as the Geršgorin discs),

$$\bigcup_{i=1}^{n} \left\{ z \in \mathbf{C} : |z - a_{ii}| \leq \sum_{k \neq i} |a_{ik}| \right\} .$$

As a consequence of *Geršgorin's Theorem*, every strictly *diagonally dominant matrix* is invertible. The latter fact is also known as the Lévy–Desplanques Theorem. *See also* ovals of Cassini.

Givens method of matrix transformation
A method for transforming a symmetric matrix M into a tridiagonal matrix $N = PMP^{-1}$. Here P is a product of two-dimensional rotation matrices.

Givens transformation   *See* Givens method of matrix transformation.

Gleason cover   The projective cover in the category of compact Hausdorff spaces and continuous maps. In this category, the projective cover of a space X always exists, and is called the *Gleason cover* of the space. It may be constructed as the Stone space (space of maximal lattice ideals) of the Boolean algebra of regular open subsets of X. A subset is *regular open* if it is equal to the interior of its closure. Alternatively, it may be constructed as the inverse limit of a family of spaces mapping epimorphically onto X. *See also* epimorphism, projective cover, Stone space.

global dimension   For an analytic set A, the number $\dim(A) = \sup_{z \in A} \dim_z(A)$, where $\dim z(A)$ is the local dimension of A at z.

global Hecke algebra   Let F be either an algebraic number field of finite degree or an algebraic function field of one variable over a finite field. Consider the general linear group $\mathrm{GL}_2(F)$ of degree 2 over F, and denote by G_A the group of rational points of $\mathrm{GL}_2(F)$ over the adele ring of F. For each place v of F, let H_v be the Hecke algebra of the standard maximal compact subgroup of the general linear group of degree 2 over the completion of F at v. Denote by ε_v the normalized Haar measure of H_v. The restricted tensor product of the local Hecke algebras H_v with respect to the family $\{\varepsilon_v\}$ is called the *global Hecke algebra* of G_A.

G-mapping   Suppose that G is a group, and X and Y are G-sets. A G-mapping is a mapping $f : X \to Y$ such that $f(gx) = gf(x)$ for each g in G and each x in X.

GNS construction   A means of constructing a cyclic representation of a C^*-algebra on a Hilbert space from a state of the C^*-algebra. The letters G, N, and S refer to Gel'fand, Naĭmark (Neumark), and Siegel (Segal), respectively.

good reduction   For a discrete valuation ring R, with quotient field K, and an Abelian variety A over K, we denote by A' the Neron minimal model of A. Thus, A' is a smooth group scheme of finite type over $\mathrm{Spec}(R)$ such that, for every scheme B' which is smooth over $\mathrm{Spec}(R)$, there is a canonical isomorphism

$$\mathrm{Hom}_{\mathrm{Spec}(R)}(B', A') \to \mathrm{Hom}_K(B'K, A) \, .$$

If A' is proper over $\mathrm{Spec}(R)$, then we say that A has a *good reduction* at R. *See* Neron minimal model.

Gorenstein ring   An algebraic ring which appears in treatments of duality in algebraic geometry. Let R be a local Artinian ring with m its maximal ideal. Then R is a *Gorenstein ring* if the annihilator of m has dimension 1 as a vector space over $K = R/m$.

graded algebra   An algebra R with a direct sum decomposition $R = R_0 \oplus R_1 \oplus \dots$ (so that the R_i are groups under addition) satisfying $R_m R_n \subseteq R_{m+n}$, for all $m, n > 0$. The elements of each R_i are called *homogeneous elements of degree i*.

graded coalgebra   A coalgebra (C, Δ, ϵ) such that there exist subspaces C_n, $n \geq 0$, such that $C = C_0 \oplus C_1 \oplus \dots$ and $\Delta(C_n) \subset \bigoplus_{i+j=n} C_i \otimes C_j$, for all $n > 0$ and such that $\epsilon(C_n) = \{0\}$, for all $n > 0$.

graded Hopf algebra   Suppose that R is a commutative ring with a unit, (A, ε) is a supplemented graded R-algebra, and $\psi : A \to A \otimes A$ is a map such that

$$\alpha_1 \circ (1_A \otimes \varepsilon) \circ \psi = 1_A = \alpha_2 \circ (\varepsilon \otimes 1_A) \circ \psi \, ,$$

with 1_A denoting the identity mapping on A, α_1 denoting an algebra isomorphism from $A \otimes R$ to A, and α_2 denoting an algebra isomorphism from $R \otimes A$ to A. Then A is called a *graded Hopf algebra*.

graded module   Let $R = R_0 \oplus R_1 \oplus \dots$ be a graded ring. A *graded R-module* is an R-module M which has a direct decomposition $M = \bigoplus_{j=-\infty}^{\infty} M_j$, such that $R_i M_j \subset R_{i+j}$, for $i \geq 0$ and $j \in \mathbf{Z}$. The elements of each M_i are called *homogeneous elements of degree i*.

graded object   An object O which can be written as a direct sum $O = \bigoplus_{a \in A} O_a$, where A is a monoid.

graded ring   A ring R with a direct sum decomposition $R = R_0 \oplus R_1 \oplus \dots$ (so that the R_i are groups under addition) satisfying $R_m R_n \subseteq R_{m+n}$ for all $m, n > 0$. The elements of each R_i are called *homogeneous elements of degree i*.

A primary example is the ring $R = k[x_1, \dots, x_n]_{\mathrm{hom}}$ of homogeneous polynomials in n variables over a coefficient ring k. In this ring, $R_n = 0$, for $n < 0$ and, for $n \geq 0$, R_n consists of the polynomials that are homogeneous of degree n.

gradient method for solving non-linear programming problem   An iterative method for solving non-linear programming problems. Suppose that the problem is to maximize the function f subject to some constraints. At each stage of the iteration, one uses the current iterate x_k, the gradient of f at x_k, and a positive number λ_k to compute the point x_{k+1} according to the formula $x_{k+1} = x_k - \lambda_k n_k$, where n_k is the unit vector in the direction of the gradient of f at the point x_k.

Gramian   For complex valued functions $f_i : [a, b] \to \mathbf{C}$, $i = 1, \dots, n$, the determinant of the $n \times n$ matrix whose i, j entry is $\int_a^b f_i \overline{f_j} \, dx$.

Gram-Schmidt process   A way of converting a basis $x_1, \dots, x_n$ for an inner product space V, $(\cdot, \cdot)$ into an orthonormal basis $v_1, \dots, v_n$, so that we have $(v_i, v_j) = 0$, if $i \neq j$ and $(v_i, v_i) = 1$, for $i, j = 1, \dots, n$. The relationship between $\{x_1, \dots, x_n\}$ and $\{v_1, \dots, v_n\}$ is given by

$$v_1 = \frac{x_1}{\|x_1\|} \, ,$$

$$v_2 = \frac{x_2 - (x_2, v_1)v_1}{\|x_2 - (x_2, v_1)v_1\|} \, ,$$

$$v_3 = \frac{x_3 - (x_3, v_1)v_1 - (x_3, v_2)v_2}{\|x_3 - (x_3, v_1)v_1 - (x_3, v_2)v_2\|} \, .$$

$$\dots$$

Here $\|x\| = \sqrt{(x, x)}$, for any $x \in V$.

Gram's Theorem   **(1)** Suppose that P is a convex polytope in Euclidean 3-space. For $k =$

0, 1, 2, denote the kth angle sum of P by $\alpha_k(P)$. Then $\alpha_0(P) - \alpha_1(P) + \alpha_2(P) = 1$. More generally, if P is a d-dimensional convex polytope in Euclidean d-space and $\alpha_k(P)$ is the kth angle sum of P for $k = 0, 1, 2, \ldots, d-1$, then

$$\sum_{i=0}^{d-1} (-1)^i \alpha_i(P) = (-1)^{d-1} \ .$$

(2) Suppose that F is a field of characteristic zero, and denote by G the general linear group of some degree over F. Consider matrix representations $\rho_1, \ldots, \rho_q$ of G over F, with n_i the degree of ρ_i, such that each ρ_i is either the rational representation of G induced by some element of G or is the contragredient map κ. Suppose that $\rho_i = \kappa$ for $i \leq s$ and $\rho_i \neq \kappa$ for $i > s$. Let $\{x_j^{(i)} : 1 \leq i \leq q, 1 \leq j \leq n_i\}$ be algebraically independent elements over F. For each $b = 1, 2, \ldots, s$, let H_b be a polynomial in $x_j^{(i)}$ for $i > s$ that is homogeneous in $x_1^{(i)}, \ldots,$ $x_{n_i}^{(i)}$ for each i, and suppose that the set V of common zeros of $H_1, \ldots, H_s$ in the affine space of dimension $n_{s+1} + \cdots + n_q$ is G-stable. Then there is a finite set C of absolute multiple covariants such that V is the set of $(\ldots, a_j^{(i)}, \ldots)$ for $i > s$ such that $(\ldots, a_m^{(b)}, \ldots, a_j^{(i)}, \ldots)$ is a zero point of C for each $a_m^{(b)}$ with $b \leq s$.

graph   A *simple graph* is a pair (V, E), where V is a set and E is a set of distinct, unordered pairs of elements of V. In a *directed graph* (or *digraph*), the elements of E are ordered pairs of distinct elements of V. Allowing multiplicities for the elements of E or allowing loops (an element of E of the form (v, v)) gives a *general graph,* also called a *graph*. The elements of V are called *vertices* and the elements of E are called *edges*. Thus a pair $(u, v) \in E$ may be visualized as a line segment joining points $u, v \in V$.

   Graphs are used widely in combinatorics and algebra since they model relationships of various kinds among sets.

graphing   *See* graph of equation, graph of function.

graph of equation   For an equation $E = E(x_1, \ldots, x_n)$ over the field K, in the variables

$x_1, \ldots, x_n$, the set $G(E) = \{(a_1, \ldots, a_n) \in K^n : E(a_1, \ldots, a_n) = 0\}$.

graph of function   For a function $f : V \to W$, where V and W are sets, the set $G(F) = \{(x, f(x)) : x \in V\}$. Thus the graph $G(f)$ is a subset of the Cartesian product $V \times W$ of the sets.

graph of inequality   For an equation $E = E(x_1, \ldots, x_n)$, over the ordered field K in the variables $x_1, \ldots, x_n$, the *graph of the inequality* $E > 0$ is the set $G(E) = \{(a_1, \ldots, a_n) \in K^n : E(a_1, \ldots, a_n) > 0\}$.

greater than   If R is an Abelian group and R^+ is a subset of R not containing 0, such that R^+ is closed under addition, $R = R^+ \cup (-R^+) \cup \{0\}$, and $R^+ \cap (-R^+)$ is empty, then, for all $a, b \in R$, we have either $a - b \in R^+$ or $a - b \in (-R^+)$ or $a - b = 0$. In the first case, we write $a > b$ and say that a is *greater than* b (or b is *less than* a); in the second case, we write $a < b$ and say that b is *greater than* a; in the third case, we say that a and b *are equal* and write $a = b$.

greater than or equal to   *See* greater than.

greatest common divisor   For integers a and b, not both zero, the unique positive integer, denoted $\gcd(a, b)$, which divides both a and b and is divisible by any other divisor of both a and b. Also called *greatest common factor* or *highest common factor.*

greatest common factor   *See* greatest common divisor.

greatest lower bound   *See* infimum.

Grössencharakter   *See* Hecke character.

Grothendieck category   A category $\mathcal{C}$ satisfying: (i.) $\mathcal{C}$ is Abelian; (ii.) $\mathcal{C}$ has a generator; (iii.) direct sums always exist in $\mathcal{C}$; and (iv.) for any object P of $\mathcal{C}$, any sub-object Q of P, and for any totally ordered family $\{R_i\}$ of sub-objects, we have $(\cup R_i) \cap Q = \cup (R_i \cap Q)$.

Grothendieck topology   Suppose that C is a category. A *Grothendieck topology* on C is a collection of families of morphisms, indexed by the objects of C (such a family that corresponds to an object S of C is called a *covering family* of S), such that

(i.) $\{\varphi : T \to S\}$ is a covering family of S whenever $\varphi : T \to S$ is an isomorphism,

(ii.) if I is an index set and $\{\varphi_i : R_i \to S\}_{i \in I}$ is a covering family of S, then, for each morphism $\varphi' : S' \to S$, the fiber product R'_i defined by $R'_i = R_i \times_S S'$ exists and the induced family $\{\varphi_i : R'_i \to S'\}$ is a covering family of S', and

(iii.) if I is an index set, A_i is an index set for each i in I, the family $\{\varphi_i : R_i \to S\}_{i \in I}$ is a covering family of S, and the family $\{s_{i,a} : S_{i,a} \to R_i\}_{a \in A_i}$ is a covering family of R_i for each i in I, then the family $\{\varphi_i \circ s_{i,a} : S_{i,a} \to S\}$ is a covering family of S.

ground field (1) If E is an extension field of a field F, then F is called the ground field (or the base field).

(2) If V is a vector space over a field F, then F is called the ground field of V (or field of scalars of V).

ground form Consider the general linear group of some degree over a field F of characteristic zero. A finite number of homogeneous forms with coefficients in F that are algebraically independent is called a set of *ground forms*. *See* general linear group.

group One of the basic structures in algebra, consisting of a set G and a (composition) map $m : G \times G \to G$, usually written as $m(g, g') = g \circ g'$, $g + g'$ or $g \cdot g'$, satisfying the following axioms:

(i.) (associativity) $g \circ (g' \circ g'') = (g \circ g') \circ g''$ for all $g, g', g'' \in G$;

(ii.) (identity) there is an element $e \in G$ such that $e \circ g = g = g \circ e$, for all $g \in G$.

(iii.) (inverses) for each $g \in G$, there is $g' \in G$ (called the *inverse* of g) such that $g \circ g' = e = g' \circ g$.

One can think of a group as describing symmetries of certain objects.

group algebra Let R be a commutative ring with identity and let G be a group with elements $\{g_\alpha\}_\alpha$. The R-group algebra is the R-algebra freely generated by the g_α, so that an element is a (formal) finite sum $r_{\alpha_1} g_{\alpha_1} + r_{\alpha_2} g_{\alpha_2} + \cdots + r_{\alpha_n} g_{\alpha_n}$. When multiplying such elements one simplifies by writing $g_\alpha g_\beta$ as g_γ, for some unique γ, and then collecting terms.

group character A representation of a group G is a homomorphism $\alpha : G \to H$, where H is a linear group over a field F. The *character* of α is the function $\mathcal{X}_\alpha : G \to F$, defined by $\mathcal{X}_\alpha(g) = \text{trace}(\alpha(g))$. In the case of finite groups G, characters characterize representations up to equivalence.

group extension The group G is an *extension* of the group H by the group F if there is a short exact sequence $1 \to H \to G \to F \to 1$. *See* exact sequence. *See also* Ext group.

grouping of terms The rearranging of terms in an expression into a form more convenient for some purpose. For example, one may group the terms involving x in the expression $5x^2 + e^y + 3x$ to produce the expression $(3x + 5x^2) + e^y$.

group inverse *See* generalized inverse.

group minimization problem Suppose that A is an $m \times n$ matrix with real entries, b a vector in Euclidean m-space $\mathbf{R}^m$, and c a vector in $\mathbf{R}^n$. Consider the vector x_B of basic variables associated with a dual feasible basis of the linear programming problem of minimizing $c^T x$ subject to the conditions that $Ax = b$ and $x_j \geq 0$ for $j = 1, \ldots, n$. Denote by S the set of vectors x such that $Ax = b$, $x_j \geq 0$ for $j = 1, \ldots, n$, and some of the coordinates of x are restricted to be integers. The problem of minimizing $c^T x$ subject to the condition that x is an element of the group generated from S by ignoring the nonnegativity constraints on the coordinates of the vector x_B is called a *group minimization problem*. One may solve this problem by finding the shortest path on a directed graph that has a special structure. If each coordinate of the vector of basic variables for the optimal solution of the group minimization problem is nonnegative, then that solution is optimal for the original linear programming problem; otherwise one may use a branch-and-bound algorithm to investigate lattice points near the optimal solution for the group minimization problem.

group of automorphisms   A one-to-one mapping $\alpha : A \rightarrow A$, where A is an algebraic object, for example an algebra or a group, such that α preserves whatever algebraic structure A has. Any such α has an inverse which also preserves the structure of A. The composition of two such automorphisms does this as well. The set of all such α forms the automorphism group of A (where the composition map in this group is a composition of functions). *See also* automorphism group.

group of classes of algebraic correspondences   Suppose that C is a nonsingular curve. Denote by G the group of divisors of the product variety $C \times C$, and denote by D the subgroup of divisors that are linearly equivalent to degenerate divisors. The quotient group G/D is called the *group of classes of algebraic correspondences*.

group of congruence classes   Let $m > 0$ be an integer. Two integers a and b are said to be *congruent* mod m (written $a \equiv b$ mod m) if $a - b$ is divisible by m. This is an equivalence relation. (*See* equivalence relation.) Denote the equivalence class of the integer a by $[a]$; thus $[a] = \{b : b \equiv a \bmod m\}$. Then the set of equivalence classes for this equivalence relation is a group under "addition" defined by $[a] + [b] = [a + b]$. This group is called the *group of congruence classes* or *group of congruences* of the integers modulo m.

group of inner automorphisms   For each element g of a group G, let $i(g) : G \rightarrow G$ be the map defined by $i(g)(g') = gg'g^{-1}$. Then we have $i(gh) = i(g)i(h)$, for all $g, h \in G$ and also $i(g^{-1}) = i(g)^{-1}$. It follows that $i(g)$ is an automorphism of G and the set of all such $i(g)$ (for all $g \in G$) is called the *group of inner automorphisms* of G. It is a normal subgroup of the group of automorphisms of G.

group of outer automorphisms   The quotient group $\mathrm{Aut}(G)/\mathrm{Inn}(G)$, where $\mathrm{Aut}(G)$ is the group of all automorphisms of G and $\mathrm{Inn}(G)$ is the normal subgroup of $\mathrm{Aut}(G)$ consisting of all inner automorphisms if G. *See* group of inner automorphisms.

group of quotients   Let S be a commutative semigroup with cancellation (so that $ab = ac$ implies $b = c$), then there is an embedding of S in a group G such that any $g \in G$ can be written as $g = st^{-1}$, where $s, t \in G$. Here G is called the *group of quotients* of S.

group of symmetries   If L is a geometric object, then a symmetry of L is a one-to-one mapping of L to itself which preserves the geometry of the object (e.g., preserves the metric, if L is a subset of a metric space). The set of all such symmetries forms a group called the *group of symmetries*.

group of twisted type   Suppose that F is a field. Denote by L a simple Lie algebra over the complex numbers that corresponds to the Dynkin diagram of some type X, and denote by C the Chevalley group of type X over F. Consider the Lie algebra $L_{\mathbf{Z}}$ spanned by Chevalley's canonical basis of L over the ring $\mathbf{Z}$ of integers. Then C is generated by linear transformations $x_a(t)$ of the Lie algebra $F \otimes_{\mathbf{Z}} L_{\mathbf{Z}}$, with a a root of L, and t in F.

(1) If X equals A_n, D_n, or E_6, then the Dynkin diagram of type X has a nontrivial symmetry τ. If $\tau(a) = b$, and if F has an automorphism σ such that the order of σ equals the order of τ, then denote by θ the automorphism of C that sends $x_a(t)$ to $x_b(t^{\sigma})$. Denote by U the θ-invariant elements of the subgroup of C generated by $\{x_a(t) : a > 0, t \in F\}$, and denote by V the θ-invariant elements of the subgroup of C generated by $\{x_b(t) : b < 0, t \in F\}$. The group generated by U and V is called a *group of twisted type*.

(2) Suppose that X equals either B_2 or F_4 and that the field F has characteristic 2, or suppose that $X = G_2$ and the field F has characteristic 3. If F has an automorphism σ such that $(t^{\sigma})^{\sigma} = t^p$ for each t in F, then one may apply a procedure similar to that described in (1) above to obtain in each case another *group of twisted type*.

groupoid   A small category in which all morphisms are invertible. *See* category.

group scheme   A *group scheme* over a scheme S is a scheme X, together with a morphism to the scheme S such that there is a section $e : S \rightarrow X$ (thought of as the identity map), a morphism $r : X \rightarrow X$ over S (thought of as the inverse map), and a morphism $m : X \times X \rightarrow X$ (thought of as the composition map) such that (i.) the composition $m \circ (\text{Id} \times r)$ is equal to the projection $X \rightarrow S$ followed by e and (ii.) the two morphisms $m \circ (\text{Id} \times r)$ and $m \circ (r \times \text{Id})$ from $X \times X \times X$ to X are the same.

Group Theorem   Suppose that R is a ring. The set of equivalence classes of fractional ideals of R that contain elements that are not zero divisors forms a group.

group-theoretic integer programming   A method that involves transforming an integer linear programming problem into a group minimization problem. If each coordinate of the vector of basic variables for the optimal solution of the corresponding group minimization problem is nonnegative, then that solution is optimal for the original programming problem; otherwise one may use a branch-and-bound algorithm to investigate lattice points near the optimal solution for the group minimization problem.

group theory   The study of groups. Group theory includes representation theory, combinatorial group theory, geometric group theory, Lie groups, finite group theory, linear groups, permutation groups, group actions, Galois theory, and more.

group variety   A variety V, together with a morphism $m : V \times V \rightarrow V$ such that the set of points given by V is a group and such that the inverse map ($v \rightarrow v^{-1}$) is also a morphism of V. Here a morphism of varieties X, Y (over a field k) is a continuous map $f : X \rightarrow Y$, such that, for every open set V in Y, and for every regular function $g : V \rightarrow k$, the function $g \circ f : f^{-1}(V) \rightarrow k$ is regular.

G-set   A set S for which there is a map $f : G \times S \rightarrow S$ such that $f(gg', s) = f(g, f(g', s))$ and $f(\text{Id}, s) = s$ for all $g, g' \in G$ and $s \in S$. Here, G is a group.

Guignard's constraint qualification   Suppose that X is a closed connected subset of real Euclidean n-space, g is a vector-valued function defined on X, and C is the set of all points in X such that $g_i(x) \leq 0$ for each $i = 1, \dots, n$. Suppose that x is a point on the boundary of X that is not an extreme point of X, denote by $\nabla g_i(x)$ the gradient of g_i at x, and denote by Y the set of vectors y for which $\nabla g_i(x) \cdot y \leq 0$ for each i such that $g_i(x) = 0$. If Y is a subset of the convex hull spanned by the vectors tangent to C at x, then g is said to satisfy *Guignard's constraint qualification* at the point x.

H

Hadamard product   Given two $m \times n$ matrices $A = (a_{ij})$ and $B = (b_{ij})$, the *Hadamard product* (or *Schur product*) of A and B is their entrywise product, usually denoted by

$$A \circ B = (a_{ij}b_{ij}) \ .$$

half-angle formulas   The trigonometric identities $\cos^2\theta = (1 + \cos(2\theta))/2$ and $\sin^2\theta = (1 + \cos(2\theta))/2$.

half-side formulas   Suppose that α, β, and γ are the angles of a spherical triangle. Denote by a the side of the triangle that is opposite α, and put $S = (\alpha + \beta + \gamma)/2$ and

$$R = \sqrt{\frac{-\cos S}{\cos(S - \alpha)\cos(S - \beta)\cos(S - \gamma)}} \ .$$

The formulas

$$\sin\left(\frac{1}{2}a\right) = \sqrt{\frac{-\cos S \cos(S - \alpha)}{\sin\beta\sin\gamma}} \ ,$$

$$\cos\left(\frac{1}{2}a\right) = \sqrt{\frac{\cos(S - \beta)\cos(S - \gamma)}{\sin\beta\sin\gamma}} \ ,$$

and

$$\tan\left(\frac{1}{2}a\right) = R\cos(S - \alpha)$$

are called *half-side formulas*.

half-spinor   An element of the representation space of either half-spin representation of the complex spinor group. *See* half-spin representation.

half-spin representation   Suppose that V is a vector space of dimension $2n$ and Q is a quadratic form on V. Write V as a direct sum of two n-dimensional isotropic spaces W and W' for Q. The representations corresponding to the sum of all even exterior powers of W and to the sum of all odd exterior powers of W are called

the *half-spin representations* of the complex orthogonal Lie algebra $so_{2n}\mathbf{C}$.

Hall subgroup   Let P be a set of prime numbers. A P-number is a number all of whose prime divisors are in P. A P'-number is a number none of whose prime divisors are in P. A finite group G is a P-group if $|G|$ is a P-number. A subgroup H of a finite group G is a *Hall P-subgroup* if H is a P-group and $[G : H]$ is a P' number. Further, H is a *Hall subgroup* if H is a P-subgroup for some P. This is equivalent to the condition $\gcd(|H|, [G : H]) = 1$.

Hamilton-Cayley Theorem   A matrix satisfies its characteristic polynomial. That is, if $p(x)$ is the characteristic polynomial of a matrix M, then $p(M) = 0$. Also called the *Cayley-Hamilton Theorem*.

Hamilton group   A non-Abelian group in which each subgroup is normal. *See* normal subgroup.

Hamilton's quaternion algebra   The non-commutative ring generated over the real numbers by $1, \mathbf{i}, \mathbf{j}, \mathbf{k}$ where 1 is the identity, $\mathbf{i}^2 = -1, \mathbf{j}^2 = -1, \mathbf{k}^2 = -1$ and $\mathbf{ijk} = -1$. Thus, an arbitrary element has the form $q = a + b\mathbf{i} + c\mathbf{j} + d\mathbf{k}$, where a, b, c, d are real numbers. The conjugate of q is the quaternion $\bar{q} = a - b\mathbf{i} - c\mathbf{j} - d\mathbf{k}$, which has the property that $q\bar{q} = a^2 + b^2 + c^2 + d^2$. This allows one to show that each non-zero q has an inverse $q^{-1} = \bar{q}/(q\bar{q})$. Thus, we have an example of a non-commutative division algebra. The set of quaternions of norm 1 ($q\bar{q} = 1$) forms a group under multiplication which is isomorphic to SU(2).

harmonic mean   Given n non-zero real numbers x_i, $i = 1, \ldots, n$, their *harmonic mean* is H, where $1/H = (1/x_1 + \cdots + 1/x_n)/n$.

harmonic motion   Simple harmonic motion is the motion of a particle subject to the differential equation

$$\frac{d^2x}{dt^2} = -n^2x \ ,$$

where n is a constant. Its solution may be written as $x = R \cos(nt + b)$, where R and b are arbitrary constants. Damped harmonic motion is the motion of a particle subject to the differential equation

$$\frac{d^2x}{dt^2} + 2p\frac{dx}{dt} + n^2x = 0 ,$$

where $2p\frac{dx}{dt}$ is a resistance proportional to the velocity and p is a constant.

harmonic progression   *See* harmonic series.

harmonic series   The series

$$1 + \frac{1}{2} + \frac{1}{3} + \cdots \frac{1}{n} + \cdots .$$

It is well known that this series diverges; a proof is given by Bernoulli. One can also see this by noting that $1/2 \geq 1/2$, $1/3+1/4 > 1/2$, $1/5+1/6 + 1/7 + 1/8 > 1/2$, $1/9 + 1/10 + \cdots + 1/16 > 1/2$, etc., showing that the series is at least as big as $n(1/2)$ for any n.

Hartshorne conjecture   If X_1 and X_2 are smooth submanifolds with ample normal bundles of a connected, projective manifold Z such that

$$\dim X_1 + \dim X_2 > \dim Z ,$$

then is $X_1 \cap X_2$ nonempty?

Hasse invariant   Let E be an elliptic curve over a perfect field k of characteristic $p > 0$ and let $F : E \to E$ be the Frobenius morphism (induced by the p-power map). Let $F^* : H^1(E, \mathcal{O}_E) \to H^1(E, \mathcal{O}_E)$ be the induced map on cohomology. If $F^* = 0$, then E has *Hasse invariant* 1; otherwise E has *Hasse invariant* 1. Here $H^1(E, \mathcal{O}_E)$ is a one-dimensional vector space over k, since E is elliptic, and $\mathcal{O}_E$ is the sheaf of regular functions on the variety E.

Hasse-Minkowski character   Let A be an $n \times n$ non-singular symmetric matrix with rational elements and let D_i ($i = 1, 2, \ldots, n$) be the leading principal minor determinant of order i in the matrix A. Suppose further that none of the D_i is zero. Then the integer

$$c_p = c_p(A) = (-1, -D_n)_p \prod_{i=1}^{n-1}(D_i, -D_{i+1})_p$$

is the Hasse-Minkowski character of A. Here p is a prime and $(a, b)_p$ is the Hilbert norm residue symbol.

Hasse-Minkowski character   Also called *Hasse-Minkowski symbol, Hasse symbol,* and *Minkowski-Hasse character.* An invariant of a quadratic form which, when considered together with the discriminant of the form and the number of variables, determines the class of a quadratic form over a local field. Let F be either a complete archimedean field or a local field of characteristic not equal to 2. Let $(a, b) = 1$ if ax^2+by^2 represents 1, otherwise let $(a, b) = -1$. If f is a nondegenerate quadratic form over F equivalent to $a_1x_1^2 + a_2x_2^2 + \cdots + a_nx_n^2$, then the *Hasse-Minkowski character* is usually defined as either

$$\chi_p(f) = \prod_{i<j}(a_i, a_j) \text{ or}$$

$$\chi_p^*(f) = \prod_{i \leq j}(a_i, a_j) .$$

Note that $\chi_p^* = \chi_p \cdot (d(f), d(f))$ where $d(f)$ is the discriminant of f. The Hasse-Minkowski character depends only on the equivalence class of the form f and not on the diagonalization used. It may also be defined in terms of the Hasse algebra of f by setting $\chi_p(f) = 1$ if the Hasse algebra associated with f splits and -1 if it does not. The Hasse-Minkowski character is sometimes called the Hasse invariant, since the Hasse invariant is either 1 or the unique element of order 2 in the Brauer group of F. *See also* Hasse invariant, Hasse-Minkowski Theorem.

Hasse-Minkowski symbol   *See* Hasse-Minkowski character.

Hasse-Minkowski Theorem   If q is a quadratic form over a global field F, with characteristic not equal to 2, then q is isotropic over F if and only if q is isotropic over all $F_\wp$, where $F_\wp$ is the completion of F at the place $\wp$.

If F and $F_\wp$ are defined as above, this theorem leads to the following statement: Two quadratic forms are equivalent over F if and only if they are equivalent over all $F_\wp$, if and only if they have the same dimension, discriminant, the same Hasse-Minkowski character for

non-archimedean $F_\wp$, and the same signature over the real completions of F. *See* Hasse-Minkowski character.

Hasse principle   Let X be a smooth projective variety over an algebraic number field k. A fundamental Diophantine problem for X is to decide whether there are any k-rational points on X and, if so, to describe them. When X is a geometrically integral quadric, the *Hasse principle* affirms that if X has rational points in every completion of k, then it has rational points in k.

Hasse's conjecture   Let m_1, m_2, and m_3 be coprime integers greater than 2. If $A(m_1, m_2, m_3)$ denotes the number of lattice points (x_1, x_2, x_3) with $(x_i, m_i) = 1$ in the tetrahedron

$$2 \max (x_i/m_i) < \sum_{i=1}^{3} x_i/m_i < 1 \, ,$$

then $A(m_1, m_2, m_3)$ is even. In 1943, Hasse made the conjecture, which arose in his investigations of class numbers of Abelian number fields.

Hasse symbol   *See* Hasse-Minkowski character.

Hasse-Witt map   *See* Hasse-Witt matrix.

Hasse-Witt matrix   Given a complete smooth algebraic variety X of dimension n defined over a perfect field k of positive characteristic p, there is a natural Frobenius morphism $F : X \to X$, whose action on the structure sheaf $\mathcal{O}_X$ of X is just as the pth power map. This action induces another action on the nth cohomology group $H^n(X, \mathcal{O}_X)$, called the *Hasse-Witt map* of X and denoted also by F. The group $H^n(X, \mathcal{O}_X)$ is a finite-dimensional k-vector space and the matrix corresponding to F is called the *Hasse-Witt matrix* of X.

Hasse zeta function   The function $\zeta(s)$, of a complex variable s, defined as follows. Suppose that V is a nonsingular complete algebraic variety defined over a finite algebraic number field F. For each prime ideal P of F, denote by V_P the reduction of V modulo P, denote by F_P the residue field of P, and denote by N_m the number

of F_P^m-rational points of V_P. Denote by Z_P the formal power series defined by

$$Z_P(u) = \exp \left(\sum_{m=1}^{\infty} \frac{N_m u^m}{m} \right) \, .$$

Denote by $\mathbf{P}$ the set of all prime ideals P of F such that V_P is defined, and denote by $N(P)$ the absolute norm of P. Then

$$\zeta(s) = \prod_{P \in \mathbf{P}} Z_P \left(N(P)^{-s}, V_P \right) \, .$$

Hausdorff space   A topological space X such that, whenever p, q are points of X, then there is an open neighborhood U of p and an open neighborhood V of q such that $U \cap V = \emptyset$. Also called T_2-space.

haversine   The complex valued function $h(z) = (1 - \cos z)/2$.

Hecke algebra   (1) Let H be a subgroup of a group G, and suppose that for each g in G, the index of $H \cap gHg^{-1}$ in H is finite. Denote by $H \backslash G/H$ the set of double cosets of G by H. If R is a commutative ring, then the module $R^{H \backslash G/H}$ has an R-algebra structure and is called the *Hecke algebra* of (G, H) over R.

(2) Let H be a subgroup of a finite group G. Suppose that F is a field, and e is an idempotent in FH such that the left ideal FHe affords an F-character. Then the subalgebra $e \cdot FG \cdot e$ is called a *Hecke algebra*.

Hecke character   Consider an algebraic number field of finite degree, denote its idele group by J and its idele class group by C. A character of J that is a character of C is called a *Hecke character* (or *Grössencharakter*). *See* idele class, idele group.

Hecke L-function   A function $L(s)$, of a complex variable s, defined as follows. Suppose that F is an algebraic number field of finite degree, and m is an integral divisor of F. Denote by χ the character of the ideal class group of F modulo m. For each integral ideal a of F, denote its absolute norm by $N(a)$. Denote by I the set of all integral ideals of F. Then

$$L(s) = \sum_{a \in I} \chi(a)/N(a)^s \, .$$

height (1) Let R be a non-trivial commutative ring with an identity. A *prime chain* of length n is a sequence $I_0 \supset I_1 \supset I_2 \supset \cdots \supset I_n$ (proper inclusions) of prime ideals (i.e., $I_i \in \mathrm{Spec}(R)$). If $I \in \mathrm{Spec}(R)$, then the supremum of the lengths of such prime chains is called the *height* of the prime ideal I. If I is a (proper) not necessarily prime ideal of R, by height(I) we will mean the minimum of the heights of the prime ideals which contain I.

(2) If $\Phi \subset R^n$ is a root system and Δ is a base for Φ, then, for $\beta \in \Phi$, we have $\beta = \sum_{\alpha \in \Phi} k_\alpha \alpha$, where the k_α are integers. Then the *height* of β is $\sum_{\alpha \in \Phi} k_\alpha$.

Heisenberg group The group of all upper-triangular integral (or sometimes real) 3×3 matrices with 1s on the diagonal. The *Heisenberg group* is nilpotent.

Held group A finite simple group of order 4,030,387,200.

Hensel's Lemma Let A be a complete local ring with m its maximal ideal. Assume that A is m-adically complete. Let $f(x) \in A[x]$ be a monic polynomial of degree n. Let $\pi : A \to A/m$ be the canonical map and extend π to $\pi : A[x] \to (A/m)[x]$. If $g_1(x)$ and $g_2(x)$ are relatively prime monic polynomials over A/m of degree r and $n - r$ (respectively), such that $\pi(f(x)) = g_1(x)g_2(x)$, then there exist $h_1(x)$, $h_2(x) \in A[x]$ having degree r and $n - r$ such that $f(x) = h_1(x)h_2(x)$ and $\pi(h_i(x)) = g_i(x)$, for $i = 1, 2$.

Herbrand quotient Let $G = \{g_1, \ldots, g_n\}$ be a finite group and let M be a G-module. Let $N : M \to M$ be the norm defined by $N(m) = g_1 m + \cdots + g_n m$. Then $N(M)$ is contained in M^G. Let $h_0 = M^G / N(M)$ and let h_1 be the first cohomology group $H^1(G, M)$. If h_0 and h_1 are finite, then the quotient of their orders is $h(M)$, the *Herbrand quotient*. The Herbrand quotient is multiplicative for short exact sequences of G-modules: if we have $1 \to M' \to M \to M'' \to 0$, then $h(M) = H(M')h(M'')$.

Herbrand's Lemma Let G be a finite group. If M' is a sub-G-module of a G-module M and the Herbrand quotient $h(M')$ exists, then $h(M)$ also exists and $h(M) = h(M')$. *See* Herbrand quotient.

Hermitian form A form $(\cdot, \cdot) : V \times V \to \mathbf{C}$, where V is a real or complex vector space, such that $(u, v) = \overline{(v, u)}$, for all $u, v \in V$. The standard Hermitian form on $\mathbf{C}^n$ is defined by $(u, v) = \sum_{i=1}^n u_i \overline{v_i}$.

Hermitian matrix A matrix M, such that its transpose and its complex conjugate (the matrix with entries equal to the complex conjugates of the entries of M) are equal. *See* transpose.

Hermitian operator An operator M, on a real or complex vector space, such that $M^* = M$, where M^* is the adjoint operator. Thus, if the Hermitian form is the standard Hermitian form, then $M^* = \overline{M^T}$, where ^{T} stands for transpose.

Hessenberg method of matrix transformation A method for transforming a matrix into a matrix (b_{ij}) by means of a triangular similarity transformation so that $b_{ij} = 0$ whenever $i - j \geq 2$.

Hessian The Jacobian of the first order derivatives of a differentiable function $f = f(x_1, \ldots, x_n)$ of n real variables; i.e., $H(f) = J(\partial f/\partial x_1, \ldots, \partial f/\partial x_n) = |\partial^2 f/\partial x_i \partial x_j|$.

Hey zeta function The function $\zeta(s)$, of a complex variable s, defined as follows. Suppose that A is a simple algebra over an algebraic number field of finite degree, o is a maximal order of A, and a is an integral left o-ideal. Denote by $N(a)$ the number of elements in o/a, and by I the set of all integral left o-ideals. Then

$$\zeta(s) = \sum_{a \in I} N(a)^{-s} .$$

higher algebra (1) Algebra at the undergraduate level.

(2) Modern, or abstract, algebra.

higher-degree Diophantine equation An algebraic equation of degree three or higher whose coefficients are integers, such that the solutions sought are to be integers. *See also* Diophantine equation.

higher differentiation   For a commutative ring R with a unit and $\mathbf{N}$ the nonnegative integers, a sequence

$$\{\delta_i : R \to R\}_{i \in \mathbf{N}}$$

of maps such that, for every x and y in R and every i and j in $\mathbf{N}$,
  (i.) $\delta_i(x+y) = \delta_i x + \delta_i y$;
  (ii.) $\delta_i(xy) = \sum_{p,q \in \mathbf{N}: p+q=i} \delta_p x \cdot \delta_q y$;
  (iii.) $\delta_i(\delta_j x) = \binom{i+j}{i}\delta_{i+j}x$; and
  (iv.) $\delta_0 x = x$.

highest common factor   *See* greatest common divisor.

highest weight   (1) Suppose that g is a complex semisimple Lie algebra, h is a Cartan subalgebra of g, and O is a lexicographic ordering on the real linear subspace of the complex-valued forms on h, spanned by the root system of g relative to h. If ρ is a representation of g, then the maximal element of the set of weights of ρ with respect to O is called the *highest weight* of ρ.

(2) If V is a vector space over the field $\mathbf{C}$ of complex numbers, g is a semisimple Lie algebra over $\mathbf{C}$, π is an irreducible finite-dimensional representation of g in V, and h is a Cartan subalgebra of g, then g has a Cartan decomposition $g = h \oplus (\oplus_\alpha g_\alpha)$, with each α being an eigenvalue of the action of h on π. Such an eigenvalue is called a *weight* of π. A nonzero vector $v \in V$ is called a *highest weight vector* of π if v is an eigenvector for the action of h on π and if $g_\alpha(v) = 0$ for each positive root α of g. The *highest weight* of π is the weight of the highest weight vector of π.

Higman-Sims group   A finite simple group of order 44,352,000.

Hilbert-Hasse norm residue symbol   For a fixed prime p, let ζ_n be a primitive p^nth root of 1 in some algebraic closure of $\mathbf{Q}_p$. Let $\mathbf{L}_n = \mathbf{Q}_p(\zeta_n)$, and let $(\cdot, \cdot)_n$ be the p^nth Hilbert norm residue symbol of $\mathbf{L}_n^*$.

Hilbert modular group   If $M \in \mathrm{PSL}(2, O_d)$ is a matrix and $\overline{M}$ stands for the matrix obtained by replacing each entry of M by its Galois conjugate, then the map $M \to (M, \overline{M})$ sends $\mathrm{PSL}(2, O_d)$ to an irreducible lattice in G (where $G = \mathrm{PSL}(2, \mathcal{R}) \times \mathrm{PSL}(2, \mathcal{R})$). This last is called a Hilbert modular group. It is known that any irreducible lattice in G is commensurable to one of the Hilbert modular groups.

Hilbert modular group   If $M \in \mathrm{PSL}(2, O_d)$ is a matrix and $\overline{M}$ stands for the matrix obtained by replacing each entry of M by its Galois conjugate, then the map $M \to (M, \overline{M})$ sends $\mathrm{PSL}(2, O_d)$ to an irreducible lattice in G (where $G = \mathrm{PSL}(2, \mathbf{R}) \times \mathrm{PSL}(2, \mathbf{R})$). This last is called a Hilbert modular group. It is known that any irreducible lattice in G is commensurable to one of the Hilbert modular groups.

Hilbert modular surface   Denote by G a Hilbert modular group, and denote by H^2 the product of the upper half-space with itself. After adding a finite number of points to H^2/G and obtaining the minimal resolution of the space, one obtains a nonsingular surface over the complex numbers. This surface is called the *Hilbert modular surface.*

Hilbert norm-residue symbol   Suppose that F is an algebraic number field that contains a primitive nth root of unity, that a and b are nonzero elements of F, and that P is a prime divisor of F. Put

$$\sigma = \left(\frac{a, F\left(\sqrt[n]{b}\right)/F}{P} \right),$$

where the symbol on the right is the norm-residue symbol. The symbol $\left(\frac{a,b}{P}\right)_n$ defined by

$$\left(\frac{a, b}{P}\right)_n = \left(\sqrt[n]{b}\right)^\sigma$$

is called the *Hilbert norm-residue symbol. See also* norm-residue symbol.

Hilbert polynomial   Let R be an Artinian ring and let $R[x_1, \ldots, x_n]$ be the (graded) polynomial ring (graded by degree). Let $M = M_0 \oplus M_1 \oplus \ldots$ be a finitely generated graded $R[x_1, \ldots, x_n]$-module. Let $L_M(n) = L(M_n)$ denote the length of the R-module. Then there is a polynomial $f(x)$ with rational entries such that $L_M(n) = f(n)$, for sufficiently large n. This is the *Hilbert polynomial* of the $R[x_1, \ldots, x_n]$-module M. *See* length of module.

Hilbert's Basis Theorem   If R is a Noetherian ring, then so is the polynomial ring $R[x]$. *See* Noetherian ring.

Hilbert scheme   Let k be an algebraically closed field. The *Hilbert scheme* parametrizes all closed subschemes of P_k^n . Here, if R is a ring, then P_k^n is the projective n space over the ring R. It satisfies the following condition: to give a closed subscheme S in P_T^n which is flat over T (for any scheme T) is the same as giving a morphism $f : T \to H$. Here the map f acts on $t \in T$ by $f(t) = $ the point of H corresponding to the fiber S_t in $P_{k(T)}^n$.

Hilbert's Irreducibility Theorem   Suppose that P is an irreducible polynomial in the n variables $x_1, \ldots, x_n$ over an algebraic number field F, and suppose that $0 < m < n$. Then there is an irreducible polynomial in the m variables x_1, $\ldots, x_m$ that is obtainable from P by assigning appropriate values in F to the $n - m$ variables $x_{m+1}, \ldots, x_n$.

Hilbert space   A linear space with a norm that is induced by a complex Hermitian inner product, and which is complete.

Hilbert-Speiser Theorem   Suppose that E/F is a finite Galois extension with Galois group G, denote by E^* and F^* the multiplicative groups of the field E and F, respectively, and denote by N the norm $N_{E/F}$. Then $\widehat{H}^0(G, F^*)$ is isomorphic to $E^*/N(F^*)$, and $H^1(G, F^*) = 0$.

Hilbert's Syzygy Theorem   (1) Let R be the graded polynomial ring in n indeterminates of degree 1 over a field, and let M be a finitely generated graded R-module. If $M_0, \ldots, M_n$ are finitely generated graded R-modules,

$$0 \to M_n \to M_{n-1} \to \cdots \to M_0 \to M \to 0$$

is an exact sequence, and $M_0, \ldots, M_{n-1}$ are free, then M_n is free.

(2) Let R be a regular local ring of dimension d, and let G be a finitely generated R-module. Then there are finitely generated free

R-modules $F_0, F_1, \ldots, F_d$ and there are R-homomorphisms $f_0 : F_0 \to G$ and $f_i : F_i \to F_{i+1}$ for $i = 1, 2, \ldots, d$ such that the sequence

$$0 \to F_d \xrightarrow{f_d} \cdots \xrightarrow{f_1} F_0 \xrightarrow{f_0} G \to 0$$

is exact.

Hilbert's Zero-point Theorem   For any field F, any finitely generated F-algebra A and any ideal I of A, the radical of I is the intersection of all the maximal ideals of A which contain I.

Hirzebruch surface   The $\mathbf{CP}^1$-bundle over $\mathbf{CP}^1$ with cross-section C where $C^2 = -n$.

Hochschild's cohomology group   Suppose that A is an algebra over a commutative ring. If M is a two-sided A-module, then consider the complex obtained from the module of all n-cochains and the coboundary operator. The cohomology of this complex is called the nth *Hochschild's cohomology group* of A relative to M. *See also* cohomology group.

Hodge's conjecture   Suppose that V is a projective nonsingular variety over a finite algebraic number field F, and denote by A the group of algebraic cycles of codimension r on $V \otimes_F \mathbf{C}$ modulo homological equivalence. Then the space of rational cohomology classes of type (r, r) on $V \otimes_F \mathbf{C}$ is spanned by A.

Hodge spectral sequence   If V is a nonsingular connected algebraic variety over the field of complex numbers, and Ω^q denotes the sheaf of germs of regular differential forms of degree q, then the spectral sequence

$$E_1^{q,p} = H^p(V, \Omega^q) \Rightarrow H^{p+q}(V, \mathbf{C})$$

is called a *Hodge spectral sequence*.

Hodge structure   Suppose that L is a lattice, and V is a finite-dimensional real vector space that contains L. Denote the complexification of V by C. A *Hodge structure* of weight m on C (or on V) is a decomposition of C as a direct sum

$$C = \bigotimes_{p,q:p+q=m} C^{p,q}$$

of complex vector spaces $C^{p,q}$ such that the complex conjugate of $C^{p,q}$ is isomorphic to $C^{q,p}$.

Hodge theory   A body of results concerning cohomology groups of manifolds. Three of these results are listed here.

(1) Suppose that X is a closed oriented Riemannian manifold. Then each element of a cohomology group of X with complex coefficients has a unique harmonic representative.

(2) Suppose that V is a compact complex manifold of dimension n that is the image of a holomorphic mapping from a compact Kähler manifold of dimension n. If $q = \max(n - p + 1, 0)$, then $H^n(V, \mathbf{C})$ carries the Hodge structure induced by the type (p, q)-decomposition of the space of differential forms on the manifold.

(3) If X is a smooth noncomplete irreducible variety, then $H^n(X, \mathbf{C})$ carries a mixed Hodge structure that is independent of the choice of complete algebraic variety $\overline{X}$ such that $\overline{X} - X$ is a subvariety.

See Hodge structure.

Hölder inequality   If $p \neq 0, 1, 1/p + 1/q = 1$ and $a_i, b_i > 0$, then we have

$$\sum_i a_i b_i \leq \left(\sum_i a_i^p\right)^{1/p} \left(\sum_i b_i^q\right)^{1/q},$$

if $p > 1$ and

$$\sum_i a_i b_i \geq \left(\sum_i a_i^p\right)^{1/p} \left(\sum_i b_i^q\right)^{1/q},$$

if $0 < p < 1$. *See also* Hölder integral inequality.

Hölder integral inequality   Suppose that f and g are measurable positive functions defined on a measurable set E, and suppose that p and q are positive real numbers such that $1/p + 1/q = 1$. If $p > 1$, then

$$\int_E fg \leq \left(\int_E f^p\right)^{1/p} \left(\int_E g^q\right)^{1/q}.$$

If $0 < p < 1$, then

$$\int_E fg \geq \left(\int_E f^p\right)^{1/p} \left(\int_E g^q\right)^{1/q}.$$

In each case, equality holds if and only if there exist two real numbers a and b such that $ab \neq 0$ and $af^p = bg^q$ almost everywhere.

Hölder's Theorem   If $n \neq 2, 6$, then the symmetric group S_n is complete. *See* symmetric group, complete group.

holomorphic function   *See* analytic function.

holomorphic functional calculus   If X is a complex Banach space (a complete normed vector space), $T : X \to X$ is a bounded linear operator and $f(z)$ is a holomorphic function on (a neighborhood of) the spectrum of T (the set of eigenvalues of T, if X is finite dimensional), then one can define $f(T)$ via a Cauchy type integral. The map $f \to f(T)$ is a homomorphism from the algebra of holomorphic functions in a neighborhood of the spectrum to T into the Banach algebra of bounded linear operators on T.

holosymmetric class   Suppose that T is an n-dimensional lattice in n-dimensional Euclidean space, and that K is a finite subgroup of the orthogonal group. Denote by A, B, and G the sets of arithmetic crystal classes, of Bravais types, and of geometric crystal classes of (T, K), respectively. There are surjective mappings $s_1 : A \to B$ and $s_2 : A \to G$, and there is an injective mapping $i : B \to A$ such that $s_1 \circ i$ is the identity mapping on B. A *holosymmetric class* is a class that belongs to the image of the mapping $s_2 \circ i$.

See arithmetic crystal class, Bravais type, geometric crystal class.

homogeneous   Of the same kind. For example, a *homogeneous polynomial* is a polynomial each of whose monomials has the same degree. *See* homogeneous polynomial. *See also* homogeneous bounded domain, homogeneous coordinate ring, homogeneous equation, homogeneous ideal, homogeneous n-chain.

homogeneous bounded domain   A bounded domain D in $\mathbf{C}^n$ such that the group of all holomorphic transformations of D acts transitively on D.

homogeneous coordinate ring Let K be a field, V a projective variety over K and $I(V)$ the ideal of $K[x_0, \ldots, x_n]$ generated by the homogeneous polynomials in $K[x_0, \ldots, x_n]$ that vanish on V. Then $K[x_0, \ldots, x_n]/I(V)$ is called the *homogeneous coordinate ring* of V.

homogeneous difference equation A linear difference equation of the form

$$x_n + c_1 x_{n-1} + \cdots + c_n x_0 = 0,$$

where $c_1, \ldots, c_n$ are given real or complex numbers and $\{x_j\}$ is an unknown infinite sequence. The notion is similar to that of homogeneous linear equation except that difference equations are often written in terms of x_n and powers of the difference operator $\Delta x_n = x_n - x_{n-1}$, so that they can be considered as analogs of differential equations.

homogeneous domain A domain with a transitive group of automorphisms. In more detail, a *domain* is a connected open subset of complex N space $\mathbf{C}^N$. A domain is *homogeneous* if it has a transitive group of analytic (holomorphic) automorphisms. This means that any pair of points z and w can be interchanged, i.e., $\phi(z) = w$, by an invertible analytic map ϕ carrying the domain onto itself. For example, the unit ball in complex N space, $\{z = (z_1, \ldots, z_N) : |z_1|^2 + \ldots + |z_N|^2 < 1\}$, is homogeneous. *See* automorphism, invertible function, invertible map. *See also* bounded homogeneous domain, Siegel domain.

homogeneous element *See* graded ring.

homogeneous equation An equation, in an unknown function f, having the form $L[f] = 0$, where L acts linearly ($L[c_1 f_1 + c_2 f_2] = c_L l[f_1] + c_2 L[f_2]$). Thus, L could be a linear differential operator (ordinary or partial). *See also* homogeneous difference equation.

homogeneous function Let F be a field. A function $f = f(x_1, \ldots, x_n) : F^n \to F$, such that, for all $c \in F$ and all $(x_1, \ldots, x_n) \in F^n$, we have $f(cx_1, \ldots, cx_n) = c^k f(x_1, \ldots, x_n)$ is called a *homogeneous function* of degree k. *See also* homogeneous polynomial.

homogeneous ideal An ideal I in a graded ring R such that I is generated by homogeneous elements. *See* graded ring.

homogeneous linear equation A linear equation in the variables $x_1, \ldots, x_n$, having the form

$$c_1 x_1 + \cdots + c_n x_n = 0,$$

where the c_i are constants.

homogeneous n-chain Let G be a group and let G act on G^{n+1} diagonally: $g(g_0, \ldots, g_n) = (gg_0, \ldots, gg_n)$. Define a boundary operator by $d(g_0, \ldots, g_n) = (g_1, \ldots, g_n) - (g_0, g_2, \ldots, g_n) + \cdots + (-1)^n(g_0, \ldots, g_{n-1})$. Then the free Abelian group with basis the elements of G^{n+1}, with this action of G on G^{n+1} and with boundary operator d is called the *group of homogeneous n-chains* of G.

homogeneous polynomial A polynomial $P(x_1, \ldots, x_n)$, which is also a homogeneous function (of some degree k). *See* homogeneous function. Thus,

$$P(x_1, \ldots, x_n)$$

$$= \sum_{m_1 + \cdots + m_n = k} c_{m_1 m_2 \cdots m_n} x_1^{m_1} \cdots x_n^{m_n}.$$

homogeneous ring *See* graded ring.

homogeneous space A smooth manifold M with a Lie group G, such that there is a transitive, smooth action of G on M.

homological algebra Originally called multilinear algebra, this branch of algebra deals with the category of left (or right) R-modules over some ring R (generally assumed to be associative and to possess an identity element). A left R-module is an Abelian group A, together with a ring homomorphism of the ring R into the ring of endomorphisms (homomorphisms) of A into itself. A right R-module is defined similarly except the ring homomorphism is replaced by a ring anti-homomorphism (i.e., the order of products is reversed). When R is commutative, a structure as a left R-module induces a structure as a right R-module and vice versa. The morphisms of this category are the

R-homomorphisms $f : A \to B$ for a pair of R-modules A and B. By an R-homomorphism we mean a group homomorphism of A into B with the property that $f(\lambda a) = \lambda f(a)$, for all $\lambda \in R$ and $a \in A$, when A and B are left R-modules and, when A and B are right R-modules, we assume that $f(a\lambda) = f(a)\lambda$. These categories have two fundamental functors, both of which take their values in the category of all Abelian groups (which is, in the terminology of this definition, the category of left Z-modules). We describe them separately.

Given two left R-modules A, B, $\mathrm{Hom}_R(A, B)$ denotes the Abelian group of all R-homomorphisms of A into B. When R is commutative, $\mathrm{Hom}_R(A, B)$ is also both a left and right R-module. This functor is covariant in the second variable and contravariant in the first. By *covariant* in the second variable we mean that if $B \to C$ is a morphism then there is an induced morphism

$$\mathrm{Hom}_R(A, B) \to \mathrm{Hom}_R(A, C) .$$

By *contravariant* in the first variable we mean that there is an induced morphism

$$\mathrm{Hom}_R(B, A) \leftarrow \mathrm{Hom}_R(C, A) .$$

This functor is left exact in both variables. By this we mean that, given any exact sequence $B \to C \to 0$, the induced group homomorphisms

$$0 \to \mathrm{Hom}_R(C, A) \to \mathrm{Hom}_R(B, A)$$

forms an exact sequence and that any exact sequence $0 \to B \to C$ induces an exact sequence $0 \to \mathrm{Hom}_R(A, B) \to \mathrm{Hom}_R(A, C)$.

When A is a right R-module and B is a left R-module then there is an Abelian group $A \otimes_R B$ called the tensor product of A and B that consists of formal sums of pairs (a, b) where a is a member of A and b is a member of B and is subject to the relations

$$((a_1 + a_2), b) \equiv (a_1, b) + (a_2, b)$$

$$(a, b_1 + b_2) \equiv (a, b_1) + (a, b_2)$$

$$(a\lambda, b) \equiv (a, \lambda b) \text{ for all } \lambda \in R .$$

$A \otimes_R B$ is covariant in both variables and right exact in both. When R is commutative, $A \otimes_R$

B can be regarded as both a left and right R-module.

Further development of these ideas leads to certain derived functors called $\mathrm{Ext}_R^n(A, B)$ in the case of the Hom functor and $\mathrm{Tor}_R^n(A, B)$ in the case of the tensor product functor. There is one such functor for each $n \geq 0$ and, in case $n = 0$, these coincide with Hom and the tensor product. This development is too lengthy and technical to describe further here. *See* Ext group, Tor.

homological dimension   Let C be the category of left R-modules where R is an associative ring with identity. We construct a projective resolution of an R-module A by first taking a projective module P_0, together with an epimorphism $P_0 \to A \to 0$ and kernel K_0. There always exists such a projective module since a free module on any set of generators for A will work. If K_0 is not projective, repeat this process with an epimorphism $P_1 \to K_0 \to 0$ and kernel K_1. Continue until a kernel K_n which is projective is obtained. The index n is called the projective dimension or homological dimension of the module A. Two facts must be proved before this notion makes sense: (i.) A has homological dimension 0 if and only if A is, itself, projective and (ii.) the index n is independent of the particular sequence of projective modules used. *See* projective module.

homological functor   A sequence of additive covariant functors of Abelian categories $H = \{H_i : \mathcal{A} \to \mathcal{A}'\}$ defined for $-\infty < i < +\infty$ with the following properties. For each short exact sequence $0 \to A' \to A \to A'' \to 0$ in $\mathcal{A}$, and morphism of exact sequences f, there exist connecting morphisms $\delta_* : H_i(A'') \to H_{i-1}(A')$ such that $\delta_* \circ H_i(f'') = H_{i-1}(f') \circ \delta_*$. The sequence

$$\cdots \to H_{i+1}(A'') \xrightarrow{\delta_*} H_i(A') \to H_i(A)$$

$$\to H_i(A'') \xrightarrow{\delta_*} H_{i-1}(A') \to \cdots$$

is a complex which is always exact.

homological mapping   A homomorphism of homology modules induced by a chain mapping. Let A be a ring with unit, and let (X, δ)

and (Y, δ') be chain complexes over A with homology modules $H(X)$ and $H(Y)$. The A-homomorphism $f_* : H(X) \to H(Y)$, of degree zero, induced by a chain mapping $f : X \to Y$, is called the *homological mapping* induced by f.

homology Let $\{A_n : n \in \mathbf{Z}\}$ be a set of R-modules over some ring R. Assume that there is a family of R-homomorphisms $f_n : A_n \to A_{n-1}$ with the property that the composite homomorphisms $f_{n-1} f_n = 0$ for all n. This data defines a chain complex. The latter condition implies that $\mathrm{Im}(f_n) \subseteq \mathrm{Ker}(f_{n-1})$, where $\mathrm{Im}(f)$ is the image of f and $\mathrm{Ker}(f)$ is the kernel of f. Then the factor module $\ker(f_{n-1})/\mathrm{Im}(f_n) = H_{n-1}$ is the $(n-1)$st homology module of the complex. The sequence of these modules is called the *homology* of the complex.

homology class The residue class of a cycle modulo a boundary in the group of chains. Let $\mathcal{A}$ be an Abelian category and let $C = (C_n, d_n)$ be a chain complex in A. Let $Z_n = \mathrm{Ker}\, d_n$ and $B_n = \mathrm{Im}\, d_{n+1}$. A residue class of Z_n / B_n is called a *homology class*.

homology group When a chain complex of Abelian groups is given, then its homology consists of a sequence of *homology groups*. *See* chain complex.

homology module When a chain complex of R-modules over some ring R is given, then its homology consists of a sequence of homology modules. *See* chain complex.

homomorphism Algebraic structures such as groups, rings, fields, and modules are all defined with one or more binary operations. Suppose that $\circ : S \times S \to S$ is one such binary operation which is, simply, a function from the Cartesian product $S \times S$ to S. We customarily write $a \circ b$ instead of the more standard functional notation $\circ(a, b)$ to describe the action of this function. Then a homomorphism is a function $f : S \to T$ from one such object to another with the property that $f(a \circ b) = f(a) \circ f(b)$ for all $a, b \in S$. Additional qualifiers such as group homomorphism, ring homomorphism, or R-homomorphism are used to describe homomorphisms defined on groups, rings, or R-modules.

Homomorphism Theorem of Groups Let G be a group and let H and N be subgroups of G such that N is normal ($aNa^{-1} \subseteq N$ for all $a \in G$.) Then (i.) $HN = NH = \{x \in G : x = nh, \text{ for some } n \in N\}$ is a subgroup of G, (ii.) N is a normal subgroup of NH, (iii.) $N \cap H$ is a normal subgroup of H, and (iv.) the factor group NH/N is isomorphic to the factor group $H/(N \cap H)$.

homotopy Let X and Y be topological spaces and let f and g be continuous functions from X to Y. We will say that f is *homotopic* to g (often written $f \sim g$) to mean that there is a continuous function $F : X \times I \to Y$ defined, on the Cartesian product of X with the unit interval $I = [0, 1]$, such that $F(x, 0) = f(x)$ and $F(x, 1) = g(x)$ for all $x \in X$. The relation of being homotopic is an equivalence relation. The function F that defines the relation is often spoken of as the *homotopy*.

When X is the unit interval (i.e., when we are talking about paths in the space Y), it is customary to make the additional assumption that $F(0, s) \equiv f(0) = g(0)$ and $F(1, s) \equiv f(1) = g(1)$.

homotopy associative Let G be a topological space carrying the structure of an H-space, defined by the continuous map $\mu : G \times G \to G$. *See* H-space. Given points $x, y, z \in G$, we can define the maps $f(x, y, z) = \mu(x, \mu(y, z))$ and $g(x, y, z) = \mu(\mu(x, y), z)$ from $G \times G \times G \to G$. We say that μ is *homotopy associative* if f and g are homotopic as maps.

homotopy commutative Let G be a topological space carrying the structure of an H-space, defined by the continuous map $\mu : G \times G \to G$. (*See* H-space.) Given points $x, y \in G$, we can define the maps $f(x, y) = \mu(x, y)$ and $g(x, y) = \mu(y, x)$ from $G \times G$ to G. We say that μ is *homotopy commutative* if f and g are homotopic as maps.

homotopy identity Let G be a topological space carrying the structure of an H-space, defined by the continuous map $\mu : G \times G \to G$.

(*See* H-space.) A constant map $e(x) = e$ is called a *homotopy identity* if the maps defined by $\mu(x, e)$ and $\mu(e, x)$ are both homotopic to the identity map i on G defined by $i(x) = x$ for all $x \in G$.

homotopy inverse   Let G be a topological space carrying the structure of an H-space, defined by the continuous map $\mu : G \times G \to G$ and having a homotopy identity e. *See* H-space, homotopy identity. A continuous map $\nu : G \to G$ defines a *homotopy inverse* if the maps defined by $\mu(x, \nu(x))$ and $\mu(\nu(x), x)$ are both homotopic to the homotopy identity e.

Hopf algebra   Two types of Hopf algebras have been defined. The first one was introduced by Hopf and is used in the study of homology and cohomology of Lie groups. The second type was introduced by Sweedler and has applications in the study of algebraic groups.

(1) A (graded) Hopf algebra (A, ϕ, ψ) over a field k is a graded algebra with multiplication ϕ which is also a graded co-algebra with comultiplication ψ such that $\phi : (A, \psi) \otimes (A, \psi) \to (A, \psi)$ is a homomorphism of graded co-algebras. This type of Hopf algebra is frequently assumed to be connected, commutative, cocommutative, and of finite type.

(2) A Hopf algebra with an antipode S is a bialgebra with antipode S. *See also* antipode.

Hopf algebra homomorphism   A bialgebra homomorphism f, between a Hopf algebra H with antipode S and a Hopf algebra H' with antipode S' such that $S'f = fS$. *See also* Hopf algebra.

Hopf comultiplication   A degree preserving linear map defined as follows. Let X be a topological space with a base point x_0 and let μ be a continuous base point preserving mapping from $X \times X$ to X. Let $\iota_1(x) = (x, x_0)$ and $\iota_2(x) = (x_0, x)$. If $\mu \circ \iota_i$ is homotopic to 1_X, then μ induces a degree preserving linear map, μ^*, from the cohomology group $H^*(X)$ to $H^*(X) \otimes H^*(X)$ via a Künneth isomorphism. In this situation μ^* is called a Hopf comultiplication and (X, μ) is called an H-space. If α is an element of $H^*(X)$, then $\mu^*(\alpha)$ is called the Hopf coproduct of α.

Hopf coproduct   The image of an element under a Hopf comultiplication. *See* Hopf comultiplication.

Horner method of solving algebraic equations An iteration method for finding the real roots of an algebraic equation. Locate a positive root between two successive integers. If a_1 is the greatest integer less than the root, use the substitution $x_1 = x - a_1$ to transform the equation into one that has a root between 0 and 1. Locate this root between successive tenths. Use the substitution $x_2 = x_1 - a_2$ to transform the equation to one that has a root between 0 and one-tenth. Continue this process. The desired root is then approximately $a_1 + a_2 + \cdots + a_n$.

Householder method of matrix transformation   A method of transforming a symmetric matrix A into a tridiagonal matrix B. The method uses a similarity transformation of the form $A \to H^{-1}AH$, where H represents an orthogonal matrix of the form $I - 2uu^*$ with $u^*u = 1$. In the above situation, I represents the identity matrix and u^* is the conjugate transpose of u.

Householder transformation   The matrix transformation $u = Hv$ where H is a symmetric and orthogonal matrix of the form $H = I - 2xx^T$, where $x^Tx = 1$. In the above situation, x^T represents the transpose of x, and I represents the identity matrix.

H-series   Since the nth factor of a Blaschke product

$$\prod \frac{\bar{a}_n}{|a_n|} \frac{a_n - z}{1 - \bar{a}_n z}$$

can be written in the form $1 + C(z, a_n)$, where $C(z, a) = (1 - |a|)/|a| - (1 - |a|^2)/|a|(1 - \bar{a}z)$, the Blaschke product converges absolutely at a point z_0 if and only if $\sum C(z_0, a_n)$ converges absolutely. Thus, in complex analysis, an H-series is a series of the form

$$\sum c_n / (1 - \bar{a}_n z) ,$$

where $0 < |a_n| < 1$ $(n = 1, 2, \cdots)$ and $\sum (1 - |a_n|) < \infty$. The set of points on C at which the H-series converges is called its set of convergence. The subject of representation theory also considers H-series.

H-series

H-space   A topological space G, together with a continuous map $\mu : G \times G \to G$, from the Cartesian product of G with itself to G. In practice, various additional conditions are imposed on such maps that make them imitate the properties normally associated with a multiplication. *See* homotopy.

hull-kernel topology   A topology on the set of primitive ideals (the structure space $\mathcal{I}$) of a Banach algebra R. Under this topology, the closure of a set $\mathcal{U}$ is the set of primitive ideals containing the intersection of the ideals in $\mathcal{U}$.

Hurwitz's relation   A relation between the Riemann matrices of two Abelian varieties T_1 and T_2 that implies the existence of a homomorphism from T_1 to T_2. Let T_1 and T_2 be Abelian varieties with Riemann matrices $\Omega_1 = (\omega_1^{(1)}, \ldots, \omega_{2n}^{(1)})$ and $\Omega_2 = (\omega_1^{(2)}, \ldots, \omega_{2m}^{(2)})$. There is a homomorphism $\lambda : T_1 \to T_2$ if and only if there is a complex matrix W, and a matrix M with integer entries, such that $W\Omega_1 = \Omega_2 M$. In particular, for every homomorphism $\lambda : T_1 \to T_2$ there is a representation matrix $W(\lambda)$ with complex coefficients, and a representation matrix $M(\lambda)$ with respect to the real coordinate systems $(\omega_1^{(1)}, \ldots, \omega_{2n}^{(1)})$ and $(\omega_1^{(2)}, \ldots, \omega_{2m}^{(2)})$, that has coefficients in $\mathbf{Z}$, such that $W(\lambda)\Omega_1 = \Omega_2 M(\lambda)$.

Hurwitz's Theorem   If every member of a normal family (of analytic functions on a connected, open, planar set Ω) is never zero on Ω, then the limit functions are either identically zero or never zero on Ω.

A family of analytic functions on a set Ω is *normal* on Ω if every sequence from the family contains a subsequence that converges uniformly on compact subsets of Ω.

Hurwitz zeta function   The function

$$\zeta(s, a) = \sum_{n=0}^{\infty} \frac{1}{(n+a)^s} , \quad 0 < a \le 1 .$$

A generalization of the Riemann zeta function considered by Hurwitz (1862).

hyperalgebra   A bialgebra, whose underlying co-algebra is co-commutative, pointed, and irreducible.

hyperbolic cosecant function   *See* hyperbolic function.

hyperbolic cosine function   *See* hyperbolic function.

hyperbolic cotangent function   *See* hyperbolic function.

hyperbolic function   The six complex functions:

hyperbolic sine: $\sinh(z) = \frac{\exp z - \exp(-z)}{2}$

hyperbolic cosine: $\cosh(z) = \frac{\exp z + \exp(-z)}{2}$

hyperbolic tangent: $\tanh(z) = \frac{\sinh(z)}{\cosh(z)}$

hyperbolic cotangent: $\coth(z) = \frac{\cosh(z)}{\sinh(z)}$

hyperbolic secant: $\operatorname{sech}(z) = \frac{1}{\cosh(z)}$

hyperbolic cosecant: $\operatorname{csch}(z) = \frac{1}{\sinh(z)}$.

Due to the fact that $\sinh(iz) = i \cdot \sin(z)$ and $\cosh(iz) = \cos(z)$, where $\sin(z)$ and $\cos(z)$ are the ordinary sine and cosine of the complex angle z, the hyperbolic functions are rarely used outside of certain specialized applications.

hyperbolic secant function   *See* hyperbolic function.

hyperbolic sine function   *See* hyperbolic function.

hyperbolic tangent function   *See* hyperbolic function.

hyperbolic transformation   A linear fractional function $f(z) = \frac{az+b}{cz+d}$ where a, b, c, and d are complex constants with $ad - bc \ne 0$ has a pair of (not necessarily distinct) fixed points. (*See* linear fractional function.) This is so because the equation $z = \frac{az+b}{cz+d}$ is linear or quadratic in z. When the two fixed points coincide, the transformation is said to be *parabolic*. When there are distinct fixed points, say α and β, then we can write the transformation in the form $\frac{w-\alpha}{w-\beta} = k\frac{z-\alpha}{z-\beta}$. When k is real (necessarily non-zero) the transformation is said to be *hyperbolic*. When $|k| = 1$, the transformation is said to be *elliptic*.

hyperbolic trigonometry One of the two classical types of non-Euclidean plane geometry, which are distinguished from each other and from Euclidean plane geometry by the form of the parallel axiom that holds. For Euclidean geometry, there is, through any point P not on a line L, exactly one line parallel to L. For elliptic (also Lobachevskian) geometry there is, through any point P not on a line L, no line parallel to L. For hyperbolic geometry there is, through any point P not on a line L, more than one (actually infinitely many) lines parallel to L.

Within the language of Riemannian geometry these three types of spaces are those of, respectively, zero, positive, and negative constant curvature. Models that exist within Euclidean geometry are given by the following constructions. For elliptic geometry, the space is the surface of a sphere and the lines are great circles on the sphere. We must regard antipodal points on the sphere as being identified in order that lines intersect in no more than one point. For hyperbolic geometry, the space is the interior of a disk and the lines are arcs of circles that intersect the boundary of the disk at right angles or are straight lines through the center of the disk. An alternative model is the upper half plane ($y > 0$) and the lines are arcs of circles centered on the real axis or straight lines orthogonal to the x-axis.

An important property of these geometries concerns the sum of the angles of a triangle. In Euclidean geometry, the sum of the angles of any triangle must be exactly equal to π (that is to say, a straight angle). In elliptic geometry the sum of the angles is always greater than π while in hyperbolic geometry the sum of the angles is always less than π.

If the non-Euclidean plane is embedded in a Euclidean space of sufficiently high dimension, then it becomes possible to measure areas of regions on the plane using the Euclidean measure of area relative to the enveloping space. It was shown by Gauss that the area of a triangle, so measured, equals the difference between π and the sum of the angles of the triangle.

hyperelliptic curve An algebraic curve, over a ground field k, is defined by an irreducible polynomial $F(X, Y)$ in the polynomial ring $k[X, Y]$. Since $F(X, Y)$ generates a maximal ideal in $k[X, Y]$, the factor ring $k[X, Y]/(F(X, Y)) = K$ is also a field and contains a canonical copy of k. If we denote the canonical images of X and Y in K by x and y, respectively, then $K = k(x, y)$ is the field extension of k generated by x and y. The field $K = k(x, y)$ is called the field of algebraic functions on the algebraic curve defined by $F(X, Y)$. A *hyperelliptic curve* is an algebraic curve of the special form $Y^2 = P(X)$, where $P(X)$ is a polynomial that has no repeated roots.

It is also usually assumed that k is algebraically closed in K. That is to say, every element of K that is not in k is transcendental over k. When this is so, k is said to be the *exact constant field* for the curve.

hyperelliptic integral Let $F(X, Y) = Y^2 - g(X)$ define a hyperelliptic curve over the field $\mathbf{C}$ of complex numbers, where the polynomial $g(X)$ is square free and of degree n. Let $K = \mathbf{C}(z, w)$ be the function field for this curve. Here, z and w are the canonical images of X and Y in the residue class field $K = \mathbf{C}[X, Y]/(F(X, Y))$. The branch points of the function field K over the projective line $\mathbf{C}(z)$ are the roots of $g(X)$ and, also, the point at infinity, when n is odd. When n is even, the point at infinity is not a branch point. We can, however, move one of the roots of $g(X)$ to ∞ by a linear change of variable in z. Thus there is no loss of generality in assuming that n is odd. The Riemann surface $\mathcal{X}$ for this curve has genus $g = 1 - 2 + \frac{1}{2}(n + 1)$, according to the standard formula for the genus in case $[K : \mathbf{C}(z)] = 2$ and there are $n + 1$ branch points of index 2. According to the Riemann–Roch Theorem, there are g linearly independent holomorphic differential forms defined on the Riemann surface. A holomorphic differential form is an expression of the form $f(z)dz$, such that, if π denotes a local parameter at a point P on the Riemann surface, then $f(z)\frac{dz}{d\pi}$, which is a member of K, has no pole at P. These are commonly referred to as differentials of the first kind. Each of them gives rise to an Abelian integral that is defined everywhere on the universal covering surface for the Riemann surface. These are *hyperelliptic integrals* of the first kind. They are more commonly referred to as Abelian integrals of the first kind. For hyperelliptic curves, these integrals can be given quite explicitly as

$$\int \frac{z^r}{\sqrt{g(z)}} dz$$

for $\leq r < g - 1$.

There are also differentials of the second and third kinds. Differentials of the second kind have poles but the residue vanishes at each of these poles. Differentials of the third kind possess at least one pole with a nonvanishing residue. Since the sum of the residues is always zero, there must be at least two poles for a differential of the third kind. To each of these is associated an Abelian integral, which, in the case of differentials of the second kind, is also defined on the entire universal covering surface for $\mathcal{X}$ although it can have poles. For differentials of the third kind, the singularities at points where there are nonvanishing residues are logarithmic singularities and are, thus, essential. The Abelian integral associated with a differential of the third kind is defined everywhere on the universal covering surface for $\mathcal{X}$, except on curves that pass through points that lie above these singularities.

hyperelliptic surface   The Riemann surface for a hyperelliptic curve defined over the complex field. *See* hyperelliptic curve.

hyperfinite factor   A factor of type II_1 which is generated, in the weak topology, by an increasing sequence $\{\mathcal{A}_n\}_{n=1}^{\infty}$ of finite dimensional $*$-subalgebras. In addition, each element in the sequence $\{\mathcal{A}_n\}_{n=1}^{\infty}$ may be assumed to be a subfactor of type I_{2^n}. Sometimes the term hyperfinite refers both to the factors defined above and to factors of type I_n, where $n < \infty$.

hypergroup   A set S with an associative multiplication that assigns to any a and b in S a nonempty subset ab of S, such that, for each a and

b in S, there exist elements x and y in S such that $b \in ax$ and $b \in ya$.

hypergroupoid   A set S, with a multiplication that associates every two elements a and b in S with a nonempty subset ab of S.

hypersurface   A term with slightly different meanings within algebraic geometry and analysis:

(**1**) Algebraic Geometry: If X denotes affine n-space over a field k, then any subvariety of dimension $n - 1$ is referred to as a hypersurface in X. The hypersurface is defined by a single irreducible polynomial in n variables. This latter fact is an expression of the algebraic fact that a polynomial ring over a field is a unique factorization domain.

(**2**) Analysis: If X denotes a manifold of dimension n, then any $n - 1$ dimensional submanifold can be referred to as a hypersurface. In many cases, the hypersurface can be represented by a single relation $f(x_1, \ldots, x_n) = 0$. However, due to the fact that the coordinate system changes from point to point, the general definition is necessarily more complicated.

hypo-Dirichlet algebra   A uniform algebra A with some properties that occur in concrete examples. Let $\Re f$ be the real part of f and let $\Re(A) = \{\Re f : f \in A\}$. Let $C_{\mathbf{R}}(X)$ be the space of all continuous real-valued functions on X and let $\log|A^{-1}| = \{\log|f| : f, f^{-1} \in A\}$. A uniform algebra A on a compact Hausdorff space X is *hypodirichlet* on X if the uniform closure of $\Re(A)$ has finite codimension in $C_{\mathbf{R}}(X)$, and the linear span of $\log|A^{-1}|$ is dense in $C_{\mathbf{R}}(X)$. A uniform algebra A is *hypodirichlet* if it is hypodirichlet on its Shilov boundary. The fundamental work on hypo-Dirichlet algebras was done by Ahern and Sarason in 1966.

I

I-adic topology   A topology that endows a ring (module) with the structure of a topological ring (module). Let I be an ideal in a ring R and let M be an R-module. The I-adic topology on M is formed by taking the cosets $x + I^n M$, $x \in M$, n a positive integer, as a base of open sets. The I-adic topology on R is formed by letting $M = R$.

icosahedral group   The group of rotations of a regular icosahedron. It is a simple group of order 60 and is isomorphic to the alternating group of degree 5.

ideal   In any ring (usually associative with identity), any subgroup J of the additive group of R is called a left ideal if $RJ \subseteq J$ and is called a right ideal if $JR \subseteq J$ and, if both of these conditions hold, is called, simply, an ideal.

ideal class   Let R be an integral domain (commutative ring with no divisors of zero). Two ideals J and K are said to be linearly equivalent if there exist nonzero elements a and b in R such that $aJ = bK$. This establishes an equivalence relation on the collection of all nonzero ideals of R which we can write as $J \approx K$. Any equivalence class under this relation is said to be an *ideal class.*

These classes respect multiplication in the sense that if $J_1 \approx J_2$ and $K_1 \approx K_2$, then $J_1 J_2 \approx K_1 K_2$. In most cases, this notion is only applied to the collection of invertible ideals of R. That is to say, those ideals J for which there is another ideal K such that JK is linearly equivalent to the unit ideal R. For this case, the collection of classes of invertible ideals forms an Abelian group referred to as the *ideal class group* of the ring.

An alternative and more modern way of defining this construction is contained in the notion of fractional ideals. A fractional ideal of an integral domain R is a nonzero R-submodule F of the quotient field of R with the property that

$aF \subseteq R$, for some nonzero $a \in R$. This includes all ordinary nonzero ideals. A fractional ideal is principal if it is of the form aR where a is some nonzero member of K. A fractional ideal F is invertible if there is another fractional ideal G such that $FG = R$. The invertible fractional ideals form an Abelian group under ordinary multiplication and this group contains the group of principal fractional ideals. The factor group is isomorphic to the ideal class group defined earlier.

ideal class group   The group of ideal classes of invertible ideals of an integral domain R or, equivalently, the group of invertible fractional ideals of R, modulo the subgroup of principal fractional ideals. *See* ideal class.

ideal class in the narrow sense   Let I_k be the Abelian group consisting of the fractional ideals of an algebraic number field k. Let P_k^+ be the subgroup of I_k consisting of all the principal ideals generated by totally positive elements of k. An *ideal class of k in the narrow sense* is a coset of I_k modulo P_k^+.

ideal group   Let K be an algebraic number field and let R be its ring of algebraic integers.

(1) The group of fractional R-ideals of K forms an Abelian group called the *ideal group* of K.

(2) Let m be an integral ideal and let $K_m = \{a/b : a, b \in R, aR \text{ and } bR \text{ relatively prime to } m\}$. Denote the group of all ideals of K that are relatively prime to m by $I_K(m)$. Let m^* be a formal product of m and a finite number of real infinite prime divisors of K, then m^* is an integral divisor of K. For $\alpha \in K_m$ let α_1 and α_2 be elements of $K_m \cap R$ such that $\alpha = \alpha_1/\alpha_2$. Let $S(m^*)$ be the group of all principal ideals generated by elements α of K_m such that $\alpha_1 \equiv \alpha_2 \pmod{m}$ and $\alpha \equiv 1 \pmod{p}$ for all the real infinite prime divisors p included in the formal product above. Any subgroup of $I_K(m)$ which contains $S(m^*)$ is called an *ideal group* modulo m^*.

(3) Sometimes the definition given above is called a congruence subgroup. In this case an *ideal group* is an equivalence class of a congruence subgroup under the following equivalence relation. Two congruence subgroups H_1 and

1-58488-052-X/01/$0.00+$.50
© 2001 by CRC Press LLC

H_2, modulo m_1^* and m_2^*, respectively, are said to be equivalent if there is a modulus n^* such that $H_1 \cap I_K(n) = H_2 \cap I_K(n)$.

idele Let K be a global field. By this we mean either an algebraic number field (a finite extension of the rational field $\mathbf{Q}$) or a field of algebraic functions in one variable over a ground field k (a finite separable extension K of the field $k(x)$ of rational functions over the ground field k, such that k is itself algebraically closed in K). These are also referred to as product formula fields since each of these types possesses a class of absolute values, which are real valued functions, defined on K with the properties that $|a| \geq 0$ for all $a \in K$, $|a| = 0$ if and only if $a = 0$ and $|a + b| \leq |a| + |b|$, for all $a, b \in K$. *See also* valuation.

Moreover this class of absolute values satisfies, for each nonzero $a \in K$, the relation $\prod_{\text{all } \wp} |a|_\wp = 1$. Consider now the direct product of copies of the multiplicative group K^* of K, indexed by absolute values. That is to say, functions from the set of absolute values to the group K^*. We will denote such a function by a, and its value at $\wp$ by $a_\wp$. An *idele* is such a function with the additional property that $|a_\wp| = 1$, for almost all $\wp$. Since each element of K^* gives rise to such a function ($a_\wp = a$ for all $\wp$) we may regard K^* as a subgroup of the group of all ideles. These are called principal ideles and the factor group is called the idele class group of the field.

idele class Members of the idele class group of a global field. *See* idele.

idele class group The group of all idele classes of a global field. *See* idele.

idele group The group of all ideles of a global field. *See* idele.

idempotent element An element a of a ring, with the property that $a^2 = a$.

idempotent subset A subset S of a ring having the property that $S^2 = S$.

Idempotent Theorem If E is an open-closed subset of the maximal ideal space of a unital

commutative Banach algebra A, then there is a unique element f of A such that $f^2 = f$ and the Gel'fand transform of f is 1 on E and 0 everywhere else. This theorem is sometimes referred to as Shilov's Idempotent Theorem.

identity (**1**) An equation. For example, a true formula such as

$$\sum_{j=1}^{N} j = \frac{N(N+1)}{2}$$

is an identity.

(**2**) That element of a group, usually denoted by the symbol 1, with the property that $1a = a1 = a$ for all a in the group (or denoted 0, with $0 + a = a + 0 = a$, in the case of an additive group).

identity character The character on a group G with the property that $\chi(g) \equiv 1$ for all $g \in G$. *See* character.

identity element A member e of a set S with a binary operation $x \circ y$, such that $x \circ e = e \circ x = x$. If the binary operation is not commutative, a left and right identity may be defined separately. A left identity e satisfies $e \circ x = x$ and a right identity e satisfies $x \circ e = x$.

identity function A function i with a domain set X such that $i(x) = x$ for all $x \in X$. *See also* identity map, identity morphism.

identity map *See* identity function.

identity matrix The identity element E in the ring of $n \times n$ matrices over some coefficient ring. The entries of E are given by the Kronecker delta-function

$$\delta_{ij} = \begin{cases} 1 & \text{if } i = j \\ 0 & \text{if } i \neq j . \end{cases}$$

identity morphism Let $\mathcal{C}$ be a category, let A be an object in $\mathcal{C}$, and let $i : A \to A$ be a morphism in $\mathcal{C}$. (A *morphism* is a generalization of a function or mapping.) i is the *identity morphism* for the object A if $f \circ i = f$ for every object B and morphism $f : A \to B$ in $\mathcal{C}$, and $i \circ g = g$ for every object C and morphism $g : C \to A$ in the category $\mathcal{C}$. One of the axioms of category theory is that every object in a category has a (necessarily unique) identity morphism.

In most familiar categories, where objects are sets with some additional structure and morphisms are particular kinds of functions, the identity morphism for an object A is simply the identity function i; that is the function i such that $i(a) = a$ for all $a \in A$. *See also* identity function, morphism.

Ihara zeta function   Let $\mathbf{R}$ be the real numbers, k_p be a p-adic field, and let Γ be a subgroup of $G = \text{PSL}_2(\mathbf{R}) \times \text{PSL}_2(k_p)$ for which the following properties hold: (i.) Γ is discrete; (ii.) the projection, Γ_R, of Γ in $\text{PSL}_2(\mathbf{R})$ is dense in $\text{PSL}_2(\mathbf{R})$; (iii.) the projection, Γ_p, of Γ in $\text{PSL}_2(k_p)$ is dense in $\text{PSL}_2(k_p)$; (iv.) the only torsion element of Γ is the identity; (v.) the quotient $\Gamma\backslash G$ is compact. Let H be the upper half plane, $\{x+iy : y > 0\}$, and, for every $z \in H$, let $\Gamma_z = \{\gamma \in \Gamma : \gamma(z) = z\}$, where Γ acts on H via Γ_R. Let $\tilde{P}(\Gamma) = \{z \in H : \Gamma_z \cong \mathbf{Z}\}$ and let $P(\Gamma) = \tilde{P}(\Gamma)/\Gamma$. If $Q \in P(\Gamma)$ is represented by $z \in H$ and γ is a generator of Γ_z, then γ is equivalent to a diagonal matrix in Γ_p whose diagonal entries are λ and λ^{-1}. Let $\deg(Q)$ be $|v(\lambda)|$ where v is the valuation of k_p. The *Ihara zeta function* of Γ is

$$Z_\Gamma(u) = \prod_{Q \in P(\Gamma)} (1 - u^{\deg(Q)})^{-1} .$$

ill-conditioned system   A system of equations for which small errors in the coefficients, or in the solving process, have a large effect on the solution.

image   For a function $f : S \to T$ from a set S into a set T, the set $\text{Im}(f)$ of all elements of T of the form $f(s)$ for some s in S. Another popular notation is $f(S)$. Also called *range*.

imaginary axis in the complex plane   Complex numbers $x + iy$, where x and y are real and $i = \sqrt{-1}$, can be identified with the points of the real plane with Cartesian coordinates (x, y). The plane is then referred to as the complex plane. The coordinate axis $x = 0$ is called the imaginary axis. Similarly, the axis $y = 0$ is referred to as the real axis.

imaginary number   Any member of the field of complex numbers of the form iy, where y is a real number and $i = \sqrt{-1}$. The field of complex numbers is the set of numbers of the form $x + iy$, where x and y are real.

imaginary part of a complex number   If $z = x + iy$ is a complex number, then the real number y is called the imaginary part of z, denoted $y = \Im z$, and the real number x is called the real part of z, denoted $x = \Re z$.

imaginary prime divisor   One of two types of infinite prime divisors on an algebraic number field K. Let $a \in K$ and let σ be any injection of K into the complex number field which does not map K into the real number field. The equivalence class of an archimedean valuation on K, given by $v(a) = |\sigma(a)|^2$, is called an imaginary (infinite) prime divisor. *See also* real prime divisor, infinite prime divisor.

imaginary quadratic field   A field K which is a quadratic (degree two) extension of the rational number field $\mathbf{Q}$ is called a quadratic number field. Since $K = \mathbf{Q}(\sqrt{m})$, for some square free integer m, we may further distinguish these fields according to whether $m < 0$ or $m > 0$. In the first case, the field is called an *imaginary quadratic field* and, in the second, a real quadratic field.

imaginary root   A root of a polynomial $f(z)$, i.e., a number r such that $f(r) = 0$, which is an imaginary number. *See* imaginary number.

imaginary unit   Any number field K (finite extension of the field $\mathbf{Q}$ of rational numbers) contains a ring of integers, denoted R. This ring consists of all roots in K of monic irreducible polynomials $f(x) \in \mathbf{Z}[x]$, where $\mathbf{Z}$ is the ring of rational integers. Alternatively, this ring is the integral closure of $\mathbf{Z}$ in K. The units of R are those elements $u \in R$ for which there is an element $v \in R$ such that $uv = 1$. In an isomorphic embedding of K into the field of complex numbers, if the image of a unit u is imaginary, then u is called an imaginary unit. Note that this depends on which embedding is used.

imperfect field   Any field that admits proper inseparable algebraic extensions. This is the opposite of perfect. The simplest example of such a field is the field $k(x)$ where $k = GF(q)$ is a finite field. The extension given by the equation $t^q - x = 0$ is inseparable. Fields of characteristic zero are perfect as are all finite fields.

implicit enumeration method of integer programming   A method of solving zero-one linear programming problems. Values are assigned to some of the variables; if the solution of the resulting linear program is either infeasible or not optimal, then all solutions containing the assigned values may be ignored.

Implicit Function Theorem   A theorem guaranteeing that an implicit equation $F(x, y) = 0$ can be solved for one of the variables. One of many different forms of the theorem is the following assertion. Let $F(x_1, \ldots, x_n, y)$ be a continuously differentiable function of $n + 1$ variables. Let P be a point in $n+1$ space and assume that $F(P) = c$ and $\frac{\partial F}{\partial y}(P) \neq 0$. Then, there is a neighborhood of P and a unique continuously differentiable function $f(x_1, \ldots, x_n)$ defined on this neighborhood, such that

$$F(x_1, \ldots, x_n, f(x_1, \ldots, x_n)) \equiv c$$

on this neighborhood.

imprimitive transitive permutation group
A permutation group G on a set X with the following two properties: (i.) for each x and y in X there is a $t \in G$ such that $t(x) = y$; (ii.) the subgroup of G consisting of all permutations which leave x fixed is not a maximal subgroup. *See also* permutation group, transitive permutation group.

improper fraction   A rational number m/n, where m and n are integers and $m > n > 0$. Such a fraction can be written in the form $\frac{m}{n} = \frac{r}{n} + q$ where $r < n$ where r and q are also integers (for simplicity we have assumed that all of these numbers are positive). The numbers r and q are obtained by long division. That is, we write $m = nq + r$ where $r < n$.

incommensurable   Two members $a, b \in S$ of a partially ordered set S such that neither $a \leq b$ nor $b \leq a$.

incomplete factorization   A method of preconditioning the system of equations $Ax = b$, in which A is approximated by LU, where L is a lower triangular matrix and U is an upper triangular matrix. In an incomplete factorization, small elements are dropped, making the approximation sparse.

inconsistent system of equations   A set of equations for which there does not exist a common solution (in an appropriate solution space).

increasing directed set   A set S, together with a binary relation $\leq$, having the properties (i.) $a \leq a$ for all $a \in S$, (ii.) $a \leq b$ and $b \leq c$ implies $a \leq c$, and (iii.) for all $a, b \in S$, there is an element $c \in S$ such that $a \leq c$ and $b \leq c$. Also called *directed set*.

indecomposable group   A group G ($G \neq \{e\}$), that is not isomorphic to the direct product of two subgroups, unless one of those subgroups is $\{e\}$.

indecomposable module   A module M over a ring R such that M is not the direct sum of two proper R-submodules.

indefinite Hermitian form   A Hermitian form equivalent to

$$\sum_{i=1}^{p} \bar{x}_i x_i - \sum_{j=1}^{q} \bar{x}_{p+j} x_{p+j}$$

in which $\bar{x}$ is the conjugate transpose and neither p nor q is zero.

indefinite quadratic form   A quadratic form over an ordered field F, which represents both positive and negative elements. If every positive number in F is a square (for example, if $F = \mathbf{R}$), then an indefinite quadratic form is equivalent to the form

$$\sum_{i=1}^{p} x_i^2 - \sum_{j=1}^{q} x_{p+j}^2 \, ,$$

in which neither p nor q is zero. The rank of the quadratic form is $p + q$ where p and q are uniquely determined by the quadratic form.

indefinite sum   A concept analogous to the indefinite integral. Given a function $f(x)$ and a fixed quantity Δx ($\Delta x \neq 0$), let $F(x)$ be a function such that $F(x + \Delta x) - F(x) = f(x) \cdot \Delta x$. If $c(x)$ is an arbitrary function, periodic, with a period Δx, then $F(x) + c(x)$ is an indefinite sum of $f(x)$.

Independence Theorem   A corollary of the Approximation Theorem. Let $e_1, e_2, \ldots, e_n$ be real numbers and let $v_1, v_2, \ldots, v_n$ be mutually nonequivalent and nontrivial multiplicative valuations of a field K. If $\prod_i v_i(a)^{e_i} = 1$ for all $a \in K \backslash \{0\}$, then $e_i = 0$ for $i = 1, 2, \ldots, n$.

independent linear equations   A system of linear equations in which deleting any equation would expand the solution set.

independent variable   A symbol that represents an arbitrary element in the domain of a function. If the domain of the function is a Cartesian product set $X_1 \times X_2 \times \cdots \times X_n$, then the independent variable may be denoted $(x_1, x_2, \ldots, x_n)$, where each x_i represents an arbitrary element in X_i. In addition, each x_i is sometimes called an *independent variable*. *See also* dependent variable.

indeterminate form   An expression of the form $0/0$, ∞/∞, $0 \cdot \infty$, $\infty - \infty$, ∞^0, 0^0, or 1^∞. Such expressions are undefined. Indeterminate forms may appear when a limit is improperly evaluated as the quotient (product, etc.) of limits and not the limit of a quotient (product, etc.). But the term may properly appear when such limits are classified, so that they can be evaluated.

indeterminate system of equations   A system of equations with an infinite number of solutions.

index   (1) Most commonly, a subscript which is used to distinguish members of a set S. Thus, in this sense, a function defined on a set (called the index set) and taking its values in the set S.

For example an infinite sequence $\{a_1, \ldots, a_n\}$ is a set indexed by the natural numbers.

(2) There are many theorems in mathematics that are referred to as "index theorems" and these generally describe properties of certain special types of indices. For example, the *index* of a linear function $T : X \to Y$ between two complex vector spaces is, by definition, $\dim \ker(T) - \dim \ker T^*$, where T^* is the adjoint, or conjugate transpose, of T. *See also* index of specialty.

index of eigenvalue   (1) The number

$$\dim \ker(A - \lambda I) - \dim \ker \left(A^* - \bar{\lambda} I \right) \, ,$$

where A is a square matrix (or a Fredholm linear operator on a Banach space), I is the identity matrix of the same size as A, and A^* is the conjugate transpose (Banach space adjoint) of A.

(2) The order of the largest (Jordan) block corresponding to λ in the *Jordan normal form* of A. The index of λ is the smallest positive integer m such that $\text{rank}(A - \lambda I)^m = \text{rank}(A - \lambda I)^{m+1}$.

index of specialty   Let D be a divisor on a complete nonsingular curve over an algebraically closed field k, and let K be a canonical divisor of X. The *index of specialty* of the divisor D is $\dim_k H^0(X, \mathcal{O}(K - D)) = \dim_k H^2(X, \mathcal{O}(D))$, where $\mathcal{O}(D)$ is the invertible sheaf associated with D. This definition also applies to divisors on nonsingular surfaces. If D is a divisor on a curve on genus g, then the Riemann-Roch Theorem may be applied to give a second formula for the index of specialty. In this case, the specialty index of D is equal to $g - \deg D + \dim |D|$, where $|D|$ is a complete linear system of D.

Index Theorem of Hodge   Let M be a compact Kähler manifold of complex dimension $2n$ and let $h^{p,q}$ denote the dimension of the space of harmonic forms of type (p, q) on M. The signature of M is $\sum_{p,q}(-1)^p h^{p,q}$. The sum may be restricted to the case where $p + q$ is even since, on a compact Kähler manifold, $h^{p,q} = h^{q,p}$. Any complex, projective, nonsingular, algebraic variety is a Kähler manifold, and the following application of the Riemann-Roch Theorem is also called the *Hodge Index Theorem*.

Let H be an ample divisor on a nonsingular projective surface X, and let D be a divisor which has a nonzero intersection number with some divisor E. If the intersection number of D and H equals zero, then the self-intersection number of D is less than zero. This implies that the induced bilinear form on the Neron-Severi group of X has only one positive eigenvalue, and that the rest of the eigenvalues are negative.

induced module An example of a scalar extension. Let K be a commutative ring, and let G be a group with a subgroup H. The canonical injection $H \rightarrow G$ induces a homomorphism of group rings $K[H] \rightarrow K[G]$. Let M be a $K[H]$ module. The *induced module* of M is the $K[G]$ module $K[G] \otimes_{K[H]} M$. *See also* induced representation.

induced representation A representation of a group G obtained by "extending" a representation of a subgroup H of G. Let K be a commutative ring; the canonical injection $H \rightarrow G$ induces a homomorphism of group rings $K[H] \rightarrow K[G]$. Let M be a $K[H]$-module, the representation of G associated with the induced $K[G]$ module $K[G] \otimes_{K[H]} M$ is the induced representation of G. It is also called the *induced representation* of the representation of H associated with M. *See also* induced module.

induced von Neumann algebra Let $\mathcal{M}$ be a von Neumann algebra which is a *-subalgebra of the set of bounded linear operators on a Hilbert space **H**. Let $\mathcal{M}'$ be the set of operators that commute with every $A \in \mathcal{M}$ and with the adjoint of every $A \in \mathcal{M}$. If E is a projection operator in $\mathcal{M}'$, then the *induced von Neumann algebra* of $\mathcal{M}$ on the subspace $E\mathbf{H}$ is the restriction of $E\mathcal{M}E = E\mathcal{M}$ to $E\mathbf{H}$.

induction The Principle of Induction is a theorem that concerns well-ordered sets. A well-ordered set is a linearly ordered set in which every nonempty subset has a smallest member. The natural numbers are a primary example of such a set. Other examples are the transfinite ordinal numbers. The principle of induction states that if S is a well-ordered set and T is a subset of S with the property: whenever $\{a \in S : a < b\} \subset T$ implies $b \in T$, then we can conclude that $T = S$. Indeed, the condition implies that the smallest member of S is in T. If the complement of T were not empty, then it would have a

smallest member. Call this member b. This is not the smallest member of S as we have seen. However, T contains all a such that $a < b$, so it must contain b as well which is a contradiction.

Certain well-ordered sets such as the natural numbers have the property that every element of S (aside from the smallest) has an immediate predecessor. The principle of induction for such sets can be rephrased as follows: any subset $T \subset S$, such that (i.) T contains the smallest element of S and (ii.) whenever T contains a then it also contains the next smallest element of S, then $T = S$. The proof is much the same since the smallest member b of the complement of T could not be the smallest member of S and, therefore, would have an immediate predecessor which is in T. Therefore, the immediate successor of b would also be in T which is, again, a contradiction. In this form the principle is called the principle of finite induction. Otherwise it is referred to as transfinite induction.

This theorem is commonly applied as a method of proof for statements that can be indexed by some well-ordered set. For example, the statements $P(n) : 1 + 2 + \cdots + n = \frac{n(n+1)}{2}$ are indexed by the natural numbers. This can be proved by finite induction. On the other hand, a theorem such as the assertion that every ideal J of a ring R is contained in a maximal ideal requires (for most rings) transfinite induction.

For most proofs that involve transfinite induction, the method is replaced by a logically equivalent method called "Zorn's Lemma" or, simply, "Zornification." In order to be applied, a basic axiom of mathematics itself (called the "Axiom of Choice") must be assumed. This has a number of formulations but the one of interest here is the one that asserts that all sets possess a well ordering. It can then be shown that every partially ordered set has a maximal chain (simply ordered subset). As an example, the class of proper ideals that contain a given ideal J of a ring R is a partially ordered set under ordinary set inclusion. By Zorn's Lemma there is a maximal chain of these and the union of the member of such a chain is a maximal and proper ideal (it does not contain the identity element) of R.

The fact that this sort of argument is logically equivalent to one using transfinite induction is not particularly trivial to prove. Both types of arguments circulated in mathematics for many years until it was realized just in this century that they were equivalent.

inequality   An expression that involves members of some partially ordered set, commonly taking the form $a < b$ or $a \leq b$. The so-called triangle inequality involving real numbers is an example and it asserts that $|a + b| \leq |a| + |b|$, for all real numbers a, b.

inequality relation   Let $\leq$ denote a relation on a set S (i.e., $a \leq b$, whenever (a, b) belongs to the relation). (*See* relation.) If it is true that (i.) $a \leq a$ for all $a \in S$, (ii.) if $a \leq b$ and $b \leq a$ then $a = b$, and (iii.) if $a \leq b$ and $b \leq c$ then $a \leq c$, then the relation is called a partial ordering or *inequality relation*. Extensions of this are the linear ordering, which is a partial ordering satisfying property (iv.) for all $a, b \in S$, $a \leq b$ or $b \leq a$, and the well ordering which is a linear ordering in which every nonempty subset of S has a smallest member.

inertia field   A valuation ring R in a field K has the property that it has exactly one maximal ideal. (*See* valuation ring.) When this ideal is principal, then every other nonzero ideal of R is a power of this ideal. Thus, a discrete valuation ring is a unique factorization domain in which there is only one irreducible element. If L is a finite extension field of K (usually assumed to be separable), then the integral closure S of the discrete valuation ring R in L has the property that it is the intersection of a finite set of discrete valuation rings of L and, equivalently, is a unique factorization domain with only a finite number of irreducible elements. The original maximal ideal of R, which is πR, has the property that π can be factored in S in the form $\pi = u \prod_{j=1}^{r} \pi_j^{e-j}$, where u is a unit of S and the elements π_j run over the finite set of distinct irreducible elements of S. Let S_j denote the discrete valuation ring that is associated with the element π_j. There are then three indices associated with this construction. The number r, above, is the decomposition index, the numbers $e_1, \ldots, e_r$ are the ramification numbers, and the numbers $f_j = [S_j/\pi_j S_j : R/\pi R]$ are the residue class degrees. When the field extensions that give rise to the residue class degrees are not separable, then certain adjustments have to be made

in these definitions. If L is a Galois extension of K with Galois group G, it can be shown that $e = e_1 = \cdots = e_r$ and $f = f_1 = \cdots = f_r$, so, in this case, $[L : K] = efr$. Now, for each j, there is a subgroup G_j of G consisting of those automorphisms σ of L over K that have the property that $(\pi_j S)^\sigma = \pi_j S$. It is easily shown that the G_j form a complete conjugacy class in G. These are called decomposition subgroups. Let Z_j denote the intermediate field that is associated with G_j under the Galois correspondence. The fields Z_j form a complete class of conjugate subfields. In the important case that G is Abelian, all of the Z_j coincide. The fields Z_j are called decomposition fields. Now, the residue class fields $S_j/\pi_j S_j = S/\pi_j S$ are also Galois over $R/\pi R$ and there is a natural homomorphism of G_j onto the Galois group of this extension. The kernel is a normal subgroup H_j of index f in G_j. The subgroups H_j are called *inertial subgroups*. The subfields T_j associated with H_j are called *inertial fields*. The inertial groups H_j are members of a complete conjugacy class as are the inertial fields T_j. Note that we have a chain of fields $K \subseteq Z_j \subseteq T_j \subseteq L$ with the relative degrees $[L : T_j] = e$, $[T_j : Z_j] = f$, and $[Z_j : K] = r$. There is a further refinement of the field extension $T_j \subseteq L$ into a chain of intermediate fields called ramification fields and these are the fields that are associated with subgroups of G_j that leave various powers of $\pi_j S$ invariant.

inertia group   (1) Let K/k be a finite Galois extension of fields and let G be its Galois group. If R is the ring of integers of K and P is a prime ideal of R, then the inertia group of P over k is $\{\sigma \in G : P^\sigma = P \text{ and } a^\sigma \equiv a \pmod{P}$ for all $a \in R\}$. This group is a subgroup of the decomposition group of P.

(2) Let k be a local field, K a normal extension of finite degree, and G be the Galois group of K/k. Let P be the valuation ideal of K and let p be a generator of P. The *inertia group* is $\{\sigma \in G : p^\sigma \equiv p \pmod{P}\}$.

inertia of (Hermitian) matrix   The ordered triple

$$i(A) = (i_+(A), i_-(A), i_0(A)) \, ,$$

where $i_+(A)$, $i_-(A)$, $i_0(A)$ are, respectively, the number of positive, negative, and zero eigenvalues (counting multiplicities), of a given Hermitian matrix A.

Sylvester's Law of Inertia states that two Hermitian matrices A and B satisfy $i(A) = i(B)$ if and only if there exists a nonsingular matrix C such that $A = CBC^*$.

The notion of inertia can be extended to arbitrary square matrices with complex entries, where $i_+(A)$, $i_-(A)$, $i_0(A)$ are, respectively, the numbers of eigenvalues with positive, negative, and zero real part.

infimum   The greatest lower bound or meet of a set of elements of a lattice. The term is most frequently used with regard to sets of real numbers. If A is a set of real numbers, the *infimum* of A is the unique real number $b = \inf A$ defined by the following two conditions: (i.) $x \geq b$ for all $x \in A$; (ii.) if $x \geq c$ for all $x \in A$, then $b \geq c$. The infimum of a set of real numbers A may not exist, that is there may be no real number b satisfying conditions (i.) and (ii.) above, but if A is bounded from below, the infimum of A is guaranteed to exist. This fact is one of the several equivalent forms of the *completeness property of the set of real numbers. See* meet. *See also* supremum.

infinite continued fraction   An expression of the form

$$a_0 + \cfrac{b_1}{a_1 + \cfrac{b_2}{a_2 + \cfrac{b_3}{a_3 + \cdots}}} \, .$$

Infinite continued fractions may be used to approximate an irrational number.

infinite Galois extension   Let K be a subfield of the field L and assume that L is an algebraic extension of K. L is a Galois extension if it is a normal, separable, algebraic extension of K. (*See* normal extension, separable extension, algebraic extension.) L is said to be normal over K if, whenever an irreducible polynomial $P(x) \in K[x]$ has a root in L, then $P(x)$ splits into linear factors in $L[x]$. Those of nonfinite degree are *infinite Galois extensions*. Alternatively, the Galois group of L over K has the property that its fixed field is precisely K. (*See* Galois group.) The fixed field is the subfield of L that is left elementwise fixed by the Galois group.

infinite height   (**1**) An element a in an Abelian p-group A is said to have *infinite height* if the equation $p^k y = a$ is solvable in A for every nonnegative integer k.

(**2**) A prime ideal p is said to have *infinite height* if there exist chains of prime ideals $p_0 < p_1 < p_2 < \cdots < p_{h-1} < p$ with arbitrarily large h. If p is a prime ideal in a Noetherian ring R, and $p \neq R$, then p does not have infinite height.

infinite matrix   A matrix with an infinite number of rows and an infinite number of columns. Such a matrix may be infinite in only two directions, as (a_{ij}), $i, j = 0, 1, 2, \ldots$:

$$\begin{pmatrix} a_{00} & a_{01} & a_{02} & \cdots \\ a_{10} & a_{11} & a_{12} & \cdots \\ a_{20} & a_{21} & a_{22} & \cdots \\ \cdot & \cdot & \cdot & \end{pmatrix},$$

or it may be infinite in all directions, as (a_{ij}), $i, j = 0, \pm 1, \pm 2, \ldots$:

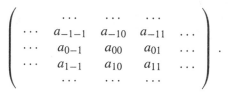

infinite prime divisor   An equivalence class of archimedean valuations on an algebraic number field.

infinitesimal automorphism   A complete holomorphic vector field on a manifold. On a Lie group the definitions of infinitesimal automorphism and derivation coincide. *See also* Killing form.

infinite solvable group   A solvable group which is not finite. There are also generalizations of the concept of solvability for infinite groups. *See* solvable group. *See also* generalized solvable group.

infinity (1) The symbol ∞, used in the notation for the sum of an infinite series of numbers

$$\sum_{n=0}^{\infty} a_n \; ;$$

an infinite interval, such as

$$[a, \infty) = \{x : x \geq a\} \; ;$$

or a limit, as x tends to ∞,

$$\lim_{x \to \infty} f(x) \; .$$

(2) The point at ∞ as, for example, the north pole of the Riemann sphere.

inflation A map of cohomology groups induced by lifting the cocycles of a factor group G/H to G. Let G be a group, M a G-module, H a normal subgroup of G, $M^H = \{m \in M : hm = m, \text{ for all } h \in H\}$, π the canonical epimorphism from G to G/H, and $i : M^H \to M$ the inclusion map. The inflation map is the map of cohomology groups

$$\inf = (\pi, i)_* : H^n \left(G/H, M^H \right)$$
$$\to H^n(G, M) \, ,$$

(for $n \geq 1$), which is achieved by lifting the homomorphism of pairs $(\pi, i) : (G/H, M^H) \to (G, M)$.

inflection point A point on a plane curve at which the curve switches from being concave to convex, relative to a fixed line; a non-singular point P on a plane curve C such that the tangent line to C at P has intersection multiplicity greater than or equal to 3. *See also* intersection multiplicity.

inhomogeneous difference equation A linear difference equation of the form

$$a_0 y_{k+n} + a_1 y_{k+n-1} + \cdots + a_n y_k = r_k \, ,$$

where r_k differs from 0 for some values of k. In the opposite case where r_k is identically equal to 0, the linear difference equation is *homogeneous*. These notions are directly analogous to the corresponding ones for linear differential equations.

inhomogeneous polarization (1) The algebraic equivalence class of a nondegenerate divisor on an Abelian variety.

(2) Let X be a proper scheme over an algebraically closed field k and let $\text{Pic}(X)$ be the Picard group of X. Let $\text{Pic}^\tau(X)$ be the subgroup of $\text{Pic}(X)$ consisting of the set of all the invertible sheaves F on X, for which there exists a nonzero integer n, such that F^n represents a point of the Picard scheme of X in the same component as the identity. A coset of $\text{Pic}^\tau(X)$ in $\text{Pic}(X)$ consisting of ample invertible sheaves is called an *inhomogeneous polarization* of X.

injection A function $i : A \to B$ such that $i(a_1) \neq i(a_2)$ whenever $a_1 \neq a_2$. Injections are also called *injective functions, injective mappings, one-to-one functions,* and *univalent functions.* The notion of an injection generalizes to the notion of a *monomorphism* or *monic morphism* in a category. *See* monomorphism. *See also* epimorphism, surjection.

injective class A class of objects in a category, each member of which is injective. The word "class" is used here rather than "set" because we cannot talk about the *set* of all injective objects in most categories without running into the sort of logical difficulties, for example Russell's paradox, connected with the "set of all sets."

injective dimension A left R module B, where R is a ring with unit, has *injective dimension n* if there is an injective resolution

$$0 \longrightarrow B \longrightarrow E_0 \longrightarrow \cdots \longrightarrow E_n \longrightarrow 0 \, ,$$

but no shorter injective resolution of B. The definition of the injective dimension of a right R module is entirely similar. *See* injective resolution. *See also* flat dimension, projective dimension.

Injective dimensions have little to do with more elementary notions of dimension, such as the dimension of a vector space, but they are related to a famous theorem called "Hilbert's Theorem on Syzygies." Suppose the ring R is commutative. Define the *global dimension* of R to be the largest injective dimension of any R module, or ∞ if there is no largest dimension. Let $D(R)$ denote the global dimension of

R, and let $R[x]$ be the ring of polynomials in x with coefficients in R. Then in modern terminology, Hilbert's Theorem on Syzygies states that $D(R[x]) = D(R) + 1$. *See also* syzygy.

injective envelope   An object E in a category $\mathcal{C}$ is the *injective envelope* of an object A if it has the following three properties: (i.) E is an injective object, (ii.) there is a monomorphism $i : A \rightarrow E$, and (iii.) there is no injective object properly between A and E. In a general category, this means that if $j : A \rightarrow E'$ and $k : E' \rightarrow E$ are monomorphisms and E' is injective, then k is actually an isomorphism. *See also* injective object, monomorphism, projective cover.

In most familiar categories, objects are sets with structure (for example, groups, topological spaces, etc.), morphisms are particular kinds of functions (for example, group homomorphisms, continuous functions, etc.), and monomorphisms are one-to-one functions of a particular kind. In these categories, property (iii.) can be phrased in terms of ordinary set containment. For example, in the category of topological spaces and continuous functions, property (iii.) reduces to the assertion that there is no injective topological space containing A as a topological subspace, and contained in E as a topological subspace, except for E itself.

injective mapping   *See* injection.

injective module   An injective object in the category of left R modules and left R module homomorphisms, where R is a ring with unit. Symmetrically, an injective object in the category of right R modules and right R module homomorphisms. *See* injective object. *See also* flat module, projective module, projective object.

In the important case where the ring R is a principal ideal domain, the injective R modules are just the divisible ones. (An R module M is *divisible* if for each element $m \in M$ and each $r \in R$, there exists $m_r \in R$ such that $r m_r = m$.) Thus, for example, the rational numbers, $\mathbf{Q}$, are an injective $\mathbf{Z}$ module, where $\mathbf{Z}$ is the ring of integers. *See also* flat module, injective object, projective module, projective object.

injective object   An object I in a category $\mathcal{C}$ satisfying the following mapping property: If $i : B \rightarrow C$ is a monomorphism in the category, and $f : B \rightarrow I$ is a morphism in the category, then there exists a (usually not unique) morphism $g : C \rightarrow I$ in the category such that $g \circ i = f$. This is summarized in the following "universal mapping diagram":

$$
\begin{array}{ccc}
B & \xrightarrow{\ i\ } & C \\
{\scriptstyle f}\searrow & & \swarrow{\scriptstyle \exists g} \\
& I &
\end{array}
$$

See also injective module, monomorphism, projective module, projective object.

In most familiar categories, objects are sets with structure (for example, groups, Banach spaces, etc.), and morphisms are particular kinds of functions (for example group homomorphisms, bounded linear transformations of norm ≤ 1, etc.), so monomorphisms are one-to-one functions (injections) of particular kinds. Here are two examples of injective objects in specific categories: (1) In the category of Abelian groups and group homomorphisms, the injective objects are just those Abelian groups which are divisible. (An Abelian group G is *divisible* if for each $g \in G$ and each integer n, there exists a group element $g_n \in G$ such that $n g_n = g$.) Thus, for example, the rational numbers $\mathbf{Q}$ are an injective object in this category. (2) The Hahn-Banach Theorem asserts that the field of complex numbers, thought of as a one-dimensional complex Banach space, is an injective object in the category of complex Banach spaces and bounded linear transformations of norm ≤ 1.

injective resolution   Let B be a left R module, where R is a ring with unit. An *injective resolution* of B is an exact sequence,

$$
0 \longrightarrow B \xrightarrow{\ \phi_0\ } E_0 \xrightarrow{\ \phi_1\ } E_1 \xrightarrow{\ \phi_2\ } \cdots,
$$

where every E_i is an injective left R module. There is a companion notion for right R modules. Injective resolutions are extremely important in homological algebra and enter into the dimension theory of rings and modules. *See also* flat resolution, injective dimension, injective module, projective resolution.

An *exact sequence* is a sequence of left R modules, such as the one above, where every ϕ_i is a left R module homomorphism (the ϕ_i are called "connecting homomorphisms"), such that $\mathrm{Im}(\phi_i) = \mathrm{Ker}(\phi_{i+1})$. Here $\mathrm{Im}(\phi_i)$ is the image of ϕ_i, and $\mathrm{Ker}(\phi_{i+1})$ is the kernel of ϕ_{i+1}. In the particular case above, because the sequence begins with 0, it is understood that the kernel of ϕ_0 is 0, that is ϕ_0 is one-to-one. There is a companion notion for right R modules.

inner automorphism   A group automorphism of the form $\phi_a(g) = aga^{-1}$. In more detail, an *automorphism* of a group G is a one-to-one mapping of G onto itself which preserves the group operation, $\phi(g_1 g_2) = \phi(g_1)\phi(g_2)$ for all g_1 and g_2 in G. If a is a fixed element of G, the mapping ϕ_a defined above is easily seen to be an automorphism. Automorphisms of this form are rather special and form a group under composition called *the group of inner automorphisms of G*.

inner derivation   A derivation of the form $D_a(x) = xa - ax$. In more detail, a *derivation* is a linear mapping D of a (possibly nonassociative) algebra A into itself, satisfying the familiar product rule for derivatives, $D(xy) = D(x)y + xD(y)$. Thus, derivations are algebraic generalizations of the derivatives of calculus. Now let A be associative. If a is a fixed element of A, the mapping D_a defined above is easily seen to be a derivation. Derivations of this form are rather special, and are called *inner derivations*.

The concept extends to Lie algebras, but in this case inner derivations are derivations of the form $D_a(x) = [x\, a]$, where $[x\, a]$ is the Lie product.

inner topology   The topology of a Lie subgroup H, as a submanifold of a Lie group G. This topology need not be the relative topology of H, viewed as a subspace of a topological space G.

inseparable element   An element of an extension field with an inseparable minimal polynomial. In more detail, let G be an extension of a field F. (This means that G is a field and $G \supseteq F$.) Let α be an algebraic element of G over F. (This means that α satisfies a polynomial equation $P(\alpha) = 0$ with coefficients in F.) Among all polynomials P with coefficients in F such that $P(\alpha) = 0$, there is one of smallest positive degree, called the *minimal polynomial* of α. The algebraic element α is *inseparable* if its minimal polynomial is inseparable. *See also* inseparable polynomial. Antonym: *separable element*.

Inseparable elements can only occur if the field F has characteristic $n \neq 0$. In particular, inseparable elements can never occur if F is the field of rational numbers, or an extension of the rationals. *See also* inseparable extension.

inseparable extension   An algebraic extension field containing an inseparable element. *See* inseparable element. Antonym: *separable extension*.

inseparable polynomial   An irreducible polynomial with coefficients in a field which factors over its splitting field with repeated factors or, more generally, a polynomial which has an inseparable polynomial among its irreducible factors.

In more detail, let F be a field and let P be a polynomial of positive degree with coefficients in F. P may or may not factor into linear factors over F (for example, $x^2 - 2$ does not factor over the rationals), but there always exists a smallest extension field of F over which P does factor (for example, $x^2 - 2$ factors as $(x - \sqrt{2})(x + \sqrt{2})$ over the field G formed by adjoining $\sqrt{2}$ to the rationals). This smallest extension field is called the *splitting field* for P.

Let $F[x]$ denote the ring of polynomials with coefficients in F. Suppose first that P is *irreducible* in $F[x]$, that is P does not factor into two or more polynomials in $F[x]$ of positive degree. P is called *separable* if its factorization over its splitting field has no repeated factors. In the general case where P is not irreducible, P is called *separable* if each irreducible factor is separable. Finally, P is called *inseparable* if it is not separable. Antonym: *separable polynomial*.

integer   (**1**) Intuitively, an *integer* is one of the signed whole numbers $0, \pm 1, \pm 2, \pm 3, \ldots$, and a *natural number* is one of the counting numbers, 1, 2, 3, ...

(**2**) Semi-formally, the *ring of integers* is the set **Z** consisting of the signed whole numbers, together with the ordinary operations of addition, $+$, and multiplication, $\cdot$. The ring of integers forms the motivating example for many of the concepts of mathematics. For example, the ring of integers, $(\mathbf{Z}, +, \cdot)$, satisfies the following properties for all x, y, and $z \in \mathbf{Z}$: (1) $x + y \in \mathbf{Z}$. (2) $x + (y + z) = (x + y) + z$. (3) $x + 0 = 0 + x$, (4) given x, there exists an element $-x$ such that $x + -x = -x + x = 0$. These are precisely the axioms for a *group,* so the integers under addition, $(\mathbf{Z}, +)$, form the first example of a group. Furthermore, addition is commutative, (5) $x + y = y + x$. Properties (1) through (5) are precisely the axioms for an *Abelian group,* so the integers under addition form the first example of an Abelian group. In addition, the integers satisfy the following additional properties for all x, y, and $z \in \mathbf{Z}$: (6) $x \cdot y \in \mathbb{Z}$. (7) $x \cdot (y + z) = x \cdot y + x \cdot z$, and $(y + z) \cdot x = y \cdot x + z \cdot z$. Properties (1) through (7) are precisely the axioms for a *ring,* so the ring of integers, $(\mathbf{Z}, +, \cdot)$, form the first, and one of the best, examples of a ring. Furthermore, the integers satisfy (8) $1 \cdot x = x \cdot 1 = x$. The number 1 is called a *unit element* because it satisfies this identity, so the ring of integers forms one of the best examples of a *ring with unit element,* or *ring with unit* for short. In addition, the ring of integers satisfies the commutative law, (9) $x \cdot y = y \cdot x$, so it forms one of the best examples of a *commutative ring.*

The ring of integers has a far richer structure than described above. For example, the ring of integers is an *integral domain,* or *domain* for short, because it satisfies property (10) $x \cdot y = 0$ implies $x = 0$ or $y = 0$ (or both). Thus, the ring of integers satisfies the familiar cancellation law: If we know $x \cdot y = x \cdot z$ for $x \neq 0$, then we know $y = z$. Finally, the integers form one of the best examples of a *Euclidean domain,* and of a *principal ideal domain. See also* Abelian group, commutative ring, Euclidean domain, principal ideal domain, ring, unit element.

Many of the most profound open (unsolved) questions in mathematics revolve around the integers. For example, a *prime number* is a (positive) integer divisible only by itself and 1. Eu-

clid proved centuries ago that there are infinitely many prime numbers. But it is still unknown whether there are infinitely many pairs of prime numbers, p_n and p_{n+1}, which differ by 2, i.e., $p_{n+1} - p_n = 2$. The conjecture that there are infinitely many such pairs of primes is called the *twin prime conjecture.* Perhaps the deepest and most important unsolved question in mathematics is the *Riemann hypothesis.* Although the Riemann hypothesis is stated in terms of the behavior of a certain analytic function called the *Riemann zeta function,* it too involves the properties of the integers at its heart. For example, if the Riemann hypothesis were true, then the twin prime conjecture (and most of the other great unsolved conjectures of number theory) would be true.

(**3**) Formally, an *integer* is an element of the *ring of integers.* The *ring of integers* is the smallest ring containing the *semi-ring of natural numbers.* The *semi-ring of natural numbers* is defined in terms of the set of natural numbers. The *set of natural numbers* is any set **N**, together with a successor function S carrying **N** to **N**, satisfying the Peano postulates:

(1) There is an element $1 \in \mathbf{N}$.

(2) $S : \mathbf{N} \to \mathbb{N}$ is a function, that is the following two properties hold: (a) given $n \in \mathbf{N}$, there is only one element $S(n)$, and (b) for each $n \in \mathbf{N}$, $S(n) \in \mathbf{N}$.

(3) For each $n \in \mathbf{N}$, $S(n) \neq 1$.

(4) S is a one-to-one function, that is if $S(m) = S(n)$ then $m = n$.

(5) (The *axiom of induction*) Suppose I is a subset of **N** satisfying the following two properties: (a) $1 \in I$, and (b) if $i \in I$ then $S(i) \in I$. Then $I = \mathbf{N}$.

Addition and multiplication, $+$ and $\cdot$, are defined inductively in terms of the successor function S, so that $S(n) = n + 1$, and the *semi-ring of natural numbers* is defined to be the set of natural numbers **N**, together with the operations of $+$ and $\cdot$.

Mathematical logic shows us that there are fundamentally different models of the natural numbers, and thus of the ring of integers. (A *model of the natural numbers* is simply a particular set **N** and successor function S satisfying the Peano postulates.) For example, if we begin by believing we understand a particular model of the natural numbers, and call this object $(\mathbf{N}, S)$, then it is possible to construct out of our pre-existing $(\mathbf{N}, S)$ a new object $(\mathbf{N}^*, S^*)$ with the following remarkable properties:

(1) $(\mathbf{N}^*, S^*)$ also satisfies the Peano postulates, and thus equally deserves to be called "the natural numbers."

(2) Every $n \in \mathbf{N}$ also belongs to $\mathbf{N}^*$.

(3) If $n \in \mathbf{N}$, then $S^*(n) = S(n)$.

(4) There exist elements of $\mathbf{N}^*$ larger than any element of $\mathbf{N}$. (These elements of $\mathbf{N}^*$ are called "infinite elements" of $\mathbf{N}^*$.)

Warning to the reader: One also has to redefine what one means by *set* and *subset* for this to work. Otherwise, (4) could not be true and $(\mathbf{N}^*, S^*)$ would simply equal $(\mathbf{N}, S)$.

This construction forms the basis of Abraham Robinson's non-standard analysis, and related constructions lie at the heart of the Gödel undecideability theorem.

(4) Other usage: In algebraic number theory, *algebraic integers* are frequently called integers, and then ordinary integers are called *rational integers*. An *algebraic integer* is an element α of an extension field of the rationals which is integral over the rational integers. *See* integral element.

integer programming   The general *linear programming* problem asks for the maximum value of a linear function L of n variables, subject to linear constraints. In other words, the problem is to maximize $L(x_1, x_2, \ldots, x_n)$ subject to the conditions $AX \leq B$, where X is the column vector formed from $x_1, \ldots, x_n$, B is a column vector, and A is a matrix. The general *integer programming* problem is the same, except the solution $(x_1, \ldots, x_n)$ is to consist of integers.

Integer programming problems frequently arise in applications, and many important combinatorial problems, such as the *travelling salesman problem,* are equivalent to integer programming problems. The formal statement of this equivalence is that integer programming is one of a class of hardest possible problems solvable quickly by inspired guessing; in other words, integer programming is NP complete. (Technically, the NP complete problem is the one of determining whether an integer vector $(x_1, \ldots, x_n)$ exists, subject to the constraints and making $L(x_1, \ldots, x_n) >$ a predetermined

constant K.) There are several efficient solution methods for particular classes of integer programming problems, but it is unlikely that there is an efficient solution method for all integer programming problems, since this would imply $NP = P$, a conjecture widely believed to be false.

integrable family of unitary representations
Let G be a topological group. A *unitary representation* of G is a group homomorphism L from G into the group of unitary operators on a Hilbert space H, which is continuous in the following sense: For each fixed h_1 and h_2 in H, the function $g \mapsto (L(g)h_1, h_2)$ is continuous. Here, $(L(g)h_1, h_2)$ denotes the inner product of $L(g)h_1$ and h_2 in the Hilbert space H. (In other words, L is a continuous map from G into the set of unitary operators endowed with the weak operator topology.) *See* unitary representation.

In the theory of unitary representations, it is frequently desirable to express a given unitary representation L as a *direct integral* or *integral direct sum* of simpler unitary representations,

$$L = \int_X l(x)\, d\mu(x)$$

where X is a set and μ is a measure on X, or, more precisely, where X is a set, $\mathcal{B}$ is a σ-field of subsets of X, and μ is a measure on $\mathcal{B}$. *See* integral direct sum. Here, $l(x)$ is a unitary representation of G for each $x \in X$. If L' is another unitary representation of G, and if L' can also be represented as a direct integral,

$$L' = \int_X l'(x)\, d\mu(x)\,,$$

and if $l'(x)$ is unitarily equivalent to $l(x)$ except possibly on a set of μ measure 0, then L' is unitarily equivalent to L. In other words, direct integrals preserve the relation of unitary equivalence. *See* unitary equivalence.

This leads to the consideration of functions from X into the set E of unitary equivalence classes of unitary representations of G. (Here, two representations are equivalent if they are unitarily equivalent.) Such a function $\mathcal{L}$ is said to be an *integrable family of unitary representations,* or an *integrable unitary representation* for short, if the following holds: There is a function l defined on X such that (1) $l(x)$ is a unitary representation in the equivalence class $\mathcal{L}(x)$ for each $x \in X$, and (2) for each pair of elements h_1 and h_2 in the Hilbert space H, and group element $g \in G$, the function $x \mapsto \langle l(x)(g)h_1, h_2 \rangle$ is μ measurable.

An integrable family of unitary representations, $\mathcal{L}$, and any of its associated functions, l, are exactly what is needed to form a new unitary representation via the direct integral,

$$L = \int_X l(x)\, d\mu(x) .$$

Because of the aforementioned preservation of unitary equivalence, one may write

$$L = \int_X \mathcal{L}\, d\mu(x)$$

instead.

integrable unitary representation   *See* integrable family of unitary representations.

integral   (**1**) Of or pertaining to the integers, as in such phrases as "integral exponent," i.e., an exponent which is an integer.

(**2**) In calculus, the anti-derivative of a continuous, real-valued function. In more detail, let f be a continuous real-valued function defined on the closed interval $[a, b]$. Let F be a function defined on $[a, b]$ such that $F'(x) = f(x)$ for all x in the interval $[a, b]$. F is called an *anti-derivative of f*, or an *indefinite integral of f*, and is denoted by $\int f(x)\, dx$. The number $F(b) - F(a)$ is called the *definite integral of f* over the interval $[a, b]$, or the *definite integral of f from a to b*, and is denoted by $\int_a^b f(x)\, dx$. By the Fundamental Theorem of Calculus, if $f(x) \geq 0$ for all x in $[a, b]$, the definite integral of f over the interval $[a, b]$ is equal to the area under the curve $y = f(x)$, $a \leq x \leq b$.

There is a sequence of rigorous and increasingly general definitions of the integral. In order of increasing generality, they are

(i.) *The Riemann integral.* Let $a = x_0 \leq x_1 \leq \cdots \leq x_n = b$ be a partition of $[a, b]$. Choose intermediate points $t_1, t_2, \ldots, t_n$ so that $x_0 \leq t_1 \leq x_1, x_1 \leq t_2 \leq x_2, \ldots, x_{n-1} \leq t_n \leq x_n$. The sum, $\sum_1^n f(t_i)(x_i - x_{i-1})$ is called a *Riemann sum*. The *Riemann integral* of f over

the interval $[a, b]$ is defined to be

$$\int_a^b f(x)\, dx = \lim_{|x_i - x_{i-1}| \to 0} \sum_{i=1}^n f(t_i)$$
$$(x_i - x_{i-1}) ;$$

in other words, the Riemann integral is defined as the limit of Riemann sums. It is non-trivial to prove that the limit exists and is independent of the particular choice of intermediate points t_i.

(ii.) There are several variants of this definition. In the most common one, $f(t_i)$ is replaced by the supremum (least upper bound) of f on the interval $[x_{i-1}, x_i]$ to obtain the *upper sum* $U(f, \mathcal{P})$, and by the infimum (greatest lower bound) of f on $[x_{i-1}, x_i]$ to obtain the *lower sum* $L(f, \mathcal{P})$. (Here, $\mathcal{P}$ refers to the partition $a = x_0 \leq x_1 \leq \ldots \leq x_n = b$.) The upper and lower integrals of f are

$$\overline{\int_a^b} f(x)\, dx = \inf_{\mathcal{P}} U(f, \mathcal{P})$$

and

$$\underline{\int_a^b} f(x)\, dx = \sup_{\mathcal{P}} L(f, \mathcal{P}) .$$

(Here, inf stands for *infimum* and sup for *supremum*. *See* infimum, supremum.) The function f is defined to be *Riemann integrable* if the upper and lower integrals of f are equal, and their common value is called the *Riemann integral* of f over the interval $[a, b]$. It is a theorem that all continuous functions are Riemann integrable. This definition has the advantage that it extends the class of integrable functions beyond the continuous ones. It is a theorem that a bounded function f is Riemann integrable if and only if it is continuous almost everywhere.

(iii.) Definitions (i.) and (ii.) generalize from intervals $[a, b]$ to suitable regions in higher dimensions.

(iv.) *The Stieltjes integral.* Everything is as in (ii.), except that $x_i - x_{i-1}$ is replaced by $\alpha(x_i) - \alpha(x_{i-1})$ in the definition of upper and lower sums, where α is a monotone increasing (and possible discontinuous) function. The Stieltjes integral of f is denoted by $\int_a^b f(x)\, d\alpha(x)$. The Stieltjes integral is more general than the Riemann integral, not in that the class of integrable functions is enlarged, but rather in that the class of things we can integrate against (the functions α) is enlarged.

(v.) *The Lebesgue integral.* Let μ be a countably additive set function on a σ-field of sets. The set function μ is called a *measure.* *See* sigma field, measure. Examples are:

(a) X is a finite set, and if A is a subset of X, then $\mu(A) =$ the number of elements in A. μ is called *counting measure* on X.

(b) X is any set, finite or infinite. The σ-field of sets is the set of all subsets of X. Let a be a fixed point of X. If A is a subset of X, then $\mu(A) = 1$ if $a \in A$, and $\mu(A) = 0$ otherwise. μ is called *a point mass at a,* or *the Dirac delta measure at a.*

(c) X is the interval $[a, b]$. The σ-field of sets is the set of Borel subsets of X. (The Borel sets include all subintervals, whether open or closed, of $[a, b]$, and many other sets besides.) If I is an interval, then $\mu(I)$ is the length of I. μ is called *Lebesgue measure* on $[a, b]$.

(d) X and the σ-field are as in (c). Let α be a monotone increasing function on $[a, b]$, and for convenience suppose it is continuous from the right. If $I = (x_1, x_2]$ is a right half closed subinterval of X, then $\mu(I) = \alpha(x_2) - \alpha(x_1)$. μ is called *a Lebesgue-Stieltjes measure* on $[a, b]$.

(e) X is an open subset of n-dimensional Euclidean space, $\mathbf{R}^n$. The σ-field is the σ-field of Borel subsets of X. If C is a small n-dimensional cube contained in X, then $\mu(C)$ is the n-dimensional volume of C. (In the familiar case $n = 2$, an n-dimensional cube is simply a square, and the n-dimensional volume is simply the area of the square. In the equally familiar case $n = 3$, an n-dimensional cube is an ordinary 3-dimensional cube, and the n-dimensional volume is the ordinary 3-dimensional volume of the cube.)

A *simple function* on X is a function which takes only finitely many values. If g is a simple function taking values $c_1, \ldots, c_n$ on the sets $A_1, \ldots, A_n$ (so $c_1, \ldots, c_n$ exhaust the finite set of values of taken by g) and $A_i = \{x \in X, g(x) = c_i\}$, then

$$\int_X g(x) \, d\mu(x) = \sum_{i=1}^{n} c_i \mu(A_i) \ .$$

(This assumes, of course, that the A_i belong to the σ-field, i.e., that the $\mu(A_i)$ are defined. Such a simple function is called *measurable.*) Now suppose f is a bounded function on X. Define the upper integral of f on X as

$$\overline{\int_X} f(x) \, d\mu(x) = \inf_{g \in \mathcal{G}} \int_X g(x) \, d\mu(x) \ ,$$

where $\mathcal{G}$ equals the set of measurable simple functions g such that $g \geq f$. Define the lower integral of f on X similarly,

$$\underline{\int_X} f(x) \, d\mu(x) = \sup_{h \in \mathcal{H}} \int_X h(x) \, dx \ ,$$

where $\mathcal{H}$ equals the set of measurable simple functions h such that $h \leq f$. The bounded function f is *Lebesgue integrable,* or simply *integrable,* if the upper and lower integrals agree, and then their common value is called the *Lebesgue integral* of f with respect to the measure μ, or simply the *integral* of f with respect to the measure μ, and is denoted by $\int_X f(x) \, dx$.

The Lebesgue integral extends both the class of integrable functions (all the way to bounded measurable functions), and the class of things we can integrate against (arbitrary measures). The theory extends to unbounded functions as well.

(vi.) *The Denjoy integral.* Similar to the Lebesgue integral, except that the upper integral is defined as the infimum of the integrals of an appropriate family of lower semi-continuous functions, and the lower integral is defined as the supremum of a family of upper semi-continuous functions. *See also* lower semi-continuous function, upper semi-continuous function.

The Denjoy integral requires extra structure on X, X must be a topological space, and the measure μ must be a Radon measure. However, the Denjoy integral is particularly well suited for dealing with certain technical difficulties connected with the integration of functions taking values in a non-separable topological vector space.

(vii.) Definition (i.) of the Riemann integral easily extends to continuous vector valued functions f taking values in a complete topological vector space. The analogous definition for the Stieltjes integral also extends to this setting. *See also* topological vector space.

(viii.) *The Pettis integral,* also called *the Dunford Pettis integral.* Let f be a vector valued function, and let μ be a measure on a space X, as in (v.) or (vi.). Suppose f takes values in a locally convex topological vector space V. The *Pettis integral* of f is defined by the conditions,

$$\left(\int_X f(x)\,d\mu(x), \lambda \right) = \int_X (f(x), \lambda)\,d\mu(x) ,$$

for all $\lambda \in V^*$. Here, V^* is the dual of V, consisting of all continuous linear functionals on V. Of course, this presupposes that the scalar valued functions $(f(x), \lambda)$ are integrable, and that the infinite system of equations,

$$(e, \lambda) = \int_X (f(x), \lambda)\,d\mu(x), \lambda \in E^* ,$$

has a solution $e \in V$. *See also* locally convex topological space.

(ix.) The definition of the Stieltjes integral extends to vector valued measures and scalar valued functions, and even to operator valued measures and vector valued functions. The Spectral Theorem is phrased in terms of such an integral. *See also* Spectral Theorem.

(x.) Recently, a seemingly minor variant of the classical Riemann integral has been discovered which has all the power of the Lebesgue integral *and more.* This new integral is variously named the *Henstock integral,* the *Kurzweil-Henstock integral,* or the *generalized Riemann integral.* The Henstock integral is defined in terms of gauges. Define a *gauge* to be a function γ which assigns to each point x of the interval $[a, b]$ a neighborhood of x. (The neighborhood may be an open interval containing x, half open if x is one of the endpoints a or b.) Let $a = x_0 < x_1 < \cdots < x_n = b$ be a partition of $[a, b]$ with intermediate points $t_1, t_2, \ldots, t_n$, as in the definition of the classical Riemann integral. Refer to such a partition as a *tagged partition*, with the intermediate points t_i as the *tags*. Define a tagged partition to be γ *fine* if $[x_{i-1}, x_i] \subseteq \gamma(t_i)$ for each i. If f is a real valued function on $[a, b]$, define the *Henstock integral* of f to be the (necessarily unique) real number L such that for each $\epsilon > 0$, there is a gauge γ such that

$$\left| L - \sum f(t_i)(x_i - x_{i-1}) \right| < \epsilon ,$$

for each *gamma* fine tagged partition. Of course, the Henstock integral of f is still denoted by

$$\int_a^b f(x)\,dx .$$

If f is positive, then Henstock integrability and Lebesgue integrability coincide, and the Henstock integral of f equals the Lebesgue integral of f. But if f varies in sign, and $|f|$ is not Henstock (and thus not Lebesgue) integrable, then the Henstock integral of f may still exist, even though the Lebesgue integral of f cannot exist under these circumstances. Thus, the Henstock integral is more general than the Lebesgue integral. The Henstock integral obtains its added power because it captures cancellation phenomena related to improper integrals that the Lebesgue integral cannot.

The Henstock integral extends to a Henstock-Stieltjes integral in a rather simple way. Henstock integration also extends to functions of several variables. However, Henstock integration on subsets of n-dimensional space, $\mathbf{R}^n$, is still an open area of investigation, as is the extension of the Henstock integral to abstract settings similar to measure spaces.

integral character   In number theory, a character which takes on only integral values.

integral closure   Let S be a commutative ring with unit, and let R be a subring of S. The *integral closure* of R in S is the set of all elements of S which are integral over R. *See* integral element. R is *integrally closed* in S if R equals its integral closure in S.

integral dependence   Let S be a commutative ring with unit, and let R be a subring of S. An element $\alpha \in S$ is *integrally dependent* over R if α is integral over R. *See also* integral element.

integral direct sum   (**1**) A representation of a Hilbert space as an L^2 space of vector valued functions. In more detail, let $(X, \mathcal{F}, \mu)$ be a measure space. Here, X is a set, $\mathcal{F}$ is a σ-field of subsets of X, and μ is a measure on $\mathcal{F}$. (*See* Hilbert space, integral, measure, sigma field.) Let $H(x)$, $x \in X$, be a family of Hilbert spaces indexed by X, and let $\mathcal{H}$ be the union of the sets $H(x)$, $x \in X$. Let L

be a set of functions f defined on X such that (a) $f(x) \in H(x)$ for all $x \in X$; (b) the function $x \mapsto \|f(x)\|$ is measurable; (c) there exists a countable family $f_1, f_2, f_3, \ldots \in L$ such that the set $\{f_1(x), f_2(x), f_3(x), \ldots\}$ is dense in $H(x)$ for each $x \in X$. Assume also the following closure property for L, (d) if g satisfies property (a) and the function $x \mapsto (f(x), g(x))$ is measurable for each $f \in L$, then $g \in L$. Here, $\|f(x)\|$ denotes the Hilbert space norm of $f(x)$ and $(f(x), g(x))$ the Hilbert space inner product in the Hilbert space $H(x)$.

Let

$$L^2(L, d\mu)$$
$$= \left\{ f \in L : \int_X \|f(x)\|^2 \, d\mu(x) < \infty \right\} .$$

The Hilbert space $L^2(L, d\mu)$ is called an *integral direct sum,* or a *direct integral,* and is frequently denoted by

$$\int_X H(x) \, d\mu(x) .$$

(2) The corresponding representation of a linear transformation T between two integral direct sums,

$$\int_X H_1(x) \, d\mu(x) \quad \text{and} \quad \int_X H_2(x) \, d\mu(x) ,$$

as an integral of bounded linear transformations $t(x)$. In more detail, let $H_1(x)$ and $H_2(x)$, $x \in X$, be two families of Hilbert spaces indexed by X as in (1) above, and let $t(x), x \in X$ be a family of linear transformations from $H_1(x)$ to $H_2(x)$ indexed by X. If f lies in $\int_X H_1(x) \, d\mu(x)$, define $T(f)$ by $T(f)(x) = t(x)(f(x))$, except possibly on a set of μ measure 0. (Of course, the domain of T will be all of $\int_X H_1(x) \, d\mu(x)$, that is, $T(f)$ will lie in $\int_X H_2(x) \, d\mu(x)$, only when the family $t(x), x \in X$ is *uniformly bounded,* except possibly on a set of μ measure 0, that is only when there is a constant K independent of x such that the norm $\|t(x)\| \leq K$, except possibly on a set of μ measure 0. In this case, the operator T will be bounded, with norm $\leq K$.) The linear transformation T is called the *integral direct sum,* or a *direct integral* of the family $t(x)$, $x \in X$, and is frequently denoted by

$$\int_X t(x) \, d\mu(x) .$$

integral divisor (1) In elementary algebra and arithmetic, a factor or divisor which is an integer, as in, for example, 3 is an integral divisor of 12.

(2) In algebraic geometry, a divisor with positive coefficients. In more detail, let X be an algebraic variety. (This simply means that X is the solution set to a system of polynomial equations. *See* algebraic variety.) A *divisor* on X is a formal sum $D = a_1 C_1 + \cdots + a_k C_k$, where the a_i are integers and the C_i are distinct irreducible subvarieties of X of codimension 1. (*Codimension* 1 means the dimension of C_i is one less than the dimension of X. *Irreducible* means C_i is not the union of two proper [i.e., strictly smaller] subvarieties. In the simplest case where X is an algebraic curve, the C_i are just points of X.) The divisor D is *integral* if all the coefficients a_i are positive. Integral divisors are also called *positive divisors* or *effective divisors.*

The notions of divisor and integral divisor extend to various related and/or more general contexts, for example to the situation where X is an analytic variety. *See* analytic variety.

Perhaps the clearest examples of integral divisors occur in elementary algebra and elementary complex analysis. Consider a polynomial or analytic function f defined in the complex plane. Let $C_1, C_2, C_3, \ldots$ be a listing of the zeros of f. The corresponding integral divisor $D = a_1 C_1 + a_2 C_2 + a_3 C_3 + \cdots$ describes the zeros of f, *counting multiplicity,* and one considers such divisors repeatedly in these subjects, for example in the statement that a polynomial of degree n has exactly n zeroes counting multiplicity. In elementary algebra and complex analysis, integral divisors are often called "sets with multiplicity," and one thinks of the coefficient a_k as meaning that the point C_k is to be counted as belonging to the set a_k times.

integral domain A commutative ring R with the property that $ab = 0$ implies $a = 0$ or $b = 0$. Here, a and b are elements of R. For example, the ring of integers is an *integral domain.*

integral element Let S be a commutative ring with unit, and let R be a subring of S. An element $\alpha \in S$ is *integral* over R if α is the solution to a polynomial equation $P(\alpha) = 0$, where the polynomial P has coefficients in R

and leading coefficient 1. (Such a polynomial is called a *monic polynomial.* For example, $x^2 + 5$ is monic, but $2x^2 + 5$ is not monic.)

The most important application of the concept of an integral element lies in algebraic number theory, where an element of an extension field of the rational numbers which is integral over the ring of integers is called an *algebraic integer,* or often just an *integer.* In this case, ordinary integers are often called *rational integers.*

If both R and S are fields, an integral element is called an *algebraic element.*

integral equivalence   (**1**) For modules, **Z**-equivalence. Here **Z** is the ring of integers. *See* R-equivalence, Z-equivalence.

(**2**) For matrices, two matrices M_1 and M_2 are *integrally equivalent* if there is an invertible matrix P, such that both P and P^{-1} have integer entries and $M_2 = P^t M_1 P$. Here, P^t denotes the transpose of P.

(**3**) For quadratic forms, two quadratic forms $Q_1(x) = x^t M_1 x$ and $Q_2(x) = x^t M_2 x$ are *integrally equivalent* if the matrices M_1 and M_2 are integrally equivalent. In other words, Q_1 and Q_2 are integrally equivalent if each can be transformed to the other by a matrix with integer entries.

integral extension   A commutative ring S with unit and containing a subring R is an *integral extension* of R if every element of S is integral over R. *See* integral element.

In the important special case where R and S are fields, an integral extension is called an *algebraic extension.*

The classic example of an integral extension is the ring S of algebraic integers in some algebraic extension field of the rational numbers. Here, the ring R is the ring of ordinary (i.e., rational) integers. *See* algebraic integer, integral element.

If S is an integral domain, there is an important relationship between being an integral extension and possessing certain finiteness conditions. Specifically, an integral domain is an integral extension of a subring R if and only if S is *module finite* over R. *Module finite* simply means that S is finitely generated as an R module. *See also* integral domain.

integral form   (**1**) A form, usually a bilinear, sesquilinear, or quadratic form, with integral coefficients. For example, the form $2x^2 - xy + 3y^2$ is integral.

(**2**) A form, usually a bilinear, sesquilinear, or quadratic form, expressed by means of integrals. For example, the inner product on the Hilbert space L^2 of square integrable functions,

$$(f, g) = \int_X f(t)\overline{g(t)} \, d\mu(t) \,,$$

is an integral (sesquilinear) form.

integral ideal   A non-zero ideal of the ring R of algebraic integers in an algebraic number field F. *See* algebraic integer, algebraic extension, ideal. *See also* integer, integral element, integral extension. In the elementary case where F is the field of rational numbers and R is the ring of ordinary integers, an integral ideal is simply a non-zero ideal of the ring of integers.

In algebraic number theory, a distinction is drawn between *fractional ideals* and *integral ideals.* An *integral ideal* is as defined above. By contrast, a *fractional ideal* is an R module lying in the algebraic number field F.

integrally closed   Let A be a subring of a ring C. Then the set of elements of C which are integral over A is called the *integral closure* of A in C. If A is equal to its integral closure, then A is said to be *integrally closed.*

integral quotient   In elementary arithmetic, a quotient in which both the numerator and the denominator are integers. For example, $3/4$ is an integral quotient, whereas $3.5/4.5$ is not, even though the latter represents (equals) the rational number $7/9$.

integral representation   (**1**) A *representation* of a group G is a homomorphism ϕ from G into a group M_n of $n \times n$ matrices. The representation is *integral* if each matrix $\phi(g)$ has integer entries.

(**2**) Any representation of a quantity by means of integrals. *See* integral.

integral ringed space   A *ringed space* is a topological space X together with a sheaf of rings $\mathcal{O}_X$ on X. This means that to each open set U of X, there is associated a ring $\mathcal{O}_X(U)$. (The remaining properties of sheaves need not concern us here.) The ringed space $(X, \mathcal{O}_X)$ is *integral* if each ring $\mathcal{O}_X(U)$ is an integral domain. For example, if X is an open subset of $\mathbf{C}^n$, complex n-space, or a complex analytic manifold, and $\mathcal{O}_X(U)$ is the ring of analytic functions defined on U, then $(X, \mathcal{O}_X)$ is an integral ringed space. *See* integral domain.

integral scheme   A *scheme* is a particular sort of ringed space. *See* scheme. An *integral scheme* is a scheme which is an integral ringed space. *See* integral ringed space.

intermediate field   Let F, G, and H be fields, with $F \subseteq G \subseteq H$. G is called an *intermediate field*.

internal product   Let G_1 and G_2 be Abelian groups, and let R be a commutative ring. A group homomorphism $\pi : G_1 \otimes G_2 \to (R, +)$, where $(R, +)$ is the underlying additive group of R, is an *internal product* if it satisfies $\pi(g_1 \otimes g_2) = \pi(g_1) \cdot \pi(g_2)$, where $\cdot$ is the multiplication in the ring R. Here, $G_1 \otimes G_2$ is the tensor product of G_1 and G_2. The notion is most frequently used in homological algebra, in which case π becomes a homomorphism of chain complexes of groups, and the relation $\pi(g_1 \otimes g_2) = \pi(g_1) \cdot \pi(g_2)$ only has to hold for cycles (or cocycles). *See also* chain complex, cocycle, cycle, tensor product.

internal symmetry   A symmetry that is an invertible mapping of a set onto itself. If the set has additional structure, then the mapping and its inverse must preserve that structure. For example, if the set is in addition a differentiable manifold, then the mapping (and automatically its inverse) must be differentiable. If the set is in addition a topological space, then the mapping and its inverse must both be continuous. If the set is in addition a metric space, then the mapping (and automatically its inverse) must be an isometry. *See also* inverse function, inverse mapping.

interpolating subset   Let $\mathcal{H}$ be a set of functions from a set X to a set Y. Let $a_1, a_2, a_3, \ldots$ be a finite or infinite sequence of elements of

X, and let $y_1, y_2, y_3, \ldots$ be a finite or infinite sequence of elements of Y. The set $A = \{a_1, a_2, a_3, \ldots\}$ is an $\mathcal{H}$ *interpolating subset* for the sequence $y_1, y_2, y_3, \ldots$ if there exists a function $h \in \mathcal{H}$ such that $h(a_1) = y_1$ for $i = 1, 2, 3, \ldots$. In this case, the sequence $a_1, a_2, a_3, \ldots$ is called an $\mathcal{H}$ *interpolating sequence* for $y_1, y_2, y_3, \ldots$.

The classic examples of interpolating sequences occur in the case where X is the field of complex numbers and $\mathcal{H}$ is the set of all polynomials with complex coefficients. The Lagrange interpolation theorem asserts that any finite sequence of complex numbers, $a_1, \ldots, a_n$, all terms of which are different, is an $\mathcal{H}$ interpolating sequence for any sequence $y_1, \ldots, y_n$ of complex numbers with the same number of terms.

Although the notion of an interpolating subset is usually reserved for discrete sets A as above, it makes sense in greater generality. Let A be an arbitrary subset of X, and let f be a function from A to Y. Then A is an $\mathcal{H}$ *interpolating subset* for the function f if there exists a function $h \in \mathcal{H}$ such that $h(a) = f(a)$ for all $a \in A$.

intersection multiplicity   A *variety* V is the set of common zeros of a set I of polynomials. In other words, $V = \{(x_1, \ldots, x_n) : P(x_1, \ldots, x_n) = 0 \text{ for all } P \in I\}$. Intuitively, the *intersection multiplicity* of varieties $V_1, \ldots, V_n$ at a point $x = (x_1, \ldots, x_n)$ where they intersect is the degree of tangency (or order of contact) of the intersection at x plus 1. For example, the parabolas

$$x_2 - x_1^2 = 0 \quad \text{and} \quad x_2 - x_1^2 - x_1 = 0$$

have intersection multiplicity 1 at $(x_1, x_2) = (0, 0)$ because they intersect transversally (are not tangent to each other) there. However,

$$x_2 - x_1^2 = 0 \quad \text{and} \quad x_2 - 5x_1^2 = 0$$

have intersection multiplicity 2 at $(x_1, x_2) = (0, 0)$ because they intersect tangentially to first order there. One must also include the possibility of x being a multiple point, in which case the intersection multiplicity should be $\geq$ the order of the multiple point.

Rigorously, if $D_1, \ldots, D_n$ are effective divisors on a smooth n-dimensional variety X and are in general position at a point $x \in X$, then the intersection multiplicity of $D_1, \ldots, D_n$ at x is

$$(D_1, \ldots, D_n)_x = \dim \left(\mathcal{O}_x / (f_1, \ldots, f_n) \right) .$$

Here is what all this means: X is *smooth* if it has no singular points. A *divisor* on X is a formal sum $D = a_1 C_1 + \cdots + a_k C_k$, where the a_i are integers and the C_i are distinct irreducible subvarieties of X of codimension $n - 1$. (*Irreducible* means C_i is not the union of two proper [i.e., strictly smaller] subvarieties. In the simplest case where X is an algebraic curve, the C_i are just points of X.) The divisor D is *effective* (also called *integral* or *positive*) if all the coefficients a_i are positive. $\mathcal{O}_x$ is the local ring of X at x. The *local ring* of X at x consists of quotients of polynomial functions, f/g, defined at and near x (i.e., on an open subset of X containing x) where $g(x) \neq 0$, and two such functions are identified if they agree on an open subset containing x. (In other words, $\mathcal{O}_x$ is the ring of terms of regular functions at x.) Locally, each divisor D_i is the divisor of a function f_i, and that is where the f_i come from. $(f_1, \ldots, f_n)$ is the ideal in $\mathcal{O}_x$ generated by $f_1, \ldots, f_n$ (by their terms actually). The quotient ring $\mathcal{O}_x / (f_1, \ldots, f_n)$ is not only a ring but also a finite dimensional vector space. The intersection multiplicity $(D_1, \ldots, D_n)$ is the dimension of this vector space. *See* general position, term, integral divisor, quotient ring.

The intersection multiplicity of effective divisors not in general position at x is defined in terms of the intersection multiplicity of equivalent divisors which are in general position. The intersection multiplicity of fewer than n effective divisors, say $D_1, \ldots, D_k$, is defined in terms of module length rather than the less general concept of vector space dimension. Let C be one of the irreducible components of the variety $\bigcap C_{i,j}$, where $D_i = \sum a_{i,j} C_{i,j}$. Then the *intersection multiplicity* of $D_1, \ldots, D_k$ *in the component C* is

$$(D_1, \ldots, D_k)_C = \ell \left(\mathcal{O}_C / (f_1, \ldots, f_k) \right)$$

where $\ell(\mathcal{O}_C / (f_1, \ldots, f_k))$ is the module length of the $\mathcal{O}_C$ module $\mathcal{O}_C / (f_1, \ldots, f_k)$ and $\mathcal{O}_C$ is the local ring of the irreducible subvariety C. *See also* local ring, module of finite length.

intersection number A *variety V* is the set of common zeros of a set I of polynomials. In other words, $V = \{(x_1, \ldots, x_n) : P(x_1, \ldots, x_n) = 0$ for all $P \in I\}$. Intuitively, the *intersection number* of varieties $V_1, \ldots, V_n$ is the number of points of intersection, counting multiplicity. Rigorously, if $D_1, \ldots, D_n$ are effective divisors on a smooth n-dimensional variety X (*see* intersection multiplicity for brief definitions of these terms), then the intersection number of $D_1, \ldots, D_n$ is

$$(D_1, \ldots, D_n) = \sum_{x \in S} (D_1, \ldots, D_n)_x ,$$

in other words, it is the sum of the intersection multiplicities over the finitely many points of intersection of the divisors $D_1, \ldots, D_n$. Here, $S = \bigcap_{i=1}^{n} S_i$, $S_i = \bigcup_j C_{i,j}$, and $D_i = \sum_j a_{i,j} C_j$.

It is also possible to define intersection numbers for fewer than n divisors, say $D_1, \ldots, D_k$. However, this definition is the culmination of an entire theory.

intersection product (1) Let $i(A, B; C)$ be the intersection multiplicity of two irreducible subvarieties A and B of an irreducible variety V, along a proper component C of $A \cap B$. The intersection product of A and B is

$$A \cdot B = \sum_n i(A, B; C_n) C_n$$

where the sum is taken over all the proper components C_n of $A \cap B$. If $X = \sum_\alpha a_\alpha A_\alpha$ and $Y = \sum_\beta b_\beta B_\beta$ are two cycles on V such that each component A_α of X intersects properly with each component B_β of Y, then the *intersection product* is

$$X \cdot Y = \sum_\alpha \sum_\beta a_\alpha b_\beta \left(A_\alpha \cdot B_\beta \right) .$$

(2) If M is an oriented n-dimensional manifold and a and b are members of the homology groups $H_p(M)$ and $H_q(M)$, then the *intersection product* of Lefschetz is $a \cdot b = D^{-1} a \frown b = D(D^{-1} a \smile D^{-1} b) \in H_{p+q-n}$ where D is the Poincaré-Lefschetz duality. *See also* intersection multiplicity, cup product, cap product, Poincaré-Lefschetz duality.

intransitive permutation group A *permutation group G* is a group of one-to-one and onto functions from a set X to itself. The group operation is understood to be a composition of functions, $\tau\sigma(x) = \tau \circ \sigma(x) = \tau(\sigma(x))$. Usually, but not always, the set X is finite. The permutation group G is *intransitive* if for some (and hence for all) $x \in X$, the set $O(x) = \{\sigma(x) : \sigma \in G\}$ is not equal to all of X. The set $O(x)$ is called the *orbit* of x, so we can say the permutation group G is intransitive if the orbit of any element $x \in X$ fails to be all of X. Synonym: *intransitive transformation group*. Antonyms: *transitive permutation group, transitive transformation group*.

invariance   As in ordinary non-technical English, the property of being unchanged with respect to some action or set of actions.

invariant   (1) Let L be a set. Let G be another set which acts on L. This means that there is a binary operation $\cdot$ so that $g \cdot l$ is an element of L for each $g \in G$ and $l \in L$. Usually, but not always, G is a group. An element $l \in L$ is *invariant under the action of G*, or a *G invariant*, if $g \cdot l = l$ for each $g \in G$.

Here are some examples:

Example (a): Let G be a group of functions from a set X to itself. Each $g \in G$ is assumed to be one-to-one and onto, and the group operation is a composition of functions, $g_1 g_2(x) = g_1 \circ g_2(x) = g_1(g_2(x))$. Let L be a set of functions from X to some set Y. The action of G on L is defined via a composition of functions: $g \cdot l = l \circ g$, i.e. $g \cdot l(x) = l(g(x))$.

Example (b): Let G equal the symmetric group on n letters. In other words, G is the permutation group of all permutations (one-to-one and onto functions) of the set $\{1, \dots, n\}$. The group operation is a composition of functions. Let L be the ring of all polynomials in n variables, $x_1, \dots, x_n$. If $g \in G$, and l is a monomial $x_1^{k_1} \cdots x_n^{k_n}$, then $g \cdot l = x_{g(1)}^{k_1} \cdots x_{g(n)}^{k_n}$. In other words, $g \cdot l$ is formed from l by rearranging the variables. If l is an arbitrary polynomial, then l is a sum of monomials, $l = \sum a_i l_i$. Define $g \cdot l$ by linearity, $g \cdot l = \sum a_i g \cdot l_i$. The polynomials which are invariant under the action of g are called *symmetric functions*. Thus, the symmetric functions are those polynomials which remain unchanged after rearranging their variables. It is a theorem that each symmet-

ric function is a sum of elementary symmetric functions, $p_1 = 1$, $p_2 =$ the sum of all pairs of variables, $p_2 = x_1^2 + x_1 x_2 + \cdots + x_1 x_n + x_2^2 + x_2 x_3 + \dots + x_2 x_n + \cdots + x_n^2$, $p_3 =$ the sum of all triples of variables, $p_n = x_1 x_2 \cdots x_n$.

Example (c): this is an important special case of example (a). G is a group of conformal transformations of the unit disk (in the complex plane) into itself. (Thus, G is a group of linear fractional transformations.) X is the unit disk, and L is the set of analytic functions defined on the unit disk. A function in L which is invariant under the action of G is called an *automorphic function*. *See* conformal transformation, linear fractional transformation.

Example (d): Let G be an Abelian group. Let K be a field and KG the group algebra over K. Let L be a left KG module. Define the action of G by module multiplication, $g \cdot l = gl$. In the theory of group representations, the set of all $l \subset L$ which are invariant under the action of G are called the *G-invariants* of L. The set of all G-invariants of L forms a left KG submodule of L which plays a key role in the theory of group representations. *See also* group algebra.

(2) Let G and L be as in (1). A subset V of L is *invariant under the action of G* if $g \cdot v \in V$ for all $v \in V$. Here is an important example: Let T be a bounded linear operator on a Hilbert space H. Let $G = \{T\}$, the set consisting of T alone. Let $L = H$. Define the action of G on L by operator application, $T \cdot l = T(l)$. A subspace of L which is invariant under the action of G is called an *invariant subspace* for T. A famous unsolved problem is the invariant subspace problem, often called the *invariant subspace conjecture:* Does every bounded linear operator on an infinite dimensional complex Hilbert space H have a proper (not $\{0\}$, not all of H) closed invariant subspace?

(3) A bilinear form f on a Lie algebra L is called an *invariant form* if $f([a\,c], b) + f(a, [b\,c]) = 0$. Here, $[a\,c]$ is the Lie algebra product of a and c.

(4) A quantity which is left unchanged under the action of a prescribed class of functions *between sets* is also called an *invariant*. For example, the Euler characteristic is a topological invariant because it is preserved under topological homeomorphisms. (A one-to-one and onto

function from one topological space to another is called a *homeomorphism* if both it and its inverse are continuous.)

(**5**) Let X be a set and let E be an equivalence relation on the set X. Let $f : X \to Y$ be a function from X to another set Y. If $f(x_1) = f(x_2)$ whenever $(x_1, x_2) \in E$, that is whenever x_1 is equivalent to x_2 modulo the equivalence relation E, then f is called an *invariant of E*. The function f is called a *complete invariant of E* if $(x_1, x_2) \in E$ if and only if $f(x_1) = f(x_2)$. Finally, a finite or infinite set F of functions on X is called a *complete system of invariants* for E if $(x_1, x_2) \in E$ if and only if $f(x_1) = f(x_2)$ for all functions $f \in F$.

Here is a well-known example: Let X be the set of all $m \times n$ matrices with coefficients in a field F. Define two such matrices M_1 and M_2 to be equivalent if there exist invertible square matrices P and Q such that $M_1 = P M_2 Q$, and let E be the resulting equivalence relation. It is a theorem of linear algebra that the *rank* of a matrix M is a complete invariant for E, that is $m \times n$ matrices M_1 and M_2 are equivalent under the above definition if and only if they have the same rank.

invariant derivation   Let D be a derivation on the function field $K(A)$ of an Abelian variety A, where K is the universal domain of A. Let T_a be translation by an element $a \in A$. If $(Df) \circ T_a = D(f \circ T_a)$, for every $f \in K(A)$, then D is called an *invariant derivation* on A.

invariant element   Let $T : X \to X$ be a function or mapping. If $x \in X$ has the property that $T(x) = x$, then x is called an *invariant* element of the operator T. This concept arises in analysis, topology, algebra, and many other branches of mathematics.

invariant element   Let $T : X \to X$ be a function or mapping. If $x \in X$ has the property that $T(x) = x$, then x is called an invariant element of the operator T. This concept arises in analysis, topology, algebra, and many other branches of mathematics.

invariant factor   Let A be an $n \times n$ matrix with distinct eigenvalues $\lambda_i,$, $i = 1, 2, \ldots, k$, and Jordan normal form J. Consider the ma-

trix B_1, obtained from J by taking the direct sum of k Jordan blocks, one for each distinct eigenvalue λ_i, having maximal order among all Jordan blocks corresponding to λ_i. Next consider B_2, obtained similarly to B_1, but from the remaining Jordan blocks in J. Continue in this manner until all Jordan blocks of J have been used, thus obtaining a sequence $B_1, B_2, \ldots, B_s$ of matrices whose sizes are non-increasing and whose direct sum is by construction permutationally similar to J.

The characteristic polynomials of the matrices B_j, $j = 1, 2, \ldots, s$ are known as the *invariant factors* of A. It is worth noting that for each $j = 1, 2, \ldots, s$, by construction, the characteristic polynomial of B_j coincides with the minimal polynomial of B_j. In particular, the minimal polynomial of B_1 is the minimal polynomial of A. It follows that two matrices are similar if and only if they have the same invariant factors.

invariant field   Let G be a group of field automorphisms of a field F. A subfield H of F is *invariant for G*, or *G-invariant*, if $g(h) = h$ for all $g \in G$ and $h \in H$; in other words, the subfield H is *G*-invariant if G fixes the elements of H. Synonym: *fixed field*.

invariant form   (**1**) A bilinear or quadratic form which is invariant under the action of a set of transformations. *See* invariant.

(**2**) A bilinear form f on a Lie algebra L such that $f([a\,c], b) + f(a, [b\,c]) = 0$. Here, $[a\,c]$ is the Lie algebra product of a and c.

invariant of group   Let G be a finite Abelian group. By the Fundamental Theorem of Abelian Groups,

$$G = \sigma(m_1) \oplus \sigma(m_2) \oplus \cdots \oplus \sigma(m_s) ,$$

where $\sigma(m_i)$ is a cyclic group of order m_i and m_i divides m_{i+1}. The numbers $m_1, m_2, \ldots, m_s$ are uniquely determined by G, are invariant under group isomorphism, and are called the *invariants of the group G*.

invariant of weight w   Let R be a ring and let L be a left R module. Let G be a set which acts on L. This means there is a binary operation $\cdot$ so that $g \cdot l$ is an element of L for each $g \in G$

and $l \in L$. Usually, G is a group. Let w be a function from G to R. An element $l \in L$ is an *invariant of weight w under the action of G,* or a *G invariant of weight w*, if $g \cdot l = w(g)l$ for each $g \in G$.

Here is an example: Let L be the set of all analytic functions in the upper half plane of the complex plane. Let G be the modular group. The *modular group* is the group of all linear fractional transformations

$$g(z) = \frac{az+b}{cz+d},$$

where a, b, c, and d are integers and the determinant $ad - bc = 1$. G acts on L by composition of functions,

$$g \cdot l(z) = l \circ g(z) = l(g(z)) = l\left(\frac{az+b}{cz+d}\right).$$

A *modular form of weight k* is a G invariant of weight w, where $w(g) = (cz+d)^k$. In other words, l is a modular form of weight k if $l \in L$ and

$$l\left(\frac{az+b}{cz+d}\right) = (cz+d)^k l(z),$$

whenever a, b, c, and d are integers and $ad - bc = 1$.

There is a companion notion of weight for right R modules. Furthermore, the notion of a G invariant of weight w extends to the situation where R and L are simply sets, and R acts on L.

inverse   Let G be a set with a binary operation $\cdot$ and an identity e. This means $g_1 \cdot g_2 \in G$ whenever g_1 and g_2 belong to G, and $g \cdot e = e \cdot g = g$ for all $g \in G$. The element $h \in G$ is an *inverse* of $g \in G$ if $g \cdot h = e$ and $h \cdot g = e$.

Often g is an element of a group G. What is very special about this case is that (a) either of the conditions $gh = e$ and $hg = e$ implies the other, and (b) the inverse of g always exists and is uniquely determined by g. The inverse of g is frequently denoted by g^{-1}. *See* group. *See also* inverse function, inverse morphism.

inverse element   An element of a set G which is the inverse of another element of G. *See* inverse.

inverse function   Let f be a function from a set X to a set Y. Diagrammatically, $f : X \to Y$. The *inverse function* to f, if it exists, is the function $g : Y \to X$ such that $f \circ g = i_Y$ and $g \circ f = i_X$. (Here, i_X and i_Y are the identity maps on X and Y, and $\circ$ denotes composition of functions.) In other words, $f(g(y)) = y$ for all $y \in Y$, and $g(f(x)) = x$ for all $x \in X$. The function f has an inverse if and only if f is one-to-one and onto, and the inverse g is defined by $g(y) =$ the unique element x such that $f(x) = y$. The inverse function to f is frequently denoted by f^{-1}. *See also* identity map, inverse mapping, inverse morphism.

Example: Let $f(x) = 10^x$, for x real. The inverse function to f is $g(x) = \log_{10} x$, for $x > 0$.

inverse limit   Suppose $\{G_\mu\}_{\mu \in I}$ is an indexed family of Abelian groups, where I is a preordered set. Suppose that there is also a family of homomorphisms $\varphi_{\mu\nu} : G_\mu \to G_\nu$, defined for all $\mu < \nu$, such that if $\mu < \nu < \kappa$, then $\varphi_{\nu\kappa} \circ \varphi_{\mu\nu} = \varphi_{\mu\kappa}$. Consider the direct product of the groups G_μ and define π_μ to be the projection onto the μth factor in this direct product. Then the inverse limit is defined to be the subgroup $G_\infty = \{x : \mu < \nu$ implies $\pi_\mu(x) = \varphi_{\nu\mu} \circ \pi_\nu(x)\}$. *See also* preordered set.

inverse mapping   *See* inverse function.

inverse matrix   The matrix B, if it exists, such that $AB = BA = I$. In more detail, let A be a square $n \times n$ matrix, and let I be the $n \times n$ identity matrix, that is the $n \times n$ matrix with entry 1 in each diagonal position and 0 elsewhere. An $n \times n$ matrix B is the *inverse* of A if $AB = I$ and $BA = I$. Either condition implies the other. The inverse of A, if it exists, is uniquely determined by A and is denoted by A^{-1}. It is a theorem of linear algebra that a matrix is invertible (i.e., has an inverse) if and only if its determinant is non-zero. There is a determinantal (involving determinants) formula for the inverse of A, $A^{-1} = \det(A)^{-1} \mathrm{adj}(A)$, where $\det(A)$ is the determinant of A and $\mathrm{adj}(A)$ is the *classical adjoint* of A, that is the transpose of the matrix of cofactors. *See* cofactor, determinant, transpose.

inverse morphism   Let $\mathcal{C}$ be a category, let A and B be objects in $\mathcal{C}$, and let $f : A \to B$ be a morphism in $\mathcal{C}$. (A *morphism* is a generalization of a function or mapping.) A morphism $g : B \to A$ is the *inverse morphism* of f if $f \circ g$ is the identity morphism on object B, and $g \circ f$ is the identity morphism on object A. The inverse morphism of f, if it exists, is uniquely determined by f.

There are also notions of *left* and *right inverse morphisms*. Morphism g is a *left inverse morphism* of f if $g \circ f$ is the identity morphism on object A, and is a *right inverse morphism* of f if $f \circ g$ is the identity morphism on B. *See also* identity morphism, inverse function.

inverse operation   *See* inverse function.

inverse proportion   Quantity a is *inversely proportional* to quantity b, or *varies inversely with* quantity b, if there is a constant k different from 0 such that $a = k/b$. For example, Newton's law of universal gravitation, "The gravitational force between two masses is directly proportional to the product of the masses and inversely proportional to the square of the distance between them," is given by the formula $F = GmM/r^2$, where G is a constant of nature called the *gravitational constant*.

inverse ratio   (**1**) The *inverse ratio* to a/b is b/a.

(**2**) *Inverse ratio* also means *inverse proportion*, as in "a varies in inverse ratio to b." *See* inverse proportion.

inverse relation   The relation formed by reversing the ordered pairs in a given relation. In more detail, a relation R is a subset of the Cartesian product $X \times Y$, where X and Y are sets. (The *Cartesian product* $X \times Y$ of X and Y is simply the set of all ordered pairs (x, y), where $x \in X$ and $y \in Y$.) The *inverse relation* to R is the relation $\{(y, x) : (x, y) \in R\}$. The inverse relation to R is usually denoted by R^{-1}. Note that if the relation R is a subset of the Cartesian product $X \times Y$, then the inverse relation R^{-1} is a subset of $Y \times X$. Example: An inverse function is a special sort of inverse relation. *See* inverse function.

inverse transformation   *See* inverse function.

inverse trigonometric function   The inverse functions to the trigonometric functions sin, cos, tan, cot, sec, and csc; that is, the arcsin, the arccosine, the arctangent, the arccotangent, the arcsecant, and the arccosecant, respectively. They are denoted by arcsin, arccos, arctan, arccot, arcsec, and arccsc, or by $\sin^{-1}$, $\cos^{-1}$, $\tan^{-1}$, $\cot^{-1}$, $\sec^{-1}$, and $\csc^{-1}$. Note that, in formulas involving trigonometric functions and inverse trigonometric functions, $\sin^{-1}(x)$ is *not* equal to the number $1/\sin(x)$, but rather to the value of the arcsin of x. Similar comments apply to $\cos^{-1}(x)$, etc.

Because the trigonometric functions sin, cos, etc. are not one-to-one, the inverse trigonometric functions are really inverse relations, although they may be thought of as multiple valued functions. Thus $\arcsin(x)$ is any angle y such that $\sin(y) = x$, and similarly with arccos (x), etc. To make the inverse trigonometric functions into single valued functions, one must specify a *branch* or, equivalently, an interval in which the trigonometric function is one-to-one. For example, the *principal branch* of the arcsin is the inverse of $\sin(x)$ restricted to the interval $-\frac{\pi}{2} \le y \le \frac{\pi}{2}$; it is denoted by Arcsin or Sin^{-1}, with a capital letter. Thus $\text{Arcsin}(x)$ is the unique angle y in the interval $-\frac{\pi}{2} \le y \le \frac{\pi}{2}$ such that $\sin(y) = x$. The principal branches of the other inverse trigonometric functions are also denoted by capital letters, $\text{Arccos} = \text{Cos}^{-1}$, $\text{Arctan} = \text{Tan}^{-1}$, etc. The principal branch of the arctan takes values in the same interval as the principal branch of the arcsin, namely $-\frac{\pi}{2} \le y \le \frac{\pi}{2}$, but the principal branch of the arccos takes values in the interval $0 \le y \le \pi$.

inverse variation   *See* inverse proportion.

inversion   The act of computing the inverse. *See* inverse.

inversion formula   Any of a number of formulas for computing the inverse of a quantity. The two most celebrated probably are:
(i.) the inversion formula for computing the inverse of a matrix, $A^{-1} = \det(A)^{-1}\text{adj}(A)$, where $\det(A)$ is the determinant of A and $\text{adj}(A)$

is the classical adjoint of A, that is the transpose of the matrix of cofactors. *See* inverse matrix.

(ii.) The Fourier inversion formula,

$$f(x) = \frac{1}{2\pi} \int_{-\infty}^{+\infty} \left[\int_{-\infty}^{+\infty} f(\lambda) e^{i(x-\lambda)\mu} \, d\lambda \right] d\mu .$$

The Fourier inversion formula is really a statement that if f has Fourier transform,

$$F(\mu) = \frac{1}{\sqrt{2\pi}} \int_{-\infty}^{+\infty} f(\lambda) e^{-i\lambda\mu} \, d\lambda ,$$

then the inverse Fourier transform is given by the inversion formula,

$$f(x) = \frac{1}{\sqrt{2\pi}} \int_{-\infty}^{+\infty} F(\mu) e^{i\mu x} \, d\mu .$$

(There have to be hypotheses on f, say $f \in L^1 \cap L^2$, for this to work.)

invertible element (1) An element with an inverse element. *See* inverse, inverse element.

(2) Let R be a ring with unit e. An element r of R, for which there exists another element a in R such that $ar = e$ and $ra = e$, is called an *invertible element* of R. This is, of course, a special case of (1) above. The element a, if it exists, is uniquely determined by r, and is denoted by r^{-1}. *See* unit.

If only the condition $ar = e$ holds, then r is said to be *left invertible*. Similarly, if only the condition $ra = e$ holds, then r is said to be *right invertible*.

invertible function A function $f : S \to T$ such that there is a function $g : T \to S$ with $f \circ g = \mathrm{id}_T$ and $g \circ f = \mathrm{id}_S$. Often the word "function" is used to specify that T is a field.

invertible map A function $f : S \to T$ such that there is a function $g : T \to S$ with $f \circ g = \mathrm{id}_T$ and $g \circ f = \mathrm{id}_S$. Often the word "map" is used to specify that T is not a field of scalars.

invertible sheaf A locally free sheaf of rank 1. In more detail, a *ringed space* is a topological space X together with a sheaf of rings $\mathcal{O}_X$ on X. This means that to each open set U of X, there is associated a ring $\mathcal{O}_X(U)$. (The remaining properties of sheaves need not concern us here.) A

sheaf of $\mathcal{O}_X$ modules is defined similarly, except that to each open subset U of X, there is associated an $\mathcal{O}_X$ module $\mathcal{F}(U)$. A sheaf $\mathcal{F}$ of $\mathcal{O}_X$ modules is *locally free of rank* 1, or *invertible,* if X can be covered by open sets U_α, $\alpha \in A$, such that $F(U_\alpha)$ is isomorphic to $\mathcal{O}_X(U_\alpha)$. *See also* ringed space, sheaf.

involution (1) A function ϕ from a set X to itself, such that $\phi^2 = \phi$. Here, $\phi^2(x) = \phi \circ \phi(x) = \phi(\phi(x))$.

(2) Let A be an algebra over the complex numbers. A function ϕ from A to itself is an *involution* if it satisfies the following four properties for all x and y in A and all complex numbers λ: (i.) $\phi(x + y) = \phi(x) + \phi(y)$, (ii.) $\phi(\lambda x) = \bar{\lambda} \phi(x)$, (iii.) $\phi(xy) = \phi(y)\phi(x)$, (iv.) $\phi((\phi(x)) = x$.

$\phi(x)$ is frequently denoted by x^*, and then the four properties take the more familiar form: (i.) $(x + y)^* = x^* + y^*$, (ii.) $(\lambda x)^* = \bar{\lambda} x^*$, (iii.) $(xy)^* = y^* x^*$, (iv.) $x^{**} = x$.

Examples: (i.) A is the complex numbers. $x^* = \bar{x}$, the complex conjugate of x. (ii.) A is the algebra of bounded linear operators on a Hilbert space. T^* is the adjoint of T.

irrational equation An equation with irrational coefficients. *See also* irrational number.

irrational exponent An exponent which is irrational. For example, the expressions 2^π and e^π involve irrational exponents. *See also* irrational number.

irrational expression An expression involving irrational numbers. *See also* irrational number. Although the word *expression* is often used loosely in elementary mathematics without a rigorous definition, it is possible to define *expression* rigorously by specifying the rules of a formal grammar.

irrational number A real number r which cannot be expressed in the form p/q, where p and q are integers. Equivalently, a real number which is not a rational number. It was the great discovery of the ancient Greek mathematician and philosopher Pythagoras that $\sqrt{2}$ is irrational. The numbers e and π are also irrational. These last two are irrational in a very strong sense, they are *transcendental,* but it took until the nineteenth century to prove this. *See also* transcendental number.

irreducible algebraic curve　An algebraic curve is a variety of dimension 1 in 2-dimensional affine or projective space. (A *variety* is the solution set to a system of polynomial equations.) An algebraic curve is *irreducible* if it is not the union of two proper (strictly smaller) subvarieties. For example, the parabola $y - x^2 = 0$ is an irreducible algebraic curve, but the pair of lines $x^2 - y^2 = 0$ is a reducible algebraic curve because it can be decomposed into the union of the two lines $x - y = 0$ and $x + y = 0$.

irreducible R-module　Let R be a ring and let M be a module over R. We say that M is irreducible over R if R has no submodules. In some contexts, we say that M is irreducible if M cannot be written as a direct sum of proper sub-modules. These are also sometimes called simple modules.

irreducible character　A character of a finite group which is not a sum of characters different from itself. Every character of a finite group is a sum of irreducible characters. *See also* character of group.

irreducible co-algebra　A co-algebra in which any two non-zero subco-algebras have non-zero intersection. A co-algebra C is irreducible if and only if C has a unique simple subco-algebra.

irreducible component　(**1**) In algebraic geometry, a variety is the solution set of a system of polynomial equations, usually in more than one variable. A variety is *irreducible* if it is not the union of two proper (strictly smaller) subvarieties. An *irreducible component* of a variety is a maximal irreducible subvariety. That is, a subvariety $W \subseteq V$ is an irreducible component of a variety V if (i.) W is irreducible, and (ii.) there is no irreducible variety W' properly between W and V ($W \subset W' \subset V$, $W \neq W'$, $W' \neq V$). Every variety is the finite union of its irreducible components. Example: $V = \{(x, y) : x^2 - y^2 = 0\}$. The irreducible com-

ponents of V are the two lines $x - y = 0$ and $x + y = 0$.

The notion of an irreducible component extends to varieties in other contexts, for example to analytic varieties.

(**2**) In combinatorial group theory, every Coxeter group can be written as the direct sum of (possibly infinitely many) irreducible Coxeter groups, called the *irreducible components* of the Coxeter group. *See also* irreducible Coxeter group.

irreducible constituent　Let $\mathcal{Z}$ denote the rational integers and $\mathcal{Q}$ the rational field. Let T be a $\mathcal{Z}$-representation of a finite group G. If T is $\mathcal{Q}$-irreducible, then T is called an irreducible constituent of the group G.

irreducible constituent　Let $\mathbf{Z}$ denote the rational integers and $\mathbf{Q}$ the rational field. Let T be a $\mathbf{Z}$-representation of a finite group G. If T is $\mathbf{Q}$-irreducible, then T is called an *irreducible* constituent of the group G.

irreducible Coxeter complex　A Coxeter complex for which the associated Coxeter group is an irreducible Coxeter group. In more detail, let (W, S) be a Coxeter group. For now, it suffices that W is a (possibly infinite) group and S is a set of generators for W. Define a *special coset* to be a coset of the form $w\langle S' \rangle$, where $w \in W$, $S' \subseteq S$, and $\langle S' \rangle$ is the group generated by S'. The *Coxeter complex* Σ associated with (W, S) is the partially ordered set of special cosets, ordered by reverse inclusion: $B \leq A$ if and only if $B \supseteq A$. Σ is an *irreducible Coxeter complex* if its Coxeter group (W, S) is an irreducible Coxeter group. *See also* irreducible Coxeter group.

Although Coxeter complexes are abstractly defined, there is a rich geometry associated with them, resembling the geometry of simplicial complexes.

irreducible Coxeter group　A Coxeter group which cannot be written as the direct sum of two other Coxeter groups. In more detail, Coxeter groups are generalizations of finite reflection groups. Let W be a (possibly infinite) group, and let S be a set of generators for W. The pair (W, S) is called a *Coxeter group* if two things are true: (i.) Each element of S has order 2 ($s^2 = s$ if $s \in S$), and (ii.) W is defined by the system of generators and relations: set of generators = S; set of relations = $\{(st)^{m(s,t)} = 1\}$, where $m(s, t)$ is the order of the element st in the group W, and there is one relation for each pair (s, t) with s and t in S and $m(s, t) < \infty$.

A Coxeter group is *irreducible* if it cannot be written as the direct sum of two other Coxeter groups. In other words, the Coxeter group (W, S) is irreducible if it cannot be written as $(W, S) = (W' \times W'', S' \cup S'')$, where (W', S') and (W'', S'') are themselves Coxeter groups. (Equivalently, a Coxeter group is irreducible if its Coxeter diagram is connected.) Every Coxeter group can be written as the direct sum of (possibly infinitely many) irreducible Coxeter groups, called the *irreducible components* of the group. *See also* Coxeter diagram, irreducible Coxeter complex.

irreducible decomposition   Informally, a decomposition of an object into irreducible components or elements. *See* irreducible component, irreducible element.

irreducible element   (**1**) An element a of a ring R with no proper factors in the ring. This means that there do not exist elements b and c in R, different from 1 and a, such that $a = bc$. Example: If R is the ring of integers, the irreducible elements are the prime numbers. *See also* prime number.

(**2**) A join or meet irreducible element of a lattice. *See* join irreducible element, meet irreducible element.

irreducible equation   A polynomial equation $P(x) = 0$, where the polynomial P is irreducible. *See also* irreducible polynomial.

irreducible fraction   A fraction a/b, where the integers a and b have no common factors other than 1 and -1. In other words, a fraction reduced to lowest terms.

irreducible homogeneous Siegel domain
Siegel domains are special kinds of domains in complex N space, $\mathbf{C}^N$. An easy way to construct new Siegel domains is to take the Cartesian product of two given Siegel domains. Thus if S_1 and S_2 are Siegel domains, $S = S_1 \times S_2 = \{s = (s_1, s_2) : s_1 \in S_1, s_2 \in S_2\}$ will also be a Siegel domain. A Siegel domain is *irreducible* if it is not the Cartesian product of two other Siegel domains. A Siegel domain is *homogeneous* if it has a transitive group of analytic (holomorphic) automorphisms. An *irreducible homogeneous Siegel domain* is a homogeneous Siegel domain which is not the Cartesian product of two other homogeneous Siegel domains. *See* homogeneous domain, Siegel domain.

irreducible linear system   A system of linear equations where no equation is a linear combination of the others. It is a theorem that a system of n linear equations in n unknowns is irreducible if and only if the determinant of the matrix of coefficients is not 0. The methods of row and column reduction provide computationally efficient tests for irreducibility. Synonym: linearly independent system of linear equations. *See also* linear combination, linearly independent elements.

irreducible matrix   *See* Frobenius normal form.

irreducible module   The module analog of a simple group. Specifically, an R module, where R is a ring, is an *irreducible module* if it contains no proper R submodules. For example, $\mathbf{Z}_p$, the integers modulo a prime number p, is an irreducible $\mathbf{Z}$ module. (Here, $\mathbf{Z}$ is the ring of integers.) In the case where the ring R is not commutative, the notion of irreducibility extends to left and right R modules.

irreducible polynomial   A polynomial with no proper factors. In greater detail, if R is a ring and $R[x]$ denotes the ring of polynomials with coefficients in R, then a polynomial P in $R[x]$ is *irreducible* if it is an irreducible element of the ring $R[x]$. *See* irreducible element. Example: If $\mathbf{R}$ is the field of real numbers and $\mathbf{C}$ is the field of complex numbers, the polynomial $P(x) = x^2 + 1$ is irreducible in $\mathbf{R}[x]$ but reducible (it factors as $(x + i)(x - i)$) in $\mathbf{C}[x]$.

irreducible projective representation   A *projective representation* of a group G is a function T from G into the group $GL(V)$ of invertible linear transformations on a vector space V, satisfying two additional axioms. (*See* projective representation.) T is *irreducible* if there does not exist a proper ($\neq 0$, $\neq V$) subspace W of V such that $T(g)(w) \in W$ for all $g \in G$ and $w \in W$. *See also* irreducible representation, irreducible unitary representation.

irreducible representation   A *representation* of a group G is a homomorphism T from G into the group GL(V) of invertible linear transformations on a vector space V. T is an *irreducible representation* if there does not exist a proper ($\neq 0$, $\neq V$) subspace W of V such that $T(g)(w) \in W$ for all $g \in G$ and $w \in W$.

The notion extends to other contexts. For example, V may be a Hilbert space and GL(V) may be replaced by the topological group of invertible bounded linear operators on V. In this case, the homomorphism T is required to be continuous, and the subspaces W are required to be closed. *See also* irreducible unitary representation.

irreducible R-module   Let R be a ring and let M be a module over R. We say that M is *irreducible* over R if R has no submodules. In some contexts, we say that M is irreducible if M cannot be written as a direct sum of proper sub-modules. These are also sometimes called simple modules.

irreducible scheme   A scheme whose underlying topological space is irreducible. In more detail, a scheme is a particular type of ringed space, $(X, \mathcal{O}_X)$. Here, X is a topological space and $\mathcal{O}_X$ is a sheaf of rings on X. The scheme $(X, \mathcal{O}_X)$ is irreducible if X is not the union of two proper ($\neq \emptyset$, $\neq X$) closed subsets. *See also* ringed space, scheme.

irreducible Siegel domain   *Siegel domains* are special kinds of domains in complex N space, $\mathbf{C}^N$. An easy way to construct new Siegel domains is to take the Cartesian product of two given Siegel domains. Thus if S_1 and S_2 are Siegel domains, $S = S_1 \times S_2 = \{s = (s_1, s_2) : s_1 \in S_1, s_2 \in S_2\}$ will also be a Siegel domain. A Siegel domain is *irreducible* if it is not the Cartesian product of two other Siegel domains.

See Siegel domain. *See also* irreducible homogeneous Siegel domain.

irreducible tensor   An element of the tensor product $V \otimes W$ of two vector spaces, which *cannot* be written as $v \otimes w$, for $v \in V$ and $w \in W$. Also called *irreducible tensor operators* or *spherical tensor operators*.

Classically, let $[a, b] = ab - ba$ and let j_x, j_y, j_z be the x-, y-, and z- components of the angular momentum $\mathbf{j}$. An *irreducible tensor* of rank k is a dynamical quantity T_q^k, where $q = k, k-1, \ldots, -k$, that satisfies the following commutation relations:

$$\left[j_z, T_q^k \right] = q T_q^k$$

$$\left[j_x \pm i j_y \right] = \sqrt{(k \mp q)(k \pm q + 1)} T_{q \mp 1}^k \ .$$

irreducible unitary representation   A *unitary representation* of a (topological) group G is a (continuous) homomorphism T from G into the group $U(H)$ of unitary operators on a Hilbert space H. T is an *irreducible unitary representation* if there does not exist a proper ($\neq 0$, $\neq H$) closed subspace W of H such that $T(g)(w) \in W$ for all $g \in G$ and $w \in W$. *See also* irreducible representation.

irreducible variety   A variety is the solution set of a system of polynomial equations (usually in several variables). An *irreducible variety* is a variety V which is not the union of two proper ($\neq \emptyset$, $\neq V$) subvarieties. For example, the parabola $y - x^2 = 0$ is an irreducible variety, but the variety $x^2 - y^2 = 0$ is reducible because it can be decomposed into the union of the two lines $x - y = 0$ and $x + y = 0$.

irredundant   In a lattice L, a representation of an element a as a join $a = a_1 \vee \cdots \vee a_n$ is *irredundant* if omitting any of the elements a_i from the join produces an element b strictly smaller than a. There is a dual notion for meets: A representation of an element a as a meet $a = a_1 \wedge \cdots \wedge a_n$ is *irredundant* if omitting any of the elements a_i from the meet produces an element b strictly larger than a. *See* join, lattice, meet.

Example: R is a Noetherian ring (a commutative ring satisfying the ascending chain condition). L is the lattice of ideals of R, ordered by inclusion. It is a theorem of ring theory, the Lasker-Noether Theorem, that every ideal in R has an irredundant representation as an intersection of primary ideals. This theorem almost completely describes the ideal theory of Noetherian rings, including such rings as the ring of polynomials in several variables with coefficients in a field, and the ring of terms of holomorphic (analytic) functions in several complex variables. *See also* Noether, Noetherian ring.

irregularity In algebraic geometry, the dimension of the Picard variety of a non-singular projective algebraic variety. *See also* Picard variety.

irregular prime A prime number p which divides the numerator of one or more of the Bernoulli numbers $B_2, B_3, \ldots, B_{p-3}$. A prime number which is not irregular is called a *regular prime*. Irregular primes were of interest because they were the class of exceptional primes for which Kummer's proof of Fermat's Last Theorem does not work. Wiles' recent proof of Fermat's Last Theorem probably makes the distinction between regular and irregular primes uninteresting, but one never knows. *See* Bernoulli number, Fermat's Last Theorem.

irregular variety A non-singular projective algebraic variety with non-zero irregularity. A variety of zero regularity is called a *regular variety*. *See also* irregularity.

isogenous Abelian varieties A pair of Abelian varieties of equal dimension, for which there is a rational group homomorphism from one variety onto the other. In more detail, an *Abelian variety* is, among other things, an algebraic variety which is also an Abelian group. A *rational group homomorphism* from one Abelian variety to another is a rational map which is also a group homomorphism. *See also* rational map.

isogenous groups A pair of topological groups (usually Lie groups) for which there is an isogeny from one to the other. *See also* isogeny.

isogeny (1) A Lie group map (a continuous, differentiable group homomorphism) $\phi : G \to H$, where G and H are Lie groups, which is

a covering space map of the underlying manifolds. The map ϕ is a *covering space map* if it is continuous and, for each $h \in H$, there exists a neighborhood U of h such that $\phi^{-1}(U)$ is a disjoint union of open sets in G mapping homeomorphically to U under ϕ.

(2) A topological group homomorphism (a continuous group homomorphism) $\phi : G \to H$, where G and H are topological groups, which is a covering space map of the underlying topological spaces.

(3) An epimorphism $\phi : G \to H$ of group schemes (over a ground scheme S) such that the kernel of ϕ is a flat, finite group scheme over S. *See also* epimorphism, scheme.

isolated component Let I be an ideal in a commutative ring R, and let $I = Q_1 \cap \cdots \cap Q_k$ be a short representation of I as an intersection of primary ideals. (*See* short representation.) Let $P_1, \ldots, P_k$ be the prime ideals belonging to $Q_1, \ldots, Q_k$. (The easiest way to specify P_i is to note that P_i is the radical of Q_i, i.e., $P_i = \{p \in R : p^n \in Q_i \text{ for some integer } n\}$.) Renumbering the primary ideals Q_i if necessary, an ideal $J = Q_1 \cap \cdots \cap Q_r$ (with $1 \le r \le k$) is an *isolated component* of I if none of the prime ideals $P_1, \ldots, P_r$ contains a prime ideal P_j not in the set $\{P_1, \ldots, P_r\}$.

Isolated components are of interest because they introduce uniqueness into the representation theory of ideals in commutative Noetherian rings. Although there are often many different short representations of an ideal I, the isolated components of I are uniquely determined. *See also* isolated primary component, Noetherian ring, primary ideal, prime ideal, radical, short representation.

isolated primary component An isolated component J of an ideal I in a commutative ring, such that J is a primary ideal. *Isolated primary components* are of interest because they must occur among the primary ideals of every short representation of I. *See also* isolated component, short representation.

isomorphic Two groups G and H are *isomorphic* if there is an isomorphism $\phi : G \to H$ between them. Isomorphic groups are regarded as being "abstractly identical," or different realizations of the same abstract group. The notion extends to other algebraic structures such as rings, to the completely general algebraic structures defined in universal algebra, and even to the theory of categories. *See also* isomorphism.

isomorphism   (1) In group theory, a mapping $\phi : G \rightarrow H$ between two groups, G and H, which is one-to-one (injective), onto (surjective), and which preserves the group operation, that is $\phi(g_1 \cdot g_2) = \phi(g_1) \cdot \phi(g_2)$. The notion extends to rings, where ϕ is required to preserve the ring addition and multiplication, to vector spaces, where ϕ is required to preserve vector addition and scalar multiplication (i.e., $\phi(\lambda_1 v_1 + \lambda_2 v_2) = \lambda_1 \phi(v_1) + \lambda_2 \phi(v_2)$), and to completely general algebraic contexts (see (2) below).

(2) In universal algebra, a mapping $\phi : \mathcal{G} \rightarrow \mathcal{H}$ between two (universal) algebras $\mathcal{G}$ and $\mathcal{H}$, which is one-to-one (injective), onto (surjective), and which preserves the operations of $\mathcal{G}$ and $\mathcal{H}$. In more detail, $\mathcal{G} = (G, F_{\mathcal{G}})$ and $\mathcal{H} = (H, F_{\mathcal{H}})$, where G is a set and $F_{\mathcal{G}}$ is a set of functions from finite Cartesian products of G with itself to G ($F_{\mathcal{G}}$ is called the set of operations on $\mathcal{G}$), and similarly for $\mathcal{H}$. Thus the functions in $F_{\mathcal{G}}$ are G-valued functions $f(g_1, \ldots, g_n)$ of n G-valued variables, and the value of n may vary with the function f. To be an isomorphism, ϕ is required to be a one-to-one and onto function from the set G to the set H, and for every function $f \in F_{\mathcal{G}}$, there must be a function $h \in F_{\mathcal{H}}$, such that $\phi(f(g_1, \ldots, g_n)) = h(\phi(g_1), \ldots, \phi((g_n))$, and *vice-versa*. (This is what is meant by "preserving the operations of $\mathcal{G}$ and $\mathcal{H}$.")

In the special case where $\mathcal{G}$ and $\mathcal{H}$ are groups, $F_{\mathcal{G}}$ equals the singleton set containing the group operation of G (the group multiplication), and similarly for $F_{\mathcal{H}}$. We thus recapture the motivating case of group isomorphisms.

(3) In category theory, a morphism $\phi : A \rightarrow B$ between two objects of a category with an inverse morphism. In other words, for ϕ to be an isomorphism, there must also be a morphism $\psi : B \rightarrow A$ (the inverse of ϕ) such that $\psi \circ \phi = \iota_A$ and $\phi \circ \psi = \iota_B$. Here, ι_A and ι_B are the identity morphisms on A and B, respectively.

The category theoretic definition captures all of cases (1) and (2) above, and also includes such examples as *isomorphisms of topological groups,* where it is required that an isomorphism ϕ be a group isomorphism *and* that ϕ and ϕ^{-1} be continuous. *See also* category, homomorphism, morphism, identity morphism.

Isomorphism Theorem of Class Field Theory
Let k be an algebraic number field. Let $I(m)$ be the multiplicative group of all fractional ideals of k which are relatively prime to a given integral divisor m of k. The Galois group of a class field K/k for an ideal group $H(m)$ is isomorphic to $I(m)/H(m)$. Therefore, every class field K/k is an Abelian extension of k.

isomorphism theorems of groups   The three standard theorems describing the relationship between homomorphisms, quotient groups, and normal subgroups. Let G and H be groups, and let $\phi : G \rightarrow H$ be a homomorphism with kernel K. (The kernel of ϕ is the set $K = \{g \in G : \phi(g) = e\}$, where e is the group identity element in H.) The *First Isomorphism Theorem* states that K is a normal subgroup of G (i.e., $gK = Kg$ for every $g \in G$), and the quotient group (factor group) G/K is isomorphic to the image of ϕ. Let S and T be subgroups of G, with T normal. The *Second Isomorphism Theorem* states that $S \cap T$ is normal in S, and $S/(S \cap T)$ is isomorphic to TS/T. Let $K \subset H \subset G$, with both K and H normal in G. The *Third Isomorphism Theorem* states that H/K is a normal subgroup of G/K, and $(G/K)/(H/K)$ is isomorphic to G/H.

There is an additional theorem which is sometimes called the *Fourth Isomorphism Theorem,* but is more commonly called *Zassenhaus's Lemma*. Let A_0, A_1, B_0, and B_1 be subgroups of G. Suppose A_0 is normal in A_1, and B_0 is normal in B_1. Zassenhaus's Lemma states that $A_0(A_1 \cap B_0)$ is normal in $A_0(A_1 \cap B_1)$, $B_0(A_0 \cap B_1)$ is normal in $B_0(A_1 \cap B_1)$, and $A_0(A_1 \cap B_1)/A_0(A_1 \cap B_0)$ is isomorphic to $B_0(A_1 \cap B_1)/B_0(A_0 \cap B_1)$. *See also* factor group, normal subgroup.

isotropic   (1) In physics and other sciences, a material or substance which responds the same way to physical forces in all directions is *isotropic*. Antonym: anisotropic.

(2) Let V be a vector space equipped with a bilinear form $(\ ,\ )$. A subspace W of V is *isotropic* (sometimes called *totally isotropic* or an *isotropy subspace*) if $W \subseteq W^\perp$, where $W^\perp$ is defined in the usual way, $W^\perp = \{v \in V : (v, w) = 0 \text{ for all } w \in W\}$. For example, if $V = \mathbf{R}^2$, 2-dimensional real space, and $(\ ,\ )$ is the Lorentz form, $((x_1, t_1), (x_2, t_2)) = x_1 x_2 - t_1 t_2$, then each of the lines forming the edge of the light cone, $\{(x, t) : x^2 - t^2 = 0\}$, is an isotropic subspace.

(3) If a differentiable manifold M has enough additional structure so that its tangent space comes equipped with a bilinear form, for example if M is a symplectic manifold, then a submanifold S of M is *isotropically embedded* if at each point $s \in S$, $T S_s$ is an isotropic subspace of $T M_s$. Here, $T S_s$ is the tangent space of S at s, and similarly for $T M_s$. *See also* symplectic manifold, tangent space.

isotropy subgroup   A group of transformations leaving a given point fixed. In more detail, let G be a group of transformations acting on a set X, and let $x_0 \in X$. The subgroup of transformations $T \in G$ leaving x_0 fixed ($T(x_0) = x_0$) is called the *isotropy subgroup* of G at the point x_0. The isotropy subgroup is also called the *stabilizer* of x_0 with respect to G.

isotropy subspace   *See* isotropic.

iteration   **(1)** Repetition; step-by-step repetition of a mathematical operation or construction.

(2) The use of loops as opposed to recursion in computer algorithms or programs.

iteration function   In numerical analysis, a function ϕ, used to compute successive approximations $x_1, x_2, x_3, \ldots$, to a quantity x, according to the formula $x_n = \phi(x_{n-1})$. For example, if we choose $\phi(x) = x - f(x)/f'(x)$ as an iteration function and then select a suitable starting point x_0, we obtain the Newton-Raphson method for approximating a zero of the function f (approximating a solution to the equation $f(x) = 0$). *See also* Newton-Raphson method of solving algebraic equations.

iteration matrix   In numerical analysis, a matrix M used to compute successive approxima-

tions $x_1, x_2, x_3, \ldots$ to a vector x, according to the formula $x_k = M x_{k-1} + c$. (Of course, one must have a conveniently chosen starting vector x_0.) For example, suppose we wish to solve the equation $Ax = b$ approximately, where A is an $n \times n$ square matrix, and x and b are n-dimensional column vectors. Write $A = L + D + U$, where L is lower triangular, D is diagonal, and U is upper triangular. If we choose the iteration matrix $M = -D^{-1}(L + U)$ and $c = D^{-1}b$, we obtain the Jacobi method for solving linear equations. On the other hand, if we choose the iteration matrix $M = -(L+D)U$ and $c = (L + D)^{-1}b$, we obtain the Gauss-Seidel method for solving linear equations. *See also* iteration function, Gauss-Seidel method for solving linear equations, Jacobi method for solving linear equations.

iterative calculation   A calculation which proceeds by means of iteration. *See* iteration.

iterative improvement   **(1)** Any one of the many algorithms for the approximate numerical solution of problems which proceed by obtaining a better approximation at each step.

(2) *See* iterative refinement.

iterative method   An algorithm or calculational process which uses iteration. A classic example is the Newton-Raphson method for computing the roots of an equation. Another classic example is the Gauss-Seidel iteration method for solving systems of linear equations. *See* iteration, iteration function, Gauss-Seidel method for solving linear equations, Newton-Raphson method of solving algebraic equations.

iterative process   *See* iterative method.

iterative refinement   **(1)** *See* iterative improvement **(1)**.

(2) In numerical analysis, a process for solving systems of linear equations which begins by obtaining a first solution using elimination (Gaussian elimination or row reduction) which is somewhat inaccurate due to roundoff errors, and then improves the accuracy of the solution using one of many iterative methods.

Iwahori subgroup   If G is a reductive group defined over a local field, then in addition to the standard BN-pair structure G has a second BN-pair structure whose associated building is Euclidean. In this case, the subgroups conjugate to B are called *Iwahori subgroups*.

Iwasawa decomposition   (1) A decomposition of a semisimple Lie algebra g over the field of real numbers as $g = k + a + n$, where k is a maximal compact subalgebra of g, a is an Abelian subalgebra of g, $a + n$ is a solvable Lie algebra, and n is a nilpotent Lie algebra.

(2) A decomposition of a connected Lie group G as $G = KAN$, where K is an (essentially) maximal compact subgroup, A is an Abelian subgroup, and N is a nilpotent subgroup. Here, G has Lie algebra g which is semisimple, $g = k + a + n$ is the Iwasawa decomposition of g as in (1) above, K, A, and N are analytic subgroups of G with Lie algebras k, a, and n, and the mapping $(x, y, z) \mapsto xyz$ is an analytic diffeomorphism of $K \times A \times N$ onto G. Furthermore, the groups A and N are simply connected.

The classic example of an Iwasawa decomposition is provided by the group $G = \mathrm{SL}(m, \mathbf{C})$, the group of $m \times m$ matrices with determinant 1 over the complex numbers. In this case, $K = \mathrm{SU}(m)$, the group of $m \times m$ unitary matrices of determinant 1, A = the group of $m \times m$ diagonal matrices of determinant 1 with positive entries on the diagonal, and N = the group of $m \times m$ upper triangular matrices with 1 in every diagonal entry. *See* Lie algebra, Lie group, semisimple Lie algebra, semisimple Lie group.

Iwasawa invariants   The integers λ, μ, and ν defined by the relation

$$\left| \mathrm{Cl}\,(k_n)_p \right| = p^{e_n}$$

where $e + n = \lambda n + \mu p^{n+\nu}$, for all sufficiently large n. Here, p is a prime, k is an algebraic number field; k_∞ is a $\mathbf{Z}_p$ extension field of k (an extension field with Galois group isomorphic to $\mathbf{Z}_p$, the integers modulo p); k_n is an intermediate field of degree p^n over k, $\mathrm{Cl}(k_n)_p$ is the pth component of the ideal class group of the field k_n, and $|\mathrm{Cl}(k_n)_p|$ is the number of elements in $\mathrm{Cl}(k_n)_p$. For cyclotomic $\mathbf{Z}_p$ extensions, the invariant $\mu = 0$.

Iwasawa's Main Conjecture   (1) A conjecture relating the characteristic polynomials of particular Galois modules to p-adic L-functions. The conjecture is an attempt to extend a classic theorem of Weil, which states that the characteristic polynomial of the Frobenius automorphism of a particular type of curve is the numerator of the zeta function of the curve. The conjecture was originally written over the field $\mathbf{Q}$, although it has been reformulated as a conjecture over any totally real field. It has been proved for real Abelian extensions of $\mathbf{Q}$ and odd primes p by Mazur and Wiles. Some work has also been done in the general case.

(2) A conjecture in number theory, relating certain Galois actions to p-adic L-functions. The conjecture asserts: $\tilde{f}_\chi(T) = g_\chi(T)$. Iwasawa's Theorem, which describes the behavior of the p-part of the class number in a $\mathbf{Z}_p$-extension, can be regarded as a local version of the Main Conjecture.

Iwasawa's Theorem   The characteristic $p \neq 0$ case of the Ado-Iwasawa Theorem: Every finite dimensional Lie algebra (over a field of characteristic p) has a faithful finite dimensional representation. The characteristic $p = 0$ case of this is Ado's Theorem. *See* Lie algebra.

J

Jacobian variety   The Picard variety of a smooth, irreducible, projective curve. *See* Picard variety.

Jacobi identity   The identity $(x \cdot y) \cdot z + (y \cdot z) \cdot x + (z \cdot x) \cdot y = 0$ satisfied by any Lie algebra. For example, if $\mathcal{A}$ is any associative algebra, and $[x, y]$ denotes the commutator, $[x, y] = xy - yx$, then the commutator satisfies the Jacobi identity $[[x, y], z] + [[y, z], x] + [[z, x], y] = 0$. *See* Lie algebra.

Jacobi method for solving linear equations
An iterative numerical method, also called the *total-step* method, for approximating the solutions to a system of linear equations. In more detail, suppose we wish to approximate the solution to the equation $Ax = b$, where A is an $n \times n$ square matrix. Write $A = L + D + U$, where L is lower triangular, D is diagonal, and U is upper triangular. The matrix D is easy to invert, so replace the exact equation $Dx = -(L+U)x+b$ by the relation $Dx_k = -(L + U)x_{k-1} + b$, and solve for x_k in terms of x_{k-1}:

$$x_k = -D^{-1}(L + U)x_{k-1} + D^{-1}b .$$

This gives us the core of the Jacobi iteration method. We choose a convenient starting vector x_0 and use the above formula to compute successive approximations $x_1, x_2, x_3, \ldots$ to the actual solution x. Under suitable conditions, the sequence of successive approximations does indeed converge to x. *See* iteration matrix.

Jacobi method of computing eigenvalues
Any of the several iterative methods for approximating all of the eigenvalues (characteristic values) of a Hermitian matrix A by constructing a finite sequence of matrices $A_0, A_1, \ldots, A_N$, where $A_0 = A$ and, for $0 < k \leq N$, $A_k = U_k^* A_{k-1} U_k$, U_k is a unitary matrix, and U_k^* denotes the conjugate transpose of U_k. The method ends with a nearly diagonal matrix A_N (all entries off the diagonal are small) with good ap-

proximations to the eigenvalues of A down the diagonal. The name is most frequently applied to the *Jacobi rotation method,* where the matrices U_k are chosen to be particularly simple unitary matrices called *planar rotation matrices. See* Jacobi rotation method.

Jacobi method of finding key matrix   A step in solving a linear system $Ax = b$ by the linear stationary iterative process. If the linear stationary iterative process is written as $x^{(k+1)} = x^{(k)} + R(b - Ax^{(k)})$, then the Jacobi method chooses R to be the inverse of the diagonal submatrix of A. *See also* linear stationary iterative process.

Jacobi rotation method   An iterative method for approximating all of the eigenvalues (characteristic values) of a Hermitian matrix A. The method begins by choosing $A_0 = A$, and then produces a sequence of matrices $A_1, A_2, \ldots, A_N$, culminating in a nearly diagonal (all entries off the diagonal are small) matrix A_N with good approximations to the eigenvalues of A down the diagonal. At each step, $A_k = U_k^* A_{k-1} U_k$, where U_k is the (unitary) matrix of a planar rotation annihilating the off diagonal entry of A_{k-1} with largest modulus, hence the name *rotation method,* and U_k^* is the conjugate transpose of U_k. The matrix $U_k = \left(u_{i,j}^{(k)}\right)$ differs from the identity matrix only in four entries, $u_{p,p}^{(k)}$, $u_{q,q}^{(k)}$, $u_{p,q}^{(k)}$, and $u_{q,p}^{(k)}$. The formulas for these entries are particularly simple in the case where the original matrix A is a real symmetric matrix: If $A_{k-1} = \left(a_{i,j}^{(k-1)}\right)$, then $u_{p,p}^{(k)} = u_{q,q}^{(k)} = \cos(\phi)$, and $u_{p,q}^{(k)} = -u_{q,p}^{(k)} = \sin(\phi)$, where

$$\tan(2\phi) = \frac{2a_{p,q}^{(k-1)}}{a_{p,p}^{(k-1)} - a_{q,q}^{(k-1)}} ,$$

and $-\frac{\pi}{4} \leq \phi \leq \frac{\pi}{4}$. In the commonly occurring case where the original matrix A has no repeated eigenvalues, the method converges quadratically. The method is named after its originator, Gustav Jacob Jacobi (1804–1851).

Jacobi's inverse problem   The problem of inverting Abelian integrals of the first kind on a compact Riemann surface $\mathcal{R}$ of genus $g \geq 1$.

Let $(\omega_1, \ldots, \omega_g)$ be a basis of Abelian differentials of the first kind on $\mathcal{R}$ and let $P_1, \ldots, P_g$ be a given set of fixed points on $\mathcal{R}$. For any given vector $(u_1, \ldots, u_g) \in \mathbf{C}^p$, the problem is to find a representation of all of the possible symmetric rational functions of $Q_1 \ldots Q_g$ as functions of $u_1, \ldots, u_g$ that satisfy

$$\sum_{j=1}^{g} \int_{P_j}^{Q_j} \omega_i = u_i .$$

In the above situation the path of integration is the same in each of the g equations. If the path is not assumed to be the same, then the system is actually a system of congruences modulo the periods of the differentials $(\omega_1, \ldots, \omega_g)$.

Jacobson radical   The set of all elements r in a ring R such that rs is quasi-regular for all $s \in R$. In more detail, let R be an arbitrary ring, possibly non-commutative, possibly without a unit element. An element $r \in R$ is *quasi-regular* if there is an element $r' \in R$ such that $r + r' + rr' = 0$. (In the special case where R has a unit element 1, this is equivalent to $(1+r)(1+r') = 1$.) For example, every nilpotent element ($r^n = 0$ for some n) is quasi-regular, but there are often other quasi-regular elements. The *Jacobson radical* of R is the set J of all elements $r \in R$ such that rs is quasi-regular for all $s \in R$.

The Jacobson radical is a two-sided ideal, and it generalizes the notion of the ordinary *radical* (the set of all nilpotent elements) of a commutative ring, though even in the special case where R is commutative, the Jacobson radical often differs from the ordinary radical. Both derive much of their importance from the following theorem: Let R be commutative. Then (i.) R is isomorphic to a subring of a direct sum of fields if and only if its radical vanishes, and (ii.) R is isomorphic to a subdirect sum of fields if and only if its Jacobson radical vanishes. *See also* radical, subdirect sum of rings, Wedderburn's Theorem.

j-algebra   A concept which reduces the study of homogeneous bounded domains to algebraic problems. Let G be a Lie algebra over $\mathbf{R}$, H a subalgebra of G, (j) a collection of linear endomorphisms of G, and ω a linear form on G. The

system $\{G, H, (j), \omega\}$ is called a *j-algebra* if the following conditions are satisfied: (i.) $j \equiv j' \bmod H$ and $jH \subset H$ for j and j' in (j), (ii.) $j^2 \equiv -1 \bmod H$, (iii.) $[h, jx] \equiv j[h, x] \bmod H$ for $h \in H$ and $x \in G$, (iv.) $[jx, jy] \equiv j[jx, y] + j[x, jy] + [x, y] \bmod H$ for x and y in G, (v.) $\omega([h, x]) = 0$ for $h \in H$, (vi.) $\omega([jx, jy]) = \omega([x, y])$, (vii.) $\omega([jx, x]) > 0$ if $x \notin H$.

Janko groups   Any of the exceptional finite simple groups J_1, J_2, J_3, and J_4. J_1 has order $2^3 \cdot 3 \cdot 5 \cdot 7 \cdot 11 \cdot 19$. J_2 has order $2^7 \cdot 3^3 \cdot 5^2 \cdot 7$ and is also called the HJ or *Hall-Janko* group. J_3 has order $2^7 \cdot 3^5 \cdot 5 \cdot 17 \cdot 19$ and J_4 has order $2^{21} \cdot 3^3 \cdot 5 \cdot 7 \cdot 11^3 \cdot 23 \cdot 29 \cdot 31 \cdot 37 \cdot 43$.

Janko-Ree group   Any member of the family of all finite simple groups for which the centralizer of every involution (element of order 2) has the form $\mathbf{Z}_2 \times \mathrm{PSL}_2(q)$, q odd. These groups consist of the Ree groups ${}^2G_2(3^n)$, for n odd, and the Janko group J_1. *See* Janko groups, Ree group.

Jensen measure   (1) A positive measure μ on the closure of an open subset Ω of $\mathbf{C}^n$ ($\mathbf{C}$ the complex numbers) such that for $x \in \Omega$,

$$\log |f(x)| \le \int \log |f(t)| \, d\mu(t) ,$$

for all f belonging to some appropriate class of holomorphic functions (such as the holomorphic functions on Ω with continuous extensions to the closure of Ω). Jensen measures are named after *Jensen's inequality,* which states that normalized Lebesgue measure on the unit circle ($1/2\pi$ times arclength measure) is a Jensen measure. In this case, Ω is taken to be the open unit disk $\{z \in \mathbf{C} : |z| < 1\}$ in the complex plane.

(2) More generally, a positive measure μ on the maximal ideal space M of a commutative Banach algebra $\mathcal{A}$ is called a *Jensen measure* for an element $\ell \in M$ if

$$\log |\ell(f)| \le \int \log |t(f)| \, d\mu(t)$$

for all $f \in \mathcal{A}$. *See also* Banach algebra, Jensen's inequality, maximal ideal space.

Jensen's inequality   (1) In complex variable theory. Let f be holomorphic on a neighborhood of the closed disc $\overline{D}(0, r)$ in the complex plane. Assume that $f(0) \neq 0$. Then Jensen's inequality is

$$\log |f(0)| \leq \frac{1}{2\pi} \int_0^{2\pi} \log |f(re^{it})| \, dt \, .$$

(2) In measure theory. Let (X, μ) be a measure space of total mass 1. Let f be a non-negative function on X. Let ϕ be a convex function of a real variable. Then Jensen's inequality is

$$\phi \left(\int_X f(x) \, d\mu(x) \right) \leq \int_X \phi \circ f(x) \, d\mu(x) \, .$$

join **(1)** In a lattice, the supremum or least upper bound of a set of elements. Specifically, if A is a subset of a lattice L, the *join* of A is the unique lattice element $b = \bigvee \{x : x \in A\}$ defined by the following two conditions: (i.) $x \leq b$ for all $x \in A$; (ii.) if $x \leq c$ for all $x \in A$, then $b \leq c$. The join of an infinite subset of a lattice may not exist; that is, there may be no element b of the lattice L satisfying conditions (i.) and (ii.) above. However, by definition, one of the axioms a lattice must satisfy is that the join of a finite subset A must always exist. The join of two elements is usually denoted by $x \vee y$.

There is a dual notion of the *meet* of a subset A of a lattice, denoted by $\bigwedge \{x : x \in A\}$, and defined by reversing the inequality signs in conditions (i.) and (ii.) above. The meet is also called the *infimum* or *greatest lower bound* of the subset A. Again, the meet of an infinite subset may fail to exist, but the meet of a finite subset always exists by the definition of a lattice.

(2) In relational database theory, the *join* (or *natural join*) of two relations is the relation formed by agreement on common attributes. Specifically, a *relation* is a set R of functions $f : A \to X$, from some set A, called the *set of attribute names,* to a set X, called the *set of possible attribute values.* (In relational database theory, the set A is always finite, so database theorists make the gloss of identifying a relation with a set of n-tuples, that is, a relation in the ordinary mathematical sense, and they then sneak the attribute names in under the table.) If R and S are two relations, with sets of attribute names A and B, respectively, and set of possible attribute values X and Y, then the *join* of R

and S is the relation $R \bowtie S$, with set of attribute names $A \cup B$ and a set of possible attribute values $X \cup Y$, defined as the set of all functions $f : A \cup B \to X \cup Y$ such that $f|_A \in R$ and $f|_B \in S$. Here, $f|_A$ is the restriction of f to A, and $f|_B$ is the restriction of f to B. Thus the formation of the natural join reduces to the familiar and ubiquitous mathematical problem of *extending* classes of functions. *See* restriction.

join irreducible element An element a of a lattice L which cannot be represented as the join of lattice elements b properly smaller than a ($b < a, b \neq a$). *See* join.

There is a dual notion of *meet irreducibility.* An element a of L is *meet irreducible* if it cannot be represented as the meet of lattice elements b properly larger than a ($a < b, a \neq b$). *See* meet.

joint proportion Quantity x is *jointly proportional to,* or *varies jointly with,* quantities y and z if there is a constant k such that $x = kyz$. *See also* direct proportion, inverse proportion.

joint spectrum Let $a_1, \dots, a_n$ be elements of a commutative Banach algebra $\mathcal{A}$, and let M be the maximal ideal space of $\mathcal{A}$. M can be identified with the space of multiplicative linear functionals on $\mathcal{A}$, that is the space of linear mappings ℓ of $\mathcal{A}$ into the complex numbers $\mathbf{C}$ such that $\ell(ab) = \ell(a)\ell(b)$. The *joint spectrum* of $a_1, \dots, a_n$ is the subset

$$\sigma(a_1, \dots, a_n) =$$
$$\{(\ell(a_1), \dots, \ell(a_n)) : \ell \in M\}$$

of $\mathbf{C}^n$. An important and useful theorem is that if $a_1, \dots, a_n$ actually generate the Banach algebra $\mathcal{A}$, then M is homeomorphic to $\sigma(a_1, \dots, a_n)$, and $\sigma(a_1, \dots, a_n)$ is polynomially convex. *See* Banach algebra, maximal ideal space, polynomial convexity, spectrum.

joint variation *See* joint proportion.

Jordan algebra A commutative, usually non-associative, algebra A satisfying the identity $(a^2 \cdot b) \cdot a = a^2 \cdot (b \cdot a)$. The model for a Jordan algebra is the algebra of $n \times n$ matrices with the multiplication $A \cdot B = \frac{1}{2}(AB + BA)$, where AB denotes the usual matrix product. The theory of Jordan algebras is somewhat analogous to the theory of Lie algebras, which is modeled on the algebra of $n \times n$ matrices with the multiplication $[A, B] = AB - BA$. *See also* Lie algebra.

Jordan canonical form   A matrix of the form

$$\begin{pmatrix} J_1 & 0 & \cdots & 0 \\ 0 & J_2 & \cdots & 0 \\ \vdots & \vdots & \ddots & \\ 0 & 0 & \cdots & J_k \end{pmatrix},$$

where each J_i is an elementary Jordan matrix, is in *Jordan canonical form*. It is a theorem of linear algebra that every $n \times n$ matrix with entries from an algebraically complete field, such as the complex numbers, is similar to a matrix in Jordan canonical form. *See also* elementary Jordan matrix.

Jordan decomposition   (**1**) The decomposition of a linear transformation $T : V \to V$, where V is a vector space, into a sum $T = T_s + T_n$, where T_s is diagonalizable, T_n is nilpotent, and the two commute. (*Nilpotent* means $T_n^k = 0$ for some integer k, and *diagonalizable* means there is a basis for V with respect to which T can be represented by a diagonal matrix.) T_s and T_n, if they exist, are uniquely determined by T. It is a theorem of linear algebra that if V is a finite dimensional vector space over an algebraically closed field, such as the complex numbers, then T_s and T_n always exist, that is T always has a Jordan decomposition. However, this need not be true if the field is, for example, the field of real numbers which is not algebraically closed. The Jordan decomposition of T is equivalent to the representation of T by a matrix in Jordan canonical form. *See* Jordan canonical form.

(**2**) The decomposition of an element a of a Lie algebra A into a sum $a = s + n$, where s is a semisimple element of A, n is a nilpotent element of A, and s and n commute. This decomposition is called the *additive Jordan decomposition* of a. The elements s and n, if they exist, are uniquely determined by a. It is a theorem that if A is a semisimple finite dimensional Lie algebra over an algebraically complete field, such as the complex numbers, then s and n always exist. *See* semisimple Lie algebra.

(**3**) The decomposition of a linear transformation T into a product $T = SU$, where S is diagonalizable, U is unipotent, and S and U commute. (*Unipotent* means that $U - I$ is nilpotent, where I is the identity transformation.) This decomposition is called the *multiplicative Jordan decomposition* of T. S and U, if they exist, are uniquely determined by T, and then the multiplicative Jordan decomposition is related to the additive Jordan decomposition $T = T_s + T_n$ by $S = T_s, U = I + S^{-1}T_n$.

(**4**) The decomposition of a linear transformation T into a product $T = EHU$, where E is elliptic, H is hyperbolic, U is unipotent, and all three commute. (*Elliptic* means E is diagonalizable and all complex eigenvalues have modulus $= 1$. *Hyperbolic* means H is diagonalizable and all complex eigenvalues have modulus < 1.) This decomposition is called the *completely multiplicative* Jordan decomposition of T. E, H, and U, if they exist, are uniquely determined by T.

(**5**) In analysis, the decomposition of a bounded additive set function μ (defined on a field of sets Σ) into the difference of two nonnegative bounded additive set functions, $\mu = \mu^+ - \mu^-$, via the formulas

$$\mu^+(E) = \sup_{F \subseteq E} \mu(F),$$

$$\mu^-(E) = - \inf_{F \subseteq E} \mu(F) ,$$

where F is restricted to belong to Σ. Here sup and inf refer to the supremum and infimum, respectively. (*See* supremum, infimum.) The set functions μ^+ and μ^- are called the *positive* or *upper variation* of μ, and the *negative* or *lower variation* of μ. The sum $|\mu| = \mu^+ + \mu^-$ is called the *total variation* of μ. *See also* additive set function.

(**6**) In analysis, the decomposition of a function of bounded variation into the difference of a monotonically increasing function and a monotonically decreasing function. (Sometimes stated *monotonically non-decreasing* and *monotonically non-increasing*.) This is a special case of (5). *See also* bounded variation, monotone function.

Jordan-Hölder Theorem   (**1**) The theorem that any two composition series of a group are equivalent. A *composition series* of a group G is a finite sequence of groups

$$G = G_0 \supset G_1 \supset \cdots \supset G_n = \{1\},$$

such that each group G_{i+1} is a maximal normal subgroup of G_i. (Equivalently, each G_{i+1} is a normal subgroup of G_i and the factor group G_i/G_{i+1} is simple and not equal to $\{1\}$.) Two composition series of G are *equivalent* if they have the same length and isomorphic factor groups. The theorem extends to groups with operators, thus to R modules, for example, and even to lattices (see (2) below). *See also* factor group, normal subgroup, simple group.

(2) The theorem that any two composition chains connecting two elements a and b in a modular lattice are equivalent. A lattice is *modular* if it satisfies the weakened distributive law,

$$x \wedge (y \vee z) = (x \wedge y) \vee (x \wedge z)$$

$$\text{whenever } x \geq y.$$

Here, $y \vee z$ denotes the lattice join or supremum (least upper bound) of y and z, and $y \vee z$ denotes the lattice meet or infimum (greatest lower bound) of y and z. (*See* join.) A *composition chain* connecting a to b is a finite sequence of lattice elements, $a = a_0 \geq a_1 \geq \cdots a_n = b$, such that there is no lattice element x strictly between a_i and a_{i+1}. (Equivalently, each interval $[a_{i+1}, a_i]$ is a two element lattice.) Two composition chains are *equivalent* if they have the same length and projective intervals. Two intervals $[w, x]$ and $[y, z]$ are *projective* if there is a finite sequence of intervals,

$$[w, x] = [w_1, x_1], [w_2, x_2], \ldots, [w_n, x_n]$$
$$= [y, z]$$

such that each pair of intervals $[w_i, x_i]$ and $[w_{i+1}, x_{i+1}]$ are transposes. Finally, two intervals are *transposes* if there are lattice elements c and d such that one interval is $[c, c \vee d]$ and the other is $[c \wedge d, d]$.

If G is a group, the lattice of its normal subgroups is a modular lattice. Thus, the Jordan-Hölder theorem for lattices gives us several Jordan-Hölder like theorems for groups, for instance for chief series and characteristic series. (The isomorphism of factor groups comes from the projectivity relation and the second isomorphism theorem for groups.) Unfortunately, the classical Jordan-Hölder theorem for composition series of groups (see (1) above) is not so easy to derive from the lattice theorem because the lattice of all subgroups may not be modular. *See also* characteristic series, isomorphism theorems of groups.

Jordan homomorphism   A mapping ϕ between Jordan algebras A and B which respects addition, scalar multiplication, and the Jordan multiplication. In other words, $\phi(a + b) = \phi(a) + \phi(b)$, $\phi(\lambda a) = \lambda\phi(a)$, and $\phi(a \cdot b) = \phi(a) \cdot \phi(b)$, for all scalars λ and for all $a, b \in A$. *See* Jordan algebra.

Jordan module   Let A be a Jordan algebra over a field of scalars K. A *Jordan A module* is a vector space V over the same field K, together with a multiplication operation $\cdot$ from $A \times V$ to V satisfying (i.) $a \cdot (v + w) = a \cdot b + a \cdot w$; (ii.) $a \cdot (\lambda v) = \lambda(a \cdot v)$, and (iii.) $(a \cdot b) \cdot v = \frac{1}{2}a \cdot (b \cdot v) + \frac{1}{2}b \cdot (a \cdot v)$, for all $a, b \in A$, $v, w \in V$, and scalars λ.

Property (iii.) seems odd; indeed the reader familiar with R modules (R a ring) would think it should be replaced by $(a \cdot b) \cdot v = a \cdot (b \cdot v)$. However, property (iii.) is easier to understand if one realizes that the Jordan multiplication $\cdot$ induces a mapping $a \mapsto T_a$ between elements a of the Jordan algebra and linear transformations T_a. Given $a \in A$, T_a is defined by $T_a(v) = a \cdot v$. Property (iii.) is chosen to guarantee that $T_{a \cdot b}$ will be the Jordan product $\frac{1}{2}(T_a T_b + T_b T_a)$ in the Jordan algebra of linear transformations of the vector space V.

In fact, the mapping $a \mapsto T_a$ is a Jordan homomorphism of A into the Jordan algebra of all linear transformations on V. A homomorphism between a Jordan algebra A and a Jordan algebra of linear transformations is called a *Jordan representation* of A. The definition of a *Jordan module* has been designed so there is a one-to-one correspondence between Jordan representations and Jordan modules. *See also* Jordan algebra, Jordan homomorphism, Jordan representation.

Jordan normal form   *See* Jordan canonical form.

Jordan representation   A Jordan homomorphism between a Jordan algebra A and a Jordan algebra of linear transformations on a vector space V, equipped with the standard Jordan product, $T \cdot S = \frac{1}{2}(TS + ST)$. There is a one-to-one correspondence between Jordan representations and Jordan modules. *See also* Jordan algebra, Jordan homomorphism, Jordan module.

Jordan-Zassenhaus theorem   Let A be a finite dimensional semisimple algebra with unit over the field of rational numbers, $\mathbf{Q}$. Let $\mathbf{Z}$ be the ring of integers, and let G be a $\mathbf{Z}$-order in A. Let L^* be a left A module, and let $\sigma(L^*)$ be the set of all left G modules L, having a finite $\mathbf{Z}$ basis, which are contained in L^*, and such that $\mathbf{Q}L = L^*$. The *Jordan-Zassenhaus Theorem* states: The set $\sigma(L^*)$ splits into a finite number of classes under $\mathbf{Z}$-equivalence.

The Jordan-Zassenhaus Theorem is a far reaching generalization of the theorem that the number of ideal classes in an algebraic number field is finite. *See* class field, ideal class, Z-basis, Z-equivalence, Z-order.

K

K3 surface   A class of algebraic surface in abstract algebraic geometry, defined in a projective space over an algebraically closed field. In projective 3-space they can be regarded as deformations of quartic surfaces. A K3 surface is characterized as a nonsingular, nonrational surface, in several ways including:

(i.) irregularity, Kodaira dimension, and the canonical divisor are zero;

(ii.) irregularity is zero and the arithmetic, geometric, and first plurigenus are all one;

(iii.) as a compact complex analytic surface, the first Chern class is zero and it has Betti numbers $b_0 = 1, b_1 = 0, b_2 = 22, b_3 = 0, b_4 = 1$.

The space of one-dimensional differential forms on a K3 surface is zero. An example of a K3 surface is any smooth surface of order four in projective three-dimensional space.

K3 surfaces were early examples of surfaces satisfying Weil's conjecture concerning the analog of the Riemann Hypothesis for algebraic varieties.

Kakeya-Eneström Theorem   Let f be a polynomial with real coefficients, say

$$f(x) = a_n x^n + a_{n-1} x^{n-1} + \cdots + a_0 ,$$

for each real number x. Suppose

$$a_n \geq a_{n-1} \geq \cdots \geq a_0 > 0 .$$

Let r be any root of the polynomial. Then $|r| \leq 1$.

Kaplansky's Density Theorem   A fundamental theorem from the theory of von Neumann algebras proved by Kaplansky in 1951.

The closure M with respect to the weak operator topology of a C^*-subalgebra A of the set of bounded linear operators on a separable Hilbert space is a von Neumann algebra. Furthermore, if A_1 is the set of elements of A with norm ≤ 1 in A (unit ball of A) and M_1 is the set of elements of M with norm ≤ 1 (unit ball of M),

then M_1 is the closure of A_1 with respect to the weak operator topology.

The theorem remains true if restated using the strong operator topology instead of the weak operator topology. Sometimes, the statement of the theorem includes the following additional information. The set of self-adjoint elements of A_1 is strongly dense in the set of self-adjoint elements of M_1, the set of positive elements of A_1 is strongly dense in M_1, and if A contains 1, the unitary group of A is strongly dense in the unitary group of M.

k-compact group   A connected algebraic group defined over a perfect field k whose k-Borel subgroups are reduced to the identity group. The name *k-anisotropic group* is used also. *See also* k-isotropic group.

k-complete scheme   Let $f : X \longrightarrow Y$ be a morphism of schemes X, Y. When f has a property, it is customary to say that X has the property over Y, or that X is a Y-(property) scheme. The property of being *complete* is connected with the property of being proper. A morphism $f : X \longrightarrow Y$ is *proper* if it is separated, of finite type, and is universally closed. Then X is called *proper* over Y. *See* separated morphism, morphism of finite type.

Now, let k be an algebraically closed field. Let X be a scheme of finite type over k which is reduced (i.e., for any element x the local ring at x has no nilpotent elements) and is irreducible (i.e., the underlying topological space is not the union of proper closed subsets). If X is *proper* over k (actually over the spectrum of k), then X is called a *k-complete scheme*.

kernel   (**1**) In algebra, where a homomorphism f is defined between two algebraic systems A and B, if the group identity of B is denoted by e, then the *kernel* of f is

$$\ker(f) = \{x \in A : f(x) = e\} .$$

Alternately, $\ker(f)$ may be denoted $f^{-1}(\{e\})$.

The *kernel* is a subset of A that usually has special properties. If A and B are groups, then the *kernel* of a homomorphism is a normal subgroup of A. If A and B are R-modules over a ring R, the *kernel* is a submodule of A. If A and B are topological linear spaces, then the *kernel* of a continuous linear operator is a closed linear subspace. The *kernel* of a semi-group homomorphism is the smallest two-sided ideal in the semi-group. Similar remarks hold for *kernels* of homomorphisms or morphisms in category theory, sheaf theory, and kernels of linear operators between spaces.

(2) In topology, for a nonempty set S in a topological space, the *kernel* of S is the largest subset T of S such that every element of T is an accumulation or cluster point of T.

(3) The word *kernel* is used in various other areas of mathematics to denote a function. In the study of integral equations, for example, the function K in the integral

$$\int_a^b K(x, y) f(y) \, dy$$

is called a *kernel*.

k-form (1) In linear algebraic group theory, a *k-form* of an algebraic group G defined over an extension field K of a field k is another algebraic group H defined over k that is K-isomorphic to G. Much work has been done in classifying the "k-forms" of various types of algebraic groups defined over K (e.g., semisimple algebraic groups or almost simple algebraic groups).

(2) More generally, if G is an algebraic group defined over k and K/k is a finite Galois extension, an algebraic group G_1 is said to be a K/k-form of G if there is a K-isomorphism from G onto G_1. For example, let k be a field and Ω a universal domain containing k. Let T be an n-dimensional algebraic k-torus with splitting field K. Then since T is K-isomorphic to the direct product of n copies of GL(1) (the multiplicative group of non-zero elements of Ω), T is a K/k-form of the n-dimensional K-split torus GL(1)^n.

See also quadratic form.

Killing form In Lie algebra theory, a symmetric bilinear form associated with the adjoint representation of a Lie algebra. Specifically, if g is a Lie algebra over a commutative ring K with 1, ρ is the adjoint (linear) representation of g and Tr denotes the trace operator, the symmetric bilinear form $B : g \times g \longrightarrow K$ given by $B(x, y) = \text{Tr}(\rho(x)\rho(y))$ is called the *Killing form*.

It is named after W. Killing who studied it in 1888. The Killing form is fundamental in the study of Killing-Cartan classification of semisimple Lie algebras over fields of characteristic 0.

k-isomorphism (1) Let k be a field and K, L extension fields of k. An isomorphism $\sigma : K \longrightarrow L$ such that $\sigma(x) = x$ for all $x \in k$ is called a *k-isomorphism*. Alternately, a *k-isomorphism* from K onto L is an isomorphism of the k-algebra K onto the k-algebra L.

(2) For other algebraic structures over a field k, a *k-isomorphism* is essentially an isomorphism (a bijective map that preserves the binary operations) and a "regular" mapping (preserving the particular structure on the sets). For example, for linear algebraic groups a *k-isomorphism* is an isomorphism that is also a birational mapping. For homogeneous k-spaces, a *k-isomorphism* is an isomorphism that is an everywhere defined pre-k-mapping.

k-isotropic group A connected algebraic group, defined over a perfect field k, whose k-Borel subgroups are nontrivial. For a reductive k-group G defined over an arbitrary field k, G is k-*isotropic* if the k-rank of G is greater than zero. *See* k-compact group. *See also* k-rank.

Kleinian group A subgroup G of the group of linear fractional functions defined on the extended complex plane $\hat{\mathbf{C}}$ such that there is an element x in $\hat{\mathbf{C}}$ which has a neighborhood U such that $g(U) \cap U = \emptyset$, for each nontrivial $g \in G$. Such groups were first studied by Klein and Poincaré in the 19th century and were named by Poincaré. *See* linear fractional function.

KMS condition A condition originally concerning finite-volume Gibbs states and later proposed for time evolution and the equilibrium states in quantum lattice systems in statistical mechanics (mathematically, within the framework of C^*-dynamical systems and a one-parameter group of automorphisms that describe the time evolution of the system). The condition was first noted by the physicists R. Kubo in 1957 and C. Martin and J. Schwinger in 1959. The letters K, M, and S are derived from their names. The equilibrium states are called *KMS states*.

Let M be a von Neumann algebra. Let ϕ be a faithful normal positive linear functional on M. Let $\{\sigma_t\}$ be a strongly continuous one-parameter group of $*$-automorphisms of M. Let S be the closed strip in the complex plane $\{z : 0 \leq \Im(z) \leq 1\}$. Then the group $\{\sigma_t\}$ will be said to satisfy the *KMS condition* if for any $x, y \in M$, there is $F : S \longrightarrow \mathbf{C}$ such that F is bounded and continuous on S, analytic in the interior of S, and satisfies the conditions

$$F(t) = \phi\left(\sigma_t(x)y\right) \quad \text{and}$$
$$F(t + i) = \phi\left(y\sigma_t(x)\right) .$$

The theory of Tomita-Takesaki shows the existence of such a group $\{\sigma_t\}$ and also that such groups are characterized by the condition. The *KMS condition* is a very important concept in the construction of type-III von Neumann algebras. *See* type-III von Neumann algebra.

Kostant's formula   A formula (named after B. Kostant) that gives the multiplicities of the weights of a finite dimensional irreducible representation constructed from a root system of a complex semisimple Lie algebra. It is a consequence of the Weyl character formula. *See* Weyl's character formula. In order to understand the (very explicit) formula, some definitions are in order. Let g be a complex semisimple Lie algebra. Let V be a $\mathbf{C}$-module and ρ a linear irreducible representation of g over V. Let Δ_+ be the set of positive roots. Let δ be the half sum of the positive roots $(\delta = \frac{1}{2}\sum_{\alpha \in \Delta_+} \alpha)$. Let W be the Weyl group of the root system. Let Λ be the highest weight of ρ. Let P be a nonnegative integer valued function (called the partition function) defined on the lattice of weights. For each weight μ, $P(\mu)$ is the number of ways μ can be expressed as a sum of positive roots. Let $m_\Lambda(\lambda)$ denote the multiplicity of a weight λ of ρ. Then *Kostant's formula* is

$$m_\Lambda(\lambda) = \sum_{w \in W} \det(w) P(w(\Lambda + \delta) - (\lambda + \delta)) .$$

This sum is very difficult to compute in practice and is thus of more theoretical than computational use. *See also* positive root, Weyl group.

k-rank   Let K be an algebraically closed field, k an arbitrary subfield of K, and G a reductive linear algebraic group defined over k.

(*See* reductive; examples of reductive groups are semisimple groups, any torus, and the general linear group.) Let T be a k-split torus in G of largest possible dimension. The dimension of T is called the *k-rank* of G. The *k-rank* is 0 if and only if G is anisotropic.

The nonzero weights of the adjoint of T are called *k-roots*. In case T is a maximal torus, *k-roots* are the usual roots of G with respect to T.

Let Z denote the centralizer of T in G; i.e.,

$$Z = \bigcap_{y \in T} \{x \in G \; : \; xy = y\} .$$

Let N denote the normalizer of T in G; i.e.,

$$N = \left\{x \in G \; : \; xTx^{-1} = T\right\} .$$

Then the finite quotient group Z/N is called the *k-Weyl group*. The group is named after the German mathematician H. Weyl (1885–1955).

k-rational divisor   (**1**) Let K be the algebraic closure (Galois extension) of a finite field k (with q elements) and let X be an algebraic curve defined over k. Then the automorphism $\sigma : k \longrightarrow k$ defined by $\sigma(x) = x^q$ defines an automorphism $\sigma : X \longrightarrow X$ defined by

$$\sigma(x_1, x_2, \ldots, x_n) = \left(x_1^q, x_2^q, \ldots, x_n^q\right)$$

that leaves all k-rational points in X fixed (called the Frobenius automorphism). Let

$$d = \sum_{x \in X} a_x x$$

be a divisor in X. (Recall that all $a_k \in \mathbf{Z}$ (integers) and all except at most a finite number are zero.) Then d is a member of the free Abelian group of divisors with base X, called Div(X). The divisor d is a *k-rational divisor* if

$$d = \sigma(d) = \sum_{x \in X} a_x \sigma(x) .$$

The set of *k-rational divisors* is a subgroup of Div(X).

(**2**) Another use of the term *k-rational divisor* involves a finite extension k of the field of rational numbers $\mathbf{Q}$. A divisor $d = \sum_{i=1}^{n} P_i$ is a *k-rational divisor* if all rational symmetric functions of the coordinates of the points P_i with coefficients in $\mathbf{Q}$ are elements of the field k.

k-rational point   The term *k-rational point* occurs in several areas of algebraic geometry, algebraic varieties, and linear algebraic groups. Examples of how the term is used follow.

(1) Let K be an algebraically closed field and k a subfield of K. Let $\mathbf{A}^2$ be the set of pairs (a, b) of elements $a, b \in K$ (the affine plane). Let f be a polynomial from $\mathbf{A}^2$ into K with coefficients from K. Recall that a plane algebraic curve is $\{(x, y) \in \mathbf{A}^2 : f(x, y) = 0\}$. Then $P = (x, y) \in \mathbf{A}^2$ is a *k-rational point* if $x, y \in k$. If $K = \mathbf{C}$ and $k = \mathbf{Q}$, then a point (x, y) is a *k-rational point* if both coordinates are rational numbers.

(2) If K is the finite field consisting of p^r elements (p a prime number) and K is the algebraic closure, then the set of *k-rational points* of a curve with coefficients in k coincides with the set of solutions of $f(x, y) = 0$, $x, y \in k$. If $r = 1$, so that k is a prime number field, then the set of *k-rational points* is equivalent to the set of solutions of the congruence $f(x, y) \equiv 0 \pmod{p}$.

(3) More generally, let k be a subfield of an arbitrary field K. Let $x = (x_1, \ldots, x_n) \in \mathbf{A}^n$. Then x is a *k-rational point* if each $x_i \in k$, $i = 1, \ldots, n$. Next, let $x = (x_1, \ldots, x_n) \in \mathbf{P}^n$ (projective space). Then x is a *k-rational point* if there is a $(n + 1)$-tuple of homogeneous coordinates $(\lambda x_0, \lambda x_1, \ldots, \lambda x_n)$, $\lambda \neq 0$ such that $\lambda x_i \in k$, for each $i = 0, 1, \ldots, n$. If $x_i \neq 0$, this is equivalent to $x_j/x_i \in k$, $\forall j = 0, 1, \ldots, n$.

(4) Let G be a linear algebraic group defined over a field k. An element p that has all of its coordinates in k is called a *k-rational point*.

(5) If X is a scheme over k, a point p of X is called a *k-rational point* of X if the residue class field (with respect to the inclusion map of k into X at p) is k.

(6) If K is an algebraically closed transcendental extension of k; V is an algebraic variety defined over k; and k' is a subfield of K, then a *k'-rational point* of V is an element of V that has all of its coordinates in k'.

Krieger's factor   The study of von Neumann algebras is carried out by studying the factors which are of type I, II, or III, with subtypes for each. (*See* factor.) *Krieger's factor* is a crossed product of a commutative von Neumann algebra with one $*$-automorphism. A *Krieger's factor* can be identified with approximately finite dimensional von Neumann algebras (over separable Hilbert spaces) of type III_0. *Krieger's factor* was named after W. Krieger who has studied them extensively.

Kronecker delta   A symbol, denoted by $\delta_{i,j}$, defined by

$$\delta_{i,j} = \begin{cases} 1 & \text{if } i = j \\ 0 & \text{if } i \neq j . \end{cases}$$

It is a special type of characteristic function defined on the Cartesian product of a set with itself. Specifically, let S be a set. Let $D = \{(x, x) : x \in S\}$. The value of the characteristic function of D is 1 if $(x, y) \in D$ (meaning that $x = y$) and is 0 otherwise.

Kronecker limit formula   (1) If ζ is the analytic continuation of the Riemann ζ-function to the complex plane $\mathbf{C}$, then

$$\lim_{s \to 1} \left[\zeta(s) - \frac{1}{s - 1} \right] =$$

$$\lim_{n \to \infty} \left[\sum_{m=1}^{n-1} \frac{1}{m + 1} + 1 - \log(n) \right] = \gamma$$

where γ is called Euler's constant. This can be expressed by saying that "near $s = 1$"

$$\zeta(s) = \frac{1}{s - 1} + \gamma + O(s - 1) .$$

This last formula is called the *Kronecker limit formula*.

(2) In the theory of elliptic integrals, modular forms, and theta functions, the function $E(z, s)$ (with $z = x + iy$) defined by the Eisenstein type series

$$\sum_{m,n} \frac{y^s}{|m(z) + n|^{2s}}$$

(where $x \in \mathbf{R}$, $y > 0$, the summation is over all pairs $(m, n) \neq (0, 0)$, and $\Re(s) > 1$) can be extended to a meromorphic function on $\mathbf{C}$ whose only pole is at $s = 1$. The *Kronecker limit formula* for $E(z, s)$ is

$$\frac{\pi}{s - 1} + 2\pi(\gamma - \log(2)) - 4\pi \log |\eta(z)| + O(s-1)$$

where

$$\eta(z) = e^{\pi i z/12} \prod_{n=1}^{\infty} \left(1 - e^{2\pi i z n}\right) .$$

The formula can be generalized to arbitrary number fields. More general Eisenstein type series lead to similar (but more complicated) limit formulas.

Kronecker product (1) Let $A = (a_{ij})$ denote an $m \times n$ matrix of complex numbers and $B = (b_{ij})$ an $r \times s$ matrix of complex numbers. Then the *Kronecker product* of the matrices A and B is defined as the $mr \times ns$ matrix described by the mn blocks C_{ij} given by

$$C_{ij} = a_{ij} B, \qquad 1 \le i \le m, 1 \le j \le n .$$

Sometimes other permutations of the $mnrs$ elements arranged in mr rows and ns columns is called the *Kronecker product*. This product is used, for example, in studying modulii spaces of Abelian varieties with endomorphism structure.

(2) Let V and W be finite dimensional vector spaces over a field k with bases $\{x_1, \ldots, x_n\}$ and $\{y_1, \ldots, y_r\}$ for V and W, respectively. Let L be a linear transformation on V and M a linear transformation on W. Let A be an $n \times n$ matrix associated with L and B an $r \times r$ matrix associated with M, determined by using the stated ordering of the basis elements. If lexicographic ordering is used for the tensor products of the basis elements in determining a basis for the tensor product of V and W, then the tensor product $V \otimes W$ of the linear transformations has the $nr \times nr$ Kronecker product matrix described in (1) as its matrix representation. *See* lexicographic linear ordering.

(3) Now let V and W be arbitrary vector spaces over a field k. Let U be a vector space whose basis vectors are elements of the Cartesian product of V and W, i.e., let the elements of U be finite sums of the form

$$\sum_{\nu=1}^{n} \alpha_\nu (x_\nu, y_\nu) .$$

Let N be the subspace of U such that

$$\sum_{\nu=1}^{n} \alpha_\nu L (x_\nu) M (y_\nu) = 0 ,$$

for each linear functional $L : V \longrightarrow k$ and each linear functional $M : W \longrightarrow k$. Define the *Kronecker product* of V and W with respect to k as the quotient space U/N. If V and W are rings with unity instead, and multiplication in U is defined componentwise, then N becomes an ideal of U and the *Kronecker product* becomes a residue class ring.

(4) Sometimes, the tensor product of algebras is referred to as their *Kronecker product*.

Kronecker's Theorem Several theorems in several fields of mathematics honor L. Kronecker (1823–1891).

(1) If f is a monic irreducible element of $k[x]$ over a field k, there is an extension field L of k containing a root c of f such that $L = k(c)$. Sometimes such an L is called a *star field*. For example, the polynomial $x^2 + 1$ over the real number field has the field of complex numbers as a star field. Further, if f has degree n, there is an extension field of k in which f factors into n linear factors, so f has exactly n roots (counting multiplicities).

(2) A field extension of the field of rational numbers which has an Abelian Galois group is a subfield of a cyclotomic field.

(3) Kronecker was among several mathematicians who studied the structure of subgroups and quotient groups of R^n generated by a finite number of elements. The following theorem was proved by him in 1884 and is also called *Kronecker's Theorem*. Let m, n be integers ≥ 1. In the following i will denote an integer between 1 and m while j will denote an integer between 1 and n. Let $a_i = (a_{i1}, a_{i2}, \ldots, a_{im})$ be m points of R^n and $b = (b_1, b_2, \ldots, b_m) \in R^n$. For every $\epsilon > 0$ there are m integers q_i and n integers p_j such that for each j

$$\left| \sum_{i=1}^{m} q_i a_{ij} - p_j - b_j \right| < \epsilon$$

if and only if for every choice of n integers r_j such that $\sum_{l=1}^{n} a_{il} r_l$ is an integer, the sum $\sum_{l=1}^{n} b_l r_l$ is an integer.

This theorem involves the closure of the subgroup of the torus T^n generated by a finite number of elements. Generalizations of the theorem have been studied in the theory of topological groups.

Kronecker symbol (1) An alternate name for the *Kronecker delta*. *See* Kronecker delta.

(2) In number theory, a generalization of the Legendre symbol, used in solving quadratic congruences

$$ax^2 + bx + c \equiv 0 \pmod{m}$$

where a, b, c are integers (members of $\mathbf{Z}$) and m is a positive integer ($m \in \mathbf{Z}^+$).

Let $d \in \mathbf{Z}$, d not a perfect square, $d \equiv 0$ or 1 (mod 4) and $m \in \mathbf{Z}^+$. For the following, recall that d is a quadratic residue of m if $x^2 = d \pmod{m}$ is solvable. Then the *Kronecker symbol* for d with respect to m, denoted by $(\frac{d}{m})$, is a mapping from $\{d\} \times \mathbf{Z}^+$ onto $\{0, {}^+1, {}^-1\}$ defined by:

$$\left(\frac{d}{1}\right) = 1 \, ,$$

$$\left(\frac{d}{2}\right) = \begin{cases} 0 & \text{if } d \text{ is even} \\ +1 & \text{if } d \equiv 1 \pmod{8} \\ -1 & \quad d \equiv 5 \pmod{8} \end{cases}$$

if m is an odd prime and $m|d$, then

$$\left(\frac{d}{m}\right) = 0 \, ,$$

if m is an odd prime and $m \nmid d$, then

$$\left(\frac{d}{m}\right) = \begin{cases} +1 & \text{if } d \text{ is a quadratic} \\ & \text{residue of } m \\ -1 & \text{if } d \text{ is not a quadratic} \\ & \text{residue of } m \end{cases}$$

if m is the product of primes $p_1, p_2, \ldots, p_r$, then

$$\left(\frac{d}{m}\right) = \left(\frac{d}{p_1}\right) \cdot \left(\frac{d}{p_2}\right) \cdot \ldots \cdot \left(\frac{d}{p_r}\right) \, .$$

The *Kronecker symbol* is used to determine the Legendre symbol, which in turn is used to find quadratic residues. For certain values of d and m the *Kronecker symbol* can be used to count the number of quadratic residues.

(3) The *Kronecker symbol* has uses and generalizations in more advanced areas of number theory as well; e.g., quadratic field theory and class field theory.

k-root *See* k-rank.

Krull-Akizuki Theorem Let R be a Noetherian integral domain, k its field of quotients, and K a finite algebraic extension of k. Let A be a subring of K containing R. Assume every non-zero prime ideal of R is maximal (i.e., assume R has Krull dimension 1). Then A is a Noetherian ring of Krull dimension 1. Also, for any ideal $g \neq (0)$ of A, A/g is a finitely generated A-module.

Here, the integral closure of a Noetherian domain of dimension one is Noetherian. This remains true for a two-dimensional Noetherian domain but not for dimension three or higher.

Krull-Azumaya Lemma Let R be a commutative ring with unit, $M \neq 0$ a finite R-module, and N an R-submodule of M. Let J be the Jacobson radical of R. The *Krull-Azumaya Lemma* is a name given to any of the following statements:

(i.) If $MJ + N = M$, then $N = M$;

(ii.) $M \neq MJ$;

(iii.) If M/MJ is spanned by a finite set $\{x_i + MJ\}$, then M is spanned by $\{x_i\}$.

This lemma is also known by the name *Nakayama's Lemma* and by the name *Azumaya-Krull-Jacobson's Lemma*. It is a basic tool in the study of non-semiprimitive rings.

Krull dimension The supremum of the lengths of chains of distinct prime ideals of a ring R. It is sometimes called the *altitude* of R. The definition was first proposed by W. Krull in 1937 and is now considered to be the "correct" definition not only for Noetherian rings but also for arbitrary rings.

With this definition any field κ has dimension zero and the polynomial ring $\kappa[x]$ has dimension one.

One reason for the acceptance of this definition for rings comes from a comparison to the situation in finite dimensional vector spaces. In a vector space of dimension n over a field κ, the largest chain of proper vector subspaces has length n. The corresponding polynomial ring $\kappa[x_1, \ldots, x_n]$ has a decreasing sequence (of length or height n) of distinct prime ideals

$$(x_1, \ldots, x_n), \ldots, (x_1), (0)$$

where the notation $(x_1, \ldots, x_n)$ denotes the ideal generated by the elements $x_1, \ldots, x_n$. This sequence is of maximal length.

Krull Intersection Theorem   Let R be a Noetherian ring, I an ideal of R, and M a finitely generated R-module. Then

(i.) there is an element $a \in I$ such that

$$(1 - a) \left(\bigcap_{j=1}^{\infty} I^j M \right) = 0 .$$

(ii.) If 0 is a prime ideal or if R is a local ring, and if I is a proper ideal of R, then

$$\bigcap_{j=1}^{\infty} I^j = 0 .$$

In some formulations, part (i.) is given as:

$$\bigcap_{j=1}^{\infty} I^j M =$$

$$\{x \in M : (1 - a)x = 0, \text{ for some } x \in I\} .$$

The theorem is important for the theory of Noetherian rings and is an application of the Artin-Rees Lemma concerning stable ring filtrations. *See* Artin-Rees Lemma.

Krull-Remak-Schmidt Theorem   The theorem arises in various branches of algebra and addresses the common length and isomorphisms between elements of the decomposition of an algebraic structure.

(**1**) In group theory. Suppose a group G satisfies the descending or ascending chain condition for normal subgroups. If $G_1 \times G_2 \times \cdots \times G_n$ and $H_1 \times H_2 \times \cdots \times H_m$ are two decompositions of G consisting of indecomposable normal subgroups, then $n = m$ and each G_i is isomorphic to some H_j.

(**2**) In ring theory. Let $A = A_1 \times A_2 \times \cdots \times A_n$ and $B = B_1 \times B_2 \times \cdots \times B_n$ be Artinian and Noetherian modules where each A_i and B_j are indecomposable modules. Then $m = n$ and each A_i is isomorphic to some B_j.

(**3**) The theorem may be formulated for other algebraic structures, e.g., for local endomorphism modules or for modular lattices.

Combinations of the names W. Krull, R. Remak, O. Schmidt, J.H.M. Wedderburn, and G. Azumaya are used to refer to this theorem. Wedderburn first stated the theorem for groups, Remak gave a proof for finite groups, Schmidt gave a proof for groups with an arbitrary system of operators, Krull extended the theorem to rings, and Azumaya found extensions of the theorem to other algebraic structures.

Krull ring   A commutative integral domain A for which there exists a family $\{v_i\}_{i \in I}$ of discrete valuations on the field of fractions K of A such that the intersection of all the valuation rings of the $\{v_i\}_{i \in I}$ is A and $v_i(x) = 0$ for all nonzero $x \in K$ and for all except (possibly) a finite number of indices $i \in I$. A Krull ring is also called a *Krull domain*. Every discrete valuation ring is a Krull ring as is a factorial ring and a principal ideal domain.

Krull rings were studied by W. Krull. They represent an attempt to get around the problem that the integral closure of a Noetherian domain is (generally) not finite. Since the valuations described above may be identified with the set of prime ideals of height one, a Krull ring may be defined alternately using prime ideal of height one.

Krull's Altitude Theorem   Also called Krull's Principal Ideal Theorem. This theorem has several forms and characterizations.

(**1**) Let R be a Noetherian ring, let $x \in R$ and let P be minimal among prime ideals of R containing x. Then the *height* of P (or the *codimension* or the *altitude* of P) is ≤ 1. *See* Krull dimension.

(**2**) Let R be a Noetherian ring containing $x_1, \ldots, x_n$. Let P be minimal among prime ideals of R containing $x_1, x_2, \ldots, x_n$. Then the height of $P \leq n$.

(**3**) Let R be a Noetherian local ring with maximal ideal m. Then the dimension of R is the minimal number n such that there exist n elements $x_1, x_2, \ldots, x_n$ not all contained in any prime ideal other than m.

Consequences of the theorem include the fact that the prime ideals in a Noetherian ring satisfy the descending chain condition so that the number of generators of a prime ideal P bounds the length of a chain of prime ideals descending from P.

Krull topology   A topology that makes the Galois group $G(L/K)$ (for the Galois extension of the field L over the field K) into a topological group. A fundamental system of neighborhoods of the field unity element of L is obtained by taking the set of all groups of the form $G(L/M)$ where M is both a subfield of L and a finite Galois extension of K.

The Krull topology is discrete if L is a finite extension of K. This topology is named after W. Krull, who extended Galois theory to infinite algebraic extensions and laid the foundations for Galois cohomology theory.

k-split   A term used in several parts of linear algebraic group theory and homology theory.

(**1**) Let k be an arbitrary field. Let Ω denote a universal domain containing k, that is, an algebraically closed field that has infinite transcendence degree over k. Let $\mathbf{G}_a$ denote the algebraic group determined by the additive group of Ω and let $\mathbf{G}_m$ denote the algebraic group determined by the multiplicative group of non-zero elements of Ω. Let G be a connected, solvable k-group. Then G is k-split if it has a composition series

$$G = G_0 \supset G_1 \supset \cdots \supset G_n = \{0\}$$

composed of connected k-subgroups with the property that each G_i/G_{i+1} is k-isomorphic to $\mathbf{G}_m$ or $\mathbf{G}_a$.

(**2**) If a k-torus T of dimension n is k-isomorphic to the direct product of n copies of $\mathbf{G}_m$, then T is said to be k-split. For a k-torus, definitions (**1**) and (**2**) are equivalent.

(**3**) Let A be a k-set, G be a k-group, M be a principal homogeneous k-space for G, and $\Gamma_{X/k}$ be the set of k-generic elements of the k-components of any k-set X. For any k-mapping h from A into G, let δh denote the k-mapping from $A \times A$ into G defined by

$$\delta h(x, y) = h(x)^{-1} h(y),$$
$$\text{for all } (x, y) \in \Gamma_{A^2/k} .$$

Recall that a one-dimensional k-cocycle f is a k-mapping from $A \times A$ into G such that

$$f(x, z) = f(x, y) f(y, z),$$
$$\text{for all } (x, y, z) \in \Gamma_{A^3/k} .$$

If f is a one-dimensional k-cocycle such that there exists a k-mapping h from $A \times A$ into M such that $f = \delta h$, then f is said to k-split in M. This definition is used in k-cohomology.

(**4**) There is a similar definition used in the cohomolgy of k-algebras involving the k-splitting of singular extensions of k-algebras bi-modules.

(**5**) If G is a connected semisimple linear algebraic group defined over a field k, then G is called k-split if there is a maximal k-split torus in G. Such a G is also said to be of *Chevalley type*.

Kummer extension   Any splitting field F of a polynomial

$$\left(x^n - a_1\right)\left(x^n - a_2\right) \ldots \left(x^n - a_r\right) ,$$

where for each $i = 1, 2, \ldots, r$ the a_i are elements of a field k that contains a primitive nth root of unity. A Kummer extension is characterized by the property of being a normal extension having an Abelian Galois group and the fact that the least common multiple of the orders of elements of the Galois group is a divisor of n.

Kummer's criterion   The German mathematician E. Kummer's attempts in the mid-19th century to prove Fermat's Last Theorem gave rise to the theory of ideals and the theory of cyclotomic fields and led to many other theories which are now of fundamental importance in several areas of mathematics. Among his many contributions to the mathematics that was developed to prove (or disprove) Fermat's Last Theorem, Kummer obtained congruences

$$B_n \left[\frac{d^{l-2n} \log(x + e^v y)}{dv^{l-2n}} \right]_{v=0} \equiv 0 \pmod{l}$$

for $n = 1, 2, \ldots, \frac{l-3}{2}$ where B_n is the nth Bernoulli number. The congruences are called *Kummer's criterion* in his honor. In 1905 the mathematician D. Mirimanoff established the equivalence of these congruences to the conditions (which also are called *Kummer's criterion*) of the following theorem.

Let x, y, z be nonzero integers. Let l be a prime > 3. Suppose x, y, z are relatively prime to each other and to l. If $\{x, y, z\}$ is a solution to the Fermat problem

$$x^l + y^l = z^l \,,$$

then

$$B_n f_{l-2n}(t) \equiv 0 \quad (\text{mod } l) \,,$$

for all $t \in \{-\frac{x}{y}, -\frac{y}{x}, -\frac{y}{z}, -\frac{z}{y}, -\frac{x}{z}, -\frac{z}{x}\}$ and for $n = 1, 2, \ldots, \frac{l-3}{2}$ where

$$f_m(t) = \sum_{i=1}^{l-1} i^{m-1} t^i \,,$$

for $m > 1$, and Bernoulli number B_n.

Kummer surface   A class of K3 surface in abstract algebraic geometry first studied by E. Kummer in 1864. It is a quartic surface which has the maximum number (16) of double points possible for a quartic surface in projective three-dimensional space. It can be described as the quotient variety of a two-dimensional Abelian variety by the automorphism subgroup generated by the sign-change automorphism $s(x) = -x$. See also K3 surface.

In projective three-dimensional space, the surface given by

$$x^4 + y^4 + z^4 + w^4 = 0$$

is a Kummer surface.

Künneth's formula   See Künneth Theorem.

Künneth Theorem   The theorem is formulated for several different areas of mathematics. Occurring in the statement of the theorem are one or more formulas concerning exact sequences known as *Künneth formulas*. The theorem itself is sometimes called Künneth formula. All Künneth-type formulas are related to studying the theory of products (e.g., tensor products) in homology and cohomology for various mathematical objects and structures. Examples follow.

(1) The first Künneth Theorem exhibits the *Künneth formula* for complexes. Let A be a ring with identity, L a complex of left A-modules and R a complex of right A-modules. Assume that for each n the boundary modules $B_n(R)$ and cycle modules $C_n(R)$ are flat. Then for each n there is a homomorphism β such that the sequence

$$0 \to \sum_{m+q=n} H_m(R) \otimes H_q(L) \xrightarrow{\alpha} H_n(R \otimes L) \xrightarrow{\beta}$$

$$\xrightarrow{\beta} \sum_{m+q=n-1} \text{Tor}_1(H_m(R) \otimes H_q(L)) \to 0$$

is exact.

(2) The next Künneth Theorem exhibits the isomorphism. Let A be a ring with identity. Let L be a complex of left A-modules and R be a complex of right A-modules. Assume that for each n the homology modules $H_n(R)$ and cycle modules $C_n(R)$ are projective (in this case $\text{Tor}_1(H_m(R) \otimes H_q(L)) = 0$). Then for each n there is an isomorphism α such that the sequence

$$\sum_{m+q=n} H_m(R) \otimes H_q(L) \cong H_n(R \otimes L) \,.$$

(3) Let G be an Abelian group. For simplicial complexes L and R and Cartesian product $L \times R$, the homology group $H_n(L \times R; G)$ splits into the direct sum

$$H_n(L \times R; G) \cong \sum_{m+q=n} H_m(L) \otimes H_q(R)$$

$$\oplus \sum_{m+q=n-1} \text{Tor}\left(H_m(L) \otimes H_q(R)\right) \,.$$

(4) Similar *Künneth formulas* and *Künneth Theorems* generalized (e.g., to spectral sequences) or stated with other criteria on the groups (e.g., torsion free groups) have been studied.

k-Weyl group   See k-rank.

L

l-adic coordinate system   Let A be an Abelian variety of dimension n over a field k of characteristic $p \geq 0$. Let l be a prime number, $\mathbf{Q}_l$ the l-adic number field, $\mathbf{Z}_l$ the group of l-adic integers, and P_l the direct product of the $2n$ quotient groups of $\mathbf{Q}_l/\mathbf{Z}_l$. Let $B_l(A)$ denote the group of points of A whose order is a power of l. If $l \neq p$, then P_l is isomorphic to $B_l(A)$. The isomorphism yields the *l-adic coordinate system* of $B_l(A)$.

l-adic representation   Let A, B be Abelian varieties of dimensions n and m, respectively, over a field k of characteristic $p \geq 0$. Let l be a prime number different from p and $\lambda : A \to B$ a rational homomorphism. Let $B_l(A)$, $B_l(B)$ denote the group of points of A, respectively B, whose order is a power of l and $\gamma : B_l(A) \to B_l(B)$ the homomorphism induced by λ. Then the $2m \times 2n$ matrix representation (with respect to the l-adic coordinate system) of γ is called the *l-adic representation* of λ.

Lagrange multiplier   To find the extrema of a function f of several variables, subject to the constraint $g = 0$, one sets $\nabla f = \lambda \cdot \nabla g$. The scalar λ is called a *Lagrange multiplier.*

Lagrange resolvent   Let k be a field of characteristic p containing the nth roots of unity such that p does not divide n. Let $k(\Theta)$ be an extension field of degree n over k with cyclic Galois group. Let σ be a generator of the Galois group. Let ζ be an nth root of unity. Then the *Lagrange resolvent,* denoted by (ζ, Θ) is defined by

$$(\zeta, \Theta) = \Theta_0 + \zeta \Theta_1 + \cdots + \zeta^{n-1} \Theta_{n-1}$$

where $\Theta_j = \sigma^j \Theta$, for each $j = 1, 2, \ldots, n-1$ and $\Theta_n = \sigma^n \Theta = \Theta_0 = \Theta$.

Since $\sigma \Theta_j = \Theta_{j+1}$, for each j, the Lagrange resolvent has the property that $\sigma(\zeta, \Theta) = \zeta^{-1}(\zeta, \Theta)$. This implies that $\sigma(\zeta, \Theta)^n = (\zeta, \Theta)^n$, meaning that $(\zeta, \Theta)^n \in K$. The Θ_j can be determined from the equations $\sum_\zeta \zeta^{-j}$

$(\zeta, \Theta) = n\Theta_j$, since p does not divide n. So, $k(\Theta)$ is generated by (ζ, Θ).

Lagrangian density   It is customary in the fields of mathematical physics and quantum mechanics to honor the Italian mathematician J.L. Lagrange (1736–1813) by using his name or the letter L to name a certain expression (involving functions of space and time variables and their derivatives) used as integrands in variational principles describing equations of motion and numerous field equations describing physical phenomena. Thus, a "Lagrangian" occurs in the statements of the "principle of least action" (first formulated by Euler and Lagrange for conservative fields and by Hamilton for nonconservative fields), as well as in Maxwell's equations of electromagnetic fields, relativity theory, electron and meson fields, gravitation fields, etc. In such systems, field equations are regarded as sets of elements describing a mechanical system with infinitely many degrees of freedom.

Specifically, let $\Omega \subset \mathbf{R}^m$, $[t_0, t_1)$ denote a time interval, and $\mathbf{f} = (f^1, f^2, \ldots, f^n) : [t_0, t_1) \longrightarrow \mathbf{R}^n$ denote a vector valued function. Let $\mathbf{L}$ be an algebraic expression featuring sums, differences, products, and quotients of the functions of $\mathbf{f}$ and their time and space derivatives. $\mathbf{L}$ may include some distributions. Define

$$L(t) = \int_\Omega \mathbf{L} \, d\mathbf{x} \,,$$

where $\mathbf{x} = (x^1, x^2, \ldots, x^m)$ and

$$V(\mathbf{f}) = \int_{t_0}^{t_1} L(t) \, dt \,.$$

Field equations are derived from considering variational problems for $V(\mathbf{f})$. In such a setup, L is called the *Lagrangian* and $\mathbf{L}$ the *Lagrangian density.*

Lanczos method of finding roots   A procedure (or attitude) used in the numerical solution (especially manual) of algebraic equations whose principal purpose is to find a first approximation of a root. Then methods (especially Newton's method) are used to improve accuracy (by iteration). For equations with real roots, the basic idea is to transform the equation into one which has a root between 0 and 1; then reduce the order of the equation by using an approximating function (e.g., a Chebychev polynomial); solve the reduced associated equation exactly for a root between 0 and 1 and convert the root found back to a root of the original equation by using the transformation functions. For complex roots the procedure is similar but finds a root of largest modulus.

As an explicit use of the ideas consider a cubic equation with real coefficients a, b, c, d (with $ad \neq 0$)

$$ax^3 + bx^2 + cx + d = 0 .$$

First change the polynomial to a monic polynomial

$$x^3 + b_1 x^2 + c_1 x + d_1 = 0$$

where $b_1 = b/a, c_1 = c/a, d_1 = d/a$. If $d_1 > 0$, transform the equation by replacing x everywhere by $-x$ obtaining

$$x^3 + b_2 x^2 + c_2 x - d_2 = 0 ,$$

with $d_2 > 0$, where $b_2 = -b_1, c_2 = c_1, d_2 = d_1$. Now, by dividing by d_2 and setting $y = x/\sqrt[3]{d_2}, b_3 = b_2/\sqrt[3]{d_2}, c_3 = c_2/(\sqrt[3]{d_2})^2$, one gets the equation

$$y^3 + b_3 y^2 + c_3 y - 1 = 0 .$$

Now $f(0) < 0$. Also, for large positive y, $f(y) > 0$. If $f(1) > 0$, there is a root between 0 and 1. If $f(1) < 0$, there is a root > 1. In this case substitute $t = 1/y$ into the equation obtaining

$$t^3 - c_3 t^2 - b_3 t - 1 = 0 .$$

In either case, substitute for t^3 or y^3 the quadratic term of the Chebychev polynomial

$$t^3 = (1/32)\left(48t^2 - 18t + 1\right)$$

and solve the resulting quadratic equation exactly. Discard any negative root and use the remaining root. Convert it back to the original equation using all the conversions. This is the *Lanczos method* for cubic equations. There are *Lanczos* methods for fourth and higher degree equations with real coefficients and ones for complex coefficients.

Lanczos method of matrix transformation

In the numerical solution of eigenvalue problems of linear differential and integral operators, matrix transformations play a fundamental role. Lanczos' method, named after C. Lanczos who introduced it in 1950, was developed as an iterative method for a nonsymmetric matrix A and involves a series of similarity transformations to reduce the matrix A to a tri-diagonal matrix T with the same (but more easily calculable) eigenvalues as A.

Specifically, given an $n \times n$ matrix A, a tri-diagonal matrix T is constructed so that $T = S^{-1}AS$. Assume that the n columns of S are denoted by $x_1, x_2, \ldots, x_n$ which will be determined so as to be linearly independent. Assume that the main diagonal of matrix $T = \{t_{ij}\}$ is denoted by $t_{ii} = b_i$ for $i = 1, 2, \ldots, n$; the superdiagonal $t_{ii+1} = c_i$ for $i = 1, 2, \ldots, n - 1$; and the subdiagonal $t_{ii-1} = d_i$, for $i = 2, 3, \ldots, n$. In this method the d_i will all be 1. The method consists of constructing a sequence $\{y_i\}$ of vectors such that $y_i^T x_i = 0$, if $i \neq j$.

Let x_0, y_0, c_0 be zero vectors and let x_1 and y_1 be chosen arbitrarily. Define $\{x_k\}$ and $\{y_k\}$ recursively, by

$$x_{k+1} = Ax_k - b_k x_k - c_{k-1} x_{k-1} ,$$
$$y_{k+1} = A^T y_k - b_k y_k - c_{k-1} y_{k-1} ,$$

with

$$b_k = \frac{y_k^T A x_k}{y_k^T x_k}$$

and

$$c_{k-1} = \frac{y_{k-1}^T A x_k}{y_{k-1}^T x_{k-1}} .$$

Assuming that $y_j^R x_j \neq 0$ for each j, it can be shown that $y_i^T x_j = 0$ if $i \neq j$, the $\{x_i\}$ are linearly independent, and that if the recursions are used for $k = n$, that the resulting $x_{n+1} = 0$. These results lead to the equations

$$Ax_1 = x_2 + b_1 x_1$$

$$Ax_k = x_{k+1} + b_k x_k + c_{k-1} x_{k-1} ,$$

for $k = 2, 3, \ldots, n - 1$ and

$$Ax_n = b_n x_n + c_{n-1} x_{n-1}$$

which yield the desired tri-diagonal matrix T.

Problems with the method involve a judicious choice of x_1 and y_1 so that $y_j^T x_j \neq 0$ for each $j = 1, 2, \ldots, n-1$. Numerically, the weakness of the method is due to a possible breakdown in this biorthogonalization of the sequences $\{x_i\}$ and $\{y_i\}$ even when the matrix A is well conditioned. The numerical procedure is likely to be unstable, for example, when $y_i^T x_i$ is small.

For symmetric matrices with $x_1 = y_1$ the sequences $\{x_i\}$ and $\{y_i\}$ are identical. In this case the numerical stability is comparable to either Givens' or Householder's methods resulting in the same tri-diagonal matrix — but with more cost in deriving it.

Theoretically, the indicated algorithm occurs in the conjugate-gradient method for solving a linear system of equations, in least square polynomial approximations to experimental data, and as one way to develop the Jordan canonical form.

Lanczos method is sometimes called the *method of minimized iterations*.

Laplace Expansion Theorem   A theorem concerning the finite sum of certain products of determinants of submatrices of a given square matrix. The theorem delineates which products yield a sum equal to the determinant of the whole matrix, so the expansions described are commonly used to evaluate determinants.

Let A be an $n \times n$ matrix with elements from a commutative ring, where it can be assumed that $n > 1$. Let A have elements a_{ij} where i denotes the row and j the column. Let $1 \leq r < n$. Suppose r rows of A, denoted by $i_1, i_2, \ldots, i_r$ are chosen where it is assumed (always) that $i_1 < i_2 < \cdots < i_r$ (lexicographic ordering). *See* lexicographic linear ordering. Let R denote this lexicographically ordered r-tuple $(i_1, i_2, \ldots, i_r)$. Suppose also that r columns $j_1, j_2, \ldots, j_r$ are chosen (lexicographically ordered). Let C be this lexicographically ordered r-tuple $(j_1, j_2, \ldots, j_r)$. By choosing elements of A from the chosen rows and columns, an $r \times r$ submatrix $A_{RC} = (b_{mn}) = (a_{i_m j_n})$ is determined. If R' denotes the lexicographically ordered $(n - r)$-tuple $i_{r+1}, i_{r+2}, \ldots, i_n$ consisting of the rows of A not chosen and C' the columns not chosen, then $A_{R'C'}$ is an $(n - r) \times (n - r)$ submatrix. Let $\lambda_{RC} = i_1 + i_2 + \cdots + i_r + j_1 + j_2 + \cdots + j_r$. Finally, let S be any lexicographically ordered r-tuple

chosen from the n numbers $1, 2, \ldots, n$. Then the *Laplace Expansion Theorem* says:

$$\sum_C (-1)^{\lambda_{RC}} \det(A_{RC}) \det(A_{S'C'}) = \det(A)$$

if $R = S$ (and $= 0$, if $R \neq S$), where the sum is taken over the $\binom{n}{r}$ selection of columns C. Also,

$$\sum_R (-1)^{\lambda_{RC}} \det(A_{RC}) \det(A_{R'S'}) = \det(A)$$

if $R = S$ (and $= 0$, if $R \neq S$), where the sum is taken over the $\binom{n}{r}$ selection of rows R. The summations resulting in the sum $\det(A)$ are called *Laplace expansions* of the determinant. If $r = 1$, the "usual" expansion of a determinant is obtained.

large semigroup algebra   Let R be a commutative ring, S a semigroup, and R^S the set of sequences of elements of R where in each sequence only a finite number of elements is different from zero. Then R^S is the direct sum of isomorphic copies of R using S as the index set. A canonical basis of R^S consists of $\{b_i\}_{i \in S}$ where for a given $i \in S$, b_i has the component indexed by i equal to 1 and the rest of the components equal to 0. The product

$$b_i b_j = b_{ij} \quad \text{for all } i, j \in S$$

determines the multiplication in R^S and makes R^S into an algebra called a *semigroup algebra* of S over R. Now suppose S has the property that for any $s \in S$ there are only a finite number of pairs $\{i, j\}$ of elements of S such that $s = ij$. If one defines multiplication of sequences $\alpha = \{\alpha_i\}$ and $\beta = \{\beta_j\}$ of R^S to be $\gamma = \{\gamma_k\}$ where

$$\gamma_k = \sum_{k=ij} \alpha_i \beta_j, \quad k \in S,$$

then the resulting algebra is called a *large semigroup algebra* and contains R^S as a subalgebra.

largest nilpotent ideal   The nilpotent ideal which is the union of all nilpotent ideals of a Lie algebra defined on a commutative ring with unit. In a left Artinian ring, the radical is nilpotent and is the *largest nilpotent ideal*. In a left Noetherian ring, the prime radical is the largest nilpotent left ideal. For a ring which has the property that every nonempty set of left ideals contains a minimal element (with respect to inclusion), the Jacobson radical is the largest nilpotent ideal.

lattice (1) A system consisting of a set S together with a partial ordering $\leq$ on S and two binary operations $\wedge$ and $\vee$ defined on S such that, for all $a, b, c \in S$,

$a \vee b \leq c$ if and only if $(a \leq c$ and $b \leq c)$,

and

$c \leq a \wedge b$ if and only if $(c \leq a$ and $c \leq b)$.

The element $a \vee b$ is called the *supremum* of a and b and the element $a \wedge b$ is called the *infimum* of a and b.

The set of all subsets of a nonempty set S ordered by the subset relation together with the operations of set union and set intersection form a *lattice*. The whole numbers, ordered by the relation *divides,* together with the operations of *greatest common divisor* and *least common multiple,* form a lattice. The real numbers, ordered by the relation *less than or equal to,* together with the operations of *maximum* and *minimum,* form a lattice.

(2) Let $\mathbf{R}^n$ denote the set of all n-tuples of real numbers. Let $\mathbf{R}^n$ be endowed with the (usual) Euclidean metric space structure. Call this metric space E^n. Let V^n denote the (usual) n-dimensional real vector space defined over $\mathbf{R}$ by adding componentwise and multiplying by scalars componentwise. Let $\mathbf{Z}$ denote the set of integers. Let σ denote the mapping between E^n and V^n that identifies points in E^n with vectors in V^n. Let L be the set of points

$$\left\{ P \in E^n : \sigma(P) = \sum_{j=1}^{n} \alpha_j v_j, \ \alpha_j \in \mathbf{Z} \right\},$$

for some basis $\{v_1, v_2, \ldots, v_n\}$ of V^n. Then L is called an n-dimensional homogeneous *lattice* in E^n.

lattice constant Let L denote the lattice group of an n-dimensional crystallographic group defined on n-dimensional Euclidean space. Then L is generated by n translations. Each inner product of two of the generating translations is called a *Lattice constant. See* lattice group.

lattice group (1) The subgroup of translations of the n-dimensional crystallographic group C defined on n-dimensional Euclidean space E^n. The *lattice group* is a commutative normal subgroup of C. Recall that C is a discrete subgroup of the group of motions on E^n that contains exactly n linearly independent translations. The lattice group is generated by these translations. In applied areas of crystallography, the group of motions is called a *space group*. Then the lattice group is called the lattice of the space group.

(2) Let L be an n-dimensional homogeneous lattice in E^n. (*See* lattice.) Then

$$\left\{ v \in V^n : v = \sigma(P), \ \text{for some } P \in L \right\}$$

is called the *lattice group* of L. This lattice group is a free module generated by the basis $\{v_1, v_2, \ldots, v_n\}$.

lattice ordered group A lattice which is also an ordered group. *See* lattice.

law A property, statement, rule, or theorem in a mathematical theory usually considered to be fundamental or intrinsic to the theory or governing the objects of the theory. Elementary examples include commutative, associative, distributive, and trichotomy laws of arithmetic, DeMorgan's laws in mathematical logic or set theory, and laws of sines, cosines, or tangents in trigonometry.

Law of Quadrants Also called *rule of species*. In spherical trigonometry, if ABC is a right spherical triangle, with right angle C and sides $a, b,$ and c (measured in terms of the angle at the center of the sphere subtended by the side), then: (i.) if a and A are both acute or both obtuse angles (said to be of *like species*), then so are b and B; (ii.) if $c < 90°$, then a and b are of like species; (iii.) if $c > 90°$, then a and b are of unlike species.

This law is used whenever one has a formula which gives two possible values for the *sine* of the side or angle, since it indicates whether the side or angle is acute or obtuse.

Law of Signs In arithmetic, a rule for simplifying expressions where two (or more) *plus* ("+") or *minus* ("−") signs appear together. The rule is to change two occurrences of the same sign to a "+" and to change two occurrences of opposite signs to a "−."

In using the *Law of Signs,* one is ignoring the different usages of the minus sign; namely, for subtraction, for additive inverse, and to denote "negative." The same is true for the plus sign; namely, for addition and to denote "positive."

Thus, $x - (-y)$ or $x - -y$ becomes $x + y$; $x + (-y)$ or $x + -y$ becomes $x - y$; $4 - {}^-3$ becomes $4 + 3$; $4 + {}^-3$ becomes $4 - 3$; $x + (+y)$ or $x + +y$ becomes $x + y$, etc.

Law of Tangents   In plane trigonometry, a ratio relationship between the lengths of sides of a triangle and tangents of its angles. For a triangle ABC, where A, B, and C represent vertices (and angles) and a, b, and c represent, respectively, the sides opposite the angles,

$$\frac{a - b}{a + b} = \frac{\tan \frac{1}{2}(A - B)}{\tan \frac{1}{2}(A + B)} .$$

Laws of Cosines   In plane and spherical trigonometry, relationships between the lengths of sides of a triangle and cosines of its angles. For a triangle ABC in plane trigonometry, where A, B, and C represent vertices (and angles) and a, b, and c represent, respectively, the sides opposite the angles,

$$c^2 = a^2 + b^2 - 2ab \cos(C) .$$

Here, the length of a side is determined by the lengths of the other two sides and the cosine of the angle included between those sides.

In spherical trigonometry, using the above designations for the angles and for the sides (measured in terms of the angle at the center of the sphere subtended by the side), there are two relationships:

$$\cos(C) = - \cos(A) \cos(B)$$
$$+ \sin(A) \sin(B) \cos(c)$$

and

$$\cos(c) = \cos(a) \cos(b) + \sin(a) \sin(b) \cos(C) .$$

Laws of Sines   In plane and spherical trigonometry, the ratio relationship between the lengths of sides of a triangle and sines of its angles. For a triangle ABC in plane trigonometry, where A, B, and C represent vertices (and angles) and a, b, and c represent, respectively, the sides opposite the angles,

$$\frac{a}{\sin(A)} = \frac{b}{\sin(B)} = \frac{c}{\sin(C)} .$$

In spherical trigonometry, using the same designations for the angles and for the sides (measured in terms of the angle at the center of the sphere subtended by the side),

$$\frac{\sin(a)}{\sin(A)} = \frac{\sin(b)}{\sin(B)} = \frac{\sin(c)}{\sin(C)} .$$

leading coefficient   The nonzero coefficient of the highest order term of a polynomial in a polynomial ring. In the expression

$$a_n x^n + a_{n-1} x^{n-1} + \cdots + a_1 x + a_0 ,$$

a_n is the leading coefficient. There are situations when it is appropriate to consider an expression when the leading coefficient is zero. In this case, such a coefficient is called a *formal leading coefficient.*

least common denominator   In arithmetic, the least common multiple of the denominators of a set of rational numbers. *See* least common multiple.

least common multiple   (1) In number theory, if a and b are two integers, a *least common multiple* of a and b, denoted by $\mathrm{lcm}\{a, b\}$, is a positive integer c such that a and b are each divisors of c (which makes c a *common multiple* of a and b), and such that c divides any other common multiple of a and b. The product of the *least common multiple* and positive greatest common divisor of two positive integers a and b is equal to the product of a and b.

The concept can be extended to a set of more than two nonzero integers. Also, the definition can be extended to a ring where a least common multiple of a finite set of nonzero elements of the ring is an element of the ring which is a common multiple of elements in the set and which is also a factor of any other common multiple.

(2) In ring theory, the ideal generated by the intersection of two ideals is known as the *least common multiple* of the ideals.

In a polynomial ring $k[x_1, \ldots, x_r]$ over a field k, the *least common multiple* of two monomials $x_1^{m_1} \cdot \ldots \cdot x_r^{m_r}$ and $x_1^{n_1} \cdot \ldots \cdot x_r^{n_r}$ is defined as $x_1^{\max(m_1, n_1)} \cdot \ldots \cdot x_r^{\max(m_r, n_r)}$.

least upper bound   *See* supremum.

Lefschetz Duality Theorem   Let X be a compact n-dimensional manifold with boundary $\dot{X}$ and orientation U over the ring R. For all R-modules G and non-negative integers q there exist isomorphisms

$$H_q(X; G) \leftarrow H_q(X \setminus \dot{X}; G) \rightarrow H^{n-q}(X, \dot{X}; G)$$

and

$$H_q(X, \dot{X}; G) \rightarrow H^{n-q}(X \setminus \dot{X}; G) \leftarrow H^{n-q}(X; G).$$

[Here $j : X \setminus \dot{X} \subset X$.]

Lefschetz fixed-point formula   One of the main properties of the l-adic cohomology. It counts the number of fixed points for a morphism on an algebraic variety or scheme.

Let X be a smooth and proper scheme of finite type of dimension n defined over an algebraically closed field k of characteristic $p \geq 0$. Let l be a prime number different from p. Let Q_l be the quotient field of the ring of l-adic integers. For each nonnegative integer i, let $H^i(X, Q_l)$ denote the l-adic cohomology of X. Let $f : X \longrightarrow X$ be a morphism that has isolated fixed points each of multiplicity one. Let f^* be the induced map on the cohomology of X. Let Tr be the trace mapping. Then the number of fixed points of f is equal to

$$\sum_{i=0}^{2n} (-1)^i \operatorname{Tr}\left(f^*; H^i(X, Q_l)\right).$$

Lefschetz number   (1) A concept applicable in several fields of mathematics, first introduced in 1923 by S. Lefschetz (1884–1972) after whom it is named. The idea is as follows. Let X be a topological space. Let $f : X \rightarrow X$

be a continuous map. For each nonnegative integer k, f induces a homomorphism f_k of the homology group $H_k(X; \mathbf{Q})$ (with coefficients in the rational number field $\mathbf{Q}$). If the ranks of the homology groups (considered as vector spaces over $\mathbf{Q}$) are finite, then there is a matrix representation for each f_k and a trace t_k of the matrix which is an invariant of f_k. The Lefschetz number of f, denoted by $L(f)$, is then given by

$$L(f) = \sum_{k=0}^{\infty} (-1)^k t_k.$$

If everything is well defined, then $L(f)$ is an integer and, if different from zero, guarantees that f has a fixed point. Everything will be well defined, for example if X is a finite complex. Alternatives for X include a chain or cochain complex of free Abelian groups (f an endomorphism of degree 0) or a finite polyhedron of degree n with integral coefficients (f continuous, sum from 0 to n). There is a similar result if X is a closed orientable manifold. Also, if f is the identity map, then $L(f)$ is the *Euler characteristic*.

(2) For complex normal Abelian varieties, the difference between the second Betti number and the Picard number is sometimes called a *Lefschetz number*.

Lefschetz pencil   Let X be a smooth complete (thus projective) algebraic variety over an infinite algebraically closed field k. In the theory of algebraic varieties it can be shown that there is a birational morphism

$$\pi : \tilde{X} \longrightarrow X$$

from a smooth complete (projective) variety $\tilde{X}$ and a mapping

$$f : \tilde{X} \longrightarrow \mathbf{P}_k^1$$

that is singular at most a finite number of points. An inverse image $f^{-1}(x)$ is called a *fiber* and contains at most one singular point which, if it exists, is an ordinary double point. The family of fibers is called a *Lefschetz pencil* of X. Sometimes the map f is called a *Lefschetz pencil*. In the theory establishing the existence of $\tilde{X}$, it is shown that fibers are hyperplane sections of X. Sometimes it is said that *Lefschetz pencils* "fiber a variety."

Lefschetz pencils are involved in the proof of the Weil conjecture. They are named in honor of S. Lefschetz (1884–1972) who studied, among many other things, nonsingular projective surfaces.

left A-module   Let A be a ring. A *left A-module* is a commutative group G, together with a mapping from $A \times G$ into G, called *scalar multiplication* and denoted here by juxtaposition, satisfying for all $x, y \in G$ and all $r, s \in A$:

$$1x = x ,$$

$$r(x + y) = rx + ry ,$$

$$(r + s)x = rx + sx ,$$

and
$$(rs)x = r(sx) .$$

Right A-modules are defined similarly. Sometimes the term *left* or *right* is omitted. If the ring is denoted by R, then the terminology is *left R-module*. If addition and scalar multiplication are just the ring operations, then the ring itself can be regarded as a left A-module. If the ring is a field, then the modules are the vector spaces over the field.

left annihilator   The *left annihilator* of an ideal S of an algebra A (over a field k) is a set

$$\{a \in A : aS = 0\} .$$

The *left annihilator* is an ideal of A. Similarly, the *left annihilator of S in R* where S is a subset of an R-module M (where R is a ring) is the set

$$\{r \in R : rS = 0\} .$$

left Artinian ring   A ring having the property that (considered as a left module over itself) every nonempty set of submodules (meaning *left ideals* in this context) has a minimal element. For such a ring, any descending chain of left ideals is finite. *Compare with* left Noetherian ring.

For a *left Artinian ring,* the *radical* is the *largest nilpotent ideal. See* largest nilpotent ideal. For a left Artinian ring, any left module is Artinian if and only if it is Noetherian.

Artinian rings are named after E. Artin (1898–1962).

left balanced functor   Let R_1, R_2, and R be rings. Let $\mathcal{C}_{R_1}$, $\mathcal{C}_{R_2}$, and $\mathcal{C}_R$ be categories of R_1-modules, R_2-modules, and R-modules, respectively. Let $T : \mathcal{C}_{R_1} \times \mathcal{C}_{R_2} \to \mathcal{C}_R$ be an additive functor that is covariant in the first variable and contravariant in the second variable. Let $A \in \mathcal{C}_{R_1}$. Let $_AT : \mathcal{C}_{R_2} \longrightarrow \mathcal{C}_R$ be the functor defined by

$$_AT(B) = T(A, B) \text{ for all } B \in \mathcal{C}_{R_2} .$$

Let $B \in \mathcal{C}_{R_2}$. Let $T_B : \mathcal{C}_{R_1} \longrightarrow \mathcal{C}_R$ be the functor defined by

$$T_B(A) = T(A, B) \text{ for all } A \in \mathcal{C}_{R_1} .$$

Then T is called *left balanced* if (i.) $_AT$ is exact for each projective module $A \in \mathcal{C}_{R_1}$ and (ii.) T_B is exact for each injective module $B \in \mathcal{C}_{R_2}$.

The definition can be extended to additive functors T of several variables (some of which are covariant and some contravariant) as follows. T is called *left balanced* if: (i.) T becomes an exact functor of the remaining variables whenever any one of the covariant variables is replaced by a projective module; and (ii.) T becomes an exact functor of the remaining variables whenever any one of the contravariant variables is replaced by an injective module.

left coset   Any set, denoted by aS, of all left multiples as of the elements s of a subgroup S of a group G and a fixed element a of G. Every *left coset* of S has the same cardinality as S. For each subgroup S of G, the group G is partitioned into its left cosets, so that for a finite group the number of elements in the group (order) is a multiple of the order of each of its subgroups.

left derived functor   Let R_1, R_2, and R be rings. Let $\mathcal{C}_{R_1}$, $\mathcal{C}_{R_2}$, and $\mathcal{C}_R$ be categories of R_1-modules, R_2-modules, and R-modules, respectively. Let $T : \mathcal{C}_{R_1} \times \mathcal{C}_{R_2} \to \mathcal{C}_R$ be an additive functor that is covariant in the first variable and contravariant in the second variable. Let X be a definite projective resolution for each module A of $\mathcal{C}_{R_1}$ and let Y be a definite injective resolution for each module B of $\mathcal{C}_{R_2}$. From homology theory, $T(X, Y)$ is a well-defined complex which can be regarded as being over $T(A, B)$. Also, because of the homotopies involved, up to natural isomorphisms, $T(X, Y)$ is independent of the chosen resolutions X and Y and depends only on A and B. Therefore, the nth homology modules $H_n(T(X, Y))$ are functions of A and B.

If $f : A \longrightarrow A'$ and $g : B' \longrightarrow B$, there exist chain transformations $F : X \longrightarrow X'$ and $G : Y' \longrightarrow Y$ over f and g. Any two such are homotopic. The induced chain transformation

$$T(f, g) : T(X, Y) \longrightarrow T(X', Y')$$

is determined up to homotopy, is independent of the choice of F and G, and depends only on f and g. This yields a well-defined homomorphism

$$H_n(T(X, Y)) \longrightarrow H_n(T(X', Y')) .$$

Define the nth *left derived functor* L_nT : $\mathcal{C}_{R_1} \times \mathcal{C}_{R_2} \to \mathcal{C}_R$ by

$$L_nT(A, B) = H_n(T(X, Y))$$

and let $L_nT(f, g)$ be the well-defined homomorphism of the last paragraph

$$L_nT(f, g) : L_nT(A, B) \longrightarrow L_nT(A', B') .$$

left global dimension   For a ring A with unity, the supremum of the set of projective (or homological) dimensions (i.e., the set of minima of the lengths of left projective resolutions) of elements of the category of left A-modules. It is also equal to the infimum of the set of injective dimensions of the category. A theorem of Auslander (1955) shows that the left global dimension is also the supremum of the set of projective dimensions of finitely generated left A-modules. As simple examples of left global dimension, the left global dimension of the integers is 1, while the left global dimension of a field is 0. *See* projective dimension, homological dimension, injective dimension.

left G-set   Let G be a group with identity e and juxtaposition denoting the group operation. Let S be a set. Let $p : G \times S \to S$ satisfy, for each $x \in S$,

$$p(ab, x) = p(a, p(b, x))$$

and

$$p(e, x) = x .$$

Then S is called a *left G-set* and G is said to act on S from the left. If juxtaposition ax is used for the value $p(a, x)$, then the two conditions become $(ab)x = a(bx)$, $ex = x$.

left hereditary ring   A ring in which every left ideal is projective. Alternately, a ring whose left global dimension is less than or equal to one. *See* left global dimension.

If a ring A is left hereditary, every submodule of a free A-module is a direct sum of modules isomorphic to left ideals of A. The converse is also true.

left ideal   A nonempty subset L of a ring R such that whenever x and y are in L and r is in R, $x - y$ and rx are in L. For example, if R is the ring of $n \times n$ real matrices, the matrices whose first column is identically zero is a left ideal. Ideals play a role in ring theory analogous to the role played by normal subgroups in group theory.

left invariant   Used in various fields to indicate not being altered or changed by a transformation, usually a left translation. For example, (in linear algebra) a subalgebra S of a linear algebra A is called *left invariant* if S contains, for any $x \in S$, all left multiples ax for each $a \in A$. If the linear algebra A has a unity element, then the left invariant subalgebra is a left ideal. In Lie group theory and transformation group theory there are left invariant measures, left invariant integrals, left invariant densities, left invariant tensor fields, etc.

left inverse element   Let S be a nonempty set on which is defined an associative binary operation, say $*$. Assume that there is a left identity element e in the set with respect to $*$. Let a denote any element of the set S. Then a *left inverse element* of a is an element b of S such that

$$b * a = e .$$

If every element of the set described above has a left inverse element, then the set is called a *group*. For a group, a left inverse element is unique and is also a *right inverse element*. For commutative groups, the inverse element of a is usually denoted by $-a$. Otherwise, the notation a^{-1} is used. An element that has a left inverse is sometimes called *left invertible*.

left Noetherian ring   A ring having the property that (considered as a left module over itself) every nonempty set of submodules (meaning *left ideals* in this context) has a maximal element. For such a ring, any ascending chain of left ideals is finite. *Compare with* left Artinian ring.

For a left Noetherian ring, the *prime radical* is the *largest nilpotent ideal. See* largest nilpotent ideal. For a left Noetherian ring, any left module is Artinian if and only if it is Noetherian.

Noetherian rings are named after A.E. Noether (1882–1935).

left order   Let g be an integral domain in which every ideal is uniquely decomposed into a product of principal ideals (i.e., a Dedekind domain). Let F be the field of quotients of g. Let A be a separable algebra of finite degree over F. Let L be a g-lattice of A. Then the set $\{x \in A : xL \subset L\}$ is an order of A called a *left order* of A. *See* order. *See also* ZG-lattice.

left projective resolution   A left resolution X of an A-module M (where A is a ring with unity) such that each X_n in the exact sequence

$$\cdots \to X_n \to X_{n-1} \to \cdots \to X_0 \to M \to 0$$

is a projective A-module, and is called a *left projective resolution* of M. *See also* left resolution.

left regular representation   Let R be a ring, M a left R-module, and $\text{Hom}_{\mathbf{Z}}(M, M)$ the ring of module endomorphisms (where M is viewed as a module over $\mathbf{Z}$). For each $r \in R$, let $\rho_r : M \longrightarrow M$ be a left translation (i.e., $\rho_r(x) = rx$ for all $x \in M$). If the ring homomorphism $\rho : R \to \text{Hom}_{\mathbf{Z}}(M, M)$, given by $\rho(r) = \rho_r$ for all $r \in M$ is an injection, then ρ is called a *left regular representation of R in $Hom_{\mathbf{Z}}(M, M)$. Left* (and *right*) *regular representations* are very important tools in ring theory with many applications throughout the theory.

left resolution   Let A be a ring with unit and M an A-module. If the sequence

$$\cdots \to X_n \xrightarrow{\partial_n} X_{n-1} \xrightarrow{\partial_{n-1}} \cdots \xrightarrow{\partial_1} X_0 \xrightarrow{\epsilon} M \to 0$$

is exact, where X is a positive chain complex of A-modules X_k, then the homomorphisms ∂_k have the property that the composition $\partial_k \partial_{k+1} = 0$ for all k, and $\epsilon : X \to M$ is a sequence of homomorphisms $\epsilon_k : X_k \to M$ such that $\epsilon_{k-1} \partial_k = 0$ for all k, then X is called a *left resolution* of M. This concept is used in homology theory in the computing of extension groups.

left satellite   Let R_1 and R be rings. Let $\mathcal{C}_{R_1}$ and $\mathcal{C}_R$ be categories of R_1-modules and R-modules, respectively. Let $T : \mathcal{C}_{R_1} \to \mathcal{C}_R$ be an additive functor. A *left satellite* of T is an additive functor defined over the same categories as T and having the same variance (contra- or co-). Once one *left satellite* of T (to be denoted by $S_1 T$) is determined, a sequence $\{S_n T\}$ of *left satellites* may be defined by iteration.

In order to define the functor $S_1 T$ by describing its action on modules and homomorphisms, some background notation and remarks on homology theory need to be presented. *Left satellites* may be defined for covariant functors and contravariant functors. Here, the procedure for creating a *left satellite* from a covariant functor will be described in some detail, with the corresponding procedure for contravariant functors only sketched. There is also a parallel development for *right satellites*.

Let A be an R_1-module. From homology theory, it is always possible to construct an exact sequence

$$0 \to M \xrightarrow{\alpha} P \xrightarrow{\beta} A \to 0$$

with P projective. For what follows, since the module $S_1 T(A)$ (to be defined) is only unique within isomorphism (with respect to the choice of P), a particular choice of exact sequence needs to be prescribed (which can always be done).

(**i.**) Suppose T is covariant. Let $A \in \mathcal{C}_{R_1}$. Let $T(\alpha) : T(M) \to T(P)$. Define $S_1 T(A)$ to be $\text{Ker}(T(\alpha))$. Then the sequence

$$0 \to S_1 T \to T(M) \to T(P)$$

is exact. Let A, $A' \in \mathcal{C}_{R_1}$. Let $g : A \to A'$. Then there are exact sequences

$$0 \to M \xrightarrow{\alpha} P \xrightarrow{\beta} A \to 0$$
$$\downarrow g$$
$$0 \to M' \xrightarrow{\alpha'} P' \xrightarrow{\beta'} A' \to 0$$

(with P and P' projective) and a homomorphism $f : P \to P'$ such that $g\beta = \beta' f$. f defines uniquely a homomorphism $f' : M \to M'$ such that $f\alpha = \alpha' f'$. One then gets the commutative diagram

$$\begin{array}{ccc} T(M) & \xrightarrow{T(\alpha)} & T(P) \\ \downarrow T(f') & & \downarrow T(f) \\ T(M') & \xrightarrow{T(\alpha')} & T(P') \, . \end{array}$$

$T(f')$ induces a homomorphism

$$\Theta(g) : \mathrm{Ker}(T(\alpha)) \to \mathrm{Ker}(T(\alpha')) \, .$$

It can be shown that $\Theta(g)$ is independent of the choice of f. Define

$$S_1 T(g) : S_1(A) \longrightarrow S_1(A')$$

by $S_1 T(g) = \Theta(g)$. With this definition, $S_1 T$ becomes a covariant functor called the *left satellite* of T.

(**ii.**) If T is contravariant, the procedure starts with an exact sequence

$$0 \to A \xrightarrow{\beta} Q \xrightarrow{\alpha} N \to 0 \, ,$$

with Q injective. Similar reasoning to the covariant case yields $S_1 T(A) = \mathrm{Ker}(T(\alpha))$ where $T(\alpha) : T(N) \longrightarrow T(Q)$ and $S_1 T(g) : S_1 T(A') \longrightarrow S_1 T(A)$. Here, $S_1 T$ is contravariant.

(**iii.**) The sequence of function $S_n T$ is defined recursively from the exact sequences

$$0 \to M \to P \to A \to 0$$

and

$$0 \to A \to Q \to N \to 0 \, ,$$

with P projective and Q injective as follows:

$$S_0 T = T \, ,$$

$$S_1 T \text{ as given above} \, .$$

If T is covariant,

$$S_{n+1} T(A) = S_n T(M)), \qquad n \geq 1 \, .$$

If T is contravariant,

$$S_{n+1} T(A) = S_n T(N)), \qquad n \geq 1 \, .$$

Left satellites may also be defined for general Abelian categories in a similar fashion.

left semihereditary ring   A ring in which every finitely generated left ideal is projective. *Compare with* left hereditary ring.

Every regular ring is left semihereditary (and right semihereditary).

left translation   A *left translation* of a group G by an element a of G is a function $f_a : G \to G$ defined by

$$f_a(x) = ax \, ,$$

for each $x \in G$.

Lehmer's method of finding roots   A method, developed in 1960 by D.H. Lehmer, for finding numerical approximations to the (real or complex, simple or multiple) roots of a polynomial using a digital computer. Lehmer's method provides a single algorithm for automatic computation applicable for all polynomials, in contrast to other methods, the choice of which, for a particular polynomial, is dependent on human judgment, experience, and intervention.

Lehmer's method includes a test for determining whether or not the given polynomial has a root inside a given circle. The test is applied at each iteration of the process (to be described next) on circles of decreasing size.

Assume that zero is not a root. Starting (in the complex plane) with the circle with center 0 and radius 1 apply the test to look for a root inside the circle. If one is found, halve the radius and retest for the smaller circle. If a root is not found, double the radius and apply the test. In a finite number of steps, an annulus

$$\{x : R < |z| < 2R\}$$

will be determined (where R is a power of 2) that contains a root in the disk of radius $2R$ but not in the disk of radius R. Next, cover the annulus with 8 overlapping disks of radius $5R/6$ and centers $(5R/3)\exp(2\pi i/8)$, for $k = 0, 1, \ldots, 7$. Test each of the 8 circles until a root is found trapped in a new annulus centered at (say) α with radii $R_1, 2R_1$

$$\{x : R_1 < |z - \alpha| < 2R_1\},$$

with $R_1 = \frac{5R}{6(2^s)}$ for some s. The new annulus is similarly covered by 8 disks and tested. After (say) k iterations of this process, one gets a circle of radius $\leq 2(5/12)^k$ which contains a root. One continues until a prescribed accuracy is reached.

The strength of Lehmer's method is its universality. The weakness is the rate of convergence of the method which in practice may be several times slower than less general methods developed for specific types of roots.

length of module   The common number of quotients M_i/M_{i+1} of submodules of a module M over a ring R in a Jordan-Hölder or composition series

$$M = M_0 \supset M_1 \supset \cdots \supset M_s = \{0\}$$

of M, if M has such a series. Any such chain has the same number of terms. A necessary and sufficient condition that M have such a series is that it be both Artinian and Noetherian. *See* composition series, Artinian module, Noetherian module.

Leopoldt's conjecture   The assertion in the study of algebraic number fields that the $\mathbf{Z}_p$-rank of the p-adic closure of the group of units of a number field is the same as the $\mathbf{Z}$-rank of the group of units. This conjecture is an open question in many cases. There are Abelian analogs of the conjecture and analogs for function fields of characteristic p. The conjecture is named after H.W. Leopoldt, an extensive contributor to the study of p-adic L-functions.

less than   *See* greater than.

less than or equal to   *See* greater than.

level structure   A notion occurring in the study of modulii spaces of Abelian varieties. Usually a letter intrinsic to the variety discussed is appended to the term.

(1) Let A be an Abelian variety of dimension n over a field k of characteristic $p \geq 0$. Let m be a positive integer that is not a multiple of p. An *m-level structure* with respect to A is a set of $2n$ points on A which form a basis for the abstract group $B_m(A)$ of points of order m on A.

(2) For polarized complex Abelian varieties, the concept of a *level structure* can be considered as a replacement of all or part of the concept of a symplectic basis. If the variety is of type D, then the D-level structure for the variety is a certain symplectic isomorphism.

There are *generalized level m-structures, orthogonal level D structures*, etc. The precise definitions are quite detailed and require considerable background and notational development. Thus, they will not be given here. The associated modulii spaces for polarized Abelian varieties with level structures are used in studying properties of geometry and arithmetic.

Levi decomposition   Let g be a finite dimensional Lie algebra over the field of real or complex numbers. Let $\text{rad}\,g$ denote the Lie subalgebra of g, called the *radical* of g. (*See* radical.) Let l be any subalgebra of g. Then the direct sum of vector spaces $\text{rad}\,g$ and l ($\text{rad}\,g \oplus l$) forms a Lie subalgebra of g. If l exists such that $\text{rad}\,g \oplus l = g$, then the sum $\text{rad}\,g \oplus l$ is called the *Levi decomposition* of g. The decomposition is named after E.E. Levi. The subalgebra l involved in the splitting is called a *Levi subalgebra* of g. A theorem of Levi (1905) states that g has such a decomposition and a theorem by Malcev (1942) establishes the uniqueness of the decomposition.

There is a similar *Levi* (product) *decomposition* in algebraic group theory given by $G = \text{Rad}\,G \times H$ where G is an algebraic group, $\text{Rad}\,G$ is a unipotent radical of G, and H is an algebraic subgroup of G called the *reductive Levi subgroup* of G. It is useful in reducing many problems to the study of *reductive* groups.

Levi subgroup   Let G be a Lie group. Let $\text{Rad}\,G$ denote the radical of G. A Lie subgroup L of G is called a *Levi subgroup* of G if (i.) the identity imbedding of L into G is a Lie group homomorphism, (ii.) $G = (\text{Rad}\,G)\,L$, (iii.) the dimension of $(\text{Rad}\,G) \cap L$ is zero. If the Lie group is connected, there is always a (connected) Levi subgroup. *See* Levi decomposition.

lexicographic linear ordering Sometimes called *dictionary order,* an ordering which treats numbers as if they were letters in a dictionary. Suppose, for example, S is a set of monomials of degree n in m variables: that is,

$$S = \left\{ x_1^{k_1} x_2^{k_2} \ldots x_m^{k_m} : \sum_{j=1}^{m} k_j = n \right\} .$$

then $x_1^{k_1} x_2^{k_2} \ldots x_m^{k_m}$ is less than or equal to ($\leq$) $x_1^{l_1} x_2^{l_2} \ldots x_m^{l_m}$ if the first nonzero difference of corresponding exponents $k_j - l_j$ satisfies $k_j - l_j \leq 0$. For example,

$$x_1 x_2 x_3^2 x_4^3 \leq x_1 x_2 x_3^3 x_4 \leq x_1^2 x_2^4 x_3 x_4 .$$

A *lexicographic ordering* for the field of complex numbers $\mathbf{C}$ is defined as follows: if $z_1 = x_1 + y_1 i$, $z_2 = x_2 + y_2 i \in \mathbf{C}$, then $z_1 \leq z_2$ if and only if

$$(x_1 \leq x_2) \text{ or } (x_1 = x_2 \text{ and } y_1 \leq y_2) .$$

Since in both of these cases the order $\leq$ is reflexive, antisymmetric, transitive, and satisfies the trichotomy law, the order is a linear order, so the order is called a *lexicographic linear order.* More generally, if $O_1, O_2, \ldots, O_m$ are ordered sets, the product set $p = \prod_{j=1}^{m} O_j$ is given a *lexicographic ordering* $<$ as follows: if $a = (a_1, a_2, \ldots, a_m)$ and $b = (b_1, b_2, \ldots, b_m) \in p$, then $a < b$ if and only if the first non-zero difference of corresponding coordinates $a_j - b_j < 0$. When each O_j is linearly ordered, so is p.

L-function Generally, a function of a complex variable that generalizes the Riemann ζ-function. *L*-functions are meromorphic on the complex numbers $\mathbf{C}$, exhibit both a Dirichlet series expansion and a Euler product expansion, and satisfy similar functional equations. Some generalizations of the ζ-function retain the ζ identifier instead of an *L*. *L-functions* are important in the analytic study of the arithmetic of objects in their corresponding mathematical structures, including rational number fields, algebraic number fields, algebraic varieties over

finite fields, representations of Galois groups, p-adic number fields, etc.

(1) The most direct generalization of the Riemann ζ-function is the Dirichlet *L-function*. Let $\mathbf{Z}$ denote the integers. Let $m \in \mathbf{Z}$ with $m > 0$. Let $\chi : \mathbf{Z} \longrightarrow \mathbf{C}$ be a Dirichlet character modulo m (i.e., $\chi \neq 0$, $\chi(a) = 0$ if a and m are not relatively prime, $\chi(ab) = \chi(a)\chi(b)$, and $\chi(a + m) = \chi(a)$). Let $s \in \mathbf{C}$ with $Re(s) > 1$. Define $L(s)$ as

$$L(s) = \sum_{n=1}^{\infty} \frac{\chi(n)}{n^s} .$$

The definition exhibits the Dirichlet series expansion which converges absolutely and makes L a holomorphic function for $\Re(s) > 1$. It is equal to the Euler product

$$\prod_p \frac{1}{1 - \frac{\chi(p)}{p^s}}$$

over primes p. It can be extended as a meromorphic function over $\mathbf{C}$. If χ is defined to be identically 1, the Riemann ζ-function itself is obtained. The two functions have many similar properties. For example, the study of the location of zeroes of the *L-function* leads to a generalized Riemann hypothesis.

(2) *Hecke L-functions* and *Hecke L-functions with grössencharakters* are generalizations of the Dirichlet *L-functions* to algebraic number fields. Other *L-functions* that can be considered generalizations of these include those of Artin and Weil involving Galois extensions of an algebraic number field. There are *L-functions* associated with algebraic varieties defined over finite fields, p-adic *L-functions* defined over the p-adic number field $\mathbf{Q}_p$, and automorphic *L-functions* defined both for the general linear group and for arbitrary reductive matrix groups. *See* Hecke L-function, Artin L-function, Weil L-function.

(3) A separate meaning of *L-function* occurs in the area of mathematical analysis where an *L-function* is defined as a continuous function $f : [a.b] \times K^n \longrightarrow K^n$ (where $[a, b]$ is a closed real interval and K^n denotes complex Euclidean n-space) that is uniformly Lipschitzian in the second variable, i.e., there is a constant L, the Lipschitz constant of f, such that, for all $t \in$

$[a, b]$, $x \in K^n$, $y \in K^n$ we have

$$\| f(t, x) - f(t, y) \| \leq L \| x - y \| \, .$$

In such a circumstance, the initial value problem

$$\frac{dx}{dt} = f(t, x), \qquad x(a) = c$$

has a unique solution for all $c \in K^n$.

Lie algebra   A theory named after the Norwegian mathematician M.S. Lie (1842–1899) who pioneered the study of what are now called Lie Groups.

Over a commutative ring with unit K, a *Lie algebra* is a K-module together with a K-module homomorphism $x \otimes y \to [x, y]$ of $L \otimes L$ into L such that, for all $x, y, z \in L$,

$$[x, x] = 0$$

and

$$[x, [y, z]] + [y, [z, x]] + [z, [x, y]] = 0 \, .$$

The product $[x, y]$ is called a *bracket product* and is an alternating bilinear function. The second equation is called the *Jacobi identity*.

An example of a Lie Algebra is obtained for an associative algebra L over K by defining

$$[x, y] = xy - yx \, .$$

Lie algebra of Lie group   There are three Lie algebras that are called a *Lie algebra of a Lie group G*. (*See* Lie algebra.)

(**1**) The Lie algebra of all left invariant derivations on G with the bracket product $[D_1, D_2]$ defined for two derivations D_1, D_2 as $D_1 D_2 - D_2 D_1$.

(**2**) The Lie algebra of all left invariant vector fields on G with the bracket product of two left invariant vector fields X and Y defined using the natural isomorphism between the vector space of left invariant vector fields and the vector space of left derivations of G.

(**3**) The Lie algebra of all tangent vectors to G at e with the bracket product between tangent vectors defined using the natural isomorphism between the space of left invariant vector fields of G and the space of tangent vectors of G at e.

The fact that the three Lie algebras defined above are isomorphic allows some flexibility in applications of the theory.

Lie group   A group G, together with the structure of a differentiable manifold over the real or complex number field, such that the functions $f : G \times G \to G$ and $g : G \to G$ defined by $f(x, y) = xy$ and $g(x) = x^{-1}$ are differentiable.

The additive groups of the fields of real or complex numbers are Lie groups. The group of invertible $n \times n$ matrices over the real or complex number fields is a Lie group.

In addition to providing a thriving branch of mathematics, Lie groups are used in many modern physical theories including that of gravitation and quantum mechanics.

Lie-Kolchin Theorem   Let V be a finite dimensional vector space over a field k. Let G be a connected solvable linear algebraic group. Let $\mathrm{GL}(V)$ be the general linear group. Let $f : G \to \mathrm{GL}(V)$ be a linear representation. Then $f(G)$ can be put into triangular form.

Alternate conclusions in the literature include the following: (i.) Then $f(G)$ leaves a flag in V invariant. (ii.) Then there exists a basis in V in which elements of $f(G)$ can be written as triangular matrices.

This theorem has been generalized by Mal'tsev.

Lie's Theorem   There are several theorems which bear the name and honor the Norwegian mathematician M.S. Lie (1842–1899) who, among other things, founded the theory of Lie groups. For solvable Lie algebras, the following version of *Lie's Theorem* is considered an important tool.

The irreducible representations of a finite dimensional, solvable Lie algebra over an algebraically closed field of characteristic zero all are one-dimensional.

Lie subalgebra   Let K be a commutative ring with unit. A *Lie subalgebra* is a subalgebra of a Lie algebra (meaning a K-submodule that is closed under the bracket product of the Lie algebra). A *Lie subalgebra* is a Lie algebra. *See* Lie algebra.

Lie subgroup   A subset of a Lie group G which is both a subgroup of G and a submanifold of G (with both structures being induced by those of G). (*See* Lie group.) As an example, an open subset which is also a subgroup of a Lie group is a Lie subgroup.

like terms   In an algebraic expression in several variables, terms with identical factors raised to identical powers (excluding values of coefficients). *Like terms* are usually collected and combined using rules of arithmetic and associative, commutative, and distributive laws.

$$3x^2y^5z \text{ and } -2x^2y^5z$$

are like terms.

limit ordinal   In the theory of ordinal numbers, a non-zero ordinal which does not have a predecessor.

line   (1) In classical Greek geometry, a *line* is an undefined term characterized by axioms and understood by intuition. A *line* is considered to be *straight* and to extend without bound. In everyday usage, the words *line* and *straight line* are used interchangeably, and although in this context the word *straight* is somewhat redundant, it is still universally used. (An archaic synonym is *right* line.) Curves that can be drawn without interruption are sometimes called *lines.*

(2) In plane analytic geometry, the graph or set of points described by a linear equation in two variables x and y, given by $ax + by = c$ (where a, b, c are real numbers) is a *line.* This equation, or alternately $ax+by+c = 0$, is called the *general equation of a line.* The graph of a real valued function f of a real variable, given (for real numbers a, b) by $f(x) = ax + b$, for each real number x, is a non-vertical line. The number a is called the *slope* of the line. With $b = 0$, there is a one-to-one correspondence between non-vertical lines and slopes. For a curve, the word *slope* is used more generally at a point on the curve as the slope of the tangent line to the curve at the point (if it exists). In this context, a line can be described as a curve with a constant slope.

The line passing through the two points (x_1, y_1) and (x_2, y_2) in the Euclidean plane is de-scribed as the set of all points (x, y) such that

$$x = x_1 \quad \text{if} \quad x_2 = x_1$$

and

$$y - y_1 = \frac{y_2 - y_1}{x_2 - x_1}(x - x_1) \quad \text{if} \quad x_2 \neq x_1.$$

The equation is called the *two point form* of the equation of a *line.* Also, a *line* has a *parametric representation* as the set of pairs (x, y) such that, for each real number t,

$$(x, y) = (1 - t)(x_1, y_1) + t(x_2, y_2).$$

(3) In affine geometry, where V is a finite dimensional vector space over a field k, a *line* is a one-dimensional affine space. If a, b are distinct points, the line joining a and b can be described as the set of all points of the form

$$(1 - t)a + tb, \quad \text{for} \quad t \in k.$$

(4) In projective geometry, a *line* is a one-dimensional projective space.

(5) Many areas of mathematics endow a *line* with properties, resulting in such concepts as parallel lines, half line, the real line, broken line, tangent line, secant line, line of curvature, long line, extended line, normal line, perpendicular line, orthogonal line, line method (in numerical analysis), etc.

linear   (1) Like or resembling a line. One of the most used terms in mathematics. Many fields of mathematics are separated into *linear* and *non-linear* subfields; e.g., the field of differential equations. The adjective is also used to specify a subfield of investigation; e.g., *linear* algebra, *linear* topological space, *linear* (or vector) space, *linear* algebraic group, general *linear* group, etc. Also, the term is used to specify a special subtype of object; e.g., *linear* combination, *linear* equation, *linear* function, *linear* order, *linear* form, etc. See the nouns modified for further explanation.

(2) If V, W are vector (*linear*) spaces over a field k, a mapping $L : V \longrightarrow W$ is called *linear* if L is additive and homogeneous. Equivalently, L is *linear* if, for each $\alpha, \beta \in k$ and $x, y \in V$,

$$L(\alpha x + \beta y) = \alpha L(x) + \beta L(y).$$

(Note that the same symbolism for the vector space operations was used for both vector spaces.)

linear algebra   (1) A set L which is a finite dimensional vector (or *linear*) space over a field F such that for each $a, b \in F$ and each $\alpha, \beta, \gamma \in L$,

$$\alpha(\beta\gamma) = (\alpha\beta)\gamma \qquad \text{(associative law)}$$

$$\alpha(a\beta + b\gamma) = a(\alpha\beta) + b(\alpha\gamma) \quad \text{(bilinearity)}$$

$$(a\alpha + b\beta)\gamma = a(\alpha\gamma) + b(\beta\gamma) .$$

(2) A mathematical theory of "linear algebras." University mathematics departments usually offer an undergraduate course where the linear algebras of matrices and linear transformations are studied.

linear algebraic group   An abstract group G, together with the structure of an affine algebraic variety, such that the product and inverse mappings given by

$$\mu : G \times G \to G, \qquad (x, y) \mapsto xy$$

$$i : G \to G, \qquad x \mapsto x^{-1}$$

are morphisms of affine algebraic varieties. Alternately, such an algebraic group is called an *affine algebraic group*. The ground field K is assumed to be an arbitrary infinite field.

Probably the most important example of a *linear algebraic group* is the *general linear group* $GL_n(K)$ or $GL_n(U)$, where U is a finite dimensional vector space over K. Since $GL_n(K)$ is an open subset (with respect to the Zariski topology) of the algebra on $n \times n$ matrices over K, it inherits the affine variety structure. Also, polynomials on $GL_n(K)$ are realized as rational functions in matrix elements with denominators being powers of determinants. It follows that the product and inverse mappings are morphisms of affine algebraic varieties.

Another definition of a *linear algebraic group* is a subgroup of the general linear group $GL_n(U)$ which is closed in the Zariski topology, where U is an algebraically closed field that has infinite transcendence degree over the prime field K. Since this subgroup is an algebraic group, it can be called an *algebraic linear group* as well. Now, there is an isomorphism between a group defined by the first definition and one defined by the second. So, the distinction is only important when the word *linear* or *affine* is omitted from the first definition, as is sometimes done in textbooks.

linear combination   Let $x_1, x_2, \ldots, x_n$ be elements of a vector space V over a field F. A *linear combination* of the elements $x_1, x_2, \ldots, x_n$ over F is an element of the form

$$a_1 x_1 + a_2 x_2 + \cdots + a_n x_n$$

where $a_1, a_2, \ldots, a_n \in F$.

Linear combination is a fundamental concept in the theory of vector spaces and is used, among other things, to define concepts of linear independence of vectors, basis, linear span, and the dimension of the vector space.

linear difference equation   An equation of the form

$$a_0 x_n + a_1 x_{n-1} + \cdots + a_k x_{n-k} = b_n$$

where $a_0, a_1, \ldots, a_k$ are scalars and $a_0 a_k \neq 0$, is called a *linear difference equation* of order k. The equation is called *non-homogeneous* if $b_n \neq 0$ and *homogeneous* if $b_n = 0$. Such recursive equations are important in many fields of numerical analysis; e.g., in finding zeros of polynomials and in determining numerical solutions of ordinary and partial differential equations. Writing the above difference equation in terms of the operator $\Delta x_n = x_n - x_{n-1}$ and its powers, one can exploit the analogy with linear differential equations.

linear disjoint extension fields   Two fields, K and L, that are extensions of another field, k, and are themselves contained in a common extension field are called *linearly disjoint* if every subset of L that is linearly independent over k is also linearly independent over K.

linear equation   Let k be a commutative ring. Let $a, b \in k$. Then an expression of the form

$$ax + b = 0$$

is called a *linear equation* in one variable x.

Let $m \geq 1$. Let $k[x_1, \ldots, x_m]$ be a polynomial ring over k. A linear equation is formed by setting a linear form equal to zero. In other words, a linear equation in m variables is given by an equation

$$a_0 + a_1 x_1 + \cdots + a_m x_m = 0 \, ,$$

where $a_1, \ldots, a_m \in k$.

In analytic geometry, where k is the real number field, a linear equation in m variables describes a line if $m = 2$, a plane if $m = 3$, and a hyperplane if $m > 3$.

linear equivalence   Any one of several similar equivalence relations defined on schemes or algebraic varieties. Generally speaking, if X is one of these structures and if the concepts of divisor and principal divisor are defined, then two divisors are related if their difference is a principal divisor. The relation is an equivalence relation, called a *linear equivalence*. The equivalence classes form a group (*See* linear equivalence class.)

More specifically, let X be a complete, nonsingular algebraic curve defined over an algebraically closed field k. Two divisors on X are said to be *linearly equivalent* if their difference is a principal divisor. A similar definition can be made for nonsingular algebraic surfaces. *See* divisor, principal divisor of functions.

More abstractly, let X be a Noetherian integral separated scheme which is nonsingular in codimension one. The divisor d_1 is called *linearly equivalent* to d_2 if their difference is a principal divisor.

The definition can be extended to general schemes by using *Cartier divisors*. *See* Cartier divisor.

linear equivalence class   An equivalence class determined by the equivalence relation called a *linear equivalence* defined on the divisors of an algebraic variety or a scheme. The equivalence classes form a group which is one of the intrinsic invariants of the variety or scheme. *See* linear equivalence.

linear extension   Let Y be a nonsingular algebraic curve over a field k. Let J denote the Jacobian variety of Y. Let ϕ be the canonical function on Y (with values in J). Let A be an Abelian variety. Then for any function f on Y into A, there exists a homomorphism h from J into A such that $f = h \circ \phi$. The function h is called the *linear extension* of f and is unique. *See* Jacobian variety. *See also* canonical function, Abelian variety.

linear form   (1) Also called *linear functional*. A linear function or transformation from a linear (vector) space V over a field F into the field F (where the field is considered to be a linear space over itself).

(2) An expression in n variables $x_1, \ldots, x_n$ of the type

$$a_1 x_1 + a_2 x_2 + \cdots + a_n x_n + b$$

where $a_1, \ldots a_n, b$ are elements of a commutative ring and at least one coefficient is not 0. If $b = 0$, the form is called *homogeneous*.

linear fractional function   A function f with domain and range either the complex numbers or the extended complex numbers and defined for numbers a, b, c, d with $ad - bc \neq 0$ by

$$f(z) = \frac{az + b}{cz + d} \, .$$

Linear fractional functions are univalent, meromorphic functions that map each circle and line in the complex plane to either a circle or a line. The class includes rigid motions, translations, rotations, and dilations. It is a subset of the class of rational functions, being those rational functions of order 1. Using function composition as the binary operation, the set of linear fractional transformations forms a group. They play an important role in the study of the geometric theory of functions.

Also called *linear fractional transformation, linear transformation, Möbius transformation, fractional linear transformation,* and *conformal automorphism. See also* linear fractional group.

linear fractional group   A group of functions formed by the set of linear fractional functions defined on the complex plane or extended complex plane using composition as the group operation. The quotient group of the full linear group modulo the subgroup of nonzero scalar matrices is isomorphic to the *linear fractional group*. Both groups are called *linear fractional* in the mathematical literature. *See* linear fractional function.

linear fractional programming   Determining the numerical solution of the problem of optimizing a function resembling the linear fractional functions of complex analysis subject to linear constraints. Linear programming methods are usually employed in the solution either by a simplex-type method or by using the solution of an associated linear programming problem with additional constraints to solve the *linear fractional* problem. Specifically, let A be an $m \times n$ matrix; let $\mathbf{a}, \mathbf{c}$ be vectors in $\mathbf{R}^n$; let $\mathbf{p}$ be a vector in $\mathbf{R}^m$; and, let $b, c \in \mathbf{R}$. Then a *linear fractional programming* problem is to minimize

$$\frac{\mathbf{a} \cdot \mathbf{x} + b}{\mathbf{c} \cdot \mathbf{x} + d}$$

subject to

$$A\mathbf{x} = \mathbf{p} \quad \text{and } \mathbf{x} \geq \mathbf{0}.$$

linear fractional transformation   Classically, a function on the complex plane having the form $\ell(z) = [az + b]/[cz + d]$ for complex constants a, b, c, d. There are analogs of this definition on any Euclidean space.

linear function   (1) A function L with domain a vector space U over a field F and range in a vector space V over F which is additive and homogeneous. This means, for each $x, y \in U$ and for each $\alpha \in F$,

$$L(x + y) = L(x) + L(y) \quad \text{(additivity)}$$

and

$$L(\alpha x) = \alpha L(x) \quad \text{(homogeneity)}.$$

The terms *linear transformation* and *linear mapping* are also used to describe a function satisfying these conditions and L is sometimes called a *linear operator*. If F is regarded as a one-dimensional vector space over itself and if $V = F$, then L is called a *linear functional* or *linear form*.

(2) If a function f has the real or complex numbers as its domain and range, it is called *linear* if for some constants a, b, $a \neq 0$, and for each x,

$$f(x) = ax + b.$$

For real valued functions of a real variable, the graph of a linear function (in this context) is a straight line. More properly, such functions should be called *affine*, but the usage *linear function* is universal.

(3) *See also* linear fractional function.

linear genus   Let X be a nonsingular algebraic surface. If X has a minimal model M, then the *linear genus* of X is an invariant equal to the arithmetic genus of a canonical divisor of M. If X is not a ruled surface, then X has a minimal model and thus a *linear genus*. *See* minimal model. *See also* arithmetic genus.

linear inequality   In analytic geometry, a *linear inequality* in m variables in Euclidean m-dimensional space is a linear form set less than, less than or equal to, greater than, or greater than or equal to zero. For example, the inequality

$$a_0 + a_1 x_1 + \cdots + a_m x_m > 0,$$

where $a_0, a_1, \ldots, a_m$ are real numbers, is a *linear inequality*.

If $m = 2$, the set of pairs of real numbers satisfying a *linear inequality* constitutes a half-plane bounded by the line given by the associated linear equation.

linear least squares problem   A problem where a linear function is to be found which best approximates a given set of data in the sense that the sum of the squares of the errors in such an approximation is minimized.

linearly dependent elements   A set of elements that satisfies some linear relation. More specifically, a set of elements $\{x_1, x_2, \ldots, x_N\}$ in a vector space is linearly dependent over a field F (such as the real or complex number systems) if there exists $\alpha_1, \ldots, \alpha_N \in F$, not all zero, with $\alpha_1 x_1 + \cdots + \alpha_N x_N = 0$. Examples include a set of two co-linear vectors or a set of three co-planar vectors.

linearly dependent function   A function which depends linearly on its input variables. More precisely, a function $f(x_1, \ldots, x_N)$ is a linearly dependent function if $f(x_1, \ldots, x_N) = \alpha_1 x_1 + \cdots + \alpha_N x_N$, where each α_i belongs to some given field, such as the real or complex number system.

linearly independent elements   A set of elements which is not linearly dependent. More precisely, a set of elements $\{x_1, x_2, \ldots, x_N\}$ in a vector space is *linearly independent* over a field F (such as the real or complex number systems) if, whenever $\alpha_1 x_1 + \cdots + \alpha_N x_N = 0$ for $\alpha_1, \ldots, \alpha_N \in F$, then necessarily all the α_i must be zero. Examples include a set of two vectors which are not co-linear or a set of three vectors which are not co-planar.

linear mapping   A function from one vector space to another $L : V \mapsto W$, which preserves the vector space operations, i.e., $L(r_1 v_1 + r_2 v_2) = r_1 L(v_1) + r_2 L(v_2)$ for all vectors v_1, v_2 $\in V$ and all scalars r_1 and r_2. Examples include multiplication of an n-dimensional vector by an $m \times n$ matrix (here V and W are $\mathbf{R}^n$ and $\mathbf{R}^m$, respectively) or the operation of differentiation (here both V and W are spaces of functions with appropriate differentiability).

Also called *linear transformation, linear operator.*

linear pencil   Given two linear transformations, or matrices, A and B, the family $A + \lambda B$, where λ is any scalar parameter, is called the (linear) pencil defined by A and B.

linear programming   The study of optimizing a linear function over a convex, linear constraint set. Linear programming arises in economics from the desire to maximize revenue or minimize cost (from some economic process) subject to constrained resources.

linear programming problem   A problem in which a linear function of one or more independent variables is maximized or minimized over a convex polygonal region. For example, to maximize the function $L(x) = 3x_1 + 2x_2 + 5x_3$ over the set $\{2x_1 + 3x_2 + 5x_3 \leq 5, x_1, x_2, x_3 \geq 0\}$. Often such a problem arises in economics, where the linear function to be maximized or minimized represents revenue or costs and the polygonal constraint region represents the limitation of resources.

linear representation   If G is a group, a linear representation of G is a group homomorphism of G into the invertible linear transformations on some vector space.

linear simple group   Denoted $L_n(q)$, where q is a non-zero power of a prime number, it is one of the classical finite simple groups. It may be identified as the projective special linear group over a finite field of order q, and denoted $\mathrm{PSL}(n, q)$. Let d be the greatest common divisor of n and $q-1$. The possible orders of a linear simple group are $q^{n(n-1)/2}(\prod_{i=2}^n (q^i - 1))/d$; where $n \geq 3$, or $n = 2$ and $q \geq 4$.

linear stationary iterative process   A method of using successive approximation to solve the system of linear equations $Ax = b$ when A is a non-singular matrix. A method is called *iterative* if the operations used to get from one approximation to the next are nearly the same. If successive steps are exactly the same, the method is called *stationary*. The linear stationary iterative process may be written as $x^{(k+1)} = x^{(k)} + R(b - Ax^{(k)})$, where R is an approximation to A^{-1} and $x^{(0)} = Rb$. With this definition $x^{(r)} = (\mathcal{I} + C + C^2 + \cdots + C^r)Rb$ where $C = \mathcal{I} - RA$ and $\mathcal{I}$ is the identity. R is sometimes referred to as the key matrix and $\mathcal{I} - RA$ is called the iteration matrix.

linear system   A finite set of simultaneous equations that depends linearly on its inputs. For example,

$$3x + 2y = 5$$

$$-x + 4y = 7$$

is a linear system of equations involving the two unknowns x and y. Another example is a linear system of differential equations, which is a system of differential equations where each one is of the form

$$L[f] = \alpha_1(x)\frac{\partial f}{\partial x_1} + \cdots + \alpha_N(x)\frac{\partial f}{\partial x_N} = \beta(x)f$$

where the α_i and β are functions of the independent variables $x = (x_1, \ldots, x_N)$ (in particular, they do not depend on the function f). The class of linear systems of differential equations is the simplest class of differential equations to understand and solve.

linear transformation   *See* linear mapping.

L-matrix   *See* sign pattern.

local class field theory   The theory of Abelian extensions of local fields; for example, the p-adic number fields. Early results were obtained using the techniques inherited from (global) class field theory. Other techniques in use include those from the theory of cohomology of groups, algebraic K-theory, and the theory of formal groups over local fields. Some of the main results of local class field theory include an isomorphism theorem and an existence theorem. Let K^* denote the multiplicative group of K. The set of all finite Abelian extensions L/K with Galois group G is in one-to-one correspondence with the norm subgroup $N = N_{L/K}(L*)$ of K^*. This correspondence yields a canonical isomorphism from G to K^*/N. The existence theorem states the converse. Any open subgroup of finite index in K^* can be realized as a norm subgroup for some Abelian extension L of K. *See also* class field theory.

local coordinates   A set of functions on a manifold M near a point p, which forms a one-to-one map of a neighborhood in M of the point p onto a neighborhood in Euclidean space (i.e., $\mathbf{R}^N$ or $\mathbf{C}^N$). Local coordinates are used to perform operations, such as differentiation or integration, to functions on a manifold by pulling them back to Euclidean space where such operations are simpler to understand.

local equation   An equation of the form $f(x) = 0$ which locally (i.e., in a neighborhood of some given point) describes a codimension one, irreducible variety.

local field   A field that is complete with respect to a discrete valuation and has a finite residue field. A *local field* with finite characteristic is isomorphic to the field of formal power series in one variable over a finite field. If the local field has characteristic 0, then it is isomorphic to a finite algebraic extension of the field of p-adic numbers ($\mathbf{Q}_p$). The term has also been applied to discretely valued fields with arbitrary residue fields. In particular, real and complex number fields have been called local fields.

local Gaussian sum   Some constants in number theory. Hecke (1918, 1920) extended the notion of character by introducing the Grössencharackter χ and defining L-functions with such characters: $L_k(s, \chi) = \sum_a \frac{\chi(a)}{N(a)}$. If we denote by $\xi_k(s, \chi)$ a certain multiple of $L_k(s, \chi)$, then $\xi_k(s, \chi)$ satisfies a functional equation $\xi_k(s, \chi) = W(\chi)\xi_k(1 - s, \bar{\chi})$ where $W(\chi)$ is a complex number with absolute value 1. Such $\xi_k(s, \chi)$ can be represented by an integral form $\xi_k(s, \chi) = c \int_{\mathbf{J}_k} \phi(t)\chi(t)V(t)^s d^*t$, where $V(t)$ is the total volume of the ideal t, c is a constant that depends on the Haar measure d^*t of the ideal group $\mathbf{J}_k$ of the given field k, and the $\phi(x)$ satisfies

$$\int_{k_p} \phi(x)\chi_p(x)V_p(x)^{-1}d^*x =$$

$$C_p N\left((\delta f)_p\right)^s \tau_p\left(\chi_p\right) \cdot \mu\left(U_{f,p}\right), \quad p|f .$$

Here $\tau_p(\chi_p)$ is a constant called the *local Gaussian sum*.

localization   The process used to construct the field of fractions from an integral domain (a ring with no non-zero zero divisors). An example is the construction of the rational numbers (fractions) from the integers.

locally constructible sheaf   Let $X_{\acute{e}t}$ be the étale site of a scheme X. A sheaf on $X_{\acute{e}t}$ which is represented by an étale covering of X is called a *locally constructible* (or *locally constant*) *sheaf*.

locally convex topological space   A topological vector space with a basis for its topology, whose members are convex.

locally finite   A term used in several different contexts. A typical example comes from the theory of coverings: Let $\mathcal{U} = \{U_\alpha\}_{\alpha \in \mathcal{U}}$ be a covering of a set E. If each $e \in E$ is contained in only finitely many of the U_α, then the covering $\mathcal{U}$ is said to be *locally finite*.

locally finite algebra   An algebra with the property that any finite number of its elements generate a finite-dimensional subalgebra.

locally Noetherian scheme   A scheme that can be covered by open affine subsets, each of which is a Noetherian ring. *See* scheme. This concept arises from the subject of algebraic geometry.

local Macaulay ring   A local ring, which, as a module over itself, has dimension equal to the number of maximal regular sequences within its maximal ideal. This class includes all Noetherian rings. This concept arises in commutative algebra.

local maximum modulus principle   A basic theorem in Banach algebras (due to Rossi) that states that the Shilov boundary of the space, A, of analytic functions on an open set $U \subset \mathbf{C}^n$ is contained in the topological boundary of U. *See* Shilov boundary.

local parameter   A mathematical quantity that is only defined on a small open subset, such as a neighborhood of a particular point. Sometimes a function, defined so as to depend upon such a quantity, can be described by a local parameter near a given point, but not a global parameter.

local ring   A ring $\neq \{0\}$ which has only one maximal ideal. Local rings are of central importance to commutative algebra.

logarithm   The logarithm of a number $N > 0$ to a given base $b > 0$ ($b \neq 1$), denoted $\log_b N$, is the unique number x with $N = b^x$. For example, $\log_2 8 = 3$ and $\log_{10} 1/100 = -2$. Of particular importance is the natural logarithm, denoted $\ln N$ or $\log N$, where the base is the number $e = \lim_{h \to \infty} (1 + 1/h)^h = 2.718 \dots$.

More rigorously, one may define logarithm to the base e by

$$\log x = \int_1^x t^{-1} \, dt$$

and $\log_b x = \log x / \log b$.

logarithmic equation   Any equation involving logarithms. For example, $\log_{10}(x + 8) + \log_{10}(x - 7) = 2$.

logarithmic function   The function $\log_b(x)$ for some given positive base b ($b \neq 1$). *See* logarithm.

The logarithm function is used to describe many physical phenomenon. For example, the Richter scale for earthquakes is defined in terms of the logarithm of the energy produced by earthquakes.

logarithm of a complex number   A logarithm of a complex number z is any number w with $e^w = z$. The logarithm of a complex number is not unique, since the exponential function is periodic with period $2\pi i$ (i.e., $e^{w+2\pi i} = e^w$), if w is a logarithm of z, then $w + 2\pi ki$ is also a logarithm of z for any integer k. The principal value of the logarithm of z is the unique logarithm whose argument is between $-\pi$ and π.

More rigorously, $\log z$ may be defined as $\log |z| + 2\pi i \, \arg(z)$, where $\log |z|$ is defined, as the logarithm of a positive real number and $\arg(z)$ is the argument of z. *See* logarithm.

long division   The long division of a number g by a number f is the process of finding unique numbers q and r with $g = fq + r$ where r (called the remainder) is less than g. Often long division involves repeating the process so that $r = fq_1 + r_1$, with $r_1 < r$. Thus, one obtains smaller and smaller remainders $r, r_1, r_2, \dots$, with r_n/f tending to 0, so that $q + q_1 + q_2 + \dots + q_n$ is a better and better approximation to g/f.

This process can be applied to elements of other rings, such as the ring of polynomials, where there is some notion of comparison (greater than or less than). In the ring of polynomials, the remainder is required to have degree smaller than q.

long multiplication   The process of multiplying two multi-digit numbers that involves arranging the intermediate products into columns. For example,

$$
\begin{array}{r}
452 \\
621 \\
\hline
452 \\
904 \\
2712 \\
\hline
280692
\end{array}
$$

loop   Any one-dimensional curve (defined as the image of a continuous function defined on the unit interval) in a topological space, which starts and ends at the same point. The space of all loops which pass through a given base point is fundamental to the field of Homotopy theory, a subfield of topology.

Lorentz group   Another name for the group $O(1, 3)$, which is the group of all 4×4 matrices A with real entries that preserve the bilinear form $F(x, y) = -x_0 y_0 + x_1 y_1 + x_2 y_2 + x_3 y_3$ (i.e., $F(Ax, Ay) = F(x, y)$). The Lorentz group (as well as all the groups $O(1, n)$) are basic to the study of differential geometry. The Lorentz group in particular is important in the study of relativity (mathematical physics).

lower bound   For a set S of real numbers, any number which is less than or equal to all elements in S. For example, any number which is less than or equal to zero is a lower bound for the set $\{x : 0 \leq x \leq 1\}$.

lower central series   A descending chain of subalgebras $D_1 G, D_2 G, \ldots$, associated to a given Lie algebra G, defined inductively as follows: $D_1 G = [G, G]$ (which means the set of all Lie brackets of an element in G with any other element in G); $D_k G = [G, D_{k-1} G]$. This concept is of fundamental importance in the classification of Lie algebras.

lower semi-continuous function   Let X be a topological space and $f : X \rightarrow \mathbf{R}$ a function. Then f is said to be *lower semi-continuous* if $f^{-1}((\beta, \infty))$ is open for every real β.

lower triangular matrix   Any square matrix $A = (a_{ij})$ with zero entries above the diagonal ($a_{ij} = 0$ for $j > i$). Any set of linear equations ($A \cdot x = b$) arising from a lower triangular matrix can be solved easily by solving the first equation $a_{11} x_1 = b_1$ for x_1 and then forward substituting in the subsequent equations to solve for $x_2, x_3, \ldots$.

lowest terms   The form of a fractional expression with no factor that is common to both numerator and denominator. For example, $3/4$ is a fraction in lowest terms, but $6/8$ is not, because of the common factor of 2.

loxodromic transformation   One of several classification schemes for linear fractional functions $w = \frac{az+b}{cz+d}$ in one complex variable. If such a transformation has two distinct finite fixed points α and β, then the transformation can be put into the following normal form:

$$\frac{w - \alpha}{w - \beta} = k \frac{z - \alpha}{z - \beta}$$

where

$$k = \frac{a - c\alpha}{a - c\beta} .$$

If $\arg k = 0$, the transformation is called hyperbolic; if $|k| = 1$ the transformation is called elliptic; otherwise the transformation is called *loxodromic*.

See linear fractional function.

Luroth's Theorem   Suppose L is the field obtained by adjoining an element α to a field F where α is transcendental over F; if E is a field with $F \subset E \subset L$, then E is obtained by adjoining some element $\beta \in E$ to F.

Lutz-Mattuck Theorem   The group of rational points of an Abelian variety of dimension n over the p-adic number field contains a subgroup of finite index that is isomorphic to n copies of the ring of p-adic integers.

The theorem is named after its authors E. Lutz (1937) and A. Mattuck (1955).

M

Macaulay local ring   *See* local Macaulay ring.

Main Theorem (class field theory)   If k is an algebraic number field, then any Abelian extension of fields K/k is a class field over k, for a suitable ideal group H.

Main Theorem (Galois Theory)   *See* Fundamental Theorem of Galois Theory.

Main Theorem (Symmetric Polynomials)
See Fundamental Theorem of Symmetric Polynomials.

mantissa   The fractional part of a number in its decimal expansion. For example, the *mantissa* of the number 3.478 is 0.478.

mapping   Another name for function. *See* function.

mathematical programming   A process of breaking down a mathematical problem into its component steps so that it can be solved by a computer.

mathematical programming problem   A mathematical problem that can be solved by programming, i.e., by breaking it down into its component steps so that it can be solved by a computer.

Mathieu group   Most non-Abelian finite simple groups can be classified into a list of infinite families. There are 26 exceptional "sporadic" groups that cannot be classified in this way. The smallest of these groups has 7920 elements and is named after its discoverer, E. Mathieu (1861).

matrix   A rectangular array of numbers or other elements:

$$\begin{bmatrix} a_{11} & a_{12} & \ldots & a_{1n} \\ a_{21} & a_{22} & \ldots & a_{2n} \\ & \ldots & \ldots & \\ a_{m1} & a_{m2} & \ldots & a_{mn} \end{bmatrix}.$$

Here n represents the number of *columns* and m the number of *rows*. These are indicated by designating the above array an $m \times n$ *matrix*.

Matrices arise in many fields, such as linear algebra. The most common use for a matrix is to represent a linear mapping from one vector space to another.

matrix group   The set of all invertible square matrices of a given dimension, under matrix multiplication. This group is one of the simplest and widely used groups where the multiplication operation is not commutative.

Subgroups of the above group may also be referred to as *matrix groups*.

matrix multiplication   The process of multiplying two matrices A and B:

$$\begin{bmatrix} a_{11} & a_{12} & \ldots & a_{1n} \\ a_{21} & a_{22} & \ldots & a_{2n} \\ & \ldots & \ldots & \\ a_{m1} & a_{m2} & \ldots & a_{mn} \end{bmatrix},$$
$$\begin{bmatrix} b_{11} & b_{12} & \ldots & b_{1r} \\ b_{21} & b_{22} & \ldots & b_{2r} \\ & \ldots & \ldots & \\ b_{s1} & b_{s2} & \ldots & b_{sr} \end{bmatrix}.$$

The number of columns of the left matrix A must be equal to the number of rows of the second matrix B ($n = s$). The ij entry of the product, $A \cdot B$ is the sum $\sum_{k=1}^{n} a_{ik}b_{kj}$.

The matrix product AB is the matrix of the composition of the linear transformations with matrices B and A.

matrix of a quadratic form   For the quadratic form on $\mathbf{R}^n$

$$Q(x) = \sum_{i,j=1}^{n} a_{ij}x_i x_j ,$$

where each a_{ij} is a real number, the matrix (a_{ij}). Usually there are further requirements, such as that this matrix must be symmetric ($a_{ij} = a_{ji}$) and positive definite ($Q(x) \geq c|x|^2$ for some positive constant c).

matrix of coefficients　　The matrix obtained from the coefficients of the variables in a system of linear equations. For example, the matrix of coefficients of the linear system

$$\left\{ \begin{array}{c} 2x + 3y = 5 \\ x + 5y = 7 \end{array} \right\}$$

is the matrix

$$\begin{pmatrix} 2 & 3 \\ 1 & 5 \end{pmatrix}.$$

The solution to a system of linear equations is usually found by manipulating its matrix of coefficients (together with the right side of the equation).

matrix representation　　(**1**) A representation of a group on a matrix group. *See* representation.
　(**2**) A matrix which is used to describe a mathematical process or object. A common example is to represent a linear map from one vector space to another by a matrix by identifying the coefficients of the expansion of the linear map in terms of given bases for the domain and range. Such a representation depends strongly on the choice of bases.

For example, if V and W are vector spaces with bases $\{v_1, \ldots, v_n\}$ and $\{w_1, \ldots, w_m\}$, the map $T : V \to W$ which satisfies $Tv_j = \sum_1^m a_{ij} w_i$ has the matrix representation

$$\begin{bmatrix} a_{11} & a_{12} & \ldots & a_{1n} \\ a_{21} & a_{22} & \ldots & a_{2n} \\ & \ldots & \ldots & \\ a_{m1} & a_{m2} & \ldots & a_{mn} \end{bmatrix}.$$

matrix unit　　The i, j *matrix unit* is an $n \times m$ matrix which has a 1 in the i, j entry and zeros elsewhere. The collection of all matrix units forms a basis for the vector space of $n \times m$ matrices. Despite its name, a matrix unit is not a unit in the algebraic sense (it does not have a multiplicative inverse).

Mauer-Cartan differential form　　A differential 1-form on a Lie group which is invariant under the group action. In more detail, for a fixed element g in the Lie group G, the mapping $T_g : G \mapsto G$ defined by $x \mapsto g \cdot x$ is a differentiable map. A Mauer-Cartan form is a 1-form v defined on G which has the property that, for any $g \in G$, the pull back of v via T_g is again v.

Mauer-Cartan system of differential equations　　The system of differential equations satisfied by a set of left-invariant 1-forms that allows one to recover the multiplication operation of the underlying Lie group.

maximal deficiency　　A geometric invariant of an algebraic surface, defined as the first cohomology group with values in the sheaf of 0-forms (smooth functions).

maximal ideal　　An ideal is maximal in a ring R if it is not properly contained in another ideal other than R itself. *See* ideal.

maximal ideal space　　The collection $\widehat{X}$ of all multiplicative linear functionals on a Banach algebra or function algebra X. $\widehat{X}$ is so called because the kernel of each such functional is a maximal ideal in X.

maximal independent system　　A linearly independent system of vectors in a vector space, with the property that any additional vector would make the system dependent. *See* linearly independent elements. For example, a set of three non-coplanar vectors in $\mathbf{R}^3$ is a maximal independent system because if any fourth vector is added, the system becomes linearly dependent.

maximal order　　The largest order of any element in a group, where the order of an element g is the smallest positive integer m with $g^m = e$, the identity element in the group.

maximal prime divisor　　(Of an ideal I of a commutative ring R.) A prime ideal P of R such that P is maximal, as an ideal of R containing I, and is disjoint from the set of elements of R which are not zero-divisors, modulo I.

maximal prime ideal   A prime ideal of a ring R which is not properly contained in a prime ideal other than R itself. *See* prime ideal.

maximal separable extension   A separable extension field of a field F which is not properly contained in any other separable extension field of F. *See* separable extension.

maximal torus   Let G be a Lie group. A *torus* H in G is a connected, compact, Abelian subgroup. If, for any other torus H' in G with $H' \supseteq H$ we have $H' = H$, then H is said to be *maximal*.

mean   Any of a variety of averages.

(**1**) The mean of a set of numbers is their sum divided by the number of entries in the set. More precisely, the mean of $x_1, \ldots, x_N$ is $(x_1 + \cdots + x_N)/N$. The mean is one of the basic statistical quantities used in analyzing data.

(**2**) The mean of a function $f(x)$, defined on a set S, is a continuous analog of the mean defined above:

$$\frac{1}{|S|} \int_S f(x)dx \, ,$$

where $|S|$ denotes the length, or volume, of S. *See also* mean of degree r.

mean of degree r   Of a function f, with respect to a weight function p, the quantity

$$\left(\frac{\int f^r \, p \, dx}{\int p \, dx} \right)^{1/r} .$$

In other words, the mean of degree r of f is the L^r norm of f with respect to the probability measure generated by the weight function p.

mean proportional   A mean proportional between two numbers a and b is the number m such that $a/m = m/b$.

mean terms of proportion   The terms b and c in the proportion $a/b = c/d$. (In older works, the proportion is sometimes written $a : b :: c : d$.)

measurable operator function   For a set X, a σ-algebra Ω on X and a Hilbert space $\mathcal{H}$, a function F from X to the set of bounded operators on $\mathcal{H}$, such that, for any vector $h \in \mathcal{H}$, the function $x \mapsto (F(x)h, h)$ is Ω measurable.

measurable vector function   The space of measurable vector functions is a set K of functions, from a measure space M to a Hilbert space H, $(\cdot, \cdot)$ with the following properties: (i.) if $x \in K$, then $||x|| = (x(\zeta), x(\zeta))^{\frac{1}{2}}$ is a measurable (scalar) function; (ii.) for any x and $y \in K$, (x, y) is measurable, and (iii.) there is a countable family $\{x_1, x_2, \ldots\}$ from K such that $\{x_1(\zeta), x_2(\zeta), \ldots\}$ is dense in H. The definition also allows for the Hilbert space H to depend on ζ.

measure   Let X be a space and let $\mathcal{A}$ be a sigma field on X. A *measure* on X is a function $\mu : \mathcal{A} \to \mathbf{R}$ that satisfies certain additivity or subadditivity properties, such as countable additivity. *See* additive set function. A measure is a device for measuring the length or the size of a set.

median   A midway point in a data set. For a discrete data set, half the data points are less than or equal to the median and half are greater than or equal to the median. More generally, the median for a distribution ρ on a probability space $\{X, p\}$ is a number m such that the probability of the event that $f(x)$ is less than or equal to m is greater than or equal to $1/2$ and the probability of the event that $f(x)$ is greater than or equal to m is greater than or equal to $1/2$.

meet   In a lattice, the infimum or greatest lower bound of a set of elements. Specifically, if A is a subset of a lattice L, the *meet* of A is the unique lattice element $b = \bigwedge\{x : x \in A\}$ defined by the following two conditions: (i.) $x \geq b$ for all $x \in A$. (ii.) If $x \geq c$ for all $x \in A$, then $b \geq c$. The meet of an infinite subset of a lattice may not exist; that is, there may be no element b of the lattice L satisfying conditions (i.) and (ii.) above. However, by definition, one of the axioms a lattice must satisfy is that the meet of a finite subset A must always exist. The meet of two elements is usually denoted by $x \wedge y$.

There is a dual notion of the *join* of a subset A of a lattice, denoted by $\bigvee\{x : x \in A\}$, and defined by reversing the inequality signs in conditions (i.) and (ii.) above. The join is also called the *supremum* or *least upper bound* of the subset A. Again, the join of an infinite subset may fail to exist, but the join of a finite subset always exists by the definition of a lattice.

meet irreducible element   An element a of a lattice L which cannot be represented as the meet of lattice elements b properly larger than a ($a < b, a \neq b$). *See* meet.

There is a dual notion of *join irreducibility*. An element a of L is *join irreducible* if it cannot be represented as the join of lattice elements b properly smaller than a ($b < a, b \neq a$). *See* join.

member of an equation   The expression on the left (or right) side of the equality sign. The member on the left side of the equals sign is called the first member of the equation, and the member on the right side is called the second member of the equation.

Meta-Abelian group   A group whose commutator subgroup (defined as the set of all $a^{-1}b^{-1}ab$, for a, b in the group) is Abelian. *See* Abelian group.

method of feasible directions   A technique in non-linear optimization for functions of several variables that reduces the problem to a series of one-dimensional optimization problems.

minimal   A term which generally means smallest but whose precise meaning depends on the context. In set theory, a minimal set with respect to a given property is a set which has the property, but, if any element is removed from the set, then the smaller set fails to have this property.

minimal basis   A basis for a vector space V with the property that, if any element is removed from the basis, it no longer forms a basis (i.e., some element of V cannot be expressed as a finite linear combination of the smaller set). *See* basis.

Often such minimality is part of the definition of basis.

minimal ideal   An ideal which does not properly contain any ideal except the zero ideal, $\{0\}$.

minimal model   A nonsingular, projective surface which is the unique relatively minimal model in its birational equivalence class. Except for rational and ruled surfaces, every non-empty birational equivalence class has a (unique) minimal model. The existence of a minimal model in the birational equivalence class of a higher dimensional variety, over the field of complex numbers, has been solved for varieties of dimension three. In the higher dimensional case, certain types of singularities must be allowed and minimal models are no longer unique.

minimal parbolic k-subgroup   A closed subgroup of a connected reductive linear algebraic group G, defined over an arbitrary ground field k, which is minimal among the parabolic subgroups of G. Any two minimal parabolic k-subgroups are conjugate to each other over k. Minimal parabolic k-subgroups play the same role for arbitrary fields that Borel subgroups play for algebraically closed fields. *See* parabolic subgroup.

minimal polynomial   For a given element T (which could be a linear transformation or an element in some field extension), a polynomial p of least degree with $p(T) = 0$.

minimal prime divisor   An ideal I in a ring R which is prime and is minimal (i.e., does not properly contain any prime ideal). *See* prime ideal.

minimal splitting field   Suppose E, F with $F \subset E$ are fields and $f(x)$ is an element of $F[x]$. The field E is a *minimal splitting field* if it is a splitting field for $f(x)$ and no proper subfield of E has this property.

minimal Weierstrass equation   A Weierstrass equation for an elliptic curve E/K is one of the form

$$y^2 + a_1 xy + a_3 y = x^3 + a_2 x^2 + a_4 x + a_6 \, .$$

We call such an equation *minimal* if, among all possible Weierstrass equations, it has least discriminant $|\mathcal{D}|$.

minimal Weierstrass equation   A Weierstrass equation for an elliptic curve E/K is one of the form

$$y^2 + a_1xy + a_3y = x^3 + a_2x^2 + a_4x + a_6 .$$

We call such an equation minimal if, among all possible Weierstrass equations, it has least discriminant $|\mathcal{D}|$.

Minkowski-Farkas Lemma   *See* Minkowski-Farkas Theorem.

Minkowski-Farkas Theorem   For every matrix $\mathbf{A}$ and vector $\mathbf{b}$ the system $\mathbf{Ax} = \mathbf{b}$ has a non-negative solution if and only if the system $\mathbf{uA} \geq 0, \mathbf{ub} < 0$ has no solution.
  This theorem is sometimes referred to as the Minkowski-Farkas Lemma. The following corollary has also been referred to as the Minkowski-Farkas Lemma. The following conditions are mutually exclusive.   Either the system of linear inequalities $\mathbf{Ax} \leq \mathbf{b}$ has a non-negative solution or the system of linear inequalities $\mathbf{uA} \geq 0, \mathbf{ub} < 0$ has a non-negative solution.

Minkowski-Hasse character   *See* Hasse-Minkowski character.

Minkowski inequality   If f and g are complex-valued, measurable functions on a measure space (X, μ) and $1 \leq p < \infty$, then

$$\left(\int_X |f(x) + g(x)|^p \, d\mu(x) \right)^{\frac{1}{p}}$$

$$\leq \left(\int_X |f(x)|^p \, d\mu(x) \right)^{\frac{1}{p}}$$

$$+ \left(\int_X |g(x)|^p \, d\mu(x) \right)^{\frac{1}{p}} .$$

The left side of the inequality is the definition of the L^p-norm of $f + g$. Thus, *Minkowski's inequality* is the statement that the L^p-norm satisfies the triangle inequality, which is one of the defining properties of a norm. Named after the German mathematician H. Minkowski (1864–1909).

Minkowski-Siegel-Tamagawa Theory   In number theory, a theory on the arithmetic of linear groups. Let S and T be the matrices of integral positive definite quadratic forms, corresponding to the lattices $\Lambda_S \subset \mathbf{R}^m$, $\Lambda_T \subset \mathbf{R}^n$, in the sense that q_S and q_T express the lengths of elements of Λ_S and Λ_T, respectively. Then an integral solution X to the equation $S[X] = T$, where $S[X] := X^t SX$, determines an isometric embedding $\Lambda_S \to \Lambda_T$. Denote by $N(S, T)$ the total number of such maps. The genus of q_S is defined to be the set of quadratic forms which are rationally equivalent to q_S. Let I be the set of these equivalence classes. One of Siegel's formulas gives the value of a certain weighted average of the numbers $N(S_x, T)$ over a set of representatives S_x for classes $x \in I$ of forms of a given genus:

$$\tilde{N}(S, T) = c_{m-n} c_m^{-1} \alpha_\infty(S, T) \prod_p \alpha_p(S, T) ,$$

where $c_1 = \frac{1}{2}$ and $c_a = 1$ for $a > 1$, $\tilde{N}(S, T)$ $= \frac{1}{\text{Mass}(S)} \sum_{x \in I} \frac{N(S_x, T)}{w(x)}$ and $w(x)$ is the order of the group of orthogonal transformations of the lattice Λ_S, and define the mass of S by $\text{Mass}(S) = \sum_{x \in I} \frac{1}{w(x)}$.
  In the special case $T = S$, we have $\tilde{N}(S, T)$ $= \frac{1}{\text{Mass}(S)}$, and the Siegel formula becomes the *Minkowski-Siegel formula:*

$$\text{Mass}(S) = c_m \alpha_\infty(S, S)^{-1} \prod_p \alpha_p(S, S)^{-1} .$$

  Siegel's formula can be deduced from an integral formula: $\tilde{N}(\varphi) = \frac{\text{vol}(g/\gamma)}{\text{vol}(G/\Gamma)} \int_{G/g} \varphi(x)dx$, where $G = O_m(\mathbf{A})$ is the locally compact group of orthogonal matrices with respect to S with coefficients in the ring of adeles $\mathbf{A}$, g and γ are closed subgroups of G with vol(g/γ) finite and φ is a continuous function with compact support on G/γ. The quantities c_{m-n} and c_m become the Tamagawa numbers $\tau(O_{m-n})$ and $\tau(O_m)$, respectively, and the *Tamagawa measure* on G can be defined.

Minkowski space   A flat space of four dimensions, designed to model the geometry of the physical universe as suggested by the special theory of relativity. It is also called *Minkowski space-time*, the *Minkowski world* or the *Minkowski universe*. There are two methods of envisioning Minkowski space. Coordinates may be written as (x, y, z, ict) where $i^2 = -1$ and c is the speed of light. In this case (x, y, z) represents the position of a point in space, and t is the time at which an event occurs at that point. The distance between two points is then

$$ds = \sqrt{(dx)^2 + (dy)^2 + (dz)^2 - c^2(dt)^2} \ .$$

It is also possible to view Minkowski space as the manifold $\mathbf{R}^4$, with a flat Lorentz metric. In this case, the coordinates are written as $(x_1, x_2, x_3, x_4) = (x, y, z, ct)$ and the space is associated with an indefinite inner product $x \cdot y = x_1 y_1 + x_2 y_2 + x_3 y_3 - x_4 y_4$. Note, in some references the time coordinate is listed first and called x_0. In addition, some references define the inner product as the negative of the one defined above. Using the definition given above, a non-zero vector x is called *time-like* or *space-like,* depending upon whether $x \cdot x$ is negative or positive. A non-zero vector x is called *null, isotropic,* or *lightlike* if $x \cdot x = 0$.

Minkowski's Theorem (1) If $K \neq \mathbf{Q}$, then $|D_K| > 1$, where K is a field, k is an algebraic number field contained in k, with $[K : k] < \infty$, and D_K is the discriminant of K.

The theorem is a consequence of the following Minkowski Lemma: let M be a lattice in $\mathbf{R}^n$, $\Delta = \mathrm{vol}(\mathbf{R}^n/M)$, and let $X \subset \mathbf{R}^n$ be a centrally symmetric convex body of finite volume $v = \mathrm{vol}(X)$. If $v > 2^n \Delta$, then there exists a nonzero $\alpha \in M \cap X$.

(2) (On convex bodies) Any convex region in n-dimensional Euclidean space which is symmetric about the origin and has a volume greater than 2^n contains another point with integral coordinates. This theorem can be generalized as follows. Let P be a convex region in n-dimensional Euclidean space which is symmetric about the origin and let Λ be a lattice with determinant Δ. If the volume of P is greater than $2^n|\Delta|$, then P contains a point of Λ other than the origin. This is one of the most important theorems in the geometry of numbers, and is one of the reasons that the geometry of numbers exists as a distinct subdivision of number theory. The following application to algebraic number fields is also called *Minkowski's Theorem.* If k is an algebraic number field of finite degree and $k \neq \mathbf{Q}$,

then the absolute value of the discriminant of k is greater than 1.

minor The determinant of an $n - 1 \times n - 1$ submatrix of an $n \times n$ matrix obtained by deleting a row and a column from the larger matrix. Minors arise as a theoretical tool for computing a determinant of a matrix (as in expansion by minors). *See* expansion of determinant.

minuend The term from which another term is to be subtracted. (The number a in $a - b$.)

minus sign The symbol "$-$" which indicates subtraction in arithmetic. The minus sign also indicates "additive inverse," so that, for example, -3 is the additive inverse of 3. This notion extends to groups (with operation $+$), where $-x$ represents the element that when added to x gives the identity in the group.

mixed decimal A number written in decimal form with an integer and a fractional part; for example, 34.587.

mixed expression A mathematical formula involving terms of more than one type. For example, $4xy + y^2 x$ is a mixed expression of the variables x and y.

mixed group A set M which can be partitioned into disjoint subsets, $M_0, M_1, \dots$ with the following properties: (i.) for $a \in M_0$ and $b \in M_i$, $i = 0, 1, \dots$, elements ab and $a \setminus b$ are defined such that $a(a \setminus b) = b$; (ii.) for $b, c \in M_i$, an element b/c of M_0 is defined such that $(b/c) \cdot c = b$; and (iii.) the associative law $(ab)c = a(bc)$ for $a, b \in M_0$ and $c \in M$ holds.

mixed ideal An ideal I of a Noetherian ring R such that there exist associated prime ideals P, Q of I such that the height of P is not equal to the height of Q. *See* height.

mixed integer programming problem A programing problem in which some of the variables are required to have integer values. In addition, all of the variables are usually required to be nonnegative. Mixed integer programming once referred only to problems that were linear in appearance. The concept has been expanded to include nonlinear programs. A mixed integer linear programming problem may be written as

$$\max\left\{cx + dy : Gx + Hy \leq b, x \in \mathbf{Z}_+^n, y \in \mathbf{R}_+^p\right\},$$

where the matrices c, d, G, H, and b have integer entries and the following dimensions: c is $1 \times n$, d is $1 \times p$, G is $m \times n$, H is $m \times p$, and b is $m \times 1$. In addition, it is assumed that m and $n + p$ are positive integers.

mixed number A number that has both an integer and a fractional part, such as $4\frac{2}{3}$.

M-matrix A matrix A that can be written as

$$A = sI - B,$$

where B is an entrywise nonnegative matrix, I is the identity matrix, and s is a positive number greater than the *spectral radius* of B. From the Perron-Frobenius Theorem for entrywise nonnegative matrices, it follows that all the eigenvalues of an M-matrix have positive real parts. M-matrices arise naturally in several areas of the mathematical sciences such as optimization, Markov chains, numerical solution of differential equations, and dynamical systems theory.

Möbius transformation *See* linear fractional function.

model A mathematical representation of a physical problem. For example, the differential equation $\frac{dy}{dt} = ky$ is a model for any physical process that is governed by exponential growth or decay, such as population growth radioactive decay.

modular arithmetic Modular arithmetic with respect to a certain number p refers to an arithmetic calculation, where at the end, only the remainder is kept after subtracting the greatest multiple of p. For example, if $p = 5$, then 3×7 (modulo 5) equals 1 (since $3 \times 7 = 21 = 4 \times 5 + 1$). *See* congruent integers, group of congruence classes.

modular automorphism A one-parameter $*$-automorphism σ_t^ϕ, of a von Neumann algebra $\mathcal{M}$, where $\sigma_t^\phi(A) \equiv \Delta_\phi^{it} A \Delta_\phi^{-it}$ for a modular operator Δ_ϕ. *See also* modular operator.

modular character A group homomorphism from the (multiplicative) group of units of the ring of integers modulo m to the multiplicative group of nonzero complex numbers. *See also* character.

modular operator A positive self-adjoint operator, defined and used in Tomita-Takesaki theory. Let ϕ be a normal semifinite faithful weight on a von Neumann algebra $\mathcal{M}$. Let $\mathcal{H}_\phi$ be the Hilbert space associated with ϕ, let $\mathcal{N}_\phi$ be the left ideal $\{A \in \mathcal{M} : \phi(A^*A)$ is finite$\}$, and let η be the associated complex linear mapping from $\mathcal{N}_\phi$ into a dense subset of $\mathcal{H}_\phi$. Let S_ϕ be the antilinear operator defined by $S_\phi\eta(A) = \eta(A^*)$ where $A \in \mathcal{N} \cap \mathcal{N}_\phi^*$ and A^* is the adjoint of A. The polar decomposition of the closure of S_ϕ defines a self-adjoint operator called a *modular operator.*

modular representation A representation of a group which is also a finite field. *See* representation.

module A nonempty set M is said to be an *R-module* (or a *module over a ring R*) if M is an Abelian group under an operation (usually denoted by +) and if, for every $r \in R$ and $m \in M$, there exists an element $rm \in M$ subject to the distributive laws: $r(m_1 + m_2) = rm_1 + rm_2$ and $(r + s)m = rm + sm$; as well as the associative law $r(sm) = (rs)m$ for $m, m_1, m_2 \in M$ and $r, s \in R$.

module of A-homomorphisms The set of all homomorphisms from an A-module M to an A-module N forms a module over the ring A called the *module of homomorphisms* from M to N, denoted $\text{Hom}_A(M, N)$. *See* homomorphism. *See also* automorphism group.

module of boundaries A concept that arises in the subject of graded modules. Regard the graded module X as a sequence of modules X_0, X_1, $X_2, \ldots$ with maps $\partial_n : X_n \mapsto X_{n-1}$ (called boundary maps) with $\partial_{n-1} \circ \partial_n = 0$. The *module of boundaries* is the graded module of images of the boundary map, i.e., $\partial_1\{X_1\}, \partial_2\{X_2\}, \ldots$ This concept arises most commonly in topology where the graded modules are chains of simplicies in a topological space X. In this case,

X_k is the free module generated (say, over the integers) by the continuous images of the standard k-dimensional simplex in Euclidean space $\{(x_1, \ldots, x_k) \in \mathbf{R}^k : x_i \geq 0, \sum_{i=1}^k x_i \leq 1\}$. The boundary map of a k-simplex is a weighted sum of (the continuous images of) its $k-1$ dimensional faces where the weights are either plus one or minus one depending on the orientation of the face. Then the module of $k-1$-boundaries is the module generated by the boundaries of k-dimensional simplicies.

module of coboundaries   A *coboundary* is the image of a cocycle under the induced boundary map. (*See* module of cocycles.) The set of coboundaries is a module over the underlying ring (often the integers or the reals).

module of cocycles   The dual of the module of cycles. More precisely, if X_0, X_1, $X_2, \ldots$ is a graded chain of cycles over a ring R (often the integers or the reals) with boundary maps $\partial_n : X_n \mapsto X_{n-1}$, then the *module of cocycles* is the graded chain $\hat{X}_0, \hat{X}_1, \hat{X}_2, \ldots$ where each $\hat{X}_n$ is the set of all ring homomorphisms from X_n to R. The induced coboundary map $\hat{\partial}_n : \hat{X}_n \mapsto \hat{X}_{n+1}$ is defined by $\hat{\partial}_n \hat{c}_n(x_{n+1}) = \hat{c}_n(\partial_n x_{n+1})$ for $\hat{c}_n \in \hat{X}_n$ and $x_{n+1} \in X_{n+1}$.

module of cycles   A concept that arises in the subject of graded modules. Regard the graded module X as a sequence of modules X_0, X_1, X_2, $\ldots$ with maps $\partial_n : X_n \mapsto X_{n-1}$ (called boundary maps) with $\partial_{n-1} \circ \partial_n = 0$. A cycle $c_n \in X_n$ is one that has zero boundary ($\partial_n c_n = 0$). The set of all cycles is a module over the underlying ring (often the integers). This concept arises most commonly in topology where the graded modules are chains of simplicies in a topological space X. In this case, X_k is the free module generated (say, over the integers) by the continuous images of the standard k-dimensional simplex in Euclidean space

$$\left\{(x_1, \ldots, x_k) \in \mathbf{R}^k : x_i \geq 0, \sum_{i=1}^k x_i \leq 1\right\}$$

(into X). The boundary map of a chain is a weighted sum of (the continuous images of) its $k-1$ dimensional faces where the weights are either plus one or minus one depending on the

orientation of the face. The cycles are the chains whose boundaries are zero. As a simple example, a circle is a cycle in $\mathbf{R}^2$ because it is the image of the one-dimensional simplex $0 \leq t \leq 1$ (via the complex exponential map, $f(t) = \exp(2\pi i t)$) and because its boundary, $f(1) - f(0)$, is zero.

module of finite length   A module M with the property that any chain of submodules of the form

$$\{0\} \subset M_0 \subset M_1 \subset M_2 \cdots \subset M_l = M$$

is finite in number. The above containments are proper.

module of finite presentation   A module M over a ring R such that there is a positive integer n and an exact sequence of R-modules $0 \to K \to R^n \to M \to 0$ where K is finitely generated.

module of finite type   **(1)** A graded module $\sum_{i \geq 0} A_i$ over a field k for which each A_i is finite dimensional.

(2) A sheaf of $\mathcal{O}$-modules (called an $\mathcal{O}$-Module) in a ringed space $(X, \mathcal{O})$, which is locally generated by a finite number of sections over $\mathcal{O}$.

(3) A finitely generated module.

module of homomorphisms   Let M and N be modules over a commutative ring R. The set of all homomorphisms of module M to module N is called the *module of homomorphisms* of M to N, with addition defined pointwise and multiplication given by $(r \cdot f)(x) = r \cdot (f(x))$, for $r \in R$, $f : M \to N$ a homomorphism and $x \in M$. The module of homomorphisms is denoted $\mathrm{Hom}_R(M, N)$.

module with operator domain   A module M, together with a set A and a map from $A \times M$ into M satisfying the following conditions. (i.) For every $a \in A$ and $x \in M$ there is a unique element $ax \in M$. (ii.) If $a \in A$ and $x, y \in M$, then $a(x + y) = ax + ay$. In this situation, M is called a *module with operator domain* A. It is also called a *module over A* or an *A-module*.

moduli of Abelian variety   Parameters or invariants which classify the set of all Abelian varieties which are equivalent under some type of equivalence relation to a given Abelian variety. The problem of finding moduli for Abelian varieties is approached both algebraically and geometrically, and involves coarse moduli schemes and inhomogeneous polarizations.

moduli scheme   *See* coarse moduli scheme.

modulus   (**1**) The modulus of a complex number $z = a + ib$ is defined as $|z| = (a^2 + b^2)^{\frac{1}{2}}$.

(**2**) Let p be a positive integer. When two integers a, b are congruent modulo p, then p is called the *modulus* of the congruence. *See* congruent integers.

modulus of common logarithm   The logarithm to the base 10 is called common logarithm. The factor by which the logarithm to a given base of any number must be multiplied to obtain the common logarithm of the same number is called modulus of the common logarithm. Because one has $\log_{10} x = \log_{10} b \log_b x$, for any base b ($b > 0$ and $b \neq 1$) and any $x > 0$, the modulus of the common logarithm for base b is seen to be $\log_{10} b$.

Moishezon space   A compact, complex, irreducible space X of (complex) dimension n whose algebraic dimension (i.e., transcendence degree of the field of meromorphic functions on X) is also equal to n.

monic polynomial   Any polynomial $p(x)$ of degree m over a ring R with leading coefficient 1_R, the unit of the ring R. *See* leading coefficient.

monoidal transformation   For any integer m with $1 < m \leq n$, we can make a quadratic transformation on the $(X_1, \ldots, X_m)$-space and "product" it with the $(X_{m+1}, \ldots, X_n)$-space to get the *monoidal transformation* of the n-space centered at the $(n - m)$-dimensional linear subspace $L : X_1 = \cdots = X_{m+1} = 0$. In greater detail, the monoidal transformation with center L sends the $(X_1, \ldots, X_n)$-space into the $(X'_1, \ldots,$ $X'_n)$-space by means of the equations

$$\begin{pmatrix} X_1 = X'_1 \\ X_2 = X'_1 X'_2 \\ \vdots \\ X_m = X'_1 X'_m \\ X_{m+1} = X'_{m+1} \\ \vdots \\ X_n = X'_n \end{pmatrix}$$

or the reverse

$$\begin{pmatrix} X'_1 = X_1 \\ X'_2 = X_2 / X_1 \\ \vdots \\ X'_m = X_m / X_1 \\ X'_{m+1} = X_{m+1} \\ \vdots \\ X'_n = X_n \end{pmatrix}$$

The origin in the $(X_1, \ldots, X_n)$-space is blown up into the linear $(m - 1)$-dimensional subspace of the $(X'_1, \ldots, X'_n)$-space given by $L' : X'_1 = X'_{m+1} = \cdots = X'_n = 0$.

monomial   A polymonial consisting of one single term such as $a_n x^n$.

monomial module   A module generated by a single generator. *See* module.

monomial representation   Let H be a subgroup of a finite group G. If ρ is a linear representation of H with representation module M, then the linear representation $\rho^G = K(G) \bigotimes_{K(H)} M$ of G is called the induced representation from ρ. (Here, K is a field.) A *monomial representation* of G is the induced representation from a degree 1 representation of a subgroup. Each monomial representation of G corresponds to a matrix representation τ such that for each $g \in G$ the matrix $\tau(g)$ has exactly one nonzero entry in each row and each column. The induced representation of G from the trivial subgroup $\{e\}$ is called the *regular representation* of G.

monomorphism   A morphism i in a category satisfying the following property: Whenever the equation $i \circ f = i \circ g$ holds for two morphisms f and g in the category, then $f = g$.

In most familiar categories, such as the category of sets and functions, a monomorphism is simply an *injective* or *one-to-one* function in the category. A function $i : A \to B$ is *one-to-one*, or *injective* if $i(a_1) \neq i(a_2)$ whenever $a_1 \neq a_2$. *See also* morphism in a category, epimorphism, injection.

monotone function   Let $I \subseteq \mathbf{R}$ be an interval and $f : I \to \mathbf{R}$ a function. If whenever $a, b \in I$ and $a < b$ it holds that $f(a) \leq f(b)$, then f is said to be *monotone increasing*. If whenever $a, b \in I$ and $a < b$ it holds that $f(a) \geq f(b)$, then f is said to be *monotone decreasing*. In both cases, f is a *monotone function*.

Moore-Penrose inverse   *See* generalized inverse.

Mordell's conjecture   Any algebraic curve of genus ≥ 2 defined over an algebraic number field k of finite degree has finitely many rational points. (Conjectured in 1922 and settled in the affirmative by G. Faltings in 1983. The original 1922 conjecture was stated for the special case $k = \mathbf{Q}$.)

Mordell-Weil Theorem   Let k be an algebraic number field of finite degree and let V be an Abelian variety of dimension n defined over k. Then the group V_k of all k-rational points on V is finitely generated.

This theorem was proved by Mordell for the special case of $n = 1$ in 1922 and by Weil for the general case in 1928.

morphism   Let $\circ$ be a binary operation on a set A, while $\circ'$ is another such operation on a set A'. A *morphism* $m : (A, \circ) \to (A', \circ')$ is a function on A to A' which preserves the operation $\circ$ on A onto the operation $\circ'$ on A', in the sense that

$$m(a \circ b) = m(a) \circ' m(b)$$

for all $a, b \in A$.

In a categorical approach to algebra, one defines a *category* $\mathcal{C}$ as a set (or, more generally, a class) of objects, together with a class of special maps, called morphisms between these objects. If $\mathcal{C}$ is the category of Abelian groups, then the morphisms would normally be group homomorphisms. If $\mathcal{C}$ is the category of topological spaces, the morphisms would be continuous maps, etc. *See also* functor.

morphism in a category   A map by means of which categorical equivalence is measured.

morphism of finite type   For R a commutative ring with 1, and X a scheme, a morphism $f : X \longrightarrow \mathrm{Spec}(R)$ is called a *morphism of finite type* provided that X has a finite open affine covering $\{U_i = \mathrm{Spec}(R_i)\}$ such that each R_i is a finitely generated R-algebra. More generally, a morphism of schemes $f : X \longrightarrow Y$ is of *finite type* if there is an affine covering $\{V_i\}$ of Y such that the restriction $f : f^{-1}(V_i) \longrightarrow V_i$ is of finite type for each i. *See* scheme, spectrum.

morphism of local ringed spaces   Let X be a topological space, and suppose that to each $x \in X$ is associated a local ring $B_{X,x}$ in a natural way. Let Y be another such space. A mapping $f : X \to Y$ is called a morphism of local ringed spaces if, whenever, $f(x) = y$, there is induced a homomorphism $B_{Y,y} \to B_{X,x}$.

morphism of local ringed spaces   Let X be a topological space, and suppose that to each $x \in X$ is associated a local ring $B_{X,x}$ in a natural way. Let Y be another such space. A mapping $f : X \to Y$ is called a *morphism of local ringed spaces* if, whenever, $f(x) = y$, there is induced a homomorphism $B_{Y,y} \to B_{X,x}$.

morphism of pointed sets   A set X with a distinguished element x^* is called a pointed set. For two pointed sets (X, x^*) and (Y, y^*), a map $f : X \longrightarrow Y$ is said to be a *morphism of pointed sets* if $f(x^*) = y^*$.

morphism of schemes   Let R be a commutative ring with 1. One obtains a sheaf of rings $\tilde{R}$ on $\mathrm{Spec}(R)$ by assigning to each point p of $\mathrm{Spec}(R)$ the ring of quotients R_p. Then $\mathrm{Spec}(R)$ is called an affine scheme when it is regarded as a local-ringed space with $\tilde{R}$ as its structure sheaf. A scheme is a local-ringed space X which is locally isomorphic to an affine scheme. A morphism of schemes X and Y is a morphism $f : X \longrightarrow Y$ as local ringed-spaces.

multilinear function   Let $V_1, V_2, \ldots, V_k, W$ be vector spaces over a field F. A function

$$f : V_1 \times V_2 \times \cdots \times V_k \to W$$

that is linear with respect to each of its arguments is called multilinear. If $k = 2$, f is called bilinear. Thus, a bilinear function $f : U \times V \to W$ satisfies

$$f(u, av_1 + bv_2) = af(u, v_1) + bf(u, v_2)$$

and

$$f(au_1 + bu_2, v) = af(u_1, v) + bf(u_2, v)$$

for all $a, b \in F$ and all $u, u_1, u_2 \in U$ and $v, v_1, v_2 \in V$.

The *determinant* of an $n \times n$ matrix A is a multilinear function when the arguments are taken to be each of the n rows (or columns) of A.

multinomial   Given n real numbers $a_1, a_2, \ldots, a_n$, and m a positive integer, then the expression $(a_1 + a_2 + \cdots + a_n)^m$ is called *multinomial*. We have

$$(a_1 + a_2 + \cdots + a_n)^m$$
$$= \sum \frac{n!}{p_1! \ldots p_m!} a_1^{p_1} \ldots a_m^{p_m}$$

where the sum is over all $p_1 + \cdots + p_m = m$. *See also* monomial.

multiobjective programming   A mathematical programming problem in which the objective function is a vector-valued function $f : \mathbf{R}^n \longrightarrow \mathbf{R}^k$, $k \geq 2$, where $\mathbf{R}^k$ is ordered in some way (e.g., lexicographic order, etc.).

multiple   Let S be a semigroup whose binary operation is multiplication. The element $a \in S$ is called a *left (right) multiple* of $b \in S$ if there exists an element $c \in S$ such that $a = cb$ ($a = bc$). Under this condition b is the left (right) divisor of a.

multiple complex   Let $\mathcal{C}$ be an Abelian category. A complex C in $\mathcal{C}$ is a family of objects $\{C^n\}_{n \in \mathbf{Z}}$ with differentials $d^n : C^n \longrightarrow C^{n+1}$ such that $d^{n+1}d^n = 0, n \in \mathbf{Z}$. A bicomplex C in $\mathcal{C}$ is a family of objects $\{C^{m;n}\}_{m,n \in \mathbf{Z}}$ and two sets of differentials $d_1^m : C^{m,n} \longrightarrow C^{m+1,n}$ and

$d_2^n : C^{m,n} \longrightarrow C^{m,n+1}$ such that $d_1^{m+1}d_1^m = d_2^{n+1}d_2^n = 0$ and $d_1^m d_2^n = d_2^n d_1^m, m, n \in \mathbf{Z}$. A multicomplex C in $\mathcal{C}$ is defined in an analogous way: A family $\{C^{n_1, n_2, \ldots, n_r}\}_{n_1, n_2, \ldots, n_r \in \mathbf{Z}}$ of objects and r sets of differentials

$$d_i^{n_i} : C^{n_1, \ldots, n_i, \ldots, n_r} \longrightarrow C^{n_1, \ldots, n_i + 1, \ldots, n_r},$$

$1 \leq i \leq r$, subject to $d_i^{n_i+1} d_i^{n_i} = 0$ and $d_i^{n_i} d_j^{n_j} = d_j^{n_j} d_i^{n_i}$, $1 \leq i, j \leq r$, $i \neq j$.

multiple covariants   Let K be a field of characteristic 0 and let $G = \text{GL}(n, K)$. Let $F = \sum C_{i_1, \ldots, i_n} m_{i_1, \ldots, i_n}$ be a homogeneous form of degree d in variables $x_1, \ldots, x_n$ with coefficients in K. (Here, $\sum i_r = d$ and

$$m_{i_1, \ldots, i_n} = \left(d! / \prod i_r! \right) x_1^{i_1} \ldots x_n^{i_n} .)$$

For each $g \in G$, we define gx_i and $(gC)_{i_1, \ldots, i_n}$ by setting, respectively,

$$(gx_1, \ldots, gx_n) = (x_1, \ldots, x_n) g^{-1}$$

and $F = \sum (gC)_{i_1, \ldots, i_n} (gm_{i_1, \ldots, i_n})$. In this case, the action of G on the polynomial ring $R = K[\ldots, C_{i_1, \ldots, i_n}, \ldots]$ is given by either a rational representation or the contragredient map $A \mapsto {}^t A^{-1}$. Then the G-invariants are called *multiple covariants*.

multiple root   A root of a polynomial, having multiplicity > 1. *See* multiplicity of root.

multiplicand   A number to be multiplied by another number.

multiplication   (1) A binary operation on a set. In group theory, and in algebra in general, it is customary to denote by $a \dot b$ or ab, the element which is associated with (a, b) under a given binary operation. The element $c = ab$ is then called the product of a and b, and the binary operation itself is called multiplication. When the term multiplication is used for a binary operation, it carries with it the implication that if a and b are in the set G, then ab is also in G.

(2) One of two binary operations on a field or ring. *See* field, ring.

multiplication by logarithms   The logarithm function $\log_b x$, especially when $b = 10$ or

$b = e$, the basis for the natural logarithm, has had extensive use in facilitating arithmetical calculations, especially before the days of computers. The rule

$$\log_b n \cdot m = \log_b n + \log_b m$$

allows one to multiply large numbers n, m by ascertaining their logarithms (from a table), adding the logarithms and then obtaining the product of m and n by another reference to the table. *See* logarithm, common logarithm, natural logarithm.

multiplication of complex numbers   Multiplication of $z_1 = a + ib$ and $z_2 = c + id$ yields $z_1 \cdot z_2 = (a+ib)(c+id) = (ac - bd) + i(ad + bc)$. Writing the complex numbers z_1 and z_2 in terms of their absolute values r_1, r_2 and their arguments ϕ_1, ϕ_2: $z_j = r_j(\cos \phi_j + i \sin \phi_j)$, yields $z_1 \cdot z_2 = r_1 r_2 [\cos(\phi_1 + {}^-\phi_2) + i \cos(\phi_1 + {}^-\phi_2)]$.

multiplication of matrices   Let F be a ring, and $A = [a_{ij}]$, $1 \le i \le m$, $1 \le j \le n$, an $m \times n$ matrix over F. Multiplication of A (on the right) by a $k \times l$ matrix $B = [b_{ij}]$ (over the same field F) is defined if $n = k$ and the product is the $m \times l$ matrix $C = [c_{ij}]$, with ij-entry is obtained by $c_{ij} = \sum_h a_{ih} b_{hj}$, $1 \le i \le m$ and $1 \le j \le l$:

$$A \cdot B = \begin{pmatrix} \sum a_{1h} b_{h1} & \cdots & \sum a_{1h} b_{hl} \\ \cdots & \sum a_{ih} b_{hj} & \cdots \\ \sum a_{mh} b_{h1} & \cdots & \sum a_{mh} b_{hl} \end{pmatrix}.$$

multiplication of polynomials   Let

$$f(x) = \sum_0^n a_k x^k, \quad g(x) = \sum_0^p b_k x^k,$$

be polynomials over an integral domain D. *Multiplication of the polynomials* gives

$$f(x) \cdot g(x) = a_0 b_0 + (a_0 b_1 + a_1 b_0) x^1$$

$$+ (a_0 b_2 + a_1 b_1 + a_2 b_0) x^2 + \dots$$

$$= \sum_0^{n+p} c_k x^k$$

with $c_k = \sum_{i=0}^k a_i b_{k-i}$.

multiplication of vectors   Any bilinear function defined on pairs of vectors. Important examples are the familiar dot product and cross product of calculus. *See also* wedge product, tensor product.

multiplicative group   (1) Any group, where the group operation is denoted by multiplication. Often a non-Abelian group is written as a multiplicative group. *See* group, Abelian group.

(2) The set F^* of all the non-zero elements of a field F. *See* field.

multiplicative identity   An element 1, in a set S with a binary operation $\cdot$, regarded as multiplication, such that $1 \cdot a = a \cdot 1 = a$, for all $a \in S$.

A multiplicative monoid is one in which the binary operation is written as multiplication and the multiplicative identity is 1 such that $a1 = 1a = a$, for all a in the monoid.

multiplicative inverse   For an element a, in a set S with a binary operation $\cdot$, considered as multiplication, and an identity element 1, an element $a^{-1} \in S$ such that $a^{-1} \cdot a = a \cdot a^{-1} = 1$. (*See also* multiplicative identity.)

A field F is a non-trival commutative ring in which every non-zero element a has a *multiplicative inverse*.

multiplicative Jordan decomposition   Let M be a free right R-module. A linear transformation τ on M is called semisimple if M has the structure of a semisimple $R[x]$ module determined by $x(m) = \tau(m)$; that is, if and only if the minimal polynomial of τ has no square factor different from the constants in $R[x]$. *The multiplicative Jordan Decomposition Theorem* states that any nonsingular linear transformation τ on M can be uniquely written as $\tau = \tau_s \tau_u$, where τ_s is a semisimple linear transformation on M and $\tau_u = 1_M + \tau_s^{-1} \tau_n$ for some nilpotent transformation τ_n on M. Here, τ_u is called the unipotent component of τ.

multiplicatively closed subset   A subset S of a ring R, which is a subsemigroup of R with respect to multiplication.

multiplicity of a point   Let f be a function defined in a neighborhood of a point p in $\mathcal{R}^N$. The multiplicity of the point p is the order to which f vanishes at p; that is, f is of order m at p if all derivatives of f up to (but not including) order m vanish, but some m^{th} order derivative does not.

In other contexts, if q is in the image of f then the order of q is the number of elements in the pre-image set of q under f.

multiplicity of a point   Let f be a function defined in a neighborhood of a point $p \in \mathbf{R}^N$. The multiplicity of the point p is the order to which f vanishes at p; that is, f is of order m at p if all derivatives of f up to (but not including) order m vanish, but some mth order derivative does not.

In other contexts, if q is in the image of f, then the order of q is the number of elements in the pre-image set of q under f.

multiplicity of root   Let D be an integral domain and $f(x)$ an element of $D[x]$. If c belongs to D and c is a root of $f(x)$ ($f(c) = 0$), then $f(x) = (x-c)^m g(x)$ where m is an integer with $0 \le m \le \deg f(x)$, $g(x) \in D[x]$ and $g(c) \ne 0$. The integer m is called the *multiplicity of the root c of $f(x)$*.

multiplicity of weight   Let g denote a complex semisimple Lie algebra, h a Cartan subalgebra of g, and R the corresponding root system. Let V be a g-module (not necessarily finite dimensional), and let $w \in h^*$ a linear form on h. We will let V^n denote the set of all $v \in V$ such that $Hv = w(H)v$, for all $H \subset h$. This is a vector subspace of V. An element of V^w is said to have weight w. The dimension of V^w is called the *multiplicity* of $w \in V$: if $V^w \ne \{0\}$, w is called a weight of V.

multiplier   Let G be a finite group and let $\mathbf{C}^* = \mathbf{C}\backslash\{0\}$. Then the second cohomology group $H^2(G, \mathbf{C}^*)$ is called the *multiplier* of G. If $H^2(G, \mathbf{C}^*) = 1$, then G is called a closed group and any projective representation of G is induced by a linear representation. *See also* Lagrange multiplier, Stokes multiplier, characteristic multiplier.

multiplier algebra   For a C^*-algebra $\mathbf{A}$, let $\mathbf{A}^{**}$ denote its enveloping von Neumann algebra. The multiplier algebra of $\mathbf{A}$ is the set $M(\mathbf{A}) = \{b \in \mathbf{A}^{**} : b\mathbf{A} + \mathbf{A}b \subseteq \mathbf{A}\}$.

multiply transitive permutation group   A permutation group G is called k-transitive (k a positive integer) if, for any k-tuples $(a_1, a_2, \ldots, a_k)$ and $(b_1, b_2, \ldots, b_k)$ of distinct elements in X, there exists a permutation $p \in G$ such that $pa_i = b_i$, for all $i = 1, \ldots, k$. If $k = 2$, G is called a *multiply transitive permutation group*.

multistage programming   A mathematical programming problem in which the objective function and the constraints have an iterative or repetitive property.

mutually associated diagrams   An $O(n)$ diagram is a Young diagram T for which the sum of the lengths of the first column and the second column is $\le n$. Two $O(n)$ diagrams T_1 and T_2 are called *mutually associated diagrams* if the lengths of their first columns sum up to n and their corresponding columns (except the first ones) have equal lengths. *See* Young diagram.

N

Nakai-Moishezon criterion   For a ringed space $(X, \mathcal{O})$, we let $\mathcal{I}$ denote a coherent sheaf of ideals of $\mathcal{O}$. When X is a k-complete scheme (where k is a field), then for any r-dimensional closed subvariety W of X and any invertible sheaf S we let $(S^r \cdot W)$ denote the intersection number of S^r with $\mathcal{O}/\mathcal{I}$. The *Nakai-Moishezon criterion* states that if $(S^r \cdot W) > 0$ for any r-dimensional closed subvariety W of X, then S is ample.

Naperian logarithm   *See* natural logarithm.

natural logarithm   The logarithm in the base e. *See* e. The natural logarithm of x is denoted $\log x$ or, in elementary textbooks, $\ln x$.

The adjective *natural* is used, due to the simplicity of the defining formula

$$\log x = \int_1^x \frac{1}{t}\, dt \; ,$$

for x real and positive.

Also called *Naperian logarithm*. *See also* logarithm, logarithmic function.

natural number   A positive integer. The system **N** of natural numbers was developed by the Italian mathematician Peano, using a few simple properties known as Peano Postulates. Let there exist a non-empty set **N** such that

Postulate I: 1 is an element in **N**.

Postulate II: For every $n \in \mathbf{N}$, there exists a unique $n^* \in \mathbf{N}$, called the successor of n.

Postulate III: 1 is not the successor of any element in **N**.

Postulate IV: If $n, m \in \mathbf{N}$ and $n^* = m^*$, then $n = m$.

Postulate V: Any subset $K \subset \mathbf{N}$ having the properties

(i.) 1 is an element of K

(ii.) $k^* \in K$ whenever $k \in K$

satisfies $K = \mathbf{N}$.

Addition on **N** is defined by:

(i.) $n + 1 = n^*$, for every $n \in \mathbf{N}$

(ii.) $n + m^* = (n + m)^*$ whenever $n + m$ is defined.

The addition satisfies the following laws:

A1 Closure Law: $n + m \in \mathbf{N}$

A2 Commutative Law: $n + m = m + n$

A3 Associative Law: $m + (n + p) = (m + n) + p$

A4 Cancellation Law: If $m + p = n + p$, then $m = n$

Multiplication on **N** is defined by

(i.) $n \cdot 1 = n$

(ii.) $n \cdot m^* = n \cdot m + n$ whenever $n \cdot m$ is defined.

M1 Closure Law: $n \cdot m \in \mathbf{N}$

M2 Commutative Law: $n \cdot m = m \cdot n$

M3 Associative Law: $m \cdot (n \cdot p) = (m \cdot n) \cdot p$

M4 Cancellation Law: If $m \cdot p = n \cdot p$, then $m = n$

Addition and multiplication are subject to the Distributive Law

D1 For all $n, m, p \in \mathbf{N}$, $m \cdot (n + p) = m \cdot n + m \cdot p$

The elements of **N** are called *natural numbers*.

natural positive cone   For a weight w on the positive elements of a von Neumann algebra $\mathcal{A}$, we set $\mathcal{N}_w = \{A \in \mathcal{A} : w(A^*A) < \infty\}$. Let $0 \le s \le \frac{1}{2}$. Then the closure W^s of the set of vectors $\Delta_w^s \eta(A)$, where A ranges over all positive elements in $\mathcal{N}_w \cap \mathcal{N}_{w^*}$, has some of the properties of the von Neumann algebra $\mathcal{A}$. (Here, Δ_w is a modular operator and η is a complex linear mapping on $\mathcal{N}_w$ defined by setting $(\eta(A_\circ), \eta(A)) = w(A^*A)$.) The special case $W^{\frac{1}{4}}$ is called the *natural positive cone*. It is a self-dual convex cone and is independent of the weight w.

negation   Given a proposition P, the proposition (not P) is called the *negation* of P and is denoted by $\sim P$ or $\neg P$.

negative   *See* negative element.

negative angle   An angle XOP, with vertex O, initial side OP and terminal side OX, such that the direction of rotation is counterclockwise.

negative cochain complex   A cochain complex C in an Abelian category such that $C^n = 0$ for $n > 0$. *See* cochain complex.

negative element   (1) Let K be an ordered field. (*See* ordered field.) If $x < 0$, we say x is *negative* and $-x$ is positive.

(2) An element $g \neq e$ of a lattice ordered group G such that $g \leq e$. See lattice ordered group.

See also negative root.

negative exponent   In a multiplicative group G with identity 1 the negative powers of an element a are a^{-1}, its multiplicative inverse, and the elements $a^{-k} = (a^{-1})^k$, for $k > 0$, where an element b^k is $bb \cdots b$ (with k factors).

negative number   A negative element of the real field. *See* negative element.

negative root   Let $\mathcal{G}$ be a complex semisimple Lie algebra. For a subalgebra $\mathcal{H}$ of $\mathcal{G}$, we set $\mathcal{H}^* = \{$all complex-valued forms on $\mathcal{H}\}$. For $h^* \in \mathcal{H}^*$, let $\mathcal{H}_{h^*} = \{g \in \mathcal{G} : \mathrm{ad}(h)g = h^*(h)g$, for all $h \in \mathcal{H}\}$. If $\mathcal{H}_{h^*} \neq 0$ and $h^* \neq 0$, then we say that h^* is a root; two roots are equivalent whenever one is a nonzero multiple of the other. The root system of $\mathcal{G}$ relative to $\mathcal{H}$ is the finite set $\Delta = \{h^* \in \mathcal{H}^* : h^* \neq 0$ and $\mathcal{H}_{h^*} \neq \{0\}\}$. Let $\mathcal{H}_{\mathbf{R}}^*$ denote the real linear subspace of $\mathcal{H}^*$ spanned by Δ. Relative to a lexicographic linear ordering of $\mathcal{H}^*$ (associated with some basis over $\mathbf{R}$), a root r is negative if $r < 0$.

Neron minimal model   Let R be a discrete valuation ring with residue field k and quotient field K. The *Neron minimal model* of an Abelian variety A over K is a smooth group scheme $\mathcal{A}$ of finite type over $\mathrm{Spec}(R)$ such that for every scheme S over $\mathrm{Spec}(R)$ there is a canonical isomorphism

$$\mathrm{Hom}_K(S_K, A) \simeq \mathrm{Hom}_{\mathrm{Spec}(R)}(S, \mathcal{A}).$$

Here, S_K is the pullback of S by $\mathrm{Spec}(K) \to \mathrm{Spec}(R)$.

Neron-Severi group   Let $\mathrm{Div}(S)$ denote the group of all divisors of a nonsingular surface S. By linearity and the Index Theorem of Hodge, there is a bilinear form $I(D \cdot D')$ on $\mathrm{Div}(S) \otimes_{\mathbf{Z}} \mathbf{Q}$ which has exactly one positive eigenvalue. (*See* Index Theorem of Hodge.) Let $J = \{D \in \mathrm{Div}(S) \otimes_{\mathbf{Z}} \mathbf{Q} : I(D \cdot D') = 0$ for all $D'\}$. Then J is a subgroup of $\mathrm{Div}(S) \otimes_{\mathbf{Z}} \mathbf{Q}$ and the quotient $X = (\mathrm{Div}(S) \otimes_{\mathbf{Z}} \mathbf{Q})/J$ is a finite dimensional vector space over $\mathbf{Q}$. This quotient group X is called the *Neron-Severi group* of the surface S and its dimension is called the Picard number of S.

network programming   A mathematical programming problem related to network flow, in which the objective function and the constraints are defined with reference to a graph.

Newton-Raphson method of solving algebraic equations   An iterative numerical method for finding an approximate value for a zero of a function f (approximating a solution to the equation $f(x) = 0$). The method begins by choosing a suitable starting value x_0. The method proceeds by constructing a sequence of successive approximations, $x_1, x_2, x_2, \ldots$, to the exact solution x, according to the formula $x_k = x_{k-1} - f(x_{k-1})/f'(x_{k-1})$. Under suitable conditions, the sequence $x_1, x_2, x_3, \ldots$ converges to x.

The name *Newton-Raphson method* is also applied to a multi-dimensional generalization of the above. Here, the object is to approximate a solution to the equation $f(x) = 0$, but this time x is a vector in an n-dimensional vector space V and f is a vector-valued function from some subset of V to V. The sequence of successive approximations is constructed according to the formula:

$$x_k = x_{k-1} - D(f(x_{k-1}))^{-1} f(x_{k-1}),$$

where Df is the Jacobian of f (the $n \times n$ matrix of partial derivatives).

Newton's formulas   (1) For interpolation:

$$f(x_0 + u\Delta x) = f(x_0) + \frac{u}{1!}\Delta_0$$

$$+ \frac{u(u-1)}{2!}\Delta_0^2 + \frac{u(u-1)(u-2)}{3!}\Delta_0^3$$

$$+ \frac{u(u-1)(u-2)(u-3)}{4!}\Delta_0^4 + \cdots,$$

where $\Delta_0 = f(x_0 + \Delta x) - f(x_0)$.

(2) For symmetric polynomials: Let $p_1, p_2, \ldots, p_r$ be the elementary symmetric polynomials in variables $X_1, X_2, \ldots, X_r$. That is, $p_1 =$

$\sum X_i$, $p_2 = \sum X_i X_j$, $\ldots$, $p_r = X_1 X_2 \ldots X_r$. For $n = 1, 2, \ldots$, let $s_n = \sum X_i^n$. The Newton formulas are: $s_n - p_1 s_{n-1} + p_2 s_{n-2} - \cdots + (-1)^{n-1} p_{n-1} s_1 + (-1)^n n p_n = 0$; and $s_n - p_1 s_{n-1} + p_2 s_{n-2} - \cdots + (-1)^k p_k s_{n-k} = 0$, for $n = k + 1, k + 2, \ldots$.

nilalgebra An algebra in which every element is nilpotent. *See* algebra, nilpotent element.

nilpotent component Let M be a linear space over a perfect field. Then any linear transformation τ of M can be uniquely written as $\tau = \tau_s + \tau_n$ (Jordan decomposition), where τ_s is semisimple and τ_n is nilpotent, and the two linear transformations τ_s and τ_n commute; they are called the *semisimple component* and the *nilpotent component* of τ, respectively. *See* Jordan decomposition.

nilpotent element An element a of a ring R such that $a^n = 0$, for some positive integer n.

nilpotent group A group G such that $C_n(G) = G$, for some n, where $1 \subset C_1(G) \subset C_2(G) \subset \cdots$ is the ascending central series of G. *See* ascending central series.

nilpotent ideal A (left, right, two-sided) ideal I of a ring R is *nil* if every element of I is nilpotent; I is a *nilpotent ideal* if $I^n = 0$ for some integer n.

nilpotent Lie algebra A Lie algebra such that there exists a number M such that any commutator of order M is zero.

nilpotent Lie group A connected Lie group whose Lie algebra is a nilpotent Lie algebra.

nilpotent matrix A square matrix A such that $A^p = 0$ where p is a positive integer. If p is the least positive integer for which $A^p = 0$, then A is said to be *nilpotent of index p*.

nilpotent radical Let $\mathcal{G}$ be a Lie algebra. The radical of $\mathcal{G}$ is the union $\mathcal{S}$ of all solvable ideals of $\mathcal{G}$; it is itself a solvable ideal of $\mathcal{G}$. The largest nilpotent ideal of $\mathcal{G}$ is the union $\mathcal{N}$ of all nilpotent ideals of $\mathcal{G}$. The ideal $\mathcal{I} = [\mathcal{S}, \mathcal{G}]$ is called the *nilpotent radical* of $\mathcal{G}$ and we have the inclusions $\mathcal{I} \subset \mathcal{N} \subset \mathcal{S} \subset \mathcal{G}$. *See* nilpotent ideal.

nilpotent subset A subset S of a ring R such that $S^n = 0$, for some positive integer n.

nilradical The ideal intersection of all the prime ideals P which contain I, where I is an ideal in a commutative ring R. The nilradical is denoted $\text{Rad} I$. If the set of prime ideals containing I is empty, then $\text{Rad} I = R$.

Noetherian domain A commutative ring R with identity, which is an integral domain and also a Noetherian ring. *See* integral domain, Noetherian ring.

Noetherian integral domain An integral domain whose set of ideals satisfies the maximum condition: every nonempty set of ideals has a maximal element.

Noetherian local ring A Noetherian ring having a unique maximal ideal.

Noetherian module A left module M, over a ring R such that, for every ascending chain $M_1 \subset M_2 \subset \cdots$ of submodules of M, there exists an integer p such that $M_k = M_p$, for all $k \geq p$.

Noetherian ring A ring R is *left* (resp. *right*) *Noetherian* if R is Noetherian as a left (resp. right) module over itself. *See* Noetherian module. R is said to be *Noetherian* if R is both left and right Noetherian.

Noetherian scheme A locally Noetherian scheme whose underlying topological space is compact. A scheme X is a *locally Noetherian scheme* if it has an affine covering $\{U_i\}$, where each U_i is the spectrum of some Noetherian ring R_i.

Noetherian semilocal ring A Noetherian ring with only a finite number of maximal ideals. *See* Noetherian ring. *See also* Noetherian local ring.

Noether's Theorem   Any ideal in a finitely generated polynomial ring is finitely generated.

non-Abelian cohomology   The cohomology $H^0(G, N)$ and $H^1(G, N)$ of a group G using non-homogeneous cochains, where N is a non-Abelian G-group.

non-Archimedian valuation   A valuation which is not Archimedian. *See* Archimedian valuation.

non-commutative field   A non-commutative ring whose nonzero elements form a (multiplicative) group.

non-convex quadratic programming   The mathematical programming problem of minimizing the objective function $f(x) = u \cdot x + (x \cdot Ax)/2$ subject to some linear equations or inequalities. Here, $f : \mathbf{R}^n \to \mathbf{R}$, $u \in \mathbf{R}^n$ is fixed, and A is a symmetric (perhaps indefinite) matrix.

non-degenerate quadratic form   A homogeneous quadratic polynomial p in $x_1, x_2, \ldots, x_n$ with coefficients in a field K is called a quadratic form. When the characteristic of K is not 2, we may write $p(x) = (x, Ax)$, where $x = (x_1, x_2, \ldots, x_n)$ and A is an $n \times n$ matrix with entries in K. We say that p is a *non-degenerate quadratic form* if the determinant $\det(A) \neq 0$.

non-degenerate representation   Let $L^1(G)$ denote the space of all complex-valued integrable functions on a locally compact group G. Then $L^1(G)$ is an algebra over $\mathbf{C}$, where multiplication is given by the convolution $(f * g)(\alpha) = \int_G f(\alpha\beta^{-1})g(\beta)d\beta$. The map $f(\alpha) \mapsto f^*(\alpha) = D(\alpha^{-1})\overline{f(\alpha^{-1})}$ is an involution of the algebra $L^1(G)$, where D is the modular function of G. For a unitary representation u of G, let $U_f = \int f(\alpha)u_\alpha d\alpha$. Then a *non-degenerate representation* of the Banach algebra $L^1(G)$ with an involution is given by the map $f \mapsto U_f$. The map $u \mapsto U$ is a bijection between the set of equivalence classes of unitary representations of G and the set of equivalence classes of non-degenerate representations of the Banach algebra $L^1(G)$ with an involution.

non-linear algebraic equation   An equation in n variables having the form $p(x_1, x_2, \ldots, x_n) = 0$, where p is a polynomial of degree > 1 with coefficients in a field.

non-linear differential equation   Any differential equation $f(t, x, x', \ldots, x^{(n)}) = 0$, where x represents an unknown function, $x' = \frac{dx}{dt}$ and $f(x_0, \ldots, x_{n+1})$ is *not* a linear function of $x_0, \ldots, x_{n+1}$.

A non-linear nth degree equation, as above, can be written in the form of a system

$$x_i' = f_i(t, x_1, \ldots, x_n), \quad i = 1, \ldots, n$$

and uniqueness can be proved, with initial conditions $x_i(a_i) = b_i$, $i = 1, \ldots, n$, under suitable conditions on the f_i.

non-linear problem   Any mathematical problem which deals with non-linear mappings or operators and their related properties.

non-linear programming   A mathematical programming problem in which the objective function or (some of) the constraints are non-linear.

non-linear transcendental equation   A non-linear analytic, but not polynomial, equation.

non-primitive character   A character of a finite group that is not a primitive character. *See* primitive character, character of group.

nonsingular matrix   A matrix having an inverse. *See* inverse matrix.

norm   (1) Let V be a vector space over $K = \mathbf{R}$ or $\mathbf{C}$. A function $\rho : V \to [0, \infty)$ is a semi-norm if it satisfies the following three conditions: (i.) $\rho(x) \geq 0$, for all $x \in V$, (ii.) $\rho(\alpha x) = |\alpha|\rho(x)$, for all $\alpha \in K$ and $x \in V$, and (iii.) $\rho(x + y) \leq \rho(x) + \rho(y)$, for all $x, y \in V$. A semi-norm ρ for which $\rho(x) = 0 \Leftrightarrow x = 0$ is called a *norm*.

(2) *See* reduced norm.

normal *-homomorphism   A *-homomorphism between two Banach algebras with involutions $\mathcal{B}_1$ and $\mathcal{B}_2$ is an algebraic homomorphism φ which preserves involution; that is,

$\varphi(x^*) = \varphi(x)^*$. A $*$-homomorphism φ is called *normal* if, for every bounded increasing net $\mathcal{N}_\alpha$ in $\mathcal{B}_1$, one has $\sup_\alpha \varphi\mathcal{N}_\alpha = \varphi(\sup_\alpha \mathcal{N}_\alpha)$. A $*$-homomorphism between two operator algebras is continuous in the strong and weak operator topologies.

normal algebraic variety An irreducible variety all of whose points are normal. *See* normal point.

normal Archimedian valuation An Archimedian valuation of an algebraic number field K of degree n which is one of the valuations v_i, $1 \leq i \leq n$, defined as follows: There are exactly n mutually distinct injections $\varphi_1, \ldots, \varphi_n$ of K into the complex number field $\mathbf{C}$. We may assume that $\varphi_i(K) \subset \mathbf{R}$ if and only if $1 \leq i \leq r_1$ and that $\varphi_i(a)$ and $\varphi_{n-i+1+r_1}(a)$ are complex conjugates for $r_1 \leq i \leq r_2 = (n - r_1)/2$. Let $v_i(a) = |\varphi_i(a)|$ for $1 \leq i \leq r_1$ and $v_i(a) = |\varphi_i(a)|^2$ for $r_1 \leq i \leq r_2 = (n - r_1)/2$. *See* Archimedian valuation.

normal basis Let L be a Galois extension of a field K and $G(L/K)$ the Galois group of L/K. (*See* Galois extension, Galois group.) If L/K is a finite Galois extension, then there is an element $l \in L$ such that the set $\{l^g : g \in G(L/K)\}$ forms a basis for L over K called a *normal basis*. *See also* Normal Basis Theorem.

The existence of a *normal basis* implies that the regular representation of $G(L/K)$ is equivalent to the K-linear representation of $G(L/K)$ by means of L.

Normal Basis Theorem Any (finite dimensional) Galois extension field L/K that has a normal basis.

normal chain of subgroups A normal chain of subgroups of a group G is a finite sequence $G = G_0 \supset G_1 \supset \cdots \supset G_n = \{e\}$ of subgroups of G with the property that each G_i is a normal subgroup of G_{i-1} for $1 \leq i \leq n$.

normal crossings Let $D \geq 0$ be a divisor (i.e., a cycle of codimension 1) on a non-singular variety V and let $x \in V$. Then D is a divisor with only *normal crossings* at x if there is a system of local coordinates $(f_1, \ldots, f_n)$ around x

such that D is defined by means of (a part of) this system of local coordinates. A divisor D is called a *divisor with only normal crossings* if it is a divisor with only normal crossings at every $x \in V$.

normal equation Let $AX = b$ be a system of m linear equations in n unknowns with $m \geq n$. The problem of finding the X which minimizes the Euclidean norm $\|b - AX\|$ is called the linear least squares problem. The normal equation for this problem is ${}^t A A X = {}^t A b$, where ${}^t A$ denotes the transpose of A. Applying the Cholesky method to this normal equation is one way of solving the linear least squares problem. *See* Cholesky method of factorization.

normal extension An extension field K of a field F, such that K is a finite extension of F and F is the fixed field of $G(K, F)$, the group of automorphism of K relative to F.

normal form A canonical choice of representatives for equivalence classes with respect to some group action. *See also* Jordan normal form.

normal function A continuous, strictly monotone function of the ordinal numbers.

Normalization Theorem for Finitely Generated Rings Let R be a finitely generated ring over an integral domain $\mathcal{I}$. Then there exist an element $\alpha \in \mathcal{I}$ (with $\alpha \neq 0$) and algebraically independent elements $Y_1, \ldots, Y_n$ of R over $\mathcal{I}$ such that the ring of quotients R_S is integral over $\mathcal{I}[\alpha^{-1}, Y_1, \ldots, Y_n]$. Here, $S = \{\alpha^i : i = 1, 2, \ldots\}$.

Normalization Theorem for Polynomial Rings Let $\mathcal{I}$ be an ideal in a polynomial ring $K[X_1, X_2, \ldots, X_n]$ in n variables over a field K. Then there exist $Y_1, Y_2, \ldots, Y_n$ in $K[X_1, X_2, \ldots, X_n]$ such that (i.) $Y_1, Y_2, \ldots, Y_n$ generate $\mathcal{I} \cap K[Y_1, Y_2, \ldots, Y_n]$ and (ii.) $K[X_1, X_2, \ldots, X_n]$ is integral over $K[Y_1, Y_2, \ldots, Y_n]$.

normalized cochain A cochain f, in Hochschild cohomology, satisfying $f(\mathfrak{l}_1, \ldots, \mathfrak{l}_n) = 0$ whenever one of the $\mathfrak{l}_i$ is 1. Here, Λ is an algebra over a commutative ring K, A is a two-sided Λ-module, an n-cochain C^n is the module of all n-linear mappings of Λ into A with $C^0 = A$. This cohomology group is defined by the n-cochains and the coboundary operator $\delta^n(f) : C^n \to C^{n+1}$ defined by

$$\delta^n(f) : (1_1, \ldots, 1_{n+1}) \mapsto \lambda_1 f(1_2, \ldots, 1_{n+1})$$

$$+ \sum_{j=1}^{n} (-1)^j f(1_1, \ldots, 1_j 1_{j+1}, \ldots, 1_{n+1})$$

$$+ (-1)^{n+1} f(1_1, \ldots, 1_n) 1_{n+1} .$$

The normalized cochains can be used to define the same cohomology group.

normalizer (1) (In ergodic theory) Let $(X, \mathcal{B}, \mu)$ be a σ-finite measure space. A map $f : X \longrightarrow X$ is called measurable if $f^{-1}(B) \in \mathcal{B}$ for every $B \in \mathcal{B}$. A bimeasurable map is a bijective measurable map whose inverse is also measurable. A non-singular bimeasurable map $f : X \longrightarrow X$ is said to be the *normalizer* for another bimeasurable map $g : X \longrightarrow X$ if for every $\varphi \in [g] = \{\varphi : \varphi$ is a non-singular measurable map of X such that, for some n, we have $\varphi(x) = g^n(x)$ for μ-almost all $x\}$ there is $\psi \in [g]$ such that $f \circ \varphi = \psi \circ f$.

(2) (In group theory) Let S be a subset of a group G. The *normalizer* of S is the subgroup $N(S) = \{g \in G : g^{-1} S g = S\}$ of G.

normal j-algebra A Lie algebra $\mathcal{G}$ over **R** such that there is a linear endomorphism j of $\mathcal{G}$ and a linear form ω on $\mathcal{G}$ satisfying the following four conditions: (i.) For every $x \in \mathcal{G}$ the eigenvalues of $\mathrm{ad}(x)$ are all real; (ii.) $[jx, jy] \equiv j[jx, y] + j[x, jy] + [x, y]$, for all $x, y \in \mathcal{G}$; (iii.) $\omega([jx, jy]) = \omega([x, y])$, for all $x, y \in \mathcal{G}$; (iv.) $\omega([jx, x]) > 0$ for $x \neq 0$.

normally flat variety Let V be a variety over a field of characteristic zero and let S be a subscheme of V defined by a sheaf of ideals $\mathcal{I}$. Let $\mathcal{O}_S$ denote the sheaf of germs of regular functions on S and let $\mathcal{O}_{S,x}$ be the stalk of $\mathcal{O}_S$ over the point $x \in S$. Then V is said to be *normally flat* along S if for any n and any $x \in S$, the quotient module $\mathcal{I}_x^n / \mathcal{I}_x^{n+1}$ is a flat $\mathcal{O}_{S,x}$-module.

normal matrix A square matrix M, with complex entries, such that $MM^* = M^*M$,

where M^* denotes the adjoint (the transpose of the complex conjugate) of M.

normal point Let V be an affine variety over an algebraically closed field k and let p be a point in V. Then p is called a *normal point* if the local ring $R_p = k[x_1, \ldots, x_n]/I(p)$ is normal. Here, $I(p)$ is the ideal of all polynomials in $k[x_1, \ldots, x_n]$ which vanish at p.

normal representation A normal $*$-homomorphism of a von Neumann algebra $\mathcal{A}$ into some operator algebra $\mathcal{B}$.

normal ring An integrally closed integral domain. A ring is integrally closed if it is equal to its integral closure in its ring of quotients. *See* integral closure.

normal series Let G be a finite group. We define p_i ($i = 1, \ldots, k$) to be the distinct primes which divide the order of G. Let $\langle P_i \rangle$ be the normal subgroup generated by the class of Sylow subgroups corresponding to the prime p_i. If r is minimal, then a collection $\langle P_1 \rangle, \langle P_2 \rangle, \ldots, \langle P_r \rangle$ which generates G is called a minimal system of Sylow classes for G; r is called the Sylow rank of G. A chain T_i ($i = 0, \ldots, r$) of normal subgroups of G ($T_0 = $ identity, $T_r = G$) such that T_{i+1} is the group generated by T_i and $\langle P_{j_i} \rangle$, where $\{j_i\}$ is a permutation of $\{1, \ldots, r\}$, is called a *normal series* for the given minimal system of Sylow classes.

normal series Let G be a finite group. We define p_i ($i = 1, \cdots, k$) to be the distinct primes which divide the order of G. Let $\langle P_i \rangle$ be the normal subgroup generated by the class of Sylow subgroups corresponding to the prime p_i. If r is minimal, then a collection $\langle P_1 \rangle, \langle P_2 \rangle, \cdots, \langle P_r \rangle$ which generates G is called a minimal system of Sylow classes for G; r is called the Sylow rank of G. A chain T_i ($i = 0, \cdots, r$) of normal subgroups of G ($T_0 = $ identity, $T_r = G$) such that T_{i+1} is the group generated by T_i and $\langle P_{j_i} \rangle$, where $\{j_i\}$ is a permutation of $\{1, \cdots, r\}$ is called a normal series for the given minimal system of Sylow classes.

normal subgroup A subgroup H of a group G such that $Hx = xH$, for all $x \in G$.

normal valuation   Let K be an algebraic number field and let $\mathcal{O}$ be the principal part of K. A p-adic valuation v of K is a *normal valuation* if, for $\alpha \in K$, one has $v(\alpha) = |p|^{-r}$, where $|p| = $ norm of $p = $ the cardinality of the set $\mathcal{O}/p$. Here, r is the degree of α with respect to p. A normal valuation of a function field is defined analogously, except that instead of $|p|$ one uses c^d, where $c > 0$ is fixed and d is the degree of the residue class field of the valuation over the base field k.

See also normal Archimedian valuation.

norm form of Diophantine equation   Let K be an algebraic number field of degree $d \geq 3$ and let $c_1, \ldots, c_n \in K$. Then the norm

$$\text{Norm} \, (c_1 x_1 + \cdots + c_n x_n)$$

$$= \prod_{j=1}^{d} \left(c_1^{(j)} x_1 + \cdots + c_n^{(j)} x_n \right)$$

is a form of degree d with rational coefficients. Here, $c_i^{(j)}$ denotes a conjugate of c_i, $1 \leq i \leq n$.

normic form   A *normic form* of order $p \geq 0$ in a field k is a homogeneous polynomial f of degree $d \geq 1$ in $n = p^d$ variables with coefficients in k such that the equation $f = 0$ has no solution in k except $(0, \ldots, 0)$.

norm-residue symbol   Let k be a finite extension of Q_p and

$$X = k\{\{t_1\}\} \ldots \{\{t_{n-1}\}\}$$

be an n-dimensional local field. *See* local field. Assume that k contains the roots of unity μ of order $q = p^\nu$, $p \neq 2$. Let ζ be a generator of μ. According to Vostokov, there is a skew-symmetric map

$$\Gamma : X^* \times \cdots \times X^* \quad \to \quad \mu$$
$$\Gamma(\alpha_1, \alpha_2, \ldots, \alpha_{n+1}) \quad = \quad \zeta^{\text{tr res}(\phi/s)}, \, \alpha_i \in X^*,$$

with the property that $\alpha_i + \alpha_j = 1 \iff \Gamma(\alpha_1, \alpha_2, \ldots, \alpha_{n+1}) = 1$ for $i \neq j$. Here tr is the trace operator of the inertia subfield of k, s is determined in X by an expansion of ζ in X, and $\phi(\alpha_1, \alpha_2, \ldots, \alpha_{n+1})$ is given by an expansion of the αs in X. Here res is the residue of ϕ/s, i.e., the coefficient of $1/t_1 t_2 \cdots t_{n-1}$.

It is known then that Γ defines a non-degenerate pairing

$$\frac{K_n(X_n)}{K_n(X_n))^q} \times \frac{X_n^*}{(X_n^*)^q} \xrightarrow{(\cdot)} \zeta^{\text{tr res}(\phi/s)}$$

satisfying the norm property, i.e.,

$$\Gamma(\alpha_1, \alpha_2, \ldots, \alpha_{n+1}) = 1$$

$$\iff \{\alpha_1, \ldots, \alpha_n\} \in K_n(X_n)$$

is a norm in $K_n(X_n(q\sqrt{\alpha_{n+1}}))$. This property gives rise to its name as the n-dimensional norm-residue symbol.

nuclear   (**1**) Let X and Y be Banach spaces. A linear operator $T \, X \to Y$ is called a *nuclear operator* if it can be written as a product $T = S_1 \Lambda S_2$ with $X \xrightarrow{S_1} l_\infty \xrightarrow{\Lambda} l_1 \xrightarrow{S_2} Y$. Here, S_1 and S_2 are bounded linear operators and Λ is multiplication by a sequence $\{\lambda_n \geq 0\}$ in l_1. Then the nuclear norm of T is given by $\|T\|_1 = \inf \|S_1\| \, \|\Lambda\|_{l_1} \|S_2\|$. When X and Y are Hilbert spaces, this norm coincides with the trace norm.

(**2**) A locally convex topological vector space Z is called a *nuclear space* if for each absolutely convex neighborhood U of 0 there is another absolutely convex neighborhood $V \subset U$ of 0 such that the natural linear mapping $\tau_{V,U} : Z(V) \longrightarrow Z(U)$ is a nuclear operator. Here, $Z(U)$ and $Z(V)$ are the normed spaces obtained from seminorms corresponding to U and V, respectively.

null sequence   Any sequence tending to 0 (in a topological space with a 0 element).

Especially, one has a *null sequence* in the $\mathcal{I}$-adic topology. Let $\mathcal{I}$ be an ideal of a ring R and let $\mathcal{M}$ be an R-module. A null sequence in $\mathcal{M}$ is a sequence which converges to zero in the $\mathcal{I}$-adic topology of $\mathcal{M}$. A base for the neighborhood system of zero in this topology is given by $\{\mathcal{I}^n \mathcal{M} : n = 1, 2, \ldots\}$.

null space   A synonym for the elements of a matrix, when the matrix is considered as a linear transformation.

number   (**1**) Any member of the real or complex numbers.

(**2**) We say that two sets have the *same number* if there is a bijection between them. This is

223

an equivalence relation on the class of sets. A *number* is an equivalence class under this relation. Examples are cardinal numbers or ordinal numbers.

number field Any subfield of the field of complex numbers.

number of irregularity The dimension of the Picard variety of an irreducible algebraic variety.

number system Any one of the fields **Q**, **R**, or **C** (rational numbers, real numbers, or complex numbers).

numerator The quantity A, in the fraction $\frac{A}{B}$ (B is called the *denominator*).

numerical Referring to an approximate calculation, commonly made with machines.

numerical radius *See* numerical range.

numerical range The *numerical range* (or *field of values*) of an $n \times n$ matrix A with entries from the complex field is the set

$$F(A) = \left\{ x^* A x \ : \ x \in \mathbf{C}^n, \ x^* x = 1 \right\} .$$

The numerical range is a compact, convex set that contains the eigenvalues of A, is invariant under unitary equivalence and is subadditive (i.e., $F(A + B) \subset F(A) + F(B)$). The numerical radius of A is defined in terms of $F(A)$ by

$$r(A) = \max\{|z| \ : \ z \in F(A)\} .$$

O

objective function   In a mathematical programming problem, the function which is to be minimized or maximized.

octahedral group   The symmetric group of degree 4. This is the group S_4 of permutations of 4 elements.

odd element   (Of a Clifford algebra.) *See* even element.

odd number   An integer that is not divisible by 2. An odd number is typically represented by $2n - 1$, where n is an integer.

omega group   Let Ω be a set and let G be a group. Then G is called an Ω-group (and Ω is called an operator domain of G) if there is a map $\Omega \times G \longrightarrow G$ such that $\omega(g_1 g_2) = \omega(g_1)\omega(g_2)$ for any $\omega \in \Omega$, $g_1, g_2 \in G$.

omega subgroup   A subgroup H of an Ω-group G is an Ω-subgroup of G if $\omega(h) \in H$ for each $\omega \in \Omega$ and $h \in H$. *See* omega group.

one   The smallest natural number. It is the identity of the multiplicative group of real numbers.

one-cycle   A collection $\mathcal{C}$ of Abelian groups $\{G_n\}_{n \in \mathbf{Z}}$ and morphisms $\partial_n : G_n \longrightarrow G_{n-1}$, $n \in \mathbf{Z}$, is called a chain complex. Any element in the kernel of ∂_n is called an n-cycle of $\mathcal{C}$. In particular, a *one-cycle* of $\mathcal{C}$ is any element of the kernel of ∂_1.

one-parameter subgroup   A one-parameter group of transformations of a smooth manifold M is a family f_t, $t \in \mathbf{R}$, of diffeomorphisms of M such that (i.) the map $\mathbf{R} \times M \longrightarrow M$ defined by $(t, x) \mapsto f_t(x)$ is smooth, and (ii.) for $s, t \in \mathbf{R}$, one has $f_s \circ f_t = f_{s+t}$. More generally, if G is a Lie group with Lie algebra $\mathcal{G}$, then for each element $X \in \mathcal{G}$ there is a continuous homomorphism $(\mathbf{R}, +) \longrightarrow G$, $t \mapsto \varphi(t)$, such that $d\varphi(d/dt) = X$. Here, d/dt is the basis of the Lie algebra of $\mathbf{R}$. Such a homomorphism is called a *one-parameter subgroup* of the Lie group G.

one-to-one function   *See* injection.

onvolution   Assume f and g are real or complex valued functions defined on $\mathbf{R}^n$ and such that f^2 and g^2 are integrable on $\mathbf{R}^n$. The *convolution* of f with g, written $f \star g$, is a function h defined on $\mathbf{R}^n$ by the formula $h(x) = \int_{-\infty}^{\infty} f(x - y)g(y)dy$. Important cases of this occur in the study of Sobolev spaces where one can choose f so that it is integrable on all closed and bounded (compact) subsets of $\mathbf{R}^n$ and choose for g a function with infinitely many derivatives ($\mathbf{C}^{\infty}$) which is zero off a compact set (i.e., it has compact support). In this special case, many smoothness properties are passed on to $f \star g$ and one can use convolutions to approximate f in various norms.

operation   Let A and B be sets. An operation of A on B is any map from (a subset of) $A \times B$ to B. Any $a \in A$ is called an operator on B and A is said to be a domain of operators on B. *See also* binary operation.

operator algebra   Let $\mathcal{B}(H)$ denote the set of bounded linear operators on a Hilbert space H. Then $\mathcal{B}(H)$ contains the identity operator and is an algebra with the operations of operator addition and operator product (composition). An *operator algebra* is any subalgebra of $\mathcal{B}(H)$.

operator domain   *See* omega group.

operator homomorphism   Let Ω be a set and let G_1 and G_2 be Ω-groups. An operator homomorphism from G_1 to G_2 is any homomorphism $\varphi : G_1 \longrightarrow G_2$ which commutes with every $\omega \in \Omega$. That is, $\varphi(\omega g_1) = \omega(\varphi g_1)$ for any $\omega \in \Omega$ and any $g_1 \in G_1$. *See* omega group.

opposite   (1) Of a non-commutative group (or ring) G, the group G^* with the operation $\star$, defined by $(a, b) \mapsto a \star b = b \circ a$, where $\circ$ is the operation of G.

(2) An oriented n-simplex σ of a simplicial complex K is an n-simplex of K with an equivalent class of total ordering of its vertices. (Two orderings are equivalent if they differ by an even permutation of the vertices.) Therefore, for every n-simplex, $n \geq 1$, there are two oriented n-simplexes called the *opposites* of one another.

optimal solution   A mathematical programming problem is the problem of finding the maximum/minimum of a given function $f : X \longrightarrow \mathbf{R}$, where X is a Banach space. A point $x^* \in X$ at which f attains its maximum/minimum is called an *optimal solution* of the problem.

orbit   A subset of elements that are related by an action. For example, let G be a Lie transformation group of a space Γ. Then a G-invariant submanifold of Γ on which G acts transitively is called an *orbit*.

order   The number of elements in a set.

ordered additive group   A commutative additive group with an ordering "$<$" such that whenever a, b, c are elements with $a < b$, then $a + c < b + c$.

ordered field   A field $\mathcal{F}$ with a total order "$<$" such that if a, b, c are elements with $a < b$, then $a + c < b + c$, and if $c > 0$ and $a < b$, then $ac < bc$.

order of equation   In a partial differential equation, the highest derivative involved in that expression.

order of field   The number of elements in a field F, often denoted $|F|$.

order of function   The *order* of an analytic function $f(z)$ at point a is the exponent of the lowest power in the Taylor Series of $f(z)$ at a.

order of polynomial   Let

$$P_n(x) = a_n x^n + a_{n-1} x^{n-1} + \cdots + a_1 x + a_0$$

be a polynomial with complex coefficients and $a_n \neq 0$. The number n is the *order* of the polynomial $P_n(x)$.

order of radical   Let A be an ideal of the ring R. The set

$$\sqrt{A} = \{x \in R : x^n \in A \text{ for some } n \in \mathbf{N}\}$$

is called a radical of A. The *order of radical* of A is the cardinal number of the set $\sqrt{A}$.

order of vanishing   Let f be a holomorphic function on an open set U in the complex plane. Suppose that $P \in U$ and that $f(P) = 0$. Let k be the least positive integer such that $f^{(k)}(P) \neq 0$ (where the superscript denotes a derivative). Then f is said to *vanish to order k at P*.

There is an analogous notion of order of vanishing for a real function. If F vanishes at a point x_0 in Euclidean space, then the order of vanishing at x_0 is the order of the first non-vanishing Taylor coefficient in the expansion about x_0.

order relation   A relation "$\leq$" on a set such that (i) $x \leq x$, (ii) if $x \leq y$ and $y \leq x$ then $x = y$, and (iii) if $x \leq y$ and $y \leq z$ then $x \leq z$.

ordinary point   A point in an analytical set around which the set has a complete analytical structure.

ordinary representation   Let G be a group and let V be a vector space. An *ordinary representation* of G is a map from G to the automorphism group $Aut(V)$ of V.

orthogonal group   The set $\mathbf{O}(n)$ of all $n \times n$ matrices with complex entries such that for each $O \in \mathbf{O}(n)$ the transpose O^t is the inverse O^{-1} of O.

orthogonality relation   The binary relation $v \cdot w = 0$ between vectors in a vector space with a scalar product.

orthogonal matrix   An element O of the orthogonal group $\mathbf{O}(n)$. *See* orthogonal group.

orthogonal measure   A measure concentrated on a set at measure 0.

orthogonal subset   Let $\mathcal{S}$ be a subset of a vector (Hilbert) space $\mathcal{V}$. The *orthogonal subset* $\mathcal{S}^\perp$ of $\mathcal{S}$ is the subset

$$\mathcal{S}^\perp = \{\vec{v} \in \mathcal{V} : \vec{v} \cdot \vec{w} = 0 \text{ for all } \vec{w} \in \mathcal{S}\}.$$

orthogonal transformation A linear transformation $A : \mathbf{R}^n \to \mathbf{R}^n$ (or, $A : \mathbf{C}^n \to \mathbf{C}^n$) that preserves the inner product, i.e., for all $\vec{v}, \vec{w} \in \mathbf{R}^n$ (or $\mathbf{C}^n$), $\vec{v} \cdot \vec{w} = A(\vec{v}) \cdot A(\vec{w})$.

outer automorphism An automorphism of a group G that is not an inner automorphism, i.e., not of the form $g \to aga^{-1}$ for some $a \in G$. The factor group of all automorphisms of G modulo the subgroup of inner automorphism is called the *group of outer automorphisms*.

ovals of Cassini The eigenvalues of an $n \times n$ matrix $A = (a_{ij})$ with entries from the complex field lie in the union of $n(n-1)/2$ ovals, known as the *ovals of Cassini*,

$$\bigcup_{i,j \in \{1,2,\dots,n\}, i \neq j} C_{ij},$$

where

$$C_{ij} = \left\{ z \in \mathbf{C} : |z - a_{ii}| \, |z - a_{jj}| \right.$$
$$\left. \leq \left(\sum_{k \neq i} |a_{ik}| \right) \left(\sum_{k \neq j} |a_{jk}| \right) \right\}.$$

See also Geršgorin's Theorem.

overfield An *overfield* of a field F is a field K containing F. This is also called an *extension field* of F.

P

p-adic numbers The completion of the rational field **Q** with respect to the p-adic valuation $|\cdot|_p$. See p-adic valuation. See also completion.

p-adic valuation For a fixed prime integer p, the valuation $|\cdot|_p$, defined on the field of rational numbers as follows. Write a rational number in the form $p^r m/n$ where r is an integer, and m, n are non-zero integers, not divisible by p. Then $|p^r m/n|_p = 1/p^r$. See valuation.

parabolic subalgebra A subalgebra of a Lie algebra **g** that contains a maximal solvable subalgebra of **g**.

parabolic subgroup A subgroup of a Lie group G that contains a maximal connected solvable Lie subgroup of G. An example is the subgroup of invertible upper triangular matrices in the group $GL_n(\mathbf{C})$ of invertible $n \times n$ matrices with complex entries.

parabolic transformation A transformation of the Riemann sphere whose fixed points are ∞ and another point.

paraholic subgroup A subgroup of a Lie group containing a Borel subgroup.

parametric equations The name given to equations which specify a curve or surface by expressing the coordinates of a point in terms of a third variable (the parameter), in contrast with a relation connecting x, y, and z, the cartesian coordinates.

partial boundary operator We call $(X_{p,q}, \partial', \partial'')$ over A a double chain complex if it is a family of left A-modules $X_{p,q}$ for $p, q \in \mathbf{Z}$ together with A-automorphisms

$$\partial'_{p,q} : X_{p,q} \to X_{p-1,q}$$

and

$$\partial''_{p,q} : X_{p,q} \to X_{p,q-1}$$

such that

$$\partial'_{p-1,q} \circ \partial'_{p,q} = \partial''_{p,q-1} \circ \partial''_{p,q}$$

$$= \partial'_{p,q-1} \circ \partial''_{p,q} + \partial''_{p-1,q} \circ \partial'_{p,q} = 0 \,.$$

We define the associated chain complex (X_n, ∂) by setting

$$X_n = \sum_{p+q=n} X_{p,q}, \quad \partial_n = \sum_{p+q=n} \partial'_{p,q} + \partial''_{p,q} \,.$$

We call ∂ the *total boundary operator*, and ∂', ∂'' the *partial boundary operators*.

partial derived functor Suppose F is a functor of n variables. If S is a subset of $\{1, \ldots, n\}$, we consider the variables whose indices are in S as active and those whose indices are in $\{1, \ldots n\} \setminus S$ as passive. By fixing all the passive variables, we obtain a functor F_S in the active variables. The partial derived functors are then defined as the derived functors $R^k F_S$. See also functor, derived functor.

partial differential The rate of change of a function of more than one variable with respect to one of the variables while holding all of the other variables constant.

partial fraction An algebraic expression of the form

$$\sum_j \sum_{m=1}^{n_j} \frac{a_{jm}}{(z - \alpha_j)^m} \,.$$

partially ordered space Let X be a set. A relation on X that satisfies the conditions:
(i.) $x \leq x$ for all $x \in X$
(ii.) $x \leq y$ and $y \leq x$ implies $x = y$
(iii.) $x \leq y$ and $y \leq z$ implies $x \leq z$
is called a partial ordering.

partial pivoting An iterative strategy, using pivots, for solving the equation $Ax = b$, where A is an $n \times n$ matrix and b is an $n \times 1$ matrix. In the method of *partial pivoting,* to obtain the matrix A^k (where $A^0 = A$), the pivot is chosen to be the entry in the kth column of A^{k-1} at or below the diagonal with the largest absolute value.

partial product   Let $\{\alpha_n\}_{n=1}^\infty$ be a given sequence of numbers (or functions defined on a common domain Ω in $\mathbf{R}^n$ or $\mathbf{C}^n$) with terms $\alpha_n \neq 0$ for all $n \in \mathbf{N}$. The formal infinite product $\alpha_1 \cdot \alpha_2 \cdots$ is denoted by $\prod_{j=1}^\infty \alpha_j$. We call

$$P_n = \prod_{j=1}^n \alpha_j$$

its nth *partial product*.

peak point   *See* peak set.

peak set   Let $\mathcal{A}$ be an algebra of functions on a domain $\Omega \subset \mathbf{C}^n$. We call $p \in \overline{\Omega}$ a *peak point* for $\mathcal{A}$ if there is a function $f \in \mathcal{A}$ such that $f(p) = 1$ and $|f(z)| < 1$ for all $z \in \overline{\Omega} \setminus \{p\}$. The set $\mathcal{P}(\mathcal{A})$ of all peak points for the algebra $\mathcal{A}$ is called the *peak set* of $\mathcal{A}$.

Peirce decomposition   Let A be a semisimple Jordan algebra over a field F of characteristic 0 and let e be an idempotent of A. For $\lambda \in F$, let $A_e(\lambda) = \{a \in A : ea = \lambda a\}$. Then

$$A = A_e(1) \oplus A_e(1/2) \oplus A_e(0) .$$

This is called the *Peirce decomposition* of A, relative to E. If 1 is the sum of idempotents e_j, let $A_{j,k} = A_{e_j}(1)$ when $j = k$ and $A_{e_j} \cap A_{e_k}$ when $j \neq k$. These are called *Peirce spaces,* and $A = \oplus_{j \leq k} A_{j,k}$. *See also* Peirce space.

Peirce's left decomposition   Let e be an idempotent element of a ring R with identity 1. Then

$$R = Re \oplus R(1 - e)$$

expresses R as a direct sum of left ideals. This is called *Peirce's left decomposition.*

Peirce space   Suppose that the unity element $1 \in K$ can be represented as a sum of the mutually orthogonal idempotents e_j. Then, putting

$$A_{j,j} = A_{e_j}(1), \quad A_{j,k} = A_{e_j}(1/2) \cap A_{e_k}(1/2),$$

we have $A = \sum_{j \leq k} \oplus A_{j,k}$. Then $A_{j,k}$ are called *Peirce spaces.*

Peirce's right decomposition   Let e be an idempotent element of a unitary ring R, then

$1 - e$ and e are orthogonal idempotent elements, and

$$R = eR + (1 - e)R$$

is the direct sum of left ideals. This is called *Peirce's right decomposition.*

Pell's equation   The Diophantine equations $x^2 - ay^2 = \pm 4$ and ± 1, where a is a positive integer, not a perfect square, are called *Pell's equations.* The solutions of such equations can be found by continued fractions and are used in the determination of the units of rings such as $\mathbf{Z}[\sqrt{a}]$. This equation was studied extensively by Gauss. It can be regarded as a starting point of modern algebraic number theory.

When $a < 0$, then *Pell's equation* has only finitely many solutions. If $a > 0$, then all solutions x_n, y_n of Pell's equation are given by

$$\pm \left(\frac{x_1 + \sqrt{a}\, y_1}{2} \right)^n = \frac{(x_n + \sqrt{a}\, y_n)}{2} ,$$

provided that the pair x_1, y_1 is a solution with the smallest $x_1 + \sqrt{a}\, y_1 > 1$. Using continued fractions, we can determine x_1, y_1 explicitly.

penalty method of solving non-linear programming problem   A method to modify a constrained problem to an unconstrained problem. In order to minimize (or maximize) a function $\phi(x)$ on a set which has constraints (such as $f_1(x) \geq 0, f_2(x) \geq 0, \ldots f_m(x) \geq 0$), a penalty or penalty function, $\psi(x, a)$, is introduced (where a is a number), where $\psi(x, a) = 0$ if $x \in X$ or $\psi(x, a) > 0$ if $x \notin X$ and ψ involves $f_1(x) \geq 0, f_2(x) \geq 0, \ldots f_m(x) \geq 0$. Then, one minimizes (or maximizes) $\phi_a(x) = \phi(x) + \psi(x, a)$ without the constraints.

percent   *Percent* means hundredths. The symbol % stands for $\frac{1}{100}$. We may write a percent as a fraction with denominator 100. For example, $31\% = \frac{31}{100}, 55\% = \frac{55}{100}, \ldots$ etc. Similarly, we may write a fraction with denominator 100 as a percent.

perfect field   A field such that every algebraic extension is separable. Equivalently, a field F is perfect if each irreducible polynomial with coefficients in F has no multiple roots (in an algebraic closure of F). Every field of characteristic 0 is perfect and so is every finite field.

perfect power   An integer or polynomial which can be written as the nth power of another integer or polynomial, where n is a positive integer. For example, 8 is a perfect cube, because $8 = 2^3$, and $x^2 + 4x + 4$ is a perfect square, because $x^2 + 4x + 4 = (x + 2)^2$.

period matrix   Let R be a compact Riemann surface of genus g. Let $\omega_1, \ldots, \omega_g$ be a basis for the complex vector space of holomorphic differentials on R and let $\alpha_1, \ldots, \alpha_{2g}$ be a basis for the 1-dimensional integral homology of R. The period matrix $\mathcal{M}$ is the $g \times 2g$ matrix whose (i, j)-th entry is the integral of ω_j over α_i. The group generated by the $2g$ columns of $\mathcal{M}$ is a lattice in $\mathbf{C}^g$ and the quotient yields a g-dimensional complex torus called the Jacobian variety of R.

period of a periodic function   Let f be a function defined on a vector space V satisfying the relation

$$f(x + \omega) = f(x)$$

for all $x \in V$ and for some $\omega \in V$. The number ω is called a *period* of $f(x)$, and $f(x)$ with a period $\omega \neq 0$ is call a *periodic function*.

period relation   Conditions on an $n \times n$ matrix which help determine when a complex torus is an Abelian manifold. In $\mathbf{C}^n$, let Γ be generated by $(1, 0, \ldots, 0), (0, 1, 0, \ldots, 0), \ldots, (0, 0, \ldots, 0, 1)$, $(a_{11}, a_{12}, \ldots, a_{1n})$, $(a_{21}, a_{22}, \ldots, a_{2n})$, $\ldots (a_{n1}, a_{n2}, \ldots, a_{nn})$. Then $\mathbf{C}^n / \Gamma$ is an Abelian manifold if there are integers $d_1, d_2, \ldots,$ $d_n \neq 0$ such that, if $A = (a_{ij})$ and $D = (\delta_{ij}d_i)$, then (i.) AD is symmetric; and (ii.) $\Im(AD)$ is positive symmetric. Conditions (i.) and (ii.) are the *period relations*.

permanent   Given an $m \times n$ matrix $A = (a_{ij})$ with $m \leq n$, the *permanent* of A is defined by

$$\text{perm} A = \sum a_{1i_1} a_{2i_2} \ldots a_{mi_m} ,$$

where the summation is taken over all m-permutations $(i_1, i_2, \ldots, i_m)$ of the set $\{1, 2, \ldots, n\}$. When A is a square matrix, the permanent therefore has an expansion similar to that of the determinant, except that the factor corresponding to the sign of the permutation is missing from each summand.

permutation group   Let A be a finite set with $\#(A) = n$. The *permutation group* on n elements is the set S_n consisting of all one-to-one functions from A onto A under the group law:

$$f \cdot g = f \circ g$$

for $f, g \in S_n$. Here $\circ$ denotes the composition of functions.

permutation matrix   An $n \times n$ matrix P, obtained from the identity matrix I_n by permutations of the rows (or columns). It follows that a *permutation matrix* has exactly one nonzero entry (equal to 1) in each row and column. There are $n!$ permutation matrices of size $n \times n$. They are orthogonal matrices, namely, $P^T P = P P^T = I_n$ (i.e., $P^T = P^{-1}$). Multiplication from the left (resp., right) by a permutation matrix permutes the rows (resp., columns) of a matrix, corresponding to the original permutation.

permutation representation   A *permutation representation* of a group G is a homomorphism from G to the group S_X of all permutations of a set X. The most common example is when $X = G$ and the permutation of G obtained from $g \in G$ is given by $x \to gx$ (or $x \to xg$, depending on whether a product of permutations is read right-to-left or left-to-right).

Peron-Frobenius Theorem   *See* Frobenius Theorem on Non-Negative Matrices.

Perron's Theorem of Positive Matrices   If A is a positive $n \times n$ matrix, A has a positive real eigenvalue λ with the following properties: (i.) λ is a simple root of the characteristic equation. (ii.) λ has a positive eigenvector $\vec{\mathbf{u}}$. (iii.) If μ is any other eigenvalue of A, then $|\mu| < \lambda$.

Peter-Weyl theory   Let G be a compact Lie group and let $C(G)$ be the commutative associative algebra of all complex valued continuous functions defined on G. The multiplicative law defined on $C(G)$ is just the usual composition of functions. Denote

$$s(G) = \left\{ f \in C(G) : \dim \sum_{g \in G} CL_g f < \infty \right\}$$

where $L_g f = f(g \cdot)$. The *Peter-Weyl theory* tells us that the subalgebra $s(G)$ is everywhere dense in $C(G)$ with respect to the uniform norm $\|f\|_\infty = \max_{g \in G} |f(g)|$.

Pfaffian differential form　The name given to the expression

$$dW = \sum_{i=1}^{n} X_i \, dx_i .$$

p-group　A group G such that the order of G is p^n, where p is a prime number and n is a non-negative integer.

Picard-Lefschetz transform　Let W be a local system attached to the monodromy representation $\varphi_p : \pi_1(U, 0) \to GL(H^p(W, \mathbf{Q}))$. For each point t_j there corresponds a cycle δ_j of $H^{n-1}(W, \mathbf{Q})$ called a vanishing cycle such that if γ_j is a loop based at 0 going once around t_j, we have for each $x \in H^{n-1}(W, \mathbf{Q})$,

$$\varphi_p\left(\gamma_j\right)(x) = x \pm \left(x, \delta_j\right) \delta_j .$$

Picard number　Let V be a complete normal variety and let $\mathcal{D}(V)$, $\mathcal{D}_a(V)$ be the group of divisors and group of divisors algebraically equivalent to zero, respectively. The rank of the quotient group $NS(V) = \mathcal{D}(V)/\mathcal{D}_a(V)$ is called the *Picard number* of V.

Picard scheme　Let $\mathcal{O}_V^*$ be the sheaf of multiplicative group of the invertible elements in $\mathcal{O}_V$. The group of linear equivalence classes of Cartier divisors can be identified with $H^1(V, \mathcal{O}_V)$. From this point of view, we can generalize the theory of the Picard variety to the case of schemes. The theory thus obtained is called the theory of *Picard schemes*.

Picard's Theorem　There are two important theorems in one complex variable proved by the French mathematician Charles Émile Picard

(1856–1941). The first Picard theorem was proved in 1879: *An entire function which is not a polynomial takes every value, with one possible exception, an infinity of times.*

The second Picard theorem was proved in 1880: *In a neighborhood of an isolated essential singularity, a single-valued, holomorphic function takes every value, with one possible exception, an infinity of times. In other words, if $f(z)$ is holomorphic for $0 < |z - z_0| < r$, and there are two unequal numbers a, b, such that $f(z) \neq a$, $f(z) \neq b$, for $|z - z_0| < r$, then z_0 is not an essential singularity.*

Picard variety　Let V be a complete normal variety. The factor group of the divisors on V, algebraically equivalent to 0 modulo the group of divisors linearly equivalent to 0, has a natural canonical structure of an Abelian variety, called the *Picard variety*.

Picard-Vessiot theory　One of two main theories of differential rings and fields. *See* Galois theory of differential fields. The *Picard-Vessiot theory* deals with linear homogeneous differential equations.

pi-group　Let π be a set of prime numbers and let π' be the set of prime numbers not in π. A π-group is a finite group whose order is a product of primes in π. A finite group is π-solvable if every Jordan-Hölder factor is either a π'-group or a solvable π-group. For a π-solvable group G, define a series of subgroups

$$1 = P_0 \subseteq N_0 \subset P_1 \subset N_1 \cdots \subset P_n \subseteq N_n = G$$

such that P_j/N_{j-1} is a maximal normal π'-subgroup of G/P_j. This is called the π-series of G and n is called the π-length of G.

pi-length　*See* pi-group.

pi-series　*See* pi-group.

pi-solvable group　A finite group G such that the order of each composition factor of G is either an element of a collection, π, of prime numbers or mutually prime to any element of π.

pivot　*See* Gaussian elimination.

pivoting   *See* Gaussian elimination.

place   A mapping $\phi : K \to \{F, \infty\}$, where K and F are fields, such that, if $\phi(a)$ and $\phi(b)$ are defined, then $\phi(a+b) = \phi(a)+\phi(b)$, $\phi(ab) = \phi(a)\phi(b)$ and $\phi(1) = 1$.

place value   The value given to a digit, depending on that digit's position in relation to the units place. For example, in 239.71, 9 represents 9 units, 3 represents 30 units, 2 represents 200 units, 7 represents $\frac{7}{10}$ units and 1 represents $\frac{1}{100}$ units.

Plancherel formula   Let G be a unimodular locally compact group and $\hat{G}$ be its quasidual. Let U be a unitary representation of G and U^* be its adjoint. For any $f, g \in L^1(G) \cap L^2(G)$, the *Plancherel formula*

$$\int_G f(x)\overline{g(x)}\, dx = \int_{\hat{G}} t(U_g^*(\xi)U_f(\xi))\, d\mu(\xi)$$

holds, where $U_f(\xi) = \int_{\hat{G}} f(x)U_x(\xi)dx$. The measure μ is called the *Plancherel measure.*

plane trigonometry   *Plane trigonometry* is related to the study of triangles, which were studied long ago by the Babylonians and ancient Greeks. The word trigonometry is derived from the Greek word for "the measurement of triangles." Today trigonometry and trigonometric functions are indispensable tools not only in mathematics, but also in many practical applications, especially those involving oscillations and rotations.

Plücker formulas   Let m be the class, n the degree, and δ, χ, i, and τ be the number of nodes, cups, inflections, tangents, and bitangents. Then

$$n(n-1) = m + 2\delta + 3\chi$$
$$m(m-1) = n + 2\tau + 3i$$
$$3n(n-2) = i + 6\delta + 8\chi$$
$$3m(m-2) = \chi + 6\tau + 8i$$
$$3(m-n) = i - \chi .$$

plurigenera   For an algebraic surface S with a canonical divisor K of S, the collection of numbers $P_i = \ell(iK)$, $(i = 2, 3, \dots)$. The plurigenera P_i, $(i = 2, 3, \dots)$ are the same for any two birationally equivalent nonsingular surfaces.

plus sign   The symbol "+" indicating the algebraic operation of addition, as in $a + b$.

Poincaré   Let R be a commutative ring with unit. Let U be an orientation over R of a compact n-manifold X with boundary. Then for all indices q and R-modules G there is an isomorphism

$$\gamma_U : H_q(X; G) \approx H^{n-q}(X; G).$$

This is called *Poincaré-Lefschetz duality*. The analogous result for a manifold X without boundary is called *Poincaré duality*.

Poincaré-Birkhoff-Witt Theorem   Let $\mathcal{G}$ be a Lie algebra over a number field K. Let $X_1, \dots, X_n$ be a basis of $\mathcal{G}$, and let $\mathcal{R} = K[Y_1, \dots, Y_n]$ be a polynomial ring on K in n indeterminates $Y_1, \dots, Y_n$. Then there exists a unique algebra homomorphism $\psi : \mathcal{R} \to \mathcal{G}$ such that $\psi(1) = 1$ and $\omega(Y_j) = X_j$, $j = 1, \dots, n$. Moreover, ψ is bijective, and the jth homogeneous component $\mathcal{R}^j$ is mapped by ψ onto $\mathcal{G}^j$. Thus, the set of monomials $\{X_1^{k_1} X_2^{k_2} \dots X_n^{k_n}\}$, $k_1, \dots, k_n \geq 0$, forms a basis of $U(\mathcal{G})$ over K. This is the so-called *Poincaré-Birkhoff-Witt Theorem*. Here $U(\mathcal{G}) = T(\mathcal{G})/J$ is the quotient associative algebra of $\mathcal{G}$ where J is the two-sided ideal of $T(\mathcal{G})$ generated by all elements of the form $X \otimes Y - Y \otimes X - [X, Y]$ and $T(\mathcal{G})$ is the tensor algebra over $\mathcal{G}$.

Poincaré differential invariant   Let $w = \alpha(z - z_\circ)/(1 - \bar{z}_\circ z)$ with $|\alpha| = 1$ and $|z_\circ| < 1$, be a conformal mapping of $|z| < 1$ onto $|w| < 1$. Then the quantity $|dw|/(1 - |w|^2) = |dz|/(1 - |z|^2)$ is called *Poincaré's differential invariant*. The disk $\{|z| < 1\}$ becomes a non-Euclidean space using any metric with $ds = |dz|/(1 - |z|^2)$.

Poincaré duality   Any theorem generalizing the following: Let M be a compact n-dimensional manifold without boundary. Then, for each p, there is an isomorphism $H^p(M; \mathbf{Z}_2) \cong H_{n-p}(M; \mathbf{Z}_2)$. If, in addition, M is assumed to be orientable, then $H^p(M) \cong H_{n-p}(M)$.

Poincaré-Lefschetz duality   Let R be a commutative ring with unit. Let U be an orientation over R of a compact n-manifold X with boundary. Then for all indices q and R-modules G there is an isomorphism

$$\gamma_U : H_q(X; G) \approx H^{n-q}(X; G) .$$

This is called *Poincaré-Lefschetz duality*. The analogous result for a manifold X without boundary is called *Poincaré duality*.

Poincaré metric   The hermitian metric

$$ds^2 = \frac{2}{(1 - |z|^2)^2} \, dz \wedge d\overline{z}$$

is called the Poincaré metric for the unit disc in the complex plane.

Poincaré's Complete Reducibility Theorem A theorem which says that, given an Abelian variety A and an Abelian subvariety X of A, there is an Abelian subvariety Y of A such that A is isogenous to $X \times Y$.

point at infinity   The point in the extended complex plane, not in the complex plane itself. More precisely, let us consider the unit sphere in $\mathbf{R}^3$:

$$S = \left\{ (x_1, x_2, x_3) \in \mathbf{R}^3 : x_1^2 + x_2^2 + x_3^2 = 1 \right\} ,$$

which we define as the extended complex numbers. Let $N = (0, 0, 1)$; that is, N is the north pole on S. We regard $\mathbf{C}$ as the plane $\{(x_1, x_2, 0) \in \mathbf{R}^3 : x_1, x_2 \in \mathbf{R}\}$ so that $\mathbf{C}$ cuts S along the equator. Now for each point $z \in \mathbf{C}$ consider the straight line in $\mathbf{R}^3$ through z and N. This intersects the sphere in exactly one point $Z \neq N$. By identifying $Z \in S$ with $z \in \mathbf{C}$, we have S identified with $\mathbf{C} \cup \{N\}$. If $|z| > 1$ then Z is in the upper hemisphere and if $|z| < 1$ then z is in the lower hemisphere; also, for $|z| = 1$, $Z = z$. Clearly Z approaches N when $|z|$ approaches ∞. Therefore, we may identify N and the point ∞ in the extended complex plane.

pointed co-algebra   Let V be a co-algebra. A nonzero subco-algebra W of V is said to be simple if W has no nonzero proper subco-algebra. The co-algebra V is called a *pointed co-algebra* if all of its simple subco-algebras are one-dimensional. *See* coalgebra.

pointed set   Denoted by (X, p), a set X where p is a member of X.

polar decomposition   Every $n \times n$ matrix A with complex entries can be written as $A = PU$, where P is a positive semidefinite matrix and U is a unitary matrix. This factorization of A is called the *polar decomposition* of the polar form of A.

polar form of a complex number   Let $z = x + iy$ be a complex number. This number has the *polar representation*

$$z = x + iy = r(\cos \theta + i \sin \theta)$$

where $r = \sqrt{x^2 + y^2}$ and $\theta = \tan^{-1}\left(\frac{y}{x}\right)$.

polarization   Let A be an Abelian variety and let X be a divisor on A. Let X' be a divisor on A such that $m_1 X \equiv m_2 X'$ for some positive integers m_1 and m_2. Let $\mathcal{X}$ be the class of all such divisors X'. When $\mathcal{X}$ contains positive nondegenerate divisors, we say that $\mathcal{X}$ determines a *polarization* on A.

polarized Abelian variety   Suppose that V is an Abelian variety. Let X be a divisor on V and let $D(X)$ denote the class of all divisors Y on V such that $mX \equiv nY$, for some integers $m, n > 0$. Further, suppose that $D(X)$ determines a polarization of V. Then the couple $(V, D(X))$ is called a *polarized Abelian variety*. *See also* Abelian variety, divisor, polarization.

pole   Let $z = a$ be an isolated singularity of a complex-valued function f. We call a a *pole* of f if

$$\lim_{z \to a} |f(z)| = \infty .$$

That is, for any $M > 0$ there is a number $\varepsilon > 0$, such that $|f(z)| \geq M$ whenever $0 < |z - a| < \varepsilon$. Usually, the function f is assumed to be holomorphic, in a punctured neighborhood $0 < |z - a| < \epsilon$.

pole divisor   Suppose X is a smooth affine variety of dimension r and suppose $Y \subset X$ is a subvariety of dimension $r - 1$. Given $f \in C(X) \setminus (0)$, let $\text{ord}_Y f < 0$ denote the order of vanishing of f on Y. Then $(f) = \sum_Y (\text{ord}_Y f) \cdot Y$ is called a pole divisor of f in Y. *See also* smooth affine variety, subvariety, order of vanishing.

polynomial   If $a_0, a_1, \ldots, a_n$ are elements of a ring R, and x does not belong to R, then

$$a_0 + a_1 x + \cdots + a_n x^n$$

is a polynomial.

polynomial convexity   Let $\Omega \subseteq C^n$ be a domain (a connected open set). If $E \subseteq \Omega$ is a subset, then define

$$\widehat{E} = \{ z \in \Omega : |p(z)| \leq \sup_{w \in E} |p(w)|$$

for all p a polynomial$\}$.

The set $\widehat{E}$ is called the *polynomially convex hull* of E in Ω. If the implication $E \subset\subset \Omega$ implies $\widehat{E} \subset\subset \Omega$ always holds, then Ω is said to be polynomially convex.

polynomial equation   An equation $P = 0$ where P is a polynomial function of one or more variables.

polynomial function   A function which is a finite sum of terms of the form $a_n x^n$, where n is a nonnegative integer and a_n is a real or complex number.

polynomial identity   An equation $P(X_1, X_2, \ldots, X_n) = 0$ where P is a polynomial in n variables with coefficients in a field K such that $P(a_1, a_2, \ldots, a_n) = 0$ for all a_i in an algebra A over K.

polynomial in m variables   A function which is a finite sum of terms $a x_1^{n_1} x_2^{n_2} \ldots x_m^{n_m}$, where $n_1, n_2, \ldots, n_m$ are nonnegative integers and a is a real or complex number. For example, $5x^2 y^3 + 3x^4 z - 2z + 3xyz$ is a polynomial in three variables.

polynomial ring   Let R be a ring. The set $R[X]$ of all polynomials in an indeterminate X with coefficients in R is a ring with respect to

the usual addition and multiplication of polynomials. The ring $R[X]$ is called the *polynomial ring* of X over R.

polynomial ring in m variables   Let R be a ring and let $X_1, X_2, \ldots, X_m$ be indeterminates. The set $R[X_1, X_2, \ldots, X_m]$ of all polynomials in $X_1, X_2, \ldots, X_m$ with coefficients in R is a ring with respect to the usual addition and multiplication of polynomials and is called the *polynomial ring in m variables* $X_1, X_2, \ldots, X_m$ over R.

Pontrjagin class   Let $\mathcal{F}$ be a complex PL sheaf over a PL manifold M. The total *Pontrjagin class* $p([\mathcal{F}]) \in H^{4*}(M; R)$ of a coset $[\mathcal{F}]$ of real PL sheaves via complexification of $[\mathcal{F}]$ satisfies these axioms:
(i.) If $[\mathcal{F}]$ is a coset of real PL sheaves of rank m on a PL manifold M, then the total Pontrjagin class $p([\mathcal{F}])$ is an element $1 + p_1([\mathcal{F}]) + \cdots + p_{[m/2]}([\mathcal{F}])$ of $H^*(M; R)$ with $p_i([\mathcal{F}]) \in H^{4i}(M; R)$;
(ii.) $p(\Xi^!{[\mathcal{F}]}) = \Xi^* p([\mathcal{F}]) \in H^{4*}(N; R)$ for any PL map $\Xi : N \to M$;
(iii.) $p([\mathcal{F}] \oplus [\mathcal{G}]) = p([\mathcal{G}])$ for any cosets $[\mathcal{F}]$ and $[\mathcal{G}]$ over M;
(iv.) If $[\mathcal{F}]$ contains a *bona fide* real vector bundle ξ over M, then $p([\mathcal{F}])$ is the classical total Pontrjagin class $p(\xi) \in H^{4*}(M; R)$.

Pontryagin multiplication   A multiplication

$$h_* : H_*(X) \otimes H_*(X) \to H_*(X) .$$

($H_*(X)$ are homology groups of the topological space X.)

Pontryagin product   The result of Pontryagin multiplication. *See* Pontryagin multiplication.

positive angle   Given a vector $\vec{v} \neq \vec{0}$ in R^n, then its direction is described completely by the *angle* α between $\vec{v}$ and $\vec{i} = (1, 0, \ldots, 0)$, the unit vector in the direction of the positive x_1-axis. If we measure the angle α counterclockwise, we say α is a *positive angle*. Otherwise, α is a negative angle.

positive chain complex   A chain complex X such that the only possible non-zero terms X_n are those X_n for which $n \geq 0$.

positive cycle   An r-cycle $A = \sum n_i A_i$ such that $n_i \geq 0$ for all i, where A_i is not in the singular locus of an irreducible variety V for all i.

positive definite function   A complex valued function f on a locally compact topological group G such that

$$\int_G f(s-t)\phi(s)\overline{\phi(t)}\,ds\,dt \geq 0$$

for every ϕ, continuous and compactly supported on G.

positive definite matrix   An $n \times n$ matrix A, such that, for all $\vec{u} \in \mathbf{R}^n$, we have

$$(A(\vec{u}), \vec{u}) \geq 0 \,,$$

with equality only when $\vec{u} = 0$.

positive divisor   A divisor that has only positive coefficients.

positive element   An element $g \in G$, where G is an ordered group, such that $g \geq e$.

positive exponent   For an expression a^b, the exponent b if $b > 0$.

positive matrix   An $n \times n$ matrix A with real entries such that $a_{jk} > 0$ for each j and k. *See also* positive definite matrix.

positive number   A real number greater than zero.

positive root   Let S be a basis of a root system ϕ in a vector space V such that each root β can be written as $\beta = \sum_{a \in S} m_a a$, where the integers m_a have the same sign. Then β is a positive root if all $m_a \geq 0$.

positive semidefinite matrix   An $n \times n$ matrix A such that, for all $\vec{u} \in \mathbf{R}^n$, we have

$$(A(\vec{u}), \vec{u}) \geq 0 \,.$$

positive Weyl chamber   The set of $\lambda \in V^*$ such that $(\beta, \lambda) > 0$ for all positive roots β, where V is a vector space over a subfield R of the real numbers.

power   Let $a_1, \ldots, a_n$ be a finite sequence of elements of a monoid M. We define the "product" of $a_1, \ldots, a_n$ by the following: we define $\prod_{j=1}^{1} a_j = a_1$, and

$$\prod_{j=1}^{k+1} a_j = \left(\prod_{j=1}^{k} a_j\right) a_{k+1} \,.$$

Then

$$\prod_{j=1}^{k} a_j \prod_{\ell=1}^{m} a_{k+\ell} = \prod_{j=1}^{k+m} a_j \,.$$

If all the $a_j = a$, we denote $a_1 \cdot a_2 \ldots a_n$ as a^n and call this the nth *power* of a.

power associative algebra   A distributive algebra A such that every element of A generates an associative subalgebra.

power method of computing eigenvalues
An iterative method for determining the eigenvalue of maximum absolute value of an $n \times n$ matrix A. Let $\lambda_1, \lambda_2, \ldots, \lambda_n$ be eigenvalues of A such that $|\lambda_1| > |\lambda_2| \geq \cdots \geq |\lambda_n|$ and let y_1 be an eigenvector such that $(\lambda_1 I - A)y_1 = 0$. Begin with a vector $x^{(0)}$ such that $(y_1, x^{(0)}) \neq 0$ and for some i_0, $x_{i_0}^{(0)} = 1$. Determine $\theta^{(0)}, \theta^{(1)}, \ldots, \theta^{(m)}, \ldots$ and $x^{(1)}, x^{(2)}, \ldots, x^{(m+1)}, \ldots$ by $Ax^{(j)} = \theta^{(j)}x^{(j+1)}$. Then $\lim_{j \to \infty} \theta^{(j)} = \lambda_1$ and $\lim_{j \to \infty} x^{(j)}$ is the eigenvector corresponding to λ_1.

power of a complex number   Let $z = x + iy = r(\cos\theta + i\sin\theta)$ be a complex number with $r = \sqrt{x^2 + y^2}$ and $\theta = \tan^{-1}\frac{y}{x}$. Let n be a positive number. The nth power of z will be the complex number $r^n(\cos n\theta + i\sin n\theta)$.

power-residue symbol   Let n be a positive integer and let K be an algebraic number field containing the nth roots of unity. Let $\alpha \in K^\times$ and let $\wp$ be a prime ideal of the ring such that $\wp$ is relatively prime to n and α. The nth power is a positive integer and let K be an algebraic number field containing the nth roots of unity.

Let $\alpha \in K^\times$ and let $\wp$ be a prime ideal of the ring of integers of K such that $\wp$ is relatively prime to n and α. The nth power residue symbol $\left(\dfrac{\alpha}{\wp}\right)_n$ is the unique nth root of unity that is congruent to $\alpha^{(N\wp-1)/n} \bmod \wp$. When $n = 2$ and $K = \mathbf{Q}$, this symbol is the usual quadratic residue symbol.

predual   Let X and Y be Banach spaces such that X is the dual of Y, $X = Y^*$. Then Y is called the *predual* of X.

preordered set   A structure space for a non-empty set R is a nonempty collection $\mathcal{X}$ of non-empty proper subsets of R given the hull-kernel topology. If there exists a binary operation $*$ on R such that $(R, *)$ is a commutative semigroup and the structure space $\mathcal{X}$ consists of prime semigroup ideals, then it is said that R has an $\mathcal{X}$-compatible operation. For $p \in R$, let $\mathcal{X}_p = \{A \in \mathcal{X} : p \notin A\}$. A preorder (reflexive and transitive relation) $\leq$ is defined on R by the rule that $a \leq b$ if and only if $\mathcal{X}_a \subseteq \mathcal{X}_b$. Then R is called a *preordered set*.

preordered set   A structure space for a non-empty set R is a nonempty collection $\mathcal{X}$ of non-empty proper subsets of R given the hull-kernel topology. If there exists a binary operation $*$ on R such that $(R, *)$ is a commutative semigroup and the structure space $\mathcal{X}$ consists of prime semigroup ideals, then it is said that R has an $\mathcal{X}$-compatible operation. For $p \in R$, let $\mathcal{X}_p = \{A \in \mathcal{X} : p \notin A\}$. A preorder (reflexive and transitive relation) $\leq$ is defined on R by the rule that $a \leq b$ if and only if $\mathcal{X}_a \subseteq \mathcal{X}_b$. Then R is called a preordered set.

presheaf   Let X be a topological space. Suppose that, for each open subset U of X, there is an Abelian group (or ring, module, etc.) $\mathcal{F}(U)$. Assume $\mathcal{F}(\phi) = 0$. In addition, suppose that whenever $U \subseteq V$ there is a homomorphism

$$\rho_{UV} : \mathcal{F}(V) \to \mathcal{F}(U)$$

such that $\rho_{UU} = $ identity and such that $\rho_{UW} = \rho_{UV}\rho_{VW}$ whenever $U \subseteq V \subseteq W$. The collection of Abelian groups along with the homomorphisms ρ_{UV} is called a *presheaf* of Abelian groups (or rings, modules, etc.). There is a standard procedure for constructing a sheaf from a presheaf.

primary Abelian group   An Abelian group in which the order of every element is a power of a fixed prime number.

primary component   Let R be a commutative ring with identity 1 and let J be an ideal of R. Assume $J = I_1 \cap \cdots \cap I_n$ with each I_i primary and with n minimal among all such representations. Then each I_i is called a primary component of J.

primary ideal   Let R be a ring with identity 1. An ideal I of R is called *primary* if $I \neq R$ and all zero divisors of R/I are nilpotent.

primary linear programming problem   A linear programming problem in which the goal is to maximize the linear function $z = \mathbf{c}x$ with the linear conditions $\sum_{j=1}^n a_{ij}x_j = b_i$ ($i = 1, 2, \ldots, m$) and $\mathbf{x} \geq 0$, where $\mathbf{x} = (x_1, x_2, \ldots, x_n)$ is the unknown vector, $\mathbf{c}$ is an $n \times 1$ vector of real numbers, b_i ($i = 1, 2, \ldots, n$) and a_{ij} ($i = 1, 2, \ldots, n, j = 1, 2, \ldots, n$) are real numbers.

primary ring   Let R be a ring and let N be the largest ideal of R containing only nilpotent elements. If R/N is nonzero and has no nonzero proper ideals, R is called *primary*.

primary submodule   Let R be a commutative ring with identity 1. Let M be an R-module. A submodule N of M is called *primary* if whenever $r \in R$ is such that there exists $m \in M/N$ with $m \neq 0$ but $rm = 0$, then $r^n(M/N) = 0$ for some integer n.

prime   A positive integer greater than 1 with the property that its only divisors are 1 and itself. The numbers $2, 3, 5, 7, 11, 13, 17, 19, 23, 27$ are the first ten primes. There are infinitely many prime numbers.

prime divisor   For an integer n, a *prime divisor* is a prime that occurs in the prime factorization of n. For an algebraic number field or for an algebraic function field of one variable (a field that is finitely generated and of transcendence degree 1 over a field K), a *prime divisor* is an equivalence class of nontrivial valuations (over K in the latter case). In the number field case, the prime divisors correspond to the nonzero prime ideals of the ring of integers and the archimedean valuations of the field.

prime element In a commutative ring R with identity 1, a *prime element* p is a nonunit such that if p divides a product ab with $a, b \in R$, then p divides at least one of a, b. When $R = \mathbf{Z}$, the prime elements are of the form $\pm p$ for prime numbers p.

prime factor A *prime factor* of an integer n is a prime number p such that n is a multiple of p.

prime field The rational numbers and the fields $\mathbf{Z}/p\mathbf{Z}$ for prime numbers p are called *prime fields*. Every field contains a unique subfield isomorphic to exactly one of these prime fields.

prime ideal Let R be a commutative ring with identity 1 and let $I \neq R$ be an ideal of R. Then I is prime if whenever $a, b \in R$ are such that $ab \in I$, then at least one of a and b is in I.

prime number A positive integer p is said to be *prime* if
(i.) $p > 1$,
(ii.) p has no positive divisors except 1 and p. The first few prime numbers are 2, 3, 5, 7, 11, 13, 17.

prime rational divisor A divisor $p = \sum n_i P_i$ on X over k satisfying the following three conditions: (i.) p is invariant under any automorphism σ of $\bar{k}/k$; (ii.) for any j, there exists an automorphism σ_j of $\bar{k}/k$ such that $P_j = P_1^{\sigma_j}$; (iii.) $n_1 = \cdots = n_t = [k(P_1) : k]_i$, where X is a nonsingular irreducible complete curve, k is a subfield of the universal domain K such that X is defined over k. Prime rational divisors generate a subgroup of the group of divisors $G(X)$, which is called a group of k-rational divisors.

primitive character Let G be a finite group and let $\mathrm{Irr}(G)$ denote the set of all irreducible

C-characters of G, where $\mathbf{C}$ denotes the field of complex numbers. Then $\chi \in \mathrm{Irr}(G)$ is called a primitive character if $\chi \neq \varphi^G$ for any character φ of a proper subgroup of G. *See also* character of group, irreducible character.

primitive element Let E be an extension field of the field F (E is a field containing F as subfield). If u is an element of E and x is an indeterminate, then we have the homomorphism $g(x) \rightarrow g(u)$ of the polynomial ring $F[x]$ into E, which is the identity on F and send $x \rightarrow u$. If the kernel is 0, then $F[u] \cong F[x]$. Otherwise, we have a monic polynomial $f(x)$ of positive degree such that the kernel is the principal ideal $(f(x))$, and then $F[u] \cong F[x]/(f(x))$. Then we say $E = F(u)$ is a simple extension of F and u a *primitive element* (= field generator of E/F).

primitive equation An equation $f(X) = 0$ such that a permutation of roots of $f(X) = 0$ is primitive, where $f(X) \in K[X]$ is a polynomial, and K is a field.

primitive hypercubic set A finite subgroup K of the orthogonal group $O(V)$ is called *fully transitive* if there is a set $S = \{e_1, \ldots, e_s\}$ that spans V on which K acts transitively and K has no invariant subspace in V. In this case, one can choose S as either
(i.) *the primitive hypercubic type:*

$$S = \{e_1, \ldots, e_n\}, \quad (e_i, e_j) = \delta_{ij} ;$$

or
(ii.) *the primitive hyperbolic type:*

$$S = \{f_1, \ldots, f_{n+1}\} ,$$

$$(f_i, f_j) = \left\{ \begin{array}{ll} 1, & i=1,\ldots n+1, i=j \\ -\frac{1}{n}, & i,j=1,\ldots,n+1, i \neq j \end{array} \right. .$$

primitive ideal Let R be a Banach algebra. A two-sided ideal I of R is *primitive* if there is a regular maximal left ideal J such that I is the set of elements $r \in R$ with $rR \subseteq I$. The regularity of J means that there is an element $u \in R$ such that $r - ru \in J$ for all $r \in R$.

primitive idempotent element An idempotent element that cannot be expressed as a sum

$a + b$ with a and b nonzero idempotents satisfying $ab = ba = 0$.

primitive permutation representation   Let G be a group acting as a group of permutations of a set X. This is called a permutation representation of G. This representation is called *primitive* if the only equivalence relations $R(x, y)$ on X such that $R(x, y)$ implies $R(gx, gy)$ for all $x, y \in X$ and all $g \in G$ are equality and the trivial relation $R(x, y)$ for all $x, y \in X$.

primitive polynomial   Let $f(x)$ be a polynomial with coefficients in a commutative ring R. When R is a unique factorization domain, $f(x)$ is called *primitive* if the greatest common divisor of the coefficients of $f(x)$ is 1. For an arbitrary ring, a slightly different definition is sometimes used: $f(x)$ is *primitive* if the ideal generated by the coefficients of $f(x)$ is R.

primitive ring   A ring R is called *left primitive* if there exists an irreducible, faithful left R-module, and R is called *right primitive* if there exists an irreducible, faithful right R-module. *See also* irreducible R-module, faithful R-module.

primitive root of unity   Let m be a positive integer and let R be a ring with identity 1. An element $\zeta \in R$ is called a primitive mth root of unity if $\zeta^m = 1$ but $\zeta^k \neq 1$ for all positive integers $k < m$.

primitive transitive permutation group   Let G be a transitive group of permutations of a set X. If the stabilizer of each element of X is a maximal subgroup of G, then G is called *primitive*.

principal adele   Let K be an algebraic number field and let $\mathbf{A}_K$ be the adeles of K. The image of the diagonal injection of K into $\mathbf{A}_K$ is the set of principal adeles.

principal antiautmorphism   A unique antiautomorphism β of a Clifford algebra $C(Q)$ such that $\beta(x) = x$, for all $x \in V$, where $C(Q) = T(V)/I(Q)$, V is an n-dimensional linear space over a field K, and Q is a quadratic form on V, $T(V)$ is the tensor algebra

over V, and $I(Q)$ is the two-sided ideal of $T(V)$ generated by elements $x \otimes x - Q(x) \cdot 1$ for $x \in V$. *Compare with* principal automorphism, i.e., the unique automorphism α of $C(Q)$ such that $\alpha(x) = -x$, for all $x \in V$.

principal automorphism   Let A be a commutative ring and let M be a module over A. Let $a \in A$. The homomorphism

$$M \ni x \mapsto ax$$

is called the principal homomorphism associated with a, and is denoted a_M. When a_M is one-to-one and onto, then we call a_M a *principal automorphism* of the module M.

principal divisor of functions   The formal sum

$$(\phi) = m_1 p_1 + \cdots + m_j p_j + n_1 q_1 + \cdots + n_k q_k$$

where $p_1, \ldots, p_j$ are the zeros and $q_1, \ldots, q_k$ are the poles of a meromorphic function ϕ, m_i is the order of p_i and n_i is the order of q_i.

principal genus   An ideal group of K formed by the set of all ideals $\mathcal{U}$ of K relatively prime to m such that $N_{K/k}(\mathcal{U})$ belongs to $H(m)$, where k is an algebraic number field, m is an integral divisor of k, $\mathcal{T}(m)$ is the multiplicative group of all fractional ideals of k which are relatively prime to m, $S(m)$ is the ray modulo m, $H(m)$ is an ideal group modulo m (i.e., a subgroup of $\mathcal{T}(m)$ containing $S(m)$), and K/k is a Galois extension.

principal H-series   An H-series Σ which is strictly decreasing and such that there exists no normal series distinct from Σ, finer than Σ, and strictly decreasing. *See also* H-series, normal series, finer.

principal ideal   Let R be a commutative ring with identity 1. A *principal ideal* is an ideal of the form $aR = \{ar | r \in R\}$ for some $a \in R$.

principal ideal domain   An integral domain in which every ideal is principal. *See* principal ideal.

principal ideal ring   A ring in which every ideal is principal. *See* principal ideal.

Principal Ideal Theorem   There are at least two results having this name:

(1) Let K be an algebraic number field and let H be the Hilbert class field of K. Every ideal of the ring of integers of K becomes principal when lifted to an ideal of the ring of integers of H. This was proved by Furtwängler in 1930.

(2) Let R be a commutative Noetherian ring with 1. If $x \in R$ and P is minimal among the prime ideals of R containing x, then the codimension of P is at most 1 (that is, there is no chain of prime ideals $P \supset P_1 \supset P_2$ (strict inclusions) in R). This was proved by Krull in 1928.

principal idele   Let K be an algebraic number field. The multiplicative group $K^\times$ injects diagonally into the group $\mathbf{I}_K$ of ideles. The image is called the set of *principal ideles*.

principal matrix   Suppose $A = [A_{ij}]$ is an $n \times n$ matrix. The principal matrices associated with A are $A(k) = [A_{ij}], 1 \le i, j \le k \le n$.

principal minor   *See* principal submatrix.

principal order   Let K be a finite extension of the rational field $\mathbf{Q}$. The ring of all algebraic integers in K is called the *principal order* of K.

principal root   A root with largest real part (if this root is unique) of the characteristic equation of a differential-difference equation.

principal series   For a semisimple Lie group, those unitary representations induced from finite dimensional unitary representations of a minimal parabolic subgroup.

principal solution   A solution $F(x)$ of the equation $\Delta F(x)/\Delta x = g(x)$, where $\Delta F(x) = F(x + \Delta x) - F(x)$. Such a solution $F(x)$ can be obtained by a formula in terms of integral, series, and limits.

principal submatrix   A submatrix of an $m \times n$ matrix A is an $(m - k) \times (n - \ell)$ matrix obtained from A by deleting certain k rows ($k < m$) and ℓ columns ($\ell < n$) of A. If $m = n$ and if the set of deleted rows coincides with the set of deleted columns, we call the submatrix obtained a *prin-*

cipal submatrix of A. Its determinant is called a principal minor of A. For example, let

$$A = \begin{pmatrix} a_{11} & a_{12} & a_{13} \\ a_{21} & a_{22} & a_{23} \\ a_{31} & a_{32} & a_{33} \end{pmatrix}.$$

Then, by deleting row 2 and column 2 we obtain the principal submatrix of A

$$\begin{pmatrix} a_{11} & a_{13} \\ a_{31} & a_{33} \end{pmatrix}.$$

Notice that the diagonal entries and A itself are principal submatrices of A.

principal value   (1) The *principal values* of arcsin, arccos, and arctan are the inverse functions of the functions $\sin x$, $\cos x$, and $\tan x$, restricted to the domains $-\frac{\pi}{2} \le x \le \frac{\pi}{2}, 0 \le x \le \pi$, and $-\frac{\pi}{2} < x < \frac{\pi}{2}$, respectively. *See* arc sine, arc cosine, arc tangent.

(2) Let $f(x)$ have a singularity at $x = c$, with $a \le c \le b$. The *Cauchy principal value* of $\int_a^b f(x)\, dx$ is

$$\lim_{\epsilon \to 0} \left(\int_a^{c-\epsilon} f(x)\, dx + \int_{c+\epsilon}^b f(x)\, dx \right).$$

The *Cauchy principal value* of an improper integral $\int_{-\infty}^\infty f(x)\, dx$ is $\lim_{c \to \infty} \int_{-c}^c f(x)\, dx$.

principle of counting constants   Let X and Y be algebraic varieties and let C be an irreducible subvariety of $X \times Y$. Let p_X and p_Y denote the projection maps onto the factors of $X \times Y$. Let $a_1 = \dim(\overline{p_X(C)})$ and $a_2 = \dim(\overline{p_Y(C)})$. There exist a nonempty open subset U_1 of $\overline{p_X(C)}$, contained in $p_X(C)$, and a nonempty open subset U_2 of $\overline{p_Y(C)})$, contained in $p_Y(C)$, such that all irreducible components of $C(x) = \{y \in Y : (x, y) \in C\}$ have the same dimension b_2 for all $x \in U_1$ and such that all irreducible components of $C^{-1}(y) = \{x \in X : (x, y) \in C\}$ have the same dimension b_2 for all $y \in U_2$. These dimensions satisfy $a_1 + b_2 = a_2 + b_1$.

principle of reflection   Two complex numbers z_1 and z_2 are said to be symmetric with respect to a circle of radius r and center z_0 if $(z_1 - z_0)(z_2 - z_0) = r^2$. The principle of reflection states that if the image of the circle under a linear fractional transformation $w = (az + b)/(cz + d)$ is again a circle (this happens unless the image is a line), then the images w_1 and w_2 of z_1 and z_2 are symmetric with respect to this new circle. *See also* linear fractional function.

The *Schwarz Reflection Principle* of complex analysis deals with the analytic continuation of an analytic function defined in an appropriate set S, to the set of reflections of the points of S.

product   A term which includes many phenomena. The most common are the following:

(1) The *product* of a set of numbers is the result obtained by multiplying them together. For an infinite product, this requires considerations of convergence.

(2) If $A_1, \ldots, A_n$ are sets, then the *product* $A_1 \times \cdots \times A_n$ is the set of ordered n-tuples $(a_1, \ldots, a_n)$ with $a_i \in A_i$ for all i. This definition can easily be extended to infinite products.

(3) Let A_1 and A_2 be objects in a category $\mathcal{C}$. A triple (P, π_1, π_2) is called the *product* of A_1 and A_2 if P is an object of $\mathcal{C}$, $\pi_i : P \to A_i$ is a morphism for $i = 1, 2$, and if whenever X is another object with morphisms $f_i : X \to A_i$, for $i = 1, 2$, then there is a unique morphism $f : X \to P$ such that $\pi_i f = f_i$ for $i = 1, 2$.

(4) *See also* bracket product, cap product, crossed product, cup product, direct product, Euler product, free product, Kronecker product, matrix multiplication, partial product, tensor product, torsion product, wedge product.

product complex   Let C_1 be a complex of right modules over a ring R and let C_2 be a complex of left R-modules. The tensor product $C_1 \otimes_R C_2$ gives a complex of Abelian groups, called the *product complex*.

product double chain complex   The double chain complex $(Z_{p,q}, \partial', \partial'')$ obtained from a chain complex X of right A-modules with boundary operator ∂_p and a chain complex Y of left A-modules with boundary operator ∂_q in

the following way:

$$Z_{p,q} = X_p \times Y_q$$
$$\partial'_{p,q} = \partial_p \times 1$$
$$\partial''_{p,q} = (-1)^p 1 \times \partial_q .$$

product formula   (1) Let K be a finite extension of the rational numbers $\mathbf{Q}$. Then $\prod_v |x|_v = 1$ for all $x \in K$, $x \neq 0$, where the product is over all the normalized absolute values (both p-adic and archimedean) of K.

(2) Let K be an algebraic number field containing the nth roots of unity, and let a and b be nonzero elements of K. For a place v of K (as in (1) above), let $(\frac{a,b}{v})_n$ be the nth norm-residue symbol. Then $\prod_v (\frac{a,b}{v})_n = 1$. *See also* norm-residue symbol.

profinite group   Any group G can be made into a topological group by defining the collection of all subgroups of finite index to be a neighborhood base of the identity. A group with this topology is called a *profinite group*.

projection matrix   A square matrix M such that $M^2 = M$.

projective algebraic variety   Let K be a field, let $\bar{K}$ be its algebraic closure, and let $\mathbf{P}^n(\bar{K})$ be n-dimensional projective space over $\bar{K}$. Let S be a set of homogeneous polynomials in $X_0, \ldots, X_n$. The set of common zeros Z of S in $\mathbf{P}^n(\bar{K})$ is called a projective algebraic variety. Sometimes, the definition also requires the set Z to be irreducible, in the sense that it is not the union of two proper subvarieties.

projective class   Let $\mathcal{A}$ be a category. A projective class is a class $\mathcal{P}$ of objects in $\mathcal{A}$ such that for each $A \in \mathcal{A}$ there is a $P \in \mathcal{P}$ and a $\mathcal{P}$-epimorphism $f : P \to A$.

projective class group   Consider left modules over a ring R with 1. Two finitely generated projective modules P_1 and P_2 are said to be equivalent if there are finitely generated free modules F_1 and F_2 such that $P_1 \oplus F_1 \simeq P_2 \oplus F_2$. The set of equivalence classes, with the operation induced from direct sums, forms a group called the *projective class group* of R.

projective cover   An object C in a category $\mathcal{C}$ is the *projective cover* of an object A if it satisfies the following three properties: (i.) C is a projective object. (ii.) There is an epimorphism $e : C \to A$. (iii.) There is no projective object properly between A and C. In a general category, this means that if $g : C' \to A$ and $f : C \to C'$ are epimorphisms and C' is projective, then f is actually an isomorphism. Thus, projective covers are simply injective envelopes "with the arrows turned around." *See also* epimorphism, injective envelope.

In most familiar categories, objects are sets with structure (for example, groups, topological spaces, etc.) and morphisms are particular kinds of functions (for example, group homomorphisms, continuous functions, etc.), and epimorphisms are onto functions (surjections) of a particular kind. Here is an example of a projective cover in a specific category: In the category of compact Hausdorff spaces and continuous maps, the projective cover of a space X always exists, and is called the *Gleason cover* of the space. It may be constructed as the Stone space (space of maximal lattice ideals) of the Boolean algebra of regular open subsets of X. A subset is *regular open* if it is equal to the interior of its closure. *See also* Gleason cover, Stone space.

projective dimension   Let R be a ring with 1 and let M be an R-module. The *projective dimension* of M is the length of the smallest projective resolution of M; that is, the projective dimension is n if there is an exact sequence $0 \to P_n \to \cdots \to P_0 \to M \to 0$, where each P_i is projective and n is minimal. If no such finite resolution exists, the projective dimension is infinite.

projective general linear group   The quotient group defined as the group of invertible matrices (of a fixed size) modulo the subgroup of scalar matrices.

projective limit   The inverse limit. *See* inverse limit.

projective module   A module M for which there exists a module N such that $M \oplus N$ is free. Equivalently, M is projective if, whenever

there is a surjection $f : A \to M$, there is a homomorphism $g : M \to A$ such that fg is the identity map of M.

projective morphism   A morphism $f : X \to Y$ of algebraic varieties over an algebraically closed field K which factors into a closed immersion $X \to \mathbf{P}^n(K) \times Y$, followed by the projection to Y. This concept can be generalized to morphisms of schemes.

projective object   An object P in a category $\mathcal{C}$ satisfying the following mapping property: If $e : C \to B$ is an epimorphism in the category, and $f : P \to B$ is a morphism in the category, then there exists a (usually not unique) morphism $g : P \to C$ in the category such that $e \circ g = f$. This is summarized in the following "universal mapping diagram":

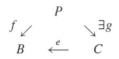

Projectivity is simply injectivity "with the arrows turned around." *See also* epimorphism, injective object, projective module.

In most familiar categories, objects are sets with structure (for example, groups, topological spaces, etc.), and morphisms are particular kinds of functions (for example, group homomorphisms, continuous maps, etc.), so epimorphisms are onto functions (surjections) of particular kinds. Here are two examples of projective objects in specific categories: (i.) In the category of Abelian groups and group homomorphisms, free groups are projective. (An Abelian group G is *free* if it is the direct sum of copies of the integers $\mathbf{Z}$.) (ii.) In the category of compact Hausdorff spaces and continuous maps, the projective objects are exactly the extremely disconnected compact Hausdorff spaces. (A compact Hausdorff space is *extremely disconnected* if the closure of every open set is again open.) *See also* compact topological space, Hausdorff space.

projective representation   A homomorphism from a group to a projective general linear group.

projective resolution   Let B be a left R module, where R is a ring with unit. A *projective* *resolution* of B is an exact sequence,

$$\cdots \xrightarrow{\phi_2} E_1 \xrightarrow{\phi_1} E_0 \xrightarrow{\phi_0} B \longrightarrow 0 \,,$$

where every E_i is a projective left R module. (We shall define *exact sequence* shortly.) There is a companion notion for right R modules. Projective resolutions are extremely important in homological algebra and enter into the dimension theory of rings and modules. *See also* flat resolution, injective resolution, projective module, projective dimension.

An *exact sequence* is a sequence of left R modules, such as the one above, where every ϕ_i is a left R module homomorphism (the ϕ_i are called "connecting homomorphisms"), such that $\mathrm{Im}(\phi_{i+1}) = \mathrm{Ker}(\phi_i)$. Here $\mathrm{Im}(\phi_{i+1})$ is the image of ϕ_{i+1}, and $\mathrm{Ker}(\phi_i)$ is the kernel of ϕ_i. In the particular case above, because the sequence ends with 0, it is understood that the image of ϕ_0 is B, that is, ϕ_0 is onto. There is a companion notion for right R modules.

projective scheme   A *projective scheme* over a scheme S is a closed subscheme of projective space over S.

projective space   Let K be a field and consider the set of $(n + 1)$-tuples $(x_0, \ldots, x_n)$ in K^{n+1} with at least one coordinate nonzero. Two tuples $(\ldots, x_i, \ldots)$ and $(\ldots, x_i', \ldots)$ are equivalent if there exists $\lambda \in K^{\times}$ such that $x_i' = \lambda x_i$ for all i. The set $\mathbf{P}^n(K)$ of all equivalence classes is called n-dimensional projective space over K. It can be identified with the set of lines through the origin in K^{n+1}. The equivalence class of $(x_0, x_1, \ldots, x_n)$ is often denoted $(x_0 : x_1 : \cdots : x_n)$. More generally, let R be a commutative ring with 1. The scheme $\mathbf{P}^n(R)$ is given as the set of homogeneous prime ideals of $R[X_0, \ldots, X_n]$ other than $(X_0, \ldots, X_n)$, with a structure sheaf defined in terms of homogeneous rational functions of degree 0. It is also possible to define projective space $\mathbf{P}^n(S)$ for a scheme S by patching together the projective spaces for appropriate rings.

projective special linear group   The quotient group defined as the group of matrices (of a fixed size) of determinant 1 modulo the subgroup of scalar matrices of determinant 1.

projective symplectic group   The quotient group defined as the group of symplectic matrices (of a given size) modulo the subgroup $\{I, -I\}$, where I is the identity matrix. *See* symplectic group.

projective unitary group   The quotient group defined as the group of unitary matrices modulo the subgroup of unitary scalar matrices. *See* unitary matrix.

proper component   Let U and V be irreducible subvarieties of an irreducible algebraic variety X. A simple irreducible component of $U \cap V$ is called *proper* if it has dimension equal to $\dim U + \dim V$ - $\dim X$.

proper equivalence   An equivalence relation R on a topological space X such that $R[K] = \{x \in X : x Rk \text{ for some } k \in K\}$ is compact for all compact sets $K \subseteq X$.

proper factor   Let a and b be elements of a commutative ring R. Then a is a *proper factor* of b if a divides b, but a is not a unit and there is no unit u with $a = bu$.

proper fraction   A positive rational number such that the numerator is less than the denominator. *See also* improper fraction.

proper intersection   Let Y and Z be subvarieties of an algebraic variety X. If every irreducible component of $Y \cap Z$ has codimension equal to $\operatorname{codim} Y + \operatorname{codim} Z$, then Y and Z are said to *intersect properly*.

proper Lorentz group   The group formed by the Lorentz transformations whose matrices have determinants greater than zero.

proper morphism of schemes   A morphism of schemes $f : X \to Y$ such that f is separated and of finite type and such that for every morphism $T \to Y$ of schemes, the induced morphism $X \times_T Y \to T$ takes closed sets to closed sets.

proper orthogonal matrix   An orthogonal matrix with determinant $+1$. *See* orthogonal matrix.

proper product Let R be an integral domain with field of quotients K and let A be an algebra over K. Let M and N be two finitely generated R-submodules of A such that $KM = KN = A$. If $\{a \in A : Ma \subseteq M\} = \{a \in A : aN \subseteq N\}$, and this is a maximal order of A, then the product MN is called a *proper product*.

proper transform Let $T : V \to W$ be a rational mapping between irreducible varieties and let V' be an irreducible subvariety of V. A proper transform of V' by T is the union of all irreducible subvarieties W' of W for which there is an irreducible subvariety of T such that V' and W' correspond.

proportion A statement equating two ratios, $\frac{a}{b} = \frac{c}{d}$, sometimes denoted by $a : b = c : d$. The terms a and d are the extreme terms and the terms b and c are the mean terms.

proportional A term in the proportion $\frac{a}{b} = \frac{c}{d}$. Given numbers a, b, and c, a number x which satisfies $\frac{a}{b} = \frac{c}{x}$ is a fourth proportional to a, b, and c. Given numbers a and b, a number x which satisfies $\frac{a}{b} = \frac{b}{x}$ is a third proportional to a and b, and a number x which satisfies $\frac{a}{x} = \frac{x}{b}$ is a mean proportional to a and b.

proportionality The state of being in proportion. *See* proportion.

Prüfer domain An integral domain R such that every nonzero ideal of R is invertible. Equivalently, the localization R_M is a valuation ring for every maximal ideal M. Another equivalent condition is that an R-module is flat if and only if it is torsion-free.

Prüfer ring An integral domain R such that all finitely generated ideals in R are invertible in the field of quotients of R. *See also* integral domain.

pseudogeometric ring Let A be a Noetherian integral domain with field of fractions K. If the integral closure of A in every finite extension of K is finitely generated over A, then A is said to satisfy the finiteness condition. A Noetherian ring R such that R/P satisfies the finite-

ness condition for all prime ideals P is called *pseudogeometric*.

pseudovaluation A map v from a ring R into the nonnegative real numbers such that (i.) $v(r) = 0$ if and only if $r = 0$, (ii.) $v(rs) \leq v(r)v(s)$, (iii.) $v(r+s) \leq v(r) + v(s)$, and (iv.) $v(-r) = v(r)$ for all $r, s \in R$.

p-subgroup A finite group whose order is a power of p is called a p-group. A p-group that is a subgroup of a larger group is called a *p-subgroup*.

p-subgroup For a finite group G and a prime integer p, a subgroup S of G such that the order of S is a power of p.

pure imaginary number An imaginary number. *See* imaginary number.

pure integer programming problem A problem similar to the primary linear programming problem in which the solution vector $\mathbf{x} = (x_1, x_2, \ldots, x_n)$ is a vector of integers: the problem is to minimize $z = \mathbf{cx}$ with the conditions $A\mathbf{x} = \mathbf{b}$, $\mathbf{x} \geq 0$, and x_j ($j = 1, 2, \ldots, n$) is an integer, where $\mathbf{c}$ is an $n \times 1$ vector of real numbers, A is an $m \times n$ matrix of real numbers, and $\mathbf{b}$ is an $m \times 1$ matrix of real numbers.

purely infinite von Neumann algebra A von Neumann algebra A which has no semifinite normal traces on A.

purely inseparable element Let L/K be an extension of fields of characteristic $p > 0$. If $\alpha \in L$ satisfies $\alpha^{p^n} \in K$ for some n, then α is called a *purely inseparable element* over K.

purely inseparable extension An extension L/K of fields such that every element of L is purely inseparable over K. *See* purely inseparable element.

purely inseparable scheme Given an irreducible polynomial $f(X)$ over a field k, if the formal derivative $df/dX = 0$, then $f(X)$ is *inseparable;* otherwise, f is *separable*. If $\text{char}(k) = 0$, every irreducible polynomial $f(X)(\neq 0)$ is separable. If $\text{char}(k) = p > 0$, an irreducible polynomial $f(X)$ is inseparable if and only if $f(X) = g(X^p)$. An algebraic element α over k is called *separable* or *inseparable* over k if the minimal polynomial of α over k is separable or inseparable. An algebraic extension of k is called *separable* if all elements of K are separable over k; otherwise, K is called *inseparable*. If α is inseparable, then k has nozero characteristic p and the minimal polynomial $f(X)$ of α can be decomposed as $f(X) = (X - \alpha_1)^{p^r}(X - \alpha_2)^{p^r} \ldots (X - \alpha_m)^{p^r}$, $r \geq 1$, where $\alpha_1, \ldots, \alpha_m$ are distinct roots of $f(X)$ in its splitting field. If $\alpha^{p^r} \in k$ for some r, we call α *purely inseparable* over k. An algebraic extension K of k is called *purely inseparable* if all elements of the field are purely inseparable over k. Let $V \subset k^n$ be a reduced irreducible affine algebraic variety. V is called *purely inseparable* if the function field $k(V)$ is purely inseparable over k. Therefore, we can define the same notion for reduced irreducible algebraic varieties. Since there is a natural equivalence between the category of algebraic varieties over k and the category of reduced, separated, algebraic k-schemes, by identifying these two categories, we define *purely inseparable* reduced, irreducible, algebraic k-schemes.

purely transcendental extension An extension of fields L/K such that there exists a set of elements $\{x_i\}_{i \in I}$, algebraically independent over K, with $L = K(\{x_i\})$. *See* algebraic independence.

pure quadratic A *quadratic* equation of the form $ax^2 + c = 0$, that is, a quadratic equation with the first degree term bx missing.

Pythagorean field A field F that contains $\sqrt{a^2 + b^2}$ for all $a, b \in F$.

Pythagorean identities The following basic identities involving trigonometric functions, resulting from the Pythagorean Theorem:

$$\sin^2(x) + \cos^2(x) = 1$$

$$\tan^2(x) + 1 = \sec^2(x)$$

$$1 + \cot^2(x) = \csc^2(x) \,.$$

Pythagorean numbers Any combination of three positive integers a, b, and c such that $a^2 + b^2 = c^2$.

Pythagorean ordered field An ordered field P such that the square root of any positive element of P is in P.

Pythagorean Theorem Consider a right triangle with legs of length a and b and hypotenuse of length c. Then $a^2 + b^2 = c^2$.

Pythagorean triple A solution in positive integers x, y, z to the equation $x^2 + y^2 = z^2$. Some examples are $(3, 4, 5)$, $(5, 12, 13)$, and $(20, 21, 29)$. If x, y, z have no common divisor greater than 1, then there are integers a, b such that $x = a^2 - b^2$, $y = 2ab$, and $z = a^2 + b^2$ (or the same equations with the roles of x and y interchanged). Also called *Pythagorean numbers*.

Q

QR method of computing eigenvalues An iterative method of finding all of the eigenvalues of an $n \times n$ matrix A. Let $A_0 = A$. Determine $A_1, A_2, \ldots, A_m, \ldots$ in the following way: If A is a real tridiagonal matrix or a complex upper Hessenberg matrix, let s_m be the eigenvalue of the 2×2 matrix closer to $a_{nn}^{(m)}$ in the lower right corner of A_m, or if A is a real upper Hessenberg matrix, let s_m and s_{m+1} be the eigenvalues of the 2×2 matrix in the lower right corner of A_m. With this (these) value(s) of s_m, write $A_m - s_m I$ as $Q_m R_m$, where Q_m is a unitary matrix and R_m is an upper Hessenberg matrix. Then define $A_{m+1} = R_m Q_m - s_m I$. Then the elements on the diagonal of $\lim_{m \to \infty} A_m$ converge to the eigenvalues of A.

quadratic Of degree 2, as, for example, a polynomial $aX^2 + bX + c$, with $a \neq 0$.

quadratic differential On a Riemann surface $\mathcal{S}$, a rule which associates to each local parameter z mapping a parametric neighborhood $U \subset \mathcal{S}$ into the extended complex plane $\overline{\mathbf{C}}$ ($z : U \to \overline{\mathbf{C}}$), a function $Q_z : z(U) \to \overline{\mathbf{C}}$ such that for any local parameter $z_1 : U_1 \to \overline{\mathbf{C}}$ and $z_2 : U_2 \to \overline{\mathbf{C}}$ with $U_1 \cap U_2 \neq \emptyset$, the following holds in this intersection:

$$\frac{Q_{z_2}\left(z_2(p)\right)}{Q_{z_1}\left(z_1(p)\right)} = \left(\frac{dz_1(p)}{dz_2(p)}\right)^2, \quad \forall p \in U_1 \cap U_2 .$$

Here $z(U)$ is the image of U in $\overline{\mathbf{C}}$ under z.

quadratic equation A polynomial equation of the form $aX^2 + bX + c = 0$. The solutions are given by the quadratic formula. *See* quadratic formula.

quadratic field A field of the form $\mathbf{Q}(\sqrt{d}) = \{a + b\sqrt{d} : a, b \in \mathbf{Q}\}$, where $\mathbf{Q}$ is the field of rational numbers and $d \in \mathbf{Q}$ is not a square.

quadratic form A polynomial of the form $Q(X_1, \ldots, X_n) = \sum_{i,j} c_{ij} X_i X_j$, where the coefficients c_{ij} lie in some ring.

quadratic formula The roots of $aX^2 + bX + c = 0$ are given by $X = \frac{-b \pm \sqrt{b^2 - 4ac}}{2a}$ (in a field of characteristic other then 2).

quadratic function A polynomial function of degree two: $ax^2 + bx + c$. For example, $8x^2 + 3x - 1$ is a quadratic function. *See also* pure quadratic.

quadratic inequality Any inequality of one of the forms:

$$
\begin{aligned}
ax^2 + bx + c &< 0 \\
ax^2 + bx + c &> 0 \\
ax^2 + bx + c &\leq 0 \\
ax^2 + bx + c &\geq 0 .
\end{aligned}
$$

quadratic polynomial A polynomial of the form $aX^2 + bX + c$.

quadratic programming Theoretical aspects and methods pertaining to minimizing or maximizing quadratic functions on sets X with constraints determined by linear equations and linear inequalities.

quadratic programming problem A quadratic programming problem in which one wants to maximize or minimize $z = \mathbf{c}^t \mathbf{x} - \frac{1}{2} \mathbf{x}^t D \mathbf{x}$ with the constraints $A\mathbf{x} \leq \mathbf{b}$ and $\mathbf{x} \geq 0$, where A is an $m \times n$ matrix of real numbers, $\mathbf{c} \in \mathbf{R}^n$, D is an $n \times n$ matrix of real numbers, and $(\ )^t$ denotes transposition.

quadratic reciprocity Let p and q be distinct odd primes. If at least one of p and q is congruent to 1 mod 4, then p is congruent to a square mod q if and only if q is congruent to a square mod p. If both p and q are congruent to 3 mod 4, then p is congruent to a square mod q if and only if q is not congruent to a square mod p.

quadratic residue An integer r is a *quadratic residue* mod n if there exists an integer x with $r \equiv x^2$ mod n. *See* modular arithmetic.

quadratic transformation   The map from the projective plane to itself given by $(x_0 : x_1 : x_2) \mapsto (x_1x_2 : x_0x_2 : x_0x_1)$. *See also* projective space.

quartic   Of degree 4, as, for example, a polynomial $aX^4 + bX^3 + cX^2 + dX + e$, with $a \neq 0$.

quartic equation   A polynomial equation of the form $aX^4 + bX^3 + cX^2 + dX + e = 0$.

quasi-affine algebraic variety   A locally closed subvariety of affine space. *See* affine space.

quasi-algebraically closed field   A field F such that, for every $d \geq 1$, every homogeneous polynomial equation of degree d in more than d variables has a nonzero solution in F. Finite fields and rational function fields $K(X)$ with K algebraically closed are quasi-algebraically closed.

quasi-coherent module   Let (X, R) be a ringed space. A sheaf of R-modules is called *quasi-coherent* if, for each $x \in X$, there exists an open set U containing x such that $M|_U$ can be expressed as A/B where A and B are free $R|_U$-modules.

quasi-dual space   The quotient space of the space of quotient representations of a locally compact group G, considered with the Borel structure subordinate to the topology of uniform convergence of matrix entries on compact sets.

quasi-equivalent representation   Two unitary representations π_1, π_2 of a group G (or symmetric representations of a symmetric algebra A) in Hilbert spaces H_1 and H_2, respectively, satisfying one of the following four equivalent conditions:
(i.) there exist unitarily equivalent representations μ_1 and μ_2 such that μ_1 is a multiple of π_1 and μ_2 is a multiple of π_2;
(ii.) the non-zero subrepresentations of π_2 are not disjoint from π_1;
(iii.) π_2 is unitarily equivalent to a subrepresentation of some multiple representation μ_2 of π_2 that has unit central support;
(iv.) there exists an isomorphism Ψ of the von Neumann algebra generated by the set $\pi_1(A)$ onto the von Neumann algebra generated by the set $\pi_2(A)$ such that

$$\Psi(\pi_1(a)) = \pi_2(a), \qquad \text{for all} \quad a \in A.$$

quasi-Frobenius algebra   Let A be an algebra over a field and let A^* be the dual of the right A-module A. Decompose A into a direct sum of indecomposable left ideals and a direct sum of indecomposable right ideals: $A = \sum_i \sum_j Ae_i^{(j)} = \sum_i \sum_j e_i^{(j)}A$, where $Ae_i^{(j)} \simeq Ae_{i'}^{(j')}$ if and only if $i = i'$, and similarly for the right ideals. A is called *quasi-Frobenius* if there exists a permutation π of the indices i such that $Ae_i^{(1)} \simeq (e_{\pi(i)}^{(1)}A)^*$.

quasi-Fuchsian group   A Fuchsian group is a group of conformal homeomorphisms of the Riemann sphere which leaves invariant a round circle in the sphere (equivalently a group of isometries of hyperbolic 3 space which leaves invariant a flat plane). A quasi-Fuchsian group is a quasiconformal deformation of a Fuchsian group: there is a quasiconformal homeomorphism of the sphere conjugating the action of the Fuchsian group to the group in question. It follows that abstractly the group is a surface group. Quasi-Fuchsian groups are extremely important for hyperbolic 3-manifolds, for instance, playing a central role in the geometrization of a large class of 3-manifolds. *See also* Kleinian group.

quasi-group   A set with a not necessarily associative law of composition $(a, b) \mapsto ab$ such that the equation $ab = c$ has a unique solution when any two of the variables are specified.

quasi-inverse   Let x, y be elements of a ring. Then y is called a *quasi-inverse* of x if $x + y - xy = x + y - yx = 0$. If 1 exists, this means that $(1 - y)$ is the inverse of $(1 - x)$. An element that has a quasi-inverse is called quasi-invertible.

quasi-inverse element   *See* quasi-inverse.

quasi-invertible element   *See* quasi-inverse.

quasi-local ring   A commutative ring with exactly one maximal ideal. Often such rings are called local rings, but some authors reserve the term local ring for Noetherian rings with unique maximal ideals.

quasi-projective algebraic variety   An open subset of a projective variety.

quasi-projective morphism   A morphism $f : X \to Y$ of varieties, or schemes, that factors as an open immersion $X \to X'$ followed by a projective morphism $X' \to Y$. *See* projective morphism.

quasi-projective scheme   An open sub-scheme of a projective scheme. *See* projective scheme.

quasi-semilocal ring   A commutative ring with only finitely many maximal ideals. Often such rings are called semilocal rings, but some authors reserve the term semilocal ring for Noetherian rings with finitely many maximal ideals.

quasi-symmetric homogeneous Siegel domain   A Siegel domain $\mathcal{S} = \mathcal{S}(U, V, \Omega, H)$ satisfying the following conditions:
(i.) Ω is homogeneous and self-dual;
(ii.) R_u is associated with T_u for all $u \in U$;
where U is a vector space over $\mathbf{R}$ of dimension m; V is a vector space over $\mathbf{C}$ of dimension n; Ω is a non-degenerate open convex cone in U with vertex at the origin; H is a Hermitian map $H : V \times V \to U_c = U \otimes_{\mathbf{R}} \mathbf{C}$; for $u \in U_c$, $R_u \in \mathrm{End}(V)$ is defined by $(u, H(v, v')) = 2h(v, R_u v')$, $v, v' \in V$; for each $u \in U$, there exists a unique element $T_u \in p(\Omega)$ such that $T_u e = u$; and $g(\Omega) = t(\Omega) + p(\Omega)$ is the Cartan decomposition. Notice that a Siegel domain $\mathcal{S}$ is *symmetric* if and only if (i.) and (ii.) above and the following condition (iii.) hold:
(iii.) R_u satisfies the relation

$$H(R(H(v'', v'))v, v'')$$
$$= H(v', R(H(v, v''))v'') .$$

quaternion   An element of the Hamiltonian quaternions. *See* Hamilton's quaternion algebra.

quaternion algebra   *See* Hamilton's quaternion algebra.

quaternion group   The group of order 8 equal to $\{\pm 1, \pm i, \pm j, \pm k\}$, with relations $i^2 = j^2 = k^2 = -1$, $ij = k$, $ji = -k$ and $(-1)a = a(-1)$ for all a. *See also* Hamilton's quaternion algebra.

quintic   Of degree 5, as, for example, a polynomial $aX^5 + bX^4 + cX^3 + dX^2 + eX + f$, with $a \neq 0$.

quintic equation   A polynomial equation of the form $aX^5 + bX^4 + cX^3 + dX^2 + eX + f = 0$.

quotient   (1) The result of dividing one number by another. *See also* numerator, denominator.

(2) The result of starting with a set (group, ring, topological space, etc.) and forming a new set consisting of equivalence classes under some equivalence relation. *See also* quotient algebra, quotient chain complex, quotient G-set, quotient group, quotient Lie algebra, quotient Lie group, quotient representation, quotient ring, quotient set.

quotient algebra   The algebra of equivalence classes of an algebra modulo a two-sided ideal.

quotient bialgebra   The quotient space H/I with the induced bialgebra structure, where I is bi-ideal of a bialgebra H, i.e., I is an ideal of an algebra (H, μ, η) which is also a co-ideal of a co-algebra (H, Δ, ϵ), and $(H, \mu, \eta, \Delta, \epsilon)$ is a bialgebra. *See also* bialgebra.

quotient bundle   Let $\pi : X \to B$ be a vector bundle and let $\pi' : X' \to B$ be a subbundle of π. The quotient bundle of π is the union of vector spaces $\pi^{-1}(b)/(\pi')^{-1}(b)$ with the quotient topology.

quotient chain complex   Let (X, ∂) be a chain complex, where $X = \sum X_n$, and let $Y = \sum Y_n$ be a subcomplex (so $\partial Y \subseteq Y$). Then X/Y with the map induced from ∂ forms the *quotient chain complex*. *See also* chain complex.

quotient co-algebra   The quotient space C/I with a co-algebra structure induced naturally from a co-algebra (C, Δ, ϵ), where I is a co-ideal of C. *See also* coalgebra.

quotient group   Let G be a group and let H be a normal subgroup. The set of all cosets forms a group, denoted G/H, where the group operation is defined by $(aH) * (bH) = (a * b)H$ (where $a * b$ denotes the group operation in G). For example, if $G = \mathbf{Z}$, the integers under addition, and $H = n\mathbf{Z}$, the multiples of a fixed integer n, then G/H is the additive group of integers modulo n.

quotient G-set   Let G be a group acting on a set X and let R be a G-compatible equivalence relation on X, that is, xRy implies $gxRgy$, for all $x, y \in X$, and all $g \in G$. The quotient set X/R, which has a natural action of G, is called a *quotient G-set*. *See also* G-set.

quotient Lie algebra   Let $\mathbf{g}$ be a Lie algebra with bracket $[\cdot, \cdot]$, and let $\mathbf{a}$ be a Lie subalgebra such that $[\mathbf{g}, \mathbf{a}] \subseteq \mathbf{a}$ (so $\mathbf{a}$ is an ideal of $\mathbf{g}$). The quotient $\mathbf{g}/\mathbf{a}$ with the bracket induced from that of $\mathbf{g}$ is called a *quotient Lie algebra*. *See also* quotient algebra.

quotient Lie group   Let G be a Lie group and let H be a closed normal subgroup. The quotient group can be given the structure of a Lie group and is called a *quotient Lie group*. *See also* quotient group.

quotient representation   Let $\rho : G \to \mathrm{GL}(V)$ be a representation of a group G as a group of linear transformations of a vector space V. Let $W \subseteq V$ be a subspace such that $\rho(G)W \subseteq W$. The quotient representation is the induced map $G \to \mathrm{GL}(V/W)$. *See* induced representation.

quotient ring   Let R be a ring and let I be a two-sided ideal. The set R/I of cosets with respect to the additive structure can be given the structure of a ring by defining $(a+I)+(b+I) = a + b + I$ and $(a + I)(b + I) = ab + I$, where $a, b \in R$. This is called a *quotient ring*. The standard example is when $R = \mathbf{Z}$, the integers, and $I = n\mathbf{Z}$, the multiples of a fixed integer n. Then R/I is the ring of integers modulo n.

quotient set   The set of equivalence classes, denoted X/R, of a set X with respect to an equivalence relation R on X.

R

Racah algebra   Let A be a dynamical quantity and let ϕ_1 and ϕ_2 be irreducible components of the state obtained by combining n angular momenta. *Racah algebra* is a systematic method of calculating the matrix element $(\phi_1, A\phi_2)$ in quantum mechanics.

Racah coefficient   Certain constants in Racah algebra. Consider irreducible representations of the rotation groups $R = \mathrm{SO}(3)$. The reduction of the tensor product of three irreducible representations can be done in two ways, $(D(j_1) \otimes D(j_2)) \otimes D(j_3)$ and $D(j_1) \otimes (D(j_2) \otimes D(j_3))$ and two corresponding sets of basis vectors. The transformation coefficient for the two ways of reduction is written in the following form:

$$\langle j_1 j_2 (j_{12}) \, j_3; j \, | \, j_1, j_2 j_3 (j_{23}) \, ; j \rangle$$

$$= \sqrt{(2j_{12} + 1)(2j_{23} + 1)} W (j_1 j_2 j j_3; j_{12} j_{23}) \cdot$$

Here $W(abcd; ef)$ are called the *Racah coefficients*.

radian   A measure of the size of an angle. If the vertex of an angle is at the center of a circle of radius 1, the length of the arc on the circle determined by the angle is the radian measure of the angle. For example, a right angle is $\pi/2$ radians. One radian equals $180/\pi$, which is approximately 57.3 degrees.

radical   A term that has many meanings:

(1) The nth root of a number.

(2) If R is a commutative ring with 1 and I is an ideal in R, the *radical* of I is the set of $r \in R$ such that $r^n \in I$ for some n (depending on r). The radical of the zero ideal is often called the *radical* of R, or the *nilradical* of R. It is the intersection of all prime ideals of R. The *Jacobson radical* of R is the intersection of all maximal ideals of R.

(3) In an arbitrary ring, the *radical* is the largest ideal that is contained in the set of quasi-invertible elements.

(4) The *radical* of an algebraic group is the largest connected closed solvable normal subgroup.

(5) The radical of a Lie algebra is the largest solvable ideal.

radical equation   An equation in which the variable is under a radical. For example, $\sqrt{x^2 + 1} - 2\sqrt[3]{x} = 3x$ is a radical equation.

radical of a group   The maximal solvable normal subgroup of a group G. In other words, a solvable normal subgroup R of G, which is not contained in a larger solvable normal subgroup of G. *See also* solvable group, normal subgroup.

radicand   In an expression $\sqrt[n]{a}$, the quantity a.

radix   The base of a system of numbers. For example, 10 is the *radix* of the decimal system.

ramification field   *See* ramification group.

ramification group   Let L/K be a Galois extension of algebraic number fields and let R be the ring of algebraic integers in L. Let P be a prime ideal of R and let Z be the decomposition group for P (the set of Galois elements g such that $gP = P$). The mth *ramification group* $V^{(m)}$ is defined to be $\{g \in Z : gx \equiv x \pmod{P^{m+1}}$ for all $x \in R\}$. The fixed field of $V^{(m)}$ is called the mth *ramification field* of P.

ramification index   (1) If S/R is a finite extension of Dedekind domains and $\wp$ is a prime ideal of R, then $\wp S = \wp_1^{e_1} \ldots \wp_g^{e_g}$, where $\wp_1, \ldots, \wp_g$ are prime ideals of S. The number e_i is called the *ramification index* of $\wp_i$.

(2) If L/K is an extension of fields and v is a (multiplicative) valuation on L, then the group index $[v(L^\times) : v(K^\times)]$ is called the *ramification index* of v.

ramification numbers   Let N be a normal subgroup of a finite group G. Suppose $\varphi \in \mathrm{Irr}N$ (the irreducible characters of N) is invariant in G. If χ is an irreducible constituent of φ^G, then the *ramification number* $e(\chi)$ is the positive integer such that $\chi_N = e(\chi)\varphi$. *See also* normal subgroup, invariant element, irreducible constituent.

Ramification Theorem   Let K be a number field (finite extension of $\mathbf{Q}$). Then a prime p ramifies in K if and only if p divides the discriminant of K.

ramified covering   A triple (X, Y, π), where X and Y are connected normal complex spaces and π is a finite surjective proper holomorphic map. For any such (X, Y, π), there exists a proper analytic subset of X, outside which π is an unramified covering.

ramified extension   An extension of algebraic number fields, for example, in which at least one prime ideal is ramified (i.e., has ramification index greater than 1). *See* ramification index.

ramified prime ideal   Let S/R be a finite extension of Dedekind domains and let P be a prime ideal of S. If the ramification index of P is greater than 1, the ideal P is said to be *ramified*. *See* ramification index.

range   *See* image.

rank of a group   Let G be a finitely generated group, having $g_1, g_2, \ldots, g_n$ as generators. Let $\tilde{G}$ have generators $\frac{g_1}{1}, \frac{g_2}{1}, \ldots, \frac{g_n}{1}$. Then the *rank of the group* G is the dimension of the finite dimensional rational vector space $\tilde{G}$.

rank of element   The number of minimal nonzero faces of an element A of a complex Δ. Here a *complex* Δ is a set with an order relation $\subset$ (read "is a face of") such that, for a given element A, the ordered subset $S(A)$ of all faces of A is isomorphic to the set of all subsets of a set.

rank of matrix   The maximal number of linearly independent columns of a rectangular matrix $A = (a_{ij})$ with entries in a field. This also equals the maximal number of linearly independent rows of A and is the size of the largest nonzero minor of A.

rank of valuation   For an additive valuation v on a field K, the Krull dimension of $R = \{x \in$ $K : v(x) \geq 0\}$; that is, the supremum of the lengths n of chains of prime ideals $\wp_0 \supset \cdots \supset \wp_n$ (strict inclusion) in R.

rational   This term often refers to a quantity equal to the ratio of two integers or polynomials. It is also used to describe a point on an algebraic variety with coordinates in a field (the point is then said to be rational over this field). *See* rational point.

rational action   The action of a matrix group G by means of a rational representation of G. *See* rational representation.

rational cohomology group   A cohomology group for transformation spaces of linear algebraic groups G over a field K introduced by Hochschild (1961). In particular, if G is a unipotent algebraic group over a field K of characteristic 0, then $H(G, A)$ is isomorphic to $H(g, A)$ where g is the Lie algebra of G and A is a g-module.

rational curve   An algebraic curve of genus 0. Such curve in n-dimensional affine space can be described by rational functions $x_1 = R_1(t), \ldots, x_n = R_n(t)$ in terms of a single parameter t.

rational equivalence   Let X_1 and X_2 be cycles on a nonsingular irreducible algebraic variety V over a field K, and let $\mathbf{A}_K^1$ be the affine line over K. Suppose there exists a cycle Z on $V \times \mathbf{A}_K^1$ such that the intersection $Z \cdot (V \times a)$ is defined for all $a \in \mathbf{A}_K^1$ and such that there are $a_1, a_2 \in \mathbf{A}_K^1$ such that $Z \cdot (V \times a_1) - Z \cdot (V \times a_2) = X_1 - X_2$. Then X_1 and X_2 are rationally equivalent.

rational expression   Any algebraic expression which can be written as a polynomial divided by a polynomial.

rational function   The quotient of two polynomials (possibly in several variables).

rational function field   A field $K(X_1, \ldots, X_n)$, where K is a given field and $X_1, \ldots, X_n$ are indeterminates.

rational homomorphism   A homomorphism $G_1 \to G_2$ of algebraic groups which is also a rational map of algebraic varieties.

rational injectivity   The notion that was used by Hochschild (1961) to introduce the *rational cohomology group* for transformation spaces of linear algebraic group G over a field k. *See* rational cohomology group.

rationalization   The process of transforming a radical expression into an expression with no radicals, or with all radicals in the numerators. *See also* rationalizing the denominator.

rationalization of denominator   Removing radicals from the denominator of an algebraic expression. For example, in an expression of the form $(a + b\sqrt{n})/(c + d\sqrt{n})$, where a, b, c, d, n are rational numbers, multiplying numerator and denominator by $c - d\sqrt{n}$. The result is of the form $(r + s\sqrt{n})/t$, where r, s, t are rational numbers. In some elementary mathematics courses, it is taught that radicals in the denominator are undesirable.

rationalization of equation   The process of transforming a radical equation into an equation with no radicals. The method is to manipulate, algebraically, the equation so only radicals are on one side of the equal sign and then raise both sides of the equation to the index of one of the radicals. The process may need to be repeated.

rationalizing the denominator   The process of removing radicals from the denominator of a fraction. The method is to multiply the numerator and the denominator of the fraction by a rationalizing factor. For example,

$$\frac{1}{\sqrt{2}} = \frac{1}{\sqrt{2}} \cdot \frac{\sqrt{2}}{\sqrt{2}} = \frac{\sqrt{2}}{2}.$$

rational map   (1) Informally, a rational function. *See* rational function.

(2) In algebraic geometry, a function from an open subset U of a variety X to another variety Y, which preserves regular functions. (*Regular functions* are functions which can be represented locally as quotients of polynomials.) In more detail, let X and Y be varieties, and

let X be irreducible. (A *variety* is the solution set of a system of polynomial equations. *Irreducible* means the variety X is not the union of two proper ($\neq 0, \neq X$) subvarieties.) A *rational map* F is an equivalence class of pairs (f, U), where U is an open subset of X and $f : U \to Y$ is a continuous function which preserves regular functions. Preserving regular functions means that if $h : W \to k$ is a regular function, where W is an open subset of Y, then $h \circ f$ is regular on the open set $\{x \in U : f(x) \in W\}$. Two such pairs (f, U) and (g, V) are deemed equivalent if $f = g$ on $U \cap V$. It is a theorem that, given a rational map F, there is a largest open set U_M such that a pair $(f, U_M) \in F$. Thus, one may think of a rational map as an actual function defined on this maximal open set U_M. The set U_M is referred to as the *set of points on which F is defined*. *See also* irreducible variety, regular function.

rational number   A number that can be expressed as the quotient of two integers m/n, with $n \neq 0$. For example, $1/2$, $-9/4$, 3, and 0 are rational numbers. The set of all rational numbers forms a field, denoted $\mathbf{Q}$. *See* field.

rational operation   Any of the four operations of addition, subtraction, multiplication, and division.

rational point   Let V be an algebraic variety defined over a field k. A point P on V is rational over k if the coordinates of P (in some affine open set containing P) lie in k. *See also* k-rational point.

rational rank of valuation   Let v be an additive valuation on a field K with values in an ordered additive group. The supremum of the cardinalities of linearly independent (over $\mathbf{Z}$) subsets of $v(K^\times)$ is called the *rational rank* of v.

rational representation   Let R be a commutative ring with 1 and let G be a subgroup of $\mathrm{GL}_n(R)$ for some n. Let $\rho : G \to \mathrm{GL}_m(R)$ for some m be a homomorphism. Suppose there exist m^2 rational functions in n^2 variables $p_{st}(\{X_{ij}\}) \in R[\{X_{ij}\}]$ with $\rho((g_{ij})) = (p_{st}(g_{uv}))$ for all $(g_{ij}) \in G$. Then ρ is called a *rational representation*.

Rational Root Theorem   Let R be a unique factorization domain with quotient field F and suppose $f(x) = a_n x^n + a_{n-1} x^{n-1} + \cdots + a_0$ is a polynomial with coefficients in R. If $c = \frac{p}{q}$ with p and q relatively prime, and c is a root of f, then p divides a_0 and q divides a_n.

This is most often used in the particular case when R is the integers and F is the rationals.

rational surface   An algebraic surface (an algebraic variety of dimension 2) that is birationally equivalent to 2-dimensional projective space. *See also* rational variety.

rational variety   A variety over a field k which, for some n, is birationally equivalent to an n-dimensional projective space over k. *See also* birational mapping, variety.

R-basis   Let R be a ring contained in a possibly larger ring S. A left S module V has an *R-basis* if V is freely generated as an R module by a family of generators, $v_\alpha, \alpha \in V$. This means (i.) each $v_\alpha \in V$; (ii.) if a finite sum $r_1 v_{\alpha_1} + \cdots + r_n v_{\alpha_n} = 0$, where $r_1, \ldots r_n \in R$, then $r_i = 0$ for $i = 1, \ldots n$; (iii.) the set of all finite sums $r_1 v_{\alpha_1} + \cdots + r_n v_{\alpha_n}$, where $r_1, \ldots r_n \in R$, is equal to V. The family $v_\alpha, \alpha \in A$, is called an *R-basis* for V.

The most important case of this occurs when R is chosen to be the ring of integers, $\mathbf{Z}$, in which case an R basis is called a $\mathbf{Z}$-basis. *See also* Z-basis.

real axis in complex plane   A complex number $p + iq$, p, q real, can be identified with the point (p, q) in Cartesian coordinates. Under this identification, numbers of the form $p + i0$, that is, real numbers, are identified with points of the form $(p, 0)$. The collection of all such points forms one of the axes in this coordinate system; this axis is called the real axis.

real closed field   A real field F such that any algebraic extension of F which is real must be equal to F. That is, F is maximal with respect to the property of reality in an algebraic closure. *See* real field.

real field   A field F such that -1 is not a sum of squares in F.

realizable linear representation   Let R and S be commutative rings with identity and let $\sigma : R \to S$ be a homomorphism. Let M be a left R-module and let M^σ denote the scalar extension of M by σ; M^σ is a left S-module. Let A be an associative R algebra and let $\rho : A \to \mathrm{End}_R(M)$ be a linear representation of A in M. Let A^σ denote the scalar extension of A and let ρ^σ denote the scalar extension of ρ. An arbitrary linear representation of the S algebra A^σ in M^σ is called *realizable* in R if it is similar to ρ^σ for some linear representation ρ over R. *See* scalar extension.

real Lie algebra   A Lie algebra over the field of real numbers. *See* Lie algebra.

real number   The real numbers, denoted $\mathbf{R}$, form a field with binary operations of $+$ (addition) and $\cdot$ (multiplication). This field is also an ordered field: There exists a subset $P \subset \mathbf{R}$ such that (i.) if $a, b \in P$ then $a + b \in P$ and $a \cdot b \in P$ and (ii.) the sets P, $-P = \{x \in \mathbf{R} : -x \in P\}$, $\{0\}$ (0 denotes the additive identity) are pairwise disjoint and their union is $\mathbf{R}$. Furthermore, the field $\mathbf{R}$ satisfies the completeness axiom: If S is a non-empty subset of $\mathbf{R}$ and if S is bounded above, then S has a least upper bound in $\mathbf{R}$. With these axioms, $\mathbf{R}$ is known as a complete ordered field. The existence of such a field is taken as an assumption by many. Others prefer to postulate the existence of the natural numbers $\mathbf{N} \subset \mathbf{R}$ by means of the Peano postulates. *See* natural number. With these postulated, the integers, $\mathbf{Z}$ are constructed from $\mathbf{N}$ and then the rational numbers $\mathbf{Q}$ are constructed as the field of quotients of $\mathbf{Z}$. To construct the real numbers from the rationals requires more work. One way is to use the method of Dedekind cuts and another involves viewing the reals as equivalence classes of Cauchy sequences. *See also* field.

real part of complex number   If $z = a + bi$, where a and b are real, is a complex number, the real part of z is a. This is denoted as $\Re(z) = a$.

real prime divisor   Let K be an algebraic number field of degree n; then there are n injections $\sigma_1, \ldots, \sigma_n$ of K into $\mathbf{C}$. We may assume that these are ordered so that $n = r_1 + 2r_2$, and so that for $1 \le j \le r_1$, σ_i maps to the real numbers, for $j \geq r_1$, σ_j maps to $\mathbf{C}$, and $\overline{\sigma}_{j+r_1+r_2} = \sigma_{j+r_1}$ for $j = 1, \ldots, r_2$. For $1 \leq j \leq r_1$ set $v_j(a) = |\sigma(a)|$, $a \in K$ and for $r_1 + 1 \leq j \leq r_2$ set $v_j(a) = |\sigma(a)|^2$. The $v_1, \ldots, v_{r_1+r_2}$ form a set of mutually nonequivalent valuations on K. Equivalence classes of the $v_1, \ldots, v_{r_1}$ are called *real prime divisors*.

real quadratic field   A field of the form $\mathbf{Q}$ $(\sqrt{m})$ where m is square free and positive.

real quadratic form   Let $A = [a_{ij}]$ be an $n \times n$ symmetric matrix. This generates a quadratic form: $Q(x) = x^T A x = \sum_{i,j=1}^{n} a_{ij} x_i x_j$. The quadratic form is called *real* if all the a_{ij} are real.

real representation   A real analytic homomorphism from a Lie group G to $\mathrm{GL}(V)$, where V is a finite dimensional vector space over $\mathbf{R}$.

real root   A root that is a real number. *See* root.

real simple Lie algebra   A Lie algebra that is real and simple. Every *real simple Lie algebra* of classical type is the Lie algebra of one of two types of subgroups of a group of invertible matrices.

real variable   A variable whose domain is understood to be either the real numbers or a subset of the real numbers. This is most often used for emphasis to distinguish the domain from the complex numbers. Functions whose domain of definition is understood to be a subset of the real numbers are called functions of a *real variable*.

reciprocal equation   A polynomial equation of the form $a_n x^n + a_{n-1} x^{n-1} + \cdots + a_1 x + a_0 = 0$ where $a_n = a_0$, $a_{n-1} = a_1$, etc. This has the property that the reciprocal of any solution is also a solution. Notice that if n is odd, then this equation has root $x = -1$. Division of this equation by $x + 1$ then yields a *reciprocal equation* of degree $n - 1$. If n is even, say $n = 2m$, then the equation may be rewritten as: $a_{2m} x^m + a_{2m-1} x^{m-1} + \cdots + a_m + \cdots + a_1 (\frac{1}{x})^{m-1} + a_0 (\frac{1}{x})^m = 0$. Because of the symmetry of the coefficients, we associate the first and last terms, the second and next-to-last terms,

etc., and consider the variable $u = x + \frac{1}{x}$, so that $x^2 + \frac{1}{x^2} = u^2 - 2$, $x^3 + \frac{1}{x^3} = u^3 + 3u$, ... and the equation reduces to an equation of degree m in u.

reciprocal linear representation   Let R be a commutative ring with identity and let M be an R module. Let A be an associative algebra over R. A homomorphism $A \to \mathrm{End}_R(M)$ is called a linear representation of A in M. An antihomomorphism (ρ is an antihomomorphism if $\rho(ab) = \rho(b)\rho(a)$) $A \to \mathrm{End}_R(M)$ is called a *reciprocal linear representation*.

reciprocal of number   The reciprocal of a number $a \neq 0$ is the number $b = \frac{1}{a}$. Then $ab = 1$, that is, the reciprocal of a number is its multiplicative inverse. Note also that the reciprocal of b is a; in other words, the reciprocal of the reciprocal of a nonzero number is the number itself.

reciprocal proportion   The reciprocal of the proportion of two numbers. If a and b are two numbers, the proportion of a to b is $\frac{a}{b}$; the reciprocal proportion is $\frac{b}{a}$.

reduced Abelian group   An Abelian group with no nontrivial divisible subgroups.

reduced algebra   Let A be a semisimple Jordan algebra over a field K. For an $\alpha \in K$ and an idempotent $e \in A$ set $A_e(\alpha) = \{x \in A : ex = \alpha x\}$. Suppose that 1 is written as a sum of idempotents e_i. If, for every i, we can write $A_{e_i}(1) = K e_i + N_i$, with N_i a nilpotent ideal of $A_{e_i}(1)$, then A is called a *reduced algebra*.

reduced basis   Let G be a discrete group of motions in n-dimensional Euclidean space and let T denote the subgroup of all translations in G. Let $\{t_1, \ldots, t_n\}$ be a basis for T and let A denote the Gram matrix $a_{ij} = (t_i, t_j)$. (Here we simply identify a translation with a vector; this we may take as the image of 0 under the translation.) If the quadratic form $q(x) = \sum a_{ij} x_i x_j$, determined by A, is reduced then $\{t_1, \ldots, t_n\}$ is called a *reduced basis* for T. (*See* reduced quadratic form.)

reduced character   Let ρ be a finite-dimensional linear representation of an associative algebra A over a field K; that is, $\rho : A \rightarrow \operatorname{End}_K(M)$ is an algebra homomorphism, where M is a K-module. Fix a basis of M and let T_a denote the matrix (relative to this basis) associated to the endomorphism of M which is the image of $a \in A$. For $a \in A$ set $\chi_\rho(a) = \operatorname{Tr} T_a$; χ_ρ is a function on A and is called the character of ρ. A character of an absolutely irreducible representation is called an absolutely irreducible character. (*See* absolutely irreducible representation.) The sum of all the absolutely irreducible characters of A is called the *reduced character* of A. *See also* reduced norm, reduced representation.

reduced Clifford group   Let K be a field, E a finite dimensional vector space over K, $q(x)$ a quadratic form on E, and $C(q)$ the Clifford algebra of the quadratic form q. Let $G = \{s \in C(q) : s$ is invertible, $s^{-1} Es = E\}$. G forms a group relative to the multiplication of $C(q)$; G is called the Clifford group of q. $C(q)$ is the direct sum of two subalgebras, $C^+(q)$ and $C^-(q)$, where $C^+(q) = K + E^2 + E^4 + \cdots$ and $C^-(q) = E + E^3 + \cdots$. The subgroup $G^+ = G \cap C^+(q)$ of G is called the special Clifford group. $C(q)$ has an antiautomorphism β which is the identity when restricted to E. This can be used to define a homomorphism N from G^+ to the multiplicative group of K by $N(s) = \beta(s) s$, $s \in G^+$. The kernel of N, denoted G_0^+, is called the *reduced Clifford group* of q. An element $s \in G$ induces a linear transformation ϕ_s on E by $\phi_s(x) = sxs^{-1}$, $x \in E$. Such a ϕ_s belongs to the orthogonal group of E relative to q and the mapping $s \rightarrow \phi_s$ is a homomorphism. This gives a representation $\phi : G \rightarrow O(q)$. Furthermore, $\phi(G^+) = SO(q)$. The subgroup $\phi(G_0^+)$ of $SO(q)$ is called the reduced orthogonal group. *See* Clifford algebra, Clifford group.

reduced dual   Let G be a unimodular locally compact group with countable base. The support of the Plancherel measure in $\widehat{G}$ is called the *reduced dual* of G.

reduced module   Let Γ be a family of locally rectifiable curves on a Riemann surface $\mathcal{R}$. We

say that a module problem is defined for Γ if there is a non-empty class P of conformally invariant metrics $\rho(z)|dz|$ on $\mathcal{R}$ such that $\rho(z)$ is square integrable in the z-plane for each local uniformizing parameter z such that

$$A_\rho(\mathcal{R}) \equiv \iint_{\mathcal{R}} \rho^2 \, dx dy$$

is defined and such that $A_\rho(\mathcal{R})$ and

$$L_\rho(\mathcal{R}) \equiv \inf_{\gamma \in \Gamma} \int_\gamma \rho \, |dz|$$

are not simultaneously 0 or ∞. We designate the quantity

$$m(\gamma) = \inf_{\rho \in R} \frac{A_\rho(\mathcal{R})}{(L_\rho(\Gamma))^2}$$

as the module of Γ.

Now let Ω be a simply connected domain of hyperbolic type in $\mathbf{C}$ with $z_0 \in \Omega$. If $r > 0$ is small, then

$$\Omega \cap \{z : |z - z_0| > r\}$$

is a doubly connected domain $\Omega(r)$. Let $\Omega(r)$ have module $m(r)$ for the class of curves separating its boundary components. The limit

$$\lim_{r \to 0} \left(m(r) + \frac{1}{2\pi} \log r \right)$$

is called the *reduced module*.

reduced module   Let Γ be a family of locally rectifiable curves on a Riemann surface $\mathcal{R}$. We say that a module problem is defined for Γ if there is a non-empty class P of conformally invariant metrics $\rho(z)|dz|$ on $\mathcal{R}$ such that $\rho(z)$ is square integrable in the z-plane for each local uniformizing parameter z such that

$$A_\rho(\mathcal{R}) \equiv \iint_{\mathcal{R}} \rho^2 \, dx dy$$

is defined and such that $A_\rho(\mathcal{R})$ and

$$L_\rho(\mathcal{R}) \equiv \inf_{\gamma \in \Gamma} \int_\gamma \rho \, |dz|$$

are not simultaneously 0 or ∞. We designate the quantity

$$m(\gamma) = \inf_{\rho \in R} \frac{A_\rho(\mathcal{R})}{(L_\rho(\Gamma))^2}$$

as the module of Γ.

Now let Ω be a simply connected domain of hyperbolic type in $\mathcal{C}$ with $z_0 \in \Omega$. If $r > 0$ is small, then

$$\Omega \cap \{z : |z - z_0| > r\}$$

is a doubly connected domain $\Omega(r)$. Let $\Omega(r)$ have module $m(r)$ for the class of curves separating its boundary components. The limit

$$\lim_{r \to 0} \left(m(r) + \frac{1}{2\pi} \log r \right)$$

is called the reduced module.

reduced norm   Let $\rho : A \to \mathrm{End}_K(M)$ be a finite dimensional linear representation of an associative algebra A over a field K in a finite dimensional K-module M. Fix a basis for M and, for $a \in A$, let T_a denote the matrix representation, with respect to this basis, of the endomorphism $\rho(a)$. For $a \in A$, define $N_\rho(a) = \det T_a$; this is called the norm of a and it depends only on the representation class of ρ. The norm is, up to a constant of the form $(-1)^r$ (where r is the degree of the representation), equal to the constant term in the characteristic polynomial of T_a. If $p_\alpha(x)$ denotes the principal polynomial of α, then the *reduced norm* of α is the constant term of $p_\alpha(x)$ multiplied by $(-1)^r$.

reduced orthogonal group   *See* reduced Clifford group.

reduced quadratic form   Let V be an n-dimensional vector space over $\mathbf{R}$ and $q(x) = \sum a_{ij}\xi_i\xi_j$ a quadratic form on V, where the a_{ij}, $i, j \in \{1, \dots, n\}$ are given elements of $\mathbf{R}$, and the $\xi_i \in \mathbf{R}$ are the coordinates of the vector $x \in V$ relative to a fixed basis of V. Set $A = [a_{ij}]$. The quadratic form q is said to be *reduced* if the matrix A is positive definite and satisfies: (i.) $a_{i\,i+1} \geq 0$ for $1 \leq i \leq n - 1$ and (ii.) $x_k^T A x_k \geq a_{kk}$ whenever x_k is a $1 \times n$ vector of the form $x_k = [y_1, \dots, y_n]^T$, where the y_i are integers such that the greatest common divisor of $y_k, y_{k+1}, \dots, y_n$ is 1.

reduced representation   (1) Suppose A is an associative algebra over a field. The direct sum of all absolutely irreducible representations of A is called the *reduced representation* of A. Its character is the reduced character. *See* absolutely irreducible representation, reduced character.

(2) Let I be an ideal in a commutative ring with identity. The ideal I is said to have a primary representation (or primary decomposition) if $I = Q_1 \cap Q_2 \cap \cdots \cap Q_n$, where each Q_i is a primary ideal. If no Q_i contains the intersection of all the others, and if the radicals of the Q_i (which are prime ideals) are distinct, the primary representation is said to be *reduced*.

reduced scheme   A scheme $(X, \mathcal{O}_X)$ such that, for every open set $U \subseteq X$, the ring $\mathcal{O}_X(U)$ has no nilpotent elements. *See also* scheme.

reduced von Neumann algebra   Let H be a Hilbert space and let $\mathcal{B}(H)$ denote the algebra of bounded operators on H. A subalgebra $\mathcal{M}$ of $\mathcal{B}(H)$ is called a $*$-subalgebra if $T^* \in \mathcal{M}$ whenever $T \in \mathcal{M}$. A $*$-subalgebra is called a von Neumann algebra if it contains the identity and is closed in the weak operator topology. Let $P \in \mathcal{B}(H)$ be a projection operator in a von Neumann algebra $\mathcal{M}$. Then the set $\mathcal{M}_P = \{PTP : T \in \mathcal{M}\}$ is a $*$-subalgebra of $\mathcal{B}(H)$ which is closed in the weak operator topology but does not contain the identity. Nevertheless, because P is a projection, the operators in this subalgebra map the Hilbert space $P(H)$ to itself and $\mathcal{M}_P$ is a von Neumann algebra on the Hilbert space $P(H)$; $\mathcal{M}_P$ is called a *reduced von Neumann algebra* on $P(H)$.

reducible component   A component of a topological space X that is not irreducible. Here a nonempty subset Y of X is *irreducible* if it cannot be expressed as the union $Y = Y_1 \cup Y_2$ of two nonempty proper subsets, each of which is closed in Y.

reducible equation   Suppose $p(x)$ is a polynomial with coefficients in a field K, i.e., $p(x) \in K[x]$. An equation of the form $p(x) = 0$ is said to be irreducible if $p(x)$ is irreducible over $K[x]$; otherwise the equation is said to be reducible. *See also* irreducible equation, reducible polynomial.

reducible linear representation (1) Let K be a commutative ring with identity and let M be a K-module. Let A be an associative K algebra. A linear representation of A in M is an algebra homomorphism, $\rho : A \to \text{End}_K(M)$ of A into the algebra of all K endomorphisms of M. Such a representation can be used to make M into a left A module by defining $am = \rho(a)m$, whenever $a \in A, m \in M$. If M is a simple A module, then ρ is said to be irreducible; otherwise it is called *reducible*.

(2) Let G be a group and K a field. A linear representation of G over K is a homomorphism ρ from G to the matrix group $\text{GL}_m(K)$. The representation is said to be *reducible* if there is another representation $\sigma : G \to \text{GL}_m(K)$ such that there exists a matrix $T \in \text{GL}_m(K)$ with $T^{-1}\rho(g)T = \sigma(g)$ for all $g \in G$ and so that $\sigma(g)$ has the form $\sigma(g) = \begin{pmatrix} \sigma_{11}(g) & \sigma_{12}(g) \\ 0 & \sigma_{22}(g) \end{pmatrix}$ for all $g \in G$.

reducible linear system Let X be a complete irreducible variety. Consider $f_1, \ldots, f_n$, elements of the function field of X, and let D be a divisor on X with $D + (f_i)$ positive for every i. (Here (f_i) denotes the divisor of f_i. *See* principal divisor of functions.) The set of all divisors of the form $(\sum k_i f_i) + D$, where the k_i are elements of the underlying field, and not all zero, is called a linear system. The linear system is irreducible if all its elements are *irreducible;* if not, then it is called *reducible*.

reducible matrix An $n \times n$ matrix A such that either (i.) $n = 1$ and $A = 0$ or (ii.) $n \geq 2$ and there is an $n \times n$ permutation matrix P and an integer m with $1 \leq m < n$ such that $P^T A P = \begin{bmatrix} C & D \\ 0 & E \end{bmatrix}$, where C is $m \times m$, E is $(n-m) \times (n-m)$, D is $m \times (n-m)$, and 0 denotes an $(n-m) \times m$ matrix of zeros.

reducible polynomial A polynomial $f(x) \in R[x]$, that is, a polynomial with coefficients in a ring R, which can be written as $f(x) = g(x)h(x)$ with $g(x)$ and $h(x)$ polynomials of lower degree which are not units in R.

reducible variety Let K be a field. If $S \subset K[x_1, \ldots, x_n]$, the set of all common zeros of S, that is, the set of all $(a_1, \ldots, a_n) \in K^n$ such that $f(a_1, \ldots, a_n) = 0$ for every $f \in S$, is called an affine K-variety. Likewise, a set in n-dimensional projective space over K consisting of common zeros of a set of homogeneous polynomials is called a projective variety. An affine or projective K-variety V is *reducible* if it can be written as $V = W_1 \cup W_2$, where each W_i is a K-variety and a proper subset of V.

reduction formulas of trigonometry The formulae: $\sin(\theta + 2\pi k) = \sin(\theta)$, $\cos(\theta + 2\pi k) = \cos(\theta)$ for k an integer; $\sin(\frac{\pi}{2} - \theta) = \cos(\theta)$, $\cos(\frac{\pi}{2} - \theta) = \sin(\theta)$ and $\sin(\theta + \pi) = -\sin(\theta)$, $\cos(\theta + \pi) = -\cos(\theta)$. These allow one to express a trigonometric function of any angle in terms of the sine and/or cosine of an angle in the interval $[0, \frac{\pi}{2})$.

reduction modulus The construction of ρ^σ from ρ, where K, L are commutative rings with identity, $\sigma : K \to L$ is an isomorphism such that $\sigma : K \to K/m$ is the canonical homomorphism, m is an ideal of K, M^σ is the scalar extension $\sigma^* M = M \otimes_K L$ of a K-module M relative to σ, and ρ is the linear representation associated with M. The linear representation ρ^σ over L associated with M^σ is the scalar extension of ρ relative to σ and is the conjugate representation of ρ relative to σ.

reduction of algebraic expression A modification of an *algebraic expression* to a simpler or more desirable form. For example, $(x+2)^2 - (x+2)(x-2)$ can be reduced to $4(x+2)$.

reductive A Lie algebra $\mathcal{G}$ is reductive if it is fully reducible under ad $\mathcal{G}_0$.

reductive action Let K be a commutative ring with identity and let $\rho : G \to \text{GL}(m, K)$ be a rational representation of a matrix group $G \subset \text{GL}(n, K)$. Suppose that R is a ring generated by elements $x_1, \ldots, x_m$ over K. An action of the group $\rho(G)$ on R can then be used to define an action of G on R by $gr = \rho(g)r$, $g \in G$, $r \in R$. This action is said to be reductive if the following condition is satisfied: Suppose M is a module generated by $y_1, \ldots, y_r \in R$ over K, which is mapped by G to itself and $f_0 \mod M$ is G-invariant. Then there is a homogeneous linear form h in $f_0, f_1, \ldots, f_r$ such that h is monic in f_0 and h is G invariant. *See* rational representation.

reductive group   Let G be an algebraic group. Let N denote the maximal connected solvable closed normal subgroup of G. If the identity element is the only unipotent element of N, then G is called a *reductive group*.

reductive Lie algebra   A Lie algebra A such that the radical of A is the center of A. This is equivalent to the condition that the adjoint representation of A is completely reducible.

The radical of A is the union of all solvable ideals of A.

redundant equation   (1) An equation containing extraneous roots that have been introduced by an algebraic operation. For example, the equation $x - 2 = \sqrt{x}$ when squared yields the equation $x^2 - 5x + 4 = 0$. This has roots 4 and 1; however, 1 is not a root of the original equation. The equation $x^2 - 5x + 4 = 0$ is said to be *redundant* because of this extraneous root.

(2) An equation in a system of equations that is a linear combination of other equations in the system. The solution to the system remains unchanged if this equation is excluded.

Ree group   Any member of the family of finite simple groups of Lie type, $^2G_2(3^n)$ and $^2F_4(2^n)$, for n odd.

refinement of chain   A chain C_1 in a partially ordered set M is a *refinement* of a chain C_2 in M if the least and greatest elements of each chain agree, and if every element of the chain C_2 belongs to C_1. C_1 is a *proper refinement* of C_2 if there is at least one element of C_1 which does not belong to C_2.

reflection   (1) The mirror image of a point or object.

A reflection in a line of a point P_1 is the point P_2 on the opposite side of the line equidistant from the line; that is, the point P_2 such that the line is the perpendicular bisector of the segment $\overline{P_1 P_2}$, joining the point and its reflection. The reflection of a geometric object in a line is just the collection of all reflections of its constituent points in the line.

If (x, y) is a point in the plane, the reflection of (x, y) in the origin is the point $(-x, -y)$; the reflection of a geometric object in the origin is the collection of reflections of its constituent points.

In a similar fashion, define the reflection of a point in a plane; for example, the reflection of (x, y, z) in the x, y plane in three dimensions is the point $(x, y, -z)$. Likewise, reflection in the origin is defined in the same way in higher dimensions.

(2) In the study of groups of symmetries of geometric objects, a group element corresponding to a transformation reflecting the object in a line or plane. In greater generality, a reflection is a mapping of an n-dimensional simply connected space of constant curvative which has an $n - 1$ dimensional hyperplane as its set of fixed points. In this case, the set of fixed points of the mapping is called the mirror of the mapping.

reflexive property   The property of a relation R on a set S:

$$(a, a) \in R \text{ for every } a \in S .$$

In other words, a relation has the reflexive property if every element is related to itself. *See* relation.

reflexive relation   A relation having the reflexive property. *See* reflexive property.

regula falsi   Literally, "rule of false position," an iterative method for finding a real solution of an equation $f(x) = 0$. Suppose that $f(x_1)$ and $f(x_2)$ have opposite sign. The method is to consider the point of intersection of the x-axis with the chord through $(x_1, f(x_1))$ and $(x_2, f(x_2))$. This point has x coordinate

$$x_3 = x_1 - f(x_1) \left(\frac{x_2 - x_1}{f(x_2) - f(x_1)} \right) .$$

The procedure is repeated with the point $(x_3, f(x_3))$ and either $(x_1, f(x_1))$ or $(x_2, f(x_2))$, depending on which of $f(x_1)$ or $f(x_2)$ has sign opposite to $f(x_3)$. This converges more slowly than Newton's method.

regular cone   A subset $C \subset V$ of a finite dimensional vector space over **R** such that C is convex, C contains no lines, and $\lambda c \in C$ whenever $c \in C$ and $\lambda > 0$.

regular extension   A field extension K/k such that k is algebraically closed in K and K is separable over k. Equivalently, the extension is regular if K is linearly disjoint from $\bar{k}$ (the algebraic closure of k) over k.

regular function   (**1**) In complex analysis, an analytic or holomorphic function. *See* analytic function, holomorphic function.

(**2**) In algebraic geometry, a function which can be represented locally as a quotient of polynomials. In more detail, let X be a variety. (A variety is the solution set of a system of polynomial equations.) Let V be an open subset of X. Let $f : V \rightarrow k$ be a function from V to a field k. The function f is *regular at the point* $p \in V$ if there is an open neighborhood U of p, contained in V, on which $f = g/h$, where g and h are polynomials and $h \neq 0$ on U. f is *regular* if it is regular at each point of its domain V.

regular ideal   A left ideal J in a ring R, for which there exists an $e \in R$ such that $r - re \in J$ for every $r \in R$. Similarly, a right ideal J, for which there exists an $e \in R$ such that $r - er \in R$ for every $r \in R$. Note that if R has an identity, then all ideals are regular.

regular local ring   Let R be a local ring and set $k = R/m$. Then k is a field. A local Noetherian ring is called *regular* if $\dim_k m/m^2 = \dim R$. (Here dim is the Krull dimension. *See* Krull dimension.) *See also* local ring.

regular mapping   (**1**) A differentiable map with nonvanishing Jacobian determinant between two manifolds of the same dimension.

(**2**) A mapping given by an n-tuple of regular functions from a variety to a variety contained in affine n-space. *See* regular function.

regular matrix   (**1**) A Markov chain is a probability model for describing a system that randomly moves among different states over successive periods of time. For such a process,

let a_{ij} denote the probability that a process currently in state j will in the next step be in state i. The matrix $A = [a_{ij}]$ is called the transition matrix of the chain. Note that the entries of A are all nonnegative and the sum of the elements in each column of A is 1. The matrix A is said to be *regular* if there is a positive integer n such that A^n has all positive entries.

(**2**) Sometimes, a synonym for non-singular or invertible matrix. *See* inverse matrix.

regular module   Let R be a commutative ring with 1 and M an R-module. An element $x \in M$ is said to be *regular* if, for every $r \in R$, there exists $t \in R$ such that $rx = rtrx$. A submodule N of M is regular if each element of N is regular. In particular, M is regular if each of its elements is regular.

regular module   Let R be a commutative ring with 1 and M an R-module. An element $x \in M$ is said to be regular if for every $r \in R$, there exists $t \in R$ such that $rx = rtrx$. A submodule N of M is regular if each element of N is regular. In particular, M is regular if each of its elements is regular.

regular polyhedral group   The group of motions which preserve a regular polygon in the plane, or the group of motions in three dimensional space that preserve a regular polyhedron. The group of motions preserving a regular n-gon in the plane is called the *dihedral group* D_n; it has order $2n$. The alternating group A_4 can be realized as the group of motions preserving a tetrahedron, and A_5 can be realized as the group of motions preserving an iscosahedron; they are called the *tetrahedral* and *iscosahedral groups*, respectively. Similarly, the symmetric group S_4 is called the *octahedral group* because it can be realized as the group of motions preserving an octahedron.

regular prime   A prime number which is not irregular. *See* irregular prime.

regular representation   (**1**) Let K be a commutative ring with identity and let A be a K algebra. An element $a \in A$ can then be viewed as an element of the algebra of K endomorphisms on A via $x \mapsto ax$ for all $x \in A$. Such a representation is called a *left regular* representation. If A has an identity, this is a faithful representation. In a similar way, define right regular representation.

(2) Let G be a group and, for $a \in G$, define a mapping $g \mapsto ag$. In this way each $a \in G$ induces a permutation on the set G. This gives a representation of G as a permutation group on G; this is called the *left regular representation* of G. In a similar way, define right regular representation. *See also* left regular representation, right regular representation.

regular ring (1) An element a in a ring R is called *regular in the sense of Von Neumann* if there exists $x \in R$ such that $axa = a$. If every element of R is regular, then R is said to be a *regular ring*. Note that every division ring is regular, whereas the ring of integers is not.

(2) A regular local ring. *See* regular local ring.

regular system of equations (1) A nonsingular system of n linear equations in n unknowns. If we denote such a system as $Ax = b$, where A is an $n \times n$ matrix, and x and b are $n \times 1$ matrices, then A has nonzero determinant and for each $n \times 1$ vector b there is a unique solution x. Also called a *Cramer system*.

(2) More generally, if $p_1(x_1, \ldots, x_m), \ldots, p_n(x_1, \ldots, x_m)$ are n polynomials in m variables over a field k, the system of equations $p_1 = 0, \ldots, p_n = 0$ is called *regular* if it has a finite number of solutions in an algebraically closed field containing k.

regular transitive permutation group A *permutation group* P is a subgroup of the symmetric group of permutations of a set S. If S has at least two elements and if, for every $\alpha, \beta \in S$, there exists a $p \in P$ such that $p\alpha = \beta$, then P is said to be *transitive*. If, in addition, no element of P, except for the identity, fixes any element of S, then P is called a *regular* transitive permutation group.

regular variety A non-singular projective algebraic variety with zero irregularity. A variety which is not regular is called an *irregular variety*. *See* irregular variety, irregularity.

regulator Let K be an algebraic number field and let $U(K)$ denote the units of K. If K has degree n, then there are n field embeddings of K into the complex numbers $\mathbf{C}$. Let r_1 be the number of real embeddings, and r_2 be the number of complex embeddings (these occur in pairs) so that $r_1 + 2r_2 = n$. Write $\sigma_1, \ldots, \sigma_{r_1}$ for the real embeddings and $\sigma_{r_1+1}, \ldots, \sigma_n$ for the complex embeddings and order the latter so that $\overline{\sigma}_{r_1+i} = \sigma_{r_1+r_2+i}$ for $i = 1, \ldots, r_2$. It is a theorem of Dirichlet that $U(K)$ is a finitely generated Abelian group of rank $r = r_1 + r_2 - 1$ and so there exists a fundamental system of units for K; these we denote by $u_1, \ldots, u_r$. For $x \in \mathbf{C}$ let $\|x\|$ denote the absolute value of x if x is real, and the square of the modulus of x if x is complex. The *regulator* of K is the absolute value of the determinant of any (all have the same value) $r \times r$ submatrix of the $r \times (r+1)$ matrix $[\ln\|\sigma_j(u_i)\|]$, $1 \le i \le r$, $1 \le j \le r+1$.

related angle Given an angle α whose terminal side does not necessarily lie in the first quadrant, the *related angle* of α is the angle in the first quadrant whose trigonometric functions have the same absolute value. For example, $\frac{\pi}{6}$ radians is the related angle of $\frac{7\pi}{6}$ radians, 70° is the related angle of 290°.

relation For A and B be sets, a *relation between A and B* is a subset R of $A \times B$. Most often considered is the case when $A = B$; in this case it is said that R is a *relation on A*. For example, $=$ is a relation on the set of real numbers; for each pair $(a, b) \in \mathbf{R} \times \mathbf{R}$ it can be determined whether or not $(a, b) \in R = \{(a, a)\}$. Similarly, $<, >, \le$, and $\ge$ are relations on the real numbers. *See also* reflexive relation, symmetric relation, transitive relation, equivalence relation.

relative algebraic number field If K is an algebraic number field and k a subfield of K, the extension of K over k is called a *relative algebraic number field*. *See* algebraic number field.

relative Bruhat decomposition The decomposition

$$(G/P)_k = G_k/P_k = \cup_{w \in_k W} \pi\left((U'_w)k\right),$$

where G is a connected reductive group defined over a field k, P is a minimal parabolic k-subgroup of G containing a maximal k-split torus S, $U = R_n(P)$ is the unipotent radical of P, $_kW = N/Z$ is the k-Weyl group of G relative to S, $Z = Z_G(S)$ is the centralizer of S in G, $N = N_G(S)$ is the normalizer of S in G, two k-subgroups U'_w and U''_w are such that $U = U'_w \times U''_w$ whenever $w \in_k W$, and π is the projection $G \to G/P$.

relative chain complex   If R is a ring, a chain complex of R-modules is a family $\{K_n, \partial_n\}$, $n \in \mathbb{Z}$, of R-modules K_n and R-module homomorphisms $\partial_n : K_n \to K_{n-1}$ such that $\partial_n \partial_{n+1} = 0$ for every n. A subcomplex of such a chain complex is a family of R-submodules $S_n \subset K_n$ such that $\partial_n S_n \subset S_{n-1}$ for every n. In this case, each ∂_n induces a well-defined homomorphism of quotient modules: $\partial'_n : K_n/S_n \to K_{n-1}/S_{n-1}$; furthermore $\partial'_n \partial'_{n+1} = 0$ for every n. $\{K_n/S_n, \partial'_n\}$ is called a relative chain complex. For a chain complex $K = \{K_n, \partial_n\}$ of R-modules, the nth homology module of the complex is defined as the quotient module $H_n(K) = \text{Ker } \partial_n/\text{Im } \partial_{n+1}$. If $S = \{S_n\}$ is a subcomplex of K, then the n-th relative homology module is defined as $H_n(K, S) = \text{Ker } \partial'_n/\text{Im } \partial'_{n+1}$. *See also* relative cochain complex.

relative cochain complex   If R is a ring, a cochain complex of R is a family $\{K_n, d_n\}$, $n \in \mathbf{Z}$, of R-modules K_n and R-module homomorphisms $d_n : K_n \to K_{n+1}$ such that $d_{n+1}d_n = 0$ for every n. (In other words, $\{K_{-n}, d_{-n}\}$ is a chain complex.) A cochain subcomplex of such a cochain complex is a sequence of R-submodules $S_n \subset K_n$ such that $d_n S_n \subset S_{n+1}$ for every n. In this case each d_n induces a well-defined homomorphism of quotient modules: $d'_n : K_n/S_n \to K_{n+1}/S_{n+1}$; furthermore, $d'_{n+1}d'_n = 0$ for every n. $\{K_n/S_n, d'_n\}$ is called a *relative cochain complex*. Cohomology and relative cohomology are then defined in a way completely analogous to the definition of homology and relative homology. *See* relative chain complex.

relative degree   Let F be a finite dimensional field extension of a field K, and denote the degree of the extension by $[F : K]$. Suppose L and

M are intermediate fields: $K \subseteq L \subseteq M \subseteq F$. The *relative degree* (or dimension) of L and M is the number $[M : L]$. This terminology allows a concise statement of the Fundamental Theorem of Galois Theory: If F is a finite dimensional Galois extension of K, then the relative degree of two intermediate fields is equal to the relative index of their corresponding subgroups of the Galois group.

relative derived functor   A homological algebra concerns a pair of algebraic categories $(\mathcal{U}, \mathcal{M})$ and a fixed functor $\Delta : \mathcal{U} \to \mathcal{M}$. A short exact sequence of objects of $\mathcal{U}$

$$0 \to A \to B \to C \to 0$$

is called *admissible* if the exact sequence

$$0 \to \Delta A \to \Delta B \to \Delta C \to 0$$

splits in $\mathcal{M}$. By means of the class $\mathcal{E}$ of admissible exact sequences, the class of $\mathcal{E}$-*projective* (resp. $\mathcal{E}$-*injective*) objects is defined as the class of objects P (resp. Q) for which the functor $\text{Hom}_{\mathcal{U}}(P, -)$ (resp. $\text{Hom}_{\mathcal{U}}(-, Q)$) is exact on the admissible short exact sequences. If $\mathcal{U}$ contains enough $\mathcal{E}$-projective or $\mathcal{E}$-injective objects, then the usual construction of homological algebra makes it possible to construct derived functors in this category, which are called *relative derived functors*. *See also* derived functor.

relative different   Let K be a relative algebraic number field over k. Let $\mathcal{D}$ and d denote the integers of K and k, respectively; that is, $\mathcal{D} = K \cap I$ and $d = k \cap I$, where I is the algebraic integers. Set $\mathcal{D}' = \{\alpha \in K : \text{Tr}_{K/k}(\alpha \mathcal{D}) \subset d\}$, where $\text{Tr}_{K/k}$ denotes the trace. $\mathcal{D}'$ is a fractional ideal in K and $(\mathcal{D}')^{-1} = \mathcal{D}_{K/k}$ is an integral ideal of K. It is called the *relative different* of K over k.

relative discriminant   Suppose K is a relative algebraic number field over k. The *relative discriminant* of K over k is the norm, with respect to K/k, of the relative different of K/k. *See* relative different.

relative homological algebra   A homological algebra associated with a pair of Abelian categories $(\mathcal{U}, \mathcal{M})$ and a fixed functor

$$\Delta : \mathcal{U} \to \mathcal{M}.$$

The functor $\Delta : \mathcal{U} \to \mathcal{M}$ is taken to be additive, exact, and faithful.

relative invariant   (1) Let K be a field and let F be an extension field of K. Let G be a group of automorphisms of F over K, and suppose that K is the fixed field of the group G. A *relative invariant* of G in F is a nonzero element $Q \in F$ such that, for each $\sigma \in G$, there exists an element $\chi(\sigma)$ in K for which $\sigma(Q) = \chi(\sigma)Q$.

(2) In algebraic geometry, a quantity invariant under birational transformation.

relatively ample   Let $(X, \mathcal{O}_X)$ be a scheme. Let F be a sheaf of $\mathcal{O}$-modules and construct a projective bundle $P(F)$ on X. A locally closed S-subscheme $f : X \to S$ of $p : P(F) \to S$ is called *quasiprojective* over S. For such a quasiprojective scheme, an invertible sheaf L on X is called *ample* over S if there exist a locally free $\mathcal{O}_S$-module of finite type on S and an immersion $\iota : X \to P(F)$ such that $\mathcal{O}_X(1) = L$. If $L^{\otimes n}$ is ample for some n, L is said to be *relatively ample* over S.

relatively minimal   *See* relatively minimal model.

relatively minimal model   A quasiprojective variety X' is said to *dominate* X if there exists a regular birational map $f : X' \to X$. A variety is a *relatively minimal variety* if it does not dominate any variety except those isomorphic to itself. A representative of a birational equivalence class is called a *model;* such a representative which is a relatively minimal variety is called a *relatively minimal model.*

relatively minimal variety   *See* relatively minimal model.

relatively prime numbers   Two integers which have no common factor other than 1 or -1.

relative norm   Let K and k be algebraic number fields with $k \subset K$ and set $n = [K : k]$. Then there are n k-linear embeddings $\phi_i : K \to \mathbf{C}$. For an ideal U of K, $U^{(i)} = \{\phi_i(u) : u \in U\}$ is an ideal of $\phi_i(K)$. Let L be the field generated by $\phi_i(K)$, $i = 1, \ldots, n$. The ideal in L gener-

ated by the $U^{(i)}$ is the extension of an ideal a of k; a is called the *relative norm* of U over k.

remainder   (1) If a is an integer and b is a positive integer, then there is a unique pair of integers q, r with $0 \leq r < b$ such that $a = qb + r$. The integer q is called the *quotient* and r is called the *remainder.*

(2) More generally, the terminology can be applied in any Euclidean domain; here, by definition, there is such a division algorithm. This can also be applied in polynomial rings: If R is an integral domain and if $a(x), b(x) \in R[X]$, then there exists $q(x), r(x) \in R[X]$ with $deg\, r(x) < deg\, b(x)$ or $r(x) = 0$ such that $a(x) = b(x)q(x) + r(x)$. Again, $q(x)$ is called the quotient of the division of $a(x)$ by $b(x)$ and $r(x)$ is called the *remainder.*

Remainder Theorem   Suppose R is a ring with identity, for example, the integers. Suppose that $f(x) = a_n x^n + a_{n-1} x^{n-1} + \cdots + a_1 x + a_0 \in R[x]$, that is, $f(x)$ is a polynomial with coefficients in R. The Remainder Theorem states that for any $c \in R$ there exists a unique $q(x) \in R[x]$ such that $f(x) = q(x)(x - c) + f(c)$.

repeating decimal   A decimal representation of a number in which a digit, or sequence of digits, is repeated without termination. For example, $\frac{1}{3} = .333\ldots$, $\frac{11}{7} = 1.571428571428\ldots$ We may consider any decimal that terminates as a repeating decimal by simply considering that it ends in an infinite sequence of zeros. With this convention, every repeating decimal is a rational number.

replica   Let V be a vector space over a field K and V^* its dual. For A in $\text{End}_K V$ (the K-endomorphisms of V) define $A^* \in \text{End}_K(V^*)$ via $A^* u(x) = u(Ax)$ for every $x \in V$ and $u \in V^*$. Let V_s^r denote the tensor product of s copies of V and r copies of V^*. Define A_s^r on V_s^r by
$$A_s^r(x_1 \otimes \cdots \otimes x_r \otimes u_1 \otimes \cdots \otimes u_r) =$$
$$\sum_{i=1}^{s} x_1 \otimes \cdots \otimes Ax_i \otimes \cdots \otimes x_s \otimes u_1 \otimes \cdots \otimes u_r$$
$$- \sum_{j=1}^{r} x_1 \otimes \cdots \otimes x_s \otimes u_1 \otimes \cdots \otimes A^* u_j \otimes \cdots \otimes u_r.$$
An element $B \in \text{End}_K(V)$ is called a *replica* of A if the nullspace of A_s^r is a subspace of the nullspace of B_s^r for every r and s.

representation   A mapping from an algebraic system to another system, usually to one which is better understood. The idea is to preserve the structure of the original system as an image and to take advantage of knowledge of the system containing this image. There are a multitude of examples of this and here we list only the more common representations.

(1) Let G be a group. Let $\rho : G \to H$ be a homomorphism of G to another group H; ρ is then called a *representation* of G. In particular, a homomorphism $\rho : G \to S_n$ of G into a permutation group S_n of permutations of a set of n elements is known as a permutation representation. Cayley's Theorem is that every finite group can be so represented using a monomorphism. A homomorphism of G into $\mathrm{GL}(n, K)$, the set of invertible $n \times n$ matrices over a commutative ring K with identity, is called a *matrix representation* of G.

(2) Let R be a commutative ring with identity and let M be a module over R. Let $\mathrm{End}_R(M)$ denote the ring of R-linear maps of M onto itself. Let A be an associative R-algebra. A *representation* of A in M is an R-algebra homomorphism $\rho : A \to \mathrm{End}_R(M)$.

(3) Let X be a Banach space and A a Banach algebra. Let $\mathcal{B}(X)$ denote the space of bounded linear operators on X. A *representation* of A in X is an algebra homomorphism $\rho : A \to \mathcal{B}(X)$ which satisfies $\|\rho(a)\| \leq \|a\|$ for every $a \in A$.

(4) Let G be a topological group and let H be a Hilbert space. A homomorphism U from G to the group of unitary linear operators on H is called a *unitary representation* of G if U is strongly continuous, that is, for every $x \in H$, $g \mapsto U_g x$ is a continuous mapping from G to H.

representation module   Let R be a commutative ring with identity and let $\rho : A \to \mathrm{End}_R(M)$ be a linear representation of an R-algebra A in an R-module M. M can be made into a left A-module, called the *representation module* of ρ, by defining $am = \rho(a)m$, for $a \in A, m \in M$.

representation space   For a representation from an algebraic structure G to the space of endomorphisms on an algebraic structure H, the *representation space* is H. *See* representation. We give several examples; all are fairly similar.

(i.) Let k be a commutative ring with identity and A an associative algebra over k. Let E be a module over k and let $\rho : A \to \mathrm{End}_k(E)$ (the space of k endomorphisms of E) be a representation. E is called the *representation space*.

(ii.) Let A be a Banach algebra and X a Banach space. For a representation of A on X, X is the *representation space*.

(iii.) Let G be a topological group and let ρ be a representation of G to the group of unitary operators on some Hilbert space H. H is called the *representation space*.

representation without multiplicity   Let U be a unitary representation of a topological group G on a Hilbert space H, that is, a homomorphism U from G to a group of unitary operators on H. A subspace N of H is called invariant if $U_g(N) \subset N$ for every $g \in G$. (For $g \in G$, U_g denotes the unitary operator that is the image of g.) For $g \in G$ let $V_g = U_g |_N$. This gives a unitary representation V of G on N; this is called the subrepresentation of U on N. If H can be written as the direct sum $H = H_1 \oplus H_2$ where the H_i are orthogonal, closed, and invariant, then U is said to be the direct sum of the subrepresentations U_1 and U_2 of U on H_1 and H_2, respectively. A unitary representation is called a *representation without multiplicity* if, whenever U is the direct sum of subrepresentations U_1 and U_2, then U_1 and U_2 have the property that no subrepresentation of U_1 is equivalent to a subrepresentation of U_2. (Two unitary representations $U_1 : G \to H_1$ and $U_2 : G \to H_2$ are said to be equivalent if there exists an isometry $T : H_1 \to H_2$ such that $T U_{1g} = U_{2g} T$ for all $g \in G$.)

representative function   Let G be a compact Lie group, and let $C(G)$ denote the algebra of continuous complex valued functions on G. For each $g \in G$, let T_g denote the operator of left translation: $T_g(x) = gx$ for $x \in G$. T_g acts on $C(G)$ by $T_g(f) = f \circ T_g$. A function $f \in C(G)$ is called a *representative function* if the vector space over $\mathbf{C}$ generated by the set of all $T_g(f)$, $g \in G$, has finite dimension. The collection of all such functions is called the *representative ring* of G.

representative of equivalence class An equivalence relation on a set S partitions the set S into disjoint sets, each consisting of related elements, that is, the set is partitioned into sets of the form $A_x = \{y \in S : x \sim y\}$, called equivalence classes. Any element of an equivalence class is called a *representative* of that equivalence class. *See* equivalence relation.

representative ring *See* representative function.

representing measure (**1**) A measure that represents a functional on a function space in the sense that the value of the functional at an element f can be realized simply as integration of f with respect to the measure. A typical theorem regarding the existence of such a measure is the Riesz representation theorem: Let X be a locally compact Hausdorff space and let T be a positive linear functional defined on the function space consisting of continuous functions on X of compact support. Then there exists a σ-algebra $\mathcal{A}$ containing the Borel sets of X and a unique positive measure μ on $\mathcal{A}$ such that $Tf = \int_X f d\mu$ for every f which is continuous and of compact support on X. In this case, μ is said to represent the functional T.

(**2**) On a function space with the property that the functionals $f \mapsto f(\zeta)$ are bounded, for ζ in some planar set Λ, the *representing measures* for the points of Λ are measures representing the evaluation functionals in the sense of (**1**). That is, representing measures are given by $\mu = \mu_\zeta$, carried on a set λ, often contained in the boundary of Λ, such that

$$f(\zeta) = \int_\lambda f(z) \, d\mu_\zeta(z) .$$

R-equivalence Let R be a ring contained in a possibly larger ring S. Let V and W be S modules with finite R bases B_V and B_W. The modules V and W are *R-equivalent* if there is an S module isomorphism between V and W which, relative to the bases B_V and B_W, has a matrix U with all entries in the ring R, and such that the inverse matrix U^{-1} also has all entries in the ring R. It is a theorem that R-equivalence is independent of the particular choice of the bases, B_V and B_W. *See also* isomorphism.

The most important case of this occurs when R is chosen to be the ring of integers, $\mathbf{Z}$, in which case R-equivalence is called $\mathbf{Z}$-*equivalence*. $\mathbf{Z}$-equivalence is related to the classical notion of *integral equivalence* for matrices. *See also* integral equivalence, Z-equivalence.

residue class Suppose R is a ring and I is an ideal in R. For $x, y \in R$, we say that x is congruent to y modulo I if $x - y \in I$. It is straightforward to check that congruence modulo I is an equivalence relation on R. A coset of I in R is called a *residue class* modulo I. If I is a principal ideal generated by an element $a \in R$, the residue class modulo I is sometimes referred to as the residue class modulo a; in particular, this is used in the case when $R = \mathbf{Z}$. *See also* residue ring.

residue class field Suppose R is a commutative ring with identity $1 \neq 0$, and M is an ideal in R. The quotient ring or residue ring, R/M, is a field if and only if M is a maximal ideal. In this case R/M is known as the *residue class field* or *residue field*. *See also* residue ring.

residue field *See* residue class field.

residue ring Let R be a ring and I an ideal of R. Define a relation $\sim$ on R by $x \sim y$ if $x - y \in I$; this is an equivalence relation. Since I is a normal subgroup of R under the operation $+$, the equivalence classes are cosets which we write additively: $a + I$. The set of all such cosets can be made into a ring with addition defined by $(a+I)+(b+I) = (a+b)+I$ and multiplication defined by $(a + I)(b + I) = ab + I$. This ring is called the *residue ring*, *residue class ring*, or *quotient ring* and is denoted R/I.

Residue Theorem (**1**) Suppose $f(z)$ is an analytic function with an isolated singularity at a point a. The residue of f at a is the coefficient of $(z-a)^{-1}$ in the Laurent expansion of f at a; we denote this by res$[f; a]$. The Residue Theorem states that if f is analytic in a simply connected domain Ω, except for isolated singularities, and if γ is a simple closed rectifiable curve lying in Ω and not passing through any singularities of

f, then

$$\frac{1}{2\pi i} \int_\gamma f(z)\, dz = \sum \text{res}\left[f; a_j\right] ,$$

where the sum is taken over the singularities a_j of f that lie inside γ.

(2) Let X be a complete nonsingular curve over an algebraically closed field k. Let K be the function field of X. Let Ω_X denote the sheaf of differentials of X over k and Ω_K be the module of differentials of K over k. For $P \in X$, let Ω_P denote its stalk at P. It can be shown that for each closed point $P \in X$, there exists a unique k-linear map $\text{res}_P : \Omega_K \to k$ which has the following properties: (i.) $\text{res}_P(\tau) = 0$ for all $\tau \in \Omega_P$, (ii.) $\text{res}_P(f^n df) = 0$ for all $n \neq -1$ and all f in the multiplicative group of K, (iii.) $\text{res}_P(f^{-1}df) = v_P(f) \cdot 1$, where v_P is the valuation associated to P. The *Residue Theorem* states that for any $\tau \in \Omega_K$,

$$\sum_{P \in X} \text{res}_P(\tau) = 0 .$$

resolution of singularities   The technique of applying birational transformations to an irreducible algebraic variety in hope of producing an equivalent projective variety without singularities.

restricted Burnside problem   A group with identity e is said to have exponent n if $g^n = e$ for every element g of the group. The *restricted Burnside problem* is the conjecture that if a group of exponent n is generated by m elements, then its order is bounded by a quantity $S(n, m)$ which depends only on n and m. For $n = p$ a prime was solved affirmatively by Kostrikin.

restricted direct product   Let I be an indexing set, and let $\{G_i\}, i \in I$ be a family of groups indexed by I. Consider the set $G = \prod_{i \in I} G_i$ and put a group structure on G by defining multiplication componentwise:

$$\{g_i\}_{i \in I}\, \{h_i\}_{i \in I} = \{g_i h_i\}_{i \in I} .$$

G is called the direct product of the groups G_i. The set of all elements $\{g_i\}_{i \in G}$ of G such that $g_i = e$ for all but a finite number of i forms a normal subgroup of G; it is called the *restricted direct product* (or *weak direct product*) of the groups G_i.

restricted Lie algebra   A Lie algebra L over a field K of characteristic $p \neq 0$ such that, for every $a \in L$, there exists an element $a^{[p]} \in L$ with the following properties: (i.) $(\lambda a)^{[p]} = \lambda^p a^{[p]}$ whenever $\lambda \in K$ and $a \in L$, (ii.) $[a, b^{[p]}] = [[\ldots[a, b], \ldots], b]$ (a p-fold commutator), and (iii.) $(a + b)^{[p]} = a^{[p]} + b^{[p]} + s(a, b)$ where $s(a, b) = s_1(a, b) + \cdots + s_{p-1}(a, b)$ and $(p - i)s_i(a, b)$ is the coefficient of λ^{p-i-1} in the expansion of the p-fold commutator

$$[[\ldots[a, \lambda a + b], \ldots], \lambda a + b] .$$

restricted minimum condition   A commutative ring R, with identity, satisfies the *restricted minimum condition* if R/I is Artinian for every nonzero ideal I of R. See Artinian ring.

restriction   Suppose A and B are sets and $f : A \to B$ is a function. Let $S \subset A$. The restriction of f to S is the function from S to B given by $a \mapsto f(a)$ for $a \in S$. This is often denoted by $f|S$ or $f|_S$.

resultant   (1) The sum of two or more vectors.

(2) A determinant formed from the coefficients of two polynomial equations. Consider the two equations

$$a_0 x^n + a_1 x^{n-1} + \cdots + a_n = 0$$

$$b_0 x^m + b_1 x^{m-1} + \cdots + b_m = 0$$

where $a_0 \neq 0$ and $b_0 \neq 0$. The resultant is the determinant of an $(m + n) \times (m + n)$ matrix

$$\begin{vmatrix} a_0 & \cdots & a_n & 0 & \cdots & \cdots & \cdots & 0 \\ 0 & a_0 & \cdots & a_n & 0 & \cdots & \cdots & 0 \\ \cdots & \cdots & \cdots & \cdots & \cdots & \cdots & \cdots & \cdots \\ 0 & \cdots & \cdots & \cdots & 0 & a_0 & \cdots & a_n \\ b_0 & \cdots & \cdots & b_m & 0 & \cdots & \cdots & 0 \\ 0 & b_0 & \cdots & \cdots & b_m & 0 & \cdots & 0 \\ \cdots & \cdots & \cdots & \cdots & \cdots & \cdots & \cdots & \cdots \\ 0 & \cdots & \cdots & 0 & b_0 & \cdots & \cdots & b_m \end{vmatrix}$$

Here the first m rows are formed with the coefficients a_i and the last n rows are formed with the coefficients b_i. The two polynomial equations have a common root if and only if their resultant is zero.

Richardson method of finding key matrix

Consider a real linear system of equations $Ax = b$, where A is an $n \times n$ matrix and x and b are $n \times 1$ matrices. One method for approximating the solution, x, involves finding a sequence of approximations x_k by means of an iterative formula $x_k = x_{k-1} + R(b - Ax_{k-1})$, where R is chosen to approximate A^{-1}. If the spectral radius of $I - RA$ is less than 1, the x_k converge to the solution for all choices of x_0. The Richardson method is the choice $R = \alpha A^T$, where $0 < \alpha < \frac{2}{\|A\|^2}$.

Riemann-Hurwitz formula

Let M_1 and M_2 be compact connected Riemann surfaces. If $f : M_1 \to M_2$ is holomorphic and $p \in M_1$, we may choose coordinate systems at p and $f(p)$ so that locally f can be expressed as $f(w) = w^n h(w)$ where $h(w)$ is holomorphic and $h(0) \neq 0$. (Here 0 is the image of both p and $f(p)$ under the respective coordinate maps.) Set $b_f(p) = n - 1$; this is called the branch number. The total branching number of f is defined as $B = \sum b_f(p)$, where the sum runs over all $p \in M$. The equidistribution property states that there exists a positive integer m such that every $q \in M_2$ is assumed m times counting multiplicity, that is, $\sum(b_f(p) + 1) = m$ for every $q \in M_2$, where the sum is taken over all $p \in f^{-1}(q)$. For $i = 1, 2$, let g_i denote the genus of M_i. The Riemann-Hurwitz formula states

$$2(g_1 - 1) - 2m(g_2 - 1) = B .$$

Riemann hypothesis

The Riemann zeta function has zeros at $-2, -4, -6, \ldots$. It is known that all other zeros of the zeta function must lie in the strip of complex numbers $\{z : 0 < \Re(z) < 1\}$. The Riemann hypothesis is the yet unproved conjecture that all of the zeros of the zeta function in this strip must lie on the line $\Re(z) = \frac{1}{2}$. Hardy proved that an infinite number of zeros lie on this line. There are numerous other equivalent formulations of this conjecture. Resolution of this conjecture would have important implications in the theory of prime numbers. *See* Riemann zeta function.

Riemann matrix

An $n \times 2n$ matrix A, for which there exists a nonsingular skew symmetric $2n \times 2n$ matrix P with integer entries so that $APA^T = 0$ and $iAPA^*$ (where A^* is the conjugate transpose matrix of A) is positive definite Hermitian.

Riemann-Roch inequality

(1) Let k be a field and let X be a nonsingular projective curve over k. A divisor on X is a formal sum $D = \sum_{p \in X} n_p P$ where each $n_p \in \mathbf{Z}$ and all but a finite number of $n_p = 0$. For such a D, the degree of D is defined as: $\deg(D) = \sum n_p$. Let K denote the function field of X over k. For a divisor D set $L(D) = \{f \in K : \mathrm{ord}_P(f) \geq -n_p$ for all P in $X\}$. (Here $\mathrm{ord}_P(f)$ denotes the order of f at P.) $L(D)$ is a finite dimensional vector space over k; set $\ell(D) = \dim_k L(D)$. The Riemann inequality (sometimes called the Riemann-Roch inequality) states that there exists a constant $g = g(X)$ such that $\ell(D) \geq \deg(D) + 1 - g$ for all divisors D.

(2) Let X be a surface. For any two divisors C and D on X we let $I(C.D)$ denote their intersection number. For a divisor D let $\mathcal{L}(D)$ denote the associated invertible sheaf and set $\ell(D) = \dim H^0(X, \mathcal{L}(D))$. Let p_a denote the arithmetic genus of X. The Riemann-Roch inequality states that if D is any divisor on X and W any canonical divisor, then $\ell(D) + \ell(W - D) \geq \frac{1}{2} I(D.(D - W)) + 1 + p_a$.

Riemann-Roch Theorem

(1) Let X be a nonsingular projective curve. Let g be the genus of X. The Riemann-Roch Theorem states that if W is a canonical divisor on X, then for any divisor D, $\ell(D) - \ell(W - D) = \deg(D) + 1 - g$. *See* Riemann-Roch inequality.

(2) Let X be a surface. For any two divisors C and D on X let $I(C.D)$ denote their intersection number. For a divisor D, let $\mathcal{L}(D)$ denote the associated invertible sheaf and set $\ell(D) = \dim H^0(X, \mathcal{L}(D))$ and $s(D) = \dim H^1(X, \mathcal{L}(D))$. Let p_a denote the arithmetic genus of X. The Riemann-Roch Theorem states that if D is any divisor on X and if W is any canonical divisor, then $\ell(D) - s(D) + \ell(W - D) = \frac{1}{2} I(D.(D - W)) + 1 + p_a$.

Riemann-Roch Theorem for the adjoint system

Let X be a surface. For any two divisors

C and D on X, let $I(C.D)$ denote their intersection number. For a divisor D let $\mathcal{L}(D)$ denote the associated invertible sheaf and set $\ell(D) = H^0(X, \mathcal{L}(D))$ and $s(D) = H^1(X, \mathcal{L}(D))$. Let p_a denote the arithmetic genus of X. The Riemann-Roch Theorem for the adjoint system states that if C is a curve with r components and K is a canonical divisor, then $\ell(C + K) = s(-C) - r + \frac{1}{2}I((K + C).C) + p_a + 2$.

Riemann's period inequality   Let $\mathcal{R}$ be a compact Riemann surface of genus g. The set of all differentials of the first kind, that is, those of the form $a(z)dz$, where $a(z)$ is holomorphic, is a vector space of dimension g over $\mathbf{C}$; let $\{\omega_1, \ldots, \omega_g\}$ denote a basis. Take normal sections $\alpha_1, \ldots, \alpha_{2g}$ as a basis for the homology group of $\mathcal{R}$, with coefficients in $\mathbf{Z}$, set $\omega_{ij} = \int_{\alpha_j} \omega_i$, and let Ω denote the $g \times 2g$ matrix whose ijth entry is ω_{ij}. Set $E = \begin{bmatrix} 0 & I_g \\ -I_g & 0 \end{bmatrix}$, where I_g denotes the $g \times g$ identity matrix. Riemann's period relation states that $\Omega E \Omega^T = 0$. *Riemann's period inequality* states that the Hermitian matrix $i\Omega E \bar{\Omega}^T$ is positive definite. *See also* Riemann matrix.

Riemann's period relation   *See* Riemann's period inequality.

Riemann surface   There are analytic "functions" (for example $w = \sqrt{z}$) that are naturally considered as multiple valued. Such functions are better understood using *Riemann surfaces*: the idea is to consider the image as a surface consisting of a suitable number of sheets so that the mapping can be considered as one-to-one from the z to w surfaces. For example, the function $w = z^{\frac{1}{3}}$ can be viewed as a three-sheeted surface (all elements of the complex plane, except zero, have three cube roots) over the complex plane. Locally, this mapping is one-to-one from the plane to the surface. More precisely, a connected Hausdorff space is called a Riemann surface if it is a complex manifold of one complex dimension. Any simply connected Riemann surface can be conformally mapped to one of the following: the finite complex plane, the complex plane together with the point at infinity (the Riemann sphere), or the interior of the unit

circle. Respectively, the surface is then called parabolic, elliptic, or hyperbolic.

Riemann theta function   Let $\mathcal{R}$ be a compact Riemann surface of genus g. Let $\{\omega_1, \ldots, \omega_g\}$ be a basis for the space of Abelian differentials of the first kind, and let $\{\alpha_1, \ldots, \alpha_{2g}\}$ be normal sections forming a basis for the homology, with coefficients in $\mathbf{Z}$, of $\mathcal{R}$. (*See* Riemann's period inequality.) Set $\Omega = (\omega_{ij})$, where $\omega_{ij} = \int_{\alpha_j} \omega_i$. It is possible to choose the ω_i so that Ω has the form $\Omega = (I_g, F)$, where I_g is a $g \times g$ identity matrix and F is a $g \times g$ complex symmetric matrix whose imaginary part is positive definite. For $u = (u_1, \ldots, u_g)$ set

$$\theta(u) = \sum_m \exp\left(2\pi i \left(mu^T + \frac{1}{2}mFm^T\right)\right)$$

where the sum is taken over all $m = (m_1, \ldots, m_g) \in \mathbf{Z}^g$. This sum converges uniformly on compact subsets. θ is the *Riemann theta function*.

Riemann zeta function   The function defined by $\zeta(z) = \sum_{n=1}^{\infty} n^{-z}$ for z in the complex plane with $\Re(z) > 1$. Euler showed that $\zeta(z) = \prod_p (1 - p^{-z})^{-1}$ where p runs over all primes. There are also various other representations of this involving integral formulas. This function can be continued analytically to a meromorphic function on the complex plane with a simple pole at $z = 1$. *See also* Riemann hypothesis.

Riesz group   An ordered Abelian group G which satisfies: (i.) whenever n is a positive integer and $g \in G$ has $g^n \geq 0$ then $g \geq 0$, and (ii.) if G_1 and G_2 are finite sets in G, and if $g_1 \leq g_2$ for every $g_1 \in G_1$ and $g_2 \in G_2$ then there exists an $h \in G$ such that $g_1 \leq h \leq g_2$ for all $g_1 \in G_1$ and $g_2 \in G_2$.

right A-module   Let A be a ring. A right A-module is an additive Abelian group G, together with a mapping $G \times A \to G$, called *scalar multiplication,* and denoted by $(x, r) \mapsto xr$, satisfying, for all $r, s \in A$ and $x, y \in G$,

$$x1 = x,$$
$$(x + y)r = xr + yr,$$
$$x(r + s) = (xr + xs),$$

and

$$x(rs) = (xr)s.$$

right annihilator   Let S be a subset of a ring R. The set $\{r \in R : Sr = 0\}$ is called the right annihilator of S. Note that the right annihilator of S is a right ideal of R and is an ideal of R if S is a right ideal.

right Artinian ring   A ring that satisfies the descending chain condition on right ideals; that is, for every chain $I_1 \supset I_2 \supset I_3 \supset \ldots$ of right ideals of the ring, there is an integer m such that $I_i = I_m$ for all $i \geq m$. *See also* right Noetherian ring, left Artinian ring, left Noetherian ring, Artinian ring, Noetherian ring.

right-balanced functor   Let T be a functor of several variables, some covariant and some contravariant, from the category of modules over a ring R to the category of modules over a ring S. T is said to be right-balanced if (i.) when any of the covariant variables of T is replaced by an injective module, T becomes an exact functor in the remaining variables, and (ii.) when any one of the contravariant variables of T is replaced by a projective module, T becomes an exact functor of the remaining variables.

right coset   Any set, denoted by Sa, of all right multiples sa of the elements s of a subgroup S of a group G and a fixed element a of G. *See* left coset.

right derived functor   *See* left derived functor.

right global dimension   For a ring A, the smallest $n \geq 0$ for which A has the property that every right A-module has projective dimension $\leq n$. This is equal to the smallest $n \geq 0$ for which each right A-module has injective dimension $\leq n$. *See* projective dimension, injective dimension. *See also* left global dimension.

right G-set   A set M, acted on, on the right, by a group G, satisfying $m(g_1 g_2) = (mg_1)g_2$ for all $g_1, g_2 \in G$ and $m \in M$, and $me = m$ for every $m \in M$, where e is the identity of G.

right hereditary ring   A ring in which every right ideal is projective. *See also* left hereditary ring, left semihereditary ring, right semihereditary ring.

right ideal   A nonempty subset I of a ring R such that, whenever x and y are in I and r is in R, $x - y$ and xr are in I. *See also* left ideal, ideal.

right injective resolution   *See* right resolution.

right invariant   (1) Let G be a locally compact topological group. A Borel measure μ on G is said to be *right invariant* if $\mu(Ea) = \mu(E)$ for every $a \in G$ and Borel set $E \subset G$.

(2) Let G be a Lie group. An element $g \in G$ defines an operator of right translation by g on G: $R_g(x) = xg$ for $x \in G$. If T is a tensor field on G and if $R_g T = T$ for every $g \in G$ (where $R_g T$ is defined in the natural way), then T is said to be *right invariant*.

right inverse element   Suppose S is a nonempty set with an associative binary operation $*$. Assume that there is a right identity element $e \in S$, with respect to $*$. If $a \in S$, a right inverse element for a is an element $b \in S$ such that

$$a * b = e.$$

If every element of such an S has an inverse element, then S is called a *group*. *See also* left inverse element.

right Noetherian ring   A ring which satisfies the ascending chain condition on right ideals. That is, for every chain $I_1 \subset I_2 \subset I_3 \subset \ldots$ of right ideals there exists an integer m such that $I_i = I_m$ for every $i \geq m$. *See also* right Artinian ring, left Noetherian ring, left Artinian ring, Noetherian ring, Artinian ring.

right order   Let g be an integral domain in which every ideal is uniquely decomposed into a product of prime ideals (i.e., a Dedekind domain). Let F be the quotient field of g. Let A be a separable algebra of finite degree over F. Let L be a g-lattice of A. For such a g-lattice L, $\{x \in A : Lx \subset L\}$ is called the *left order* of A.

A g-lattice L is a g-submodule of A that is finitely generated over g and satisfies $A = FL$.

right regular representation (1) Let R be a commutative ring with identity and let M be an R-algebra. For $r \in R$, consider the mapping $\rho_r : M \to M$ given by right translation (i.e., $\rho_r(x) = xr$, for $x \in M$). The mapping $\rho : R \to \mathrm{Hom}_{\mathbf{Z}}(M, M)$ given by $\rho(r) = \rho_r$ is an antihomomorphism; it is called the *right regular representation* of M. If M has an identity, this is a faithful representation.

(2) Let G be a group, and for $a \in G$ define a mapping $g \to ga$. In this way, each $a \in G$ induces a permutation on the set G. This gives a representation of G as a permutation group on G (again, the representation is an antihomomorphism): this is called the right regular representation. *See also* regular representation, left regular representation.

right resolution Consider a cochain complex (X, δ) of A-modules X^n, $n \in \mathbf{Z}$, and A module homomorphisms $\delta_n : X^n \to X^{n+1}$ with $\delta^{n+1}\delta^n = 0$. Suppose $\varepsilon : M \to X^0$ is an augmentation of X over M, that is, an A module homomorphism from an A module M to X^0 such that the composition $M \xrightarrow{\varepsilon} X^0 \xrightarrow{\delta^0} X^1$ is trivial. If $0 \xrightarrow{\varepsilon} M \xrightarrow{\delta^0} X^0 \xrightarrow{\delta^0} \cdots \longrightarrow X^n \xrightarrow{\delta^n} \cdots$ is exact, then X is called a *right resolution* of M. If, in addition, each X^n is an injective A-module, then the cochain complex X is called a right injective resolution of M.

right satellite Let $\mathcal{A}$ be an Abelian category and let $\mathcal{R}$ be a selective Abelian category. If $\mathcal{A}$ has enough proper injectives, then a $\mathcal{P}$ connected pair (T, E_*, S) of covariant functors is right universal if and only if each proper short exact sequence in $\mathcal{A}$, $0 \to C \to J \to K \to 0$, with J proper injective, induces a right exact sequence $T(J) \to T(K) \to S(C) \to 0$ in $\mathcal{R}$. Given T, the S with this property is uniquely determined. It is called the *right satellite* of T. *See also* left satellite.

right semihereditary ring A ring in which every finitely generated right ideal is projective. *See also* left semihereditary ring, left hereditary ring, right hereditary ring.

right translation The function $f_a : G \to G$, on a group G, defined by $f_a(x) = xa$, for $x \in G$. If g is a function on G, a right translation of g by a is the function g_a defined by $g_a(x) = g(xa^{-1})$.

right triangle A triangle with a right angle. The side opposite the right angle is called the hypotenuse and the sides adjacent the right angle are often called the legs of the triangle. The Pythagorean Theorem states that in a *right triangle* the sum of the squares of the lengths of the legs is equal to the square of the length of the hypotenuse.

ring A nonempty set R, with two binary operations, usually denoted by $+$ and $\cdot$, which satisfy the following axioms: (i.) With respect to $+$, R is an Abelian group. (ii.) $\cdot$ is associative: $a \cdot (b \cdot c) = (a \cdot b) \cdot c$ for all $a, b, c \in R$. (iii.) $+$ and $\cdot$ satisfy the distributive laws: $a \cdot (b + c) = a \cdot b + a \cdot c$ and $(b + c) \cdot a = b \cdot a + c \cdot a$ for all $a, b, c \in R$. The operation $+$ is called addition and the operation $\cdot$ is called multiplication. If multiplication is commutative, the ring is called a commutative ring. If there is an identity for multiplication, that is, an element 1 which has $x \cdot 1 = 1 \cdot x = x$ for all $x \in R$, the ring is called a ring with identity. The identity for the additive Abelian group is denoted as 0. A division ring is a ring whose nonzero elements form a group under multiplication. A commutative division ring is called a field.

ringed space A topological space X, together with a sheaf of rings $\mathcal{O}_X$ on X. The main import of this statement is that to each open set U of X, there is associated a ring $\mathcal{O}_X(U)$. Example: Let X be an open subset of $\mathbf{C}^n$, or, more generally, a complex analytic manifold. For each open subset U, let $\mathcal{O}_X(U) =$ the ring of analytic (holomorphic) functions defined on U. Then the pair $(X, \mathcal{O}_X)$ is an important example of a ringed space. The sheaf $\mathcal{O}_X$ is called the *Oka sheaf*. *See* sheaf.

ring homomorphism   Let R and S be rings, each with binary operations denoted by $+$ and $\cdot$. A *ring homomorphism* from R to S is a mapping $f : R \to S$ which satisfies: (i.) $f(a + b) = f(a) + f(b)$ and (ii.) $f(a \cdot b) = f(a) \cdot f(b)$ for all $a, b \in R$.

ring isomorphism   A ring homomorphism that is both injective and surjective. *See* ring homomorphism.

ring of differential polynomials   Let K be a field and $\delta_1, \ldots, \delta_n$ a set of commuting derivations on K. An extension field L of K is called a differential extension field of K if the δ_i extend to derivations (which we again denote by δ_i) on L such that $\delta_i(k) \in K$ for every $k \in K$. Let $x_1, \ldots, x_m$ be elements of a differential extension field of K with derivations $\delta_1, \ldots, \delta_n$ and suppose that $\delta_1^{r_1} \delta_2^{r_2} \cdots \delta_n^{r_n} x_i$ (where the r_i are nonnegative integers and $1 \leq i \leq m$) are algebraically independent over K. The collection of all polynomials (over K) in these expressions is called the *ring of differential polynomials* in the variables $x_1, \ldots, x_m$.

ring of fractions   Let R be a commutative ring with identity, and S a multiplicative subset of R, that is, $1 \in S$ and S has the property that $ab \in S$ whenever $a, b \in S$. Define an equivalence relation on the set $R \times S$ by $(r, s) \sim (r', s') \Leftrightarrow s_1(rs' - r's) = 0$ for some $s_1 \in S$. Note that in the case when S contains no zero divisors and $0 \notin S$, then $(r, s) \sim (r', s') \Leftrightarrow rs' - r's = 0$. The equivalence class of (r, s) is usually denoted as $\frac{r}{s}$ and the set of all equivalence classes defined in this way is denoted $S^{-1}R$. Note that if $0 \in S$, then $S^{-1}R$ consists only of the element $\frac{0}{1}$. It can be shown that $S^{-1}R$ is a commutative ring with identity, where addition and multiplication are defined by $\frac{r}{s} + \frac{r'}{s'} = \frac{(rs'+r's)}{ss'}$ and $(\frac{r}{s})(\frac{r'}{s'}) = \frac{rr'}{ss'}$. The ring $S^{-1}R$ is called the *ring of fractions* of R by S.

ring of polynomials   Let R be a ring and x an indeterminate. The *ring of polynomials* over R, denoted $R[x]$, is the set of all polynomials in the variable x, with coefficients in R. This may be formally defined by considering the set of all sequences $(a_0, a_1, a_2, \ldots)$ of elements of R such that $a_i = 0$ for all but a finite number of indices i and defining addition and multiplication by: $(a_0, a_1, \ldots) + (b_0, b_1, \ldots) = (a_0+b_0, a_1+b_1, \ldots)$ and $(a_0, a_1, \ldots)(b_0, b_1, \ldots) = (c_0, c_1, \ldots)$ where $c_n = \sum_{i=0}^{n} a_{n-i} b_i$. With these operations, it is easy to see that $R[x]$ is a ring. Usually, however, the element $(a_0, \ldots, a_n, 0, 0, \ldots)$ is denoted by $a_0 + a_1 x + \cdots + a_n x^n$. Polynomial rings in several variables can be defined inductively: $R[x, y] = R[x][y]$.

ring of scalars   If M is a module over a ring R, R is called the *ring of scalars*. *See* module.

ring of total quotients   Let R be a commutative ring with identity and let U be the multiplicative subset of R consisting of all elements of R that are not zero divisors. The ring of total quotients is the ring of fractions $U^{-1}R$. *See* ring of fractions.

ring operations   The two binary operations that occur in the definition of ring. These are usually denoted by $+$ and $\cdot$ and are called addition and multiplication. *See* ring.

ring theory   The study of rings and their properties. *See* ring.

Ritt's Basis Theorem   In the theory of differential rings, a polynomial generated by the derivatives of a finite set S of elements is called a differential polynomial of the elements of S. *Ritt's Basis Theorem* states that any set A of differential polynomials contains a finite subset B such that, if u is any element of A, then there exists an integer power of u that is a linear combination of the elements of B and their derivatives. This theorem on differential polynomials is the analog of Hilbert's Basis Theorem in the ring of ordinary polynomials. *See also* Hilbert's Basis Theorem.

root   (**1**) (Of an equation) A number which satisfies the equation (makes the equation true when it is substituted in for the variable).

(**2**) (Of a polynomial $p(x)$) A root of the *equation* $p(x) = 0$.

(**3**) (Of a number) A number which satisfies $x^n = b$. This number is called the nth root of b, and is denoted by $\sqrt[n]{b}$.

(4) (Of a complex number) A number which satisfies $z^n = a + bi$. This number is called the nth root of $a + bi$, and is denoted by $\sqrt[n]{a + bi}$.

Also, the elements of a root system Φ. *See* root system.

root extraction   An algorithm to find the nth root of a number.

roots of unity   If n is a positive integer, then any complex z such that $z^n = 1$ is called an nth *root of unity*. There are n distinct nth roots of unity. A theorem that is useful for calculating roots of unity is *De Moivre's Theorem,* which states that for a complex number z in polar form $z = r[\cos\theta + i\sin\theta]$,

$$[r(\cos\theta + i\sin\theta)]^n = r^n(\cos n\theta + i\sin n\theta).$$

root subspace   Let E be a subfield of a field K, and let $p(x)$ be a polynomial with coefficients in E. If c is a root of $p(x)$ (i.e., $p(c) = 0$), the smallest subfield of K that contains both E and c is called a *root subspace.*

root system   A finite subset Φ of a real vector space V which satisfies three conditions: (i.) $0 \in \Phi$ and Φ generates V; (ii.) for each $\alpha \in \Phi$ there exists an element α^* in the dual space V^* of V such that the reflection map $r_\alpha : \beta \to \beta - \alpha^*(\beta)\alpha$ stabilizes Φ and such that $\langle \alpha, \alpha^* \rangle = 2$, where $\langle x, \alpha^* \rangle = \alpha^*(x)$; and (iii.) if $\alpha, \beta \in \Phi$, then $\eta_{\beta,\alpha} \equiv \langle \alpha, \beta^* \rangle \equiv \beta^*(\alpha)$ is an integer.

R-order   Let R be a ring. Let A be a finite dimensional algebra with unit element e over a field F. A subring G of A is called an *R-order* in A if it satisfies (i.) $e \in G$, (ii.) G contains a basis of A as a vector space over the field F (an F-basis of A), and (iii.) G is a finitely generated R module.

The most important case of this occurs when R is chosen to be the ring of integers, $\mathbf{Z}$, in which case an R-order is called a $\mathbf{Z}$-order. *See also* $\mathbf{Z}$-order.

Rosen's gradient projection method   An algorithm which seeks to minimize a convex differentiable function p on a non-empty convex set F. Starting with some u_0 in F, if $p(u_0)$ is not minimal, then p decreases in the direction of the gradient of $-p$. Because p is a convex differentiable function, $p(x) < p(u_0)$ for each point along the gradient of $-p$ at u_0. Map u_0 to the point u, where that gradient intersects the boundary of F. Repeat with the gradient of $-p$ at u_1, u_2, etc. We have $p(u_0) > p(u_1) > p(u_2) \ldots$ If the process converges, p is minimized at the limit of the sequence $u_0, u_1, \ldots$.

rotation   Rigid motion about a fixed axis wherein every point not on the axis is rotated through the same angle about the axis. A rotated point set retains its shape. A rotation of Euclidian n-space is a linear transformation that preserves distances and the orientation of the space.

rotation group   The group formed by the set of all rotations of $\mathbf{R}^n$, also called the *orthogonal group* of $\mathbf{R}^n$. The *rotation group* of 3-space appears prominently in theoretical mechanics.

Roth's Theorem   A result in the theory of rational approximations of irrational numbers. If $k > 2$ and a is a real algebraic number, then there are only a finite number of integer pairs (x, y) with $x > 0$ such that

$$\left| a - \frac{y}{x} \right| < \frac{1}{x^k}.$$

See algebraic number.

Rouché's Theorem   Let C be a simple rectifiable curve in the complex plane and suppose that the functions F and f are analytic in and on C. If $|F| > |f|$ at each point on C, then the functions F and $f + F$ have the same number of zeros in the finite region bounded by C. The theorem is helpful for proving the existence of and locating zeros of a complex function.

rounding of number   A method of approximating numbers. To round a decimal to the nth place, one sets all of the digits after the nth place to zero, and changes the nth place according to the following rule: If the first digit after the nth place is greater than or equal to 5, the nth place is increased by one. If the first digit after the nth place is less than five, the nth place is unaffected by the rounding. For example, π rounded to the hundredths place is 3.14; 35560 rounded to the nearest thousand is 36000.

row finite matrix   An infinite matrix wherein there are a finite number of non-zero entries in each row.

row nullity   For A an $m \times n$ matrix, the dimension of the row null space (the space of solutions described by $Ax = 0$). If A is of rank r, the row nullity is equal to $m - r$.

row of matrix   A single horizontal sequence of elements, extending across a matrix. For example, in the matrix

$$\begin{bmatrix} a_{11} & a_{12} & \cdots & a_{1n} \\ a_{21} & a_{22} & \cdots & a_{2n} \\ \cdot & \cdot & \cdot & \cdot \\ a_{m1} & a_{m2} & \cdots & a_{mn} \end{bmatrix},$$

any of the arrays $a_{j1} a_{j2} \ldots a_{jn}$ is a row.

row vector   A horizontal array of elements; a $1 \times n$ matrix.

r-ple point on a curve   Suppose C is a plane curve, P is a point on C, and F is a field. Write $C = C_r + C_{r+1} + \cdots + C_{r+s}, r, s \in \mathbf{Z}^+$, where each C_i is a form of degree i in $F[X, Y]$ and $C_r \neq 0$. Then P is called an r-ple point on the curve C. *See also* multiplicity of a point.

ruled surface   A surface S that contains, for each point p in S, a line that contains p. Some simple examples of ruled surfaces are planes and hyperboloids of one sheet.

Rule of Three   A philosophy of mathematical pedagogy first stressed by the Calculus Consortium based at Harvard University, in which graphical, numerical, and analytical approaches are used to aid in conceptualizations of problems and solutions. The same group has recently coined the term *Rule of Four* to imply the addition of a verbal approach to calculus. The term was used later by the University of Alabama in instructing probability through mathematics, simulation, and data analysis.

S

saddle point   A point at which the first two partial derivatives of a function $f(x, y)$ in both variables are zero, yet which is not a local extremum. Also called a hyperbolic point. The term *saddle* hails from the appearance of the plotted surface, wherein the saddle point appears as a mountain pass between two hills. Similar saddle phenomena occur with three hills; they are sometimes called *monkey saddles,* with an extra place for a monkey's tail.

Satake diagram   Used in the theory of Lie algebras to describe the classification of a non-compact real simple Lie Algebra L that arises from the comparison of the conjugation operation of L with the complexification of L.

satisfy   Meet specified conditions, as in an equation or set of equations. Any values that reduce an equation to an identity *satisfy* that equation. Also, to meet a set of hypotheses, such as satisfying the conditions of a theorem.

Sato's conjecture   A conjecture asserting

$$\lim_{x \to \infty} \frac{\text{number of primes } p \le x : \theta_p \in [\alpha, \beta]}{\text{number of primes} < x}$$

$$= \frac{2}{\pi} \int_\alpha^\beta \sin^2 \theta \, d\theta, \ 0 \prec \alpha < \beta < \pi \ ,$$

where E is an elliptic curve over the rational number field **Q** such that E does not have complex multiplication, N is its conductor, $L(s, E) =$

$$\prod_{p|N} \left(1 - \epsilon_p p^{-s}\right)^{-1} \prod_{p \nmid N} \left(1 - a_p p^{-s} + p^{1-2s}\right)^{-1}$$

is the L-function of E over **Q**, $\epsilon_p = 0$ or ± 1, $1 - a_p u + p u^2 = p_1(u, E \bmod p) = 1 - a_p u + p u^2 = (1 - \pi_p u)(1 - \bar{\pi}_p u)$ with $\pi_p = \sqrt{p} e^{i \theta_p}$ ($0 < \theta_p < \pi$). When E has complex multiplication, the distribution of θ_p for half of p is uniform in $[0, \pi]$, and θ_p is $\frac{\pi}{2}$ for the remaining half of p.

scalar extension   Suppose $\varphi : R_1 \to R_2$ is a ring homomorphism and suppose M is an R_2-module. Consider M to be an R_1-module by scalar restriction. Define the R_1-module $M_{R_2} = R_2 \otimes_{R_1} M$. Then M_{R_2} may be considered to be an R_2-module by defining $r_2 \cdot (r_2' \otimes m)$ to be $r_2 \cdot r_2' \otimes m$ for all $r_2, r_2' \in R_2$ and for all $m \in M$. This R_2-module M_{R_2} is called a *scalar extension. See also* scalar restriction.

scalar matrix   A square matrix, wherein all of the elements on the main diagonal are equal, i.e., all a_{ii} are the same, and all other elements are 0, i.e., $a_{ij} = 0$, for $i \neq j$.

scalar multiple   (1) (In a linear space) The product, ax, of an element, x, in a set L and an element, a, in a field K; K is the scalar field.

(2) (Of a linear operator) The product of a scalar a and a linear operator T: $(aT)x = a(Tx)$.

(3) (In a module) The product ax of an element x in a module M and an element a in a ring A (where M is a module over A); A is the ring of scalars.

(4) (Of a vector) The product $c\vec{v}$ of a vector $\vec{v}$ and a real number c. The vector $c\vec{v}$ is on the same line as the line containing $\vec{v}$, and the ratio of $c\vec{v}$ to $\vec{v}$ is c (if $\vec{v}$ is the zero vector, then $c\vec{v}$ is also the zero vector).

scalar multiplication   *See* vector space.

scalar quantity   When considering a vector space over a field k, an element of k. *See* vector space.

scalar restriction   Suppose $\varphi : R_1 \to R_2$ is a ring homomorphism and suppose M is an R_2-module. Then M can be made into an R_1-module by defining $r \cdot m$ to be $\varphi(r) \cdot m$ for all $r \in R_1$ and for all $m \in M$. This R_1-module M is called a *scalar restriction. See also* ring homomorphism, module.

scheme   A ringed space X for which there is an affine scheme Spec(A) such that Spec(A) and X are locally isomorphic.

Schmidt's Theorem   If a field F is perfect, then every extension E/F is separable.

See also perfect field, separable extension.

Schottky group   Suppose $C_1, C_{-1}, \ldots, C_k$, C_{-k} are $2k$ circles in the complex plane. Denote the interior of C_i by I_i and the exterior, including the point at infinity, by E_i, $i = \pm 1, \ldots, \pm k$. Let $\mu_1, \ldots, \mu_k$ be Möbius transformations satisfying $\mu_i(C_i) = C_{-i}$, $\mu_i(E_i) = I_{-j}$, $\mu_i(I_i) = E_{-i}$. Further, μ_{-i} is defined to be μ_i^{-1}, $i = 1, \ldots, k$. Then the group generated by $\mu_1, \ldots, \mu_k$ is called a *Schottky group*. *See also* Möbius transformation.

Schreier conjecture   A conjecture which states that for an arbitrary finite simple group, G, the outer automorphisms of G form a solvable group. The conjecture is true for all known finite simple groups.

Schur index   Let G be a finite group, $F \subset S$ where S is a splitting field for G, and $\varphi \in$ Irr$_S(G)$. If Ψ is an irreducible S-representation which affords φ and Ξ is an irreducible F-representation such that Ψ is a constituent of Ξ^S, then the multiplicity of Ψ as a constituent of Ξ^S is called the *Schur index* of φ over F. *See also* splitting field, irreducible representation, constituent.

Schur product   *See* Hadamard product.

Schur's Lemma   (1) Let M and N be simple A-modules, and let $f : M \to N$ be an A-homomorphism. Then f is either an isomorphism or f is the zero homomorphism.

(2) Let A be a ring and let M be a simple A-module. Let $D = \mathcal{E}_A(M)$ be the endomorphism ring of M. Then D is a skew field.

Schur subgroup   The subgroup of the Brauer group $Br(k)$ consisting of those algebra classes that contain a Schur algebra A over k, where k is a field of characteristic 0.

Schur-Zassenhaus Theorem   Suppose N is a normal subgroup of a finite group G such that $|N|$ and $|G : N|$ are relatively prime. Then G contains subgroups of order $|G : N|$ and any two of these subgroups are conjugate in G.

See also normal subgroup, conjugate subgroup.

Schwarz inequality   The inequality which states that for complex numbers $x_1, \ldots, x_n$ and $y_1, \ldots, y_n$,

$$\left| \sum_{i=1}^{n} x_i \bar{y}_i \right|^2 \le \sum_{i=1}^{n} |x_i|^2 \sum_{i=1}^{n} |y_i|^2 .$$

The inequality is commonly generalized in abstract inner product spaces to mean $|(\mathbf{x}, \mathbf{y})| \le \|\mathbf{x}\| \|\mathbf{y}\|$ for vectors $\mathbf{x}$ and $\mathbf{y}$, where $(\cdot, \cdot)$ is the inner product and $\| \cdot \|$ is its norm. Also known as the *Cauchy-Schwarz* or *Bunyakovskiĭ* inequality, it is a special case of the Hölder integral inequality. *See* Hölder's Theorem.

scientific notation   The convention in applied mathematics wherein real numbers are represented as decimals (between one and ten) multiplied by powers of ten. For example, 36000 is 3.6×10^4 and 0.00031 is 3.1×10^{-4} in *scientific notation*. The notation is particularly useful in dealing with numbers that are very large or very small in magnitude.

secant function   One of the fundamental trigonometric functions, denoted sec x. By definition, $\sec x = \frac{1}{\cos x}$, and hence the *secant function* is (i.) periodic, satisfying $\sec(x + 2\pi) = \sec x$; (ii.) bounded below, satisfying $1 \le |\sec x|$, for all real x, and (iii.) undefined where the cosine is 0 ($x = \pm \frac{\pi}{2}, \pm \frac{3\pi}{2}, \ldots$). Many other properties of the secant function can be derived from those of the cosine and sine. *See* cosine function. *See also* secant of an angle.

secant method   The iterative sequence of approximate solutions of the equation $f(x) = 0$, given by

$$x_{n+1} = x_n - f(x_n) \cdot \frac{x_n - x_{n+1}}{f(x_n) - f(x_{n+1})}$$

for $f(x_n) \ne f(x_{n+1})$, $n \ge 1$.

secant of an angle   Written $\sec \alpha$, the reciprocal of the x-coordinate of the point where the terminal ray of the angle α whose initial ray lies along the positive x-axis intersects the unit circle. If $0 < \alpha < \frac{\pi}{2}$ (α in radians) so that the angle is one of the angles in a right triangle with adjacent side a, opposite side b, and hypotenuse c, then $\sec \alpha = \frac{c}{a}$.

second difference   In difference equations, where the difference of y at x is $\Delta y(x) = y(x + \Delta x) - y(x)$, the second difference is defined as $\Delta^2 y(x) = \Delta(\Delta y(x)) = y(x + 2\Delta x) - 2y(x + \Delta x) + y(x)$.

second factor of class number   *See* first factor of class number.

secular equation   The characteristic equation of a square matrix A. *See* characteristic equation. For a square symmetric matrix of real elements, all of the solutions to the secular equation (the eigenvalues of A) are real.

sedenion   Any element of A_4, where A_n is the Clifford algebra over **R** with $\lambda_{ii} = -1$. A_4 is important in spinor theory and the theory of Dirac's equation.

Selberg trace formula   Let G be a connected semisimple Lie group and Γ a discrete subgroup. Let T be the regular representation of G on $\Gamma \backslash G$ defined by $(T_g f)(x) = f(xy)$, $f \in L^2(\Gamma \backslash G)$. When $\Gamma \backslash G$ is compact, one has $T = \sum_{k=1}^{\infty} T^{(k)}$ as the irreducible decomposition of T. Let χ_k be the character of the irreducible unitary representation $T^{(k)}$. Then one has the *trace formula*

$$\sum_{k=1}^{\infty} \int_G f(g) \chi_k(g) dg = \sum_{\{\gamma\}} \int_{\mathcal{D}_\gamma} f(x^{-1} \gamma x) \, dx$$

where $\{\gamma\}$ is the conjugate class of γ in Γ and $\mathcal{D}_\gamma$ is the quotient space of the centralizer G_γ of γ in G by the centralizer Γ_γ of γ in Γ. When $\Gamma \backslash G$ is not compact, the irreducible decomposition of T on $\Gamma \backslash G$ contains not only the discrete sum but also the direct integral (continuous spectrum). Selberg showed that even in this case, there are explicit examples in which the trace formula holds for the part with discrete spectrum. Also the part with continuous spectrum can be described by the generalized Eisenstein series.

Selberg zeta function   The zeta function

$$Z_\Gamma(s, M) = \prod_i \prod_{n=0}^{\infty} \det\left(I - M(\gamma_i) N(\gamma_i)^{-s-n}\right),$$

where $\Gamma \subset SL(2, \mathbf{R})$ is a Fuchsian group, operating on the complex upper half-plane **H**, P_1,

$P_2, \ldots$ are the conjugacy classes of primitive hyperbolic elements of Γ, $\gamma_j \in P_j$ are their representatives, $\gamma \mapsto M(g)$ is a matrix representation of Γ, $N(\gamma) = \xi_2^2$ the norm of hyperbolic γ where ξ_1, ξ_2 are two eigenvalues of γ. Here, when the two eigenvalues of $\gamma \in \Gamma$ are distinct real numbers ξ_1, ξ_2 ($\xi_1 \xi_2 = 1$, $\xi_1 < \xi_2$), we call γ *hyperbolic*. When γ is hyperbolic, $\gamma^n (n = 1, 2, \ldots)$ is also hyperbolic. When $\pm\gamma$ is not a positive power of another hyperbolic element, γ is called a *primitive hyperbolic element*.

self-dual regular cone   A cone which is self-dual and regular. Let X be a vector space with an inner product $(\cdot, \cdot)$. A subset $C \subset X$ is called a *cone* if (i.) $x, y \in C$ implies $x + y \in C$; (ii.) $x \in C$ and $t > 0$ imply $tx \in C$; (iii.) if x and $-x$ are in C, then $x = 0$. A cone is called *regular* if the relations $x_1 \leq x_2 \leq \cdots \leq z$ imply that $\{x_n\}$ is norm-convergent. A subset $C^d := \{x \in X : (x, y) \geq 0, \text{ for all } y \in C\} \subset X$ is called a *dual cone*. C is called *self-dual* if $C = C^d$.

self-intersection number   On an algebraic surface, the intersection number (or Kronecker index) $I(d_1 \cdot d_2)$ is defined for any divisors d_1 and d_2. *See* intersection number. The self-intersection number is the case in which $d_1 = d_2$, i.e., the number $I(d \cdot d) = (d^2)$ is the *self-intersection number* of the divisor d.

semidefinite Hermitian form   A Hermitian form H with n variables such that the signature of H is either $(r, 0)$ or $(0, r)$, with $1 \leq r < n$. *See* semidefinite quadratic form.

semidefinite quadratic form   A real quadratic form Q with n variables such that the signature of Q is either $(r, 0)$ or $(0, r)$, with $1 \leq r < n$. *See* semidefinite Hermitian form.

semidefinite von Neumann algebra   A von Neumann algebra A which has an exact semifinite normal trace on A.

semidirect product   A group G which can be written as the product of a normal subgroup N of G and a subgroup H of G where $N \cap H = \{1\}$.

semifinite function   A measurable function f with the property that $\{x : |f(x)| = \infty\}$ is σ-

finite (the countable union of sets of finite measure).

semigroup   Generalization of a group in which a binary operation is defined and associative on a set. The existence of an identity and multiplicative inverses are not guaranteed in a semigroup as they are in a group.

semigroup algebra   An algebra over a field K which has a multiplicative semigroup G as its basis.

semigroup bialgebra   An algebra $\Psi(A)$ over a field Ψ with a basis A that is at the same time a multiplicative semigroup.

semi-invariant   (1) A common eigenvector of a family of endomorphisms of a vector space or module.

(2) A numerical characteristic of random variables related to the concept of moment of higher order. If $\mathbf{v} = (v_1, \ldots, v_k) \in \mathbf{R}^k$ is a random vector, $\phi_{\mathbf{v}}(\mathbf{u}) = E(e^{i(\mathbf{u}, \mathbf{v})})$ is its moment generating function, where $\mathbf{u} = (u_1, \ldots, u_k) \in \mathbf{R}^k$ and $(\mathbf{u}, \mathbf{v}) = \sum_{j=1}^{n} v_j u_j$. If for some $n \geq 1$ the nth moment $E(|v_j|^n) < \infty$, $j = 1, \ldots, k$, then the (mixed) moments

$$m_{\mathbf{v}}^{(\alpha_1, \ldots, \alpha_k)} = E\left(v_1^{\alpha_1} \cdots v_k^{\alpha_k}\right)$$

exist for all multi-indices $(\alpha_1, \ldots, \alpha_k) \in (\mathbf{Z}_+)^k$ such that $|\alpha| = \alpha_1 + \cdots + \alpha_k \leq n$. Under these conditions,

$$\phi_{\mathbf{v}}(\mathbf{u}) =$$

$$\sum_{|\alpha| \leq n} \frac{i^{\alpha_1 + \cdots + \alpha_k}}{\alpha_1! \cdots \alpha_k!} m_{\mathbf{v}}^{(\alpha_1, \ldots, \alpha_k)} u_1^{\alpha_1} \cdots u_k^{\alpha_k} + o\left(|\mathbf{u}|^n\right),$$

where $|\mathbf{u}| = |u_1| + \cdots + |u_k|$ and for sufficiently small $|\mathbf{u}|$, the principal value of $\log \phi_{\mathbf{v}}(\mathbf{u})$ can be expressed by Taylor's expansion as

$$\log \phi_{\mathbf{v}}(\mathbf{u}) =$$

$$\sum_{|\alpha| \leq n} \frac{i^{\alpha_1 + \cdots + \alpha_k}}{\alpha_1! \cdots \alpha_k!} a_{\mathbf{v}}^{(\alpha_1, \ldots, \alpha_k)} u_1^{\alpha_1} \cdots u_k^{\alpha_k} + o\left(|\mathbf{u}|^n\right)$$

where the coefficients $a_{\mathbf{v}}^{(\alpha_1, \ldots, \alpha_k)}$ are called (mixed) *semi-invariants,* or *cumulants,* of order $\alpha = (\alpha_1, \ldots, \alpha_k)$ of the vector $\mathbf{v} = (v_1, \ldots, v_k)$.

semilinear mapping   A $\mathbf{Z}$-linear mapping $\phi : E \to E'$, where E and E' are modules, such that $\phi(ax) = \bar{a}\phi(x)$ for all $x \in E$.

semilocal ring   A ring R with finitely many maximal ideals.

semiprimary ring   A ring R such that, for any ideal A and any positive number $n \in \mathbf{N}$ with $A^n = 0$, we have $A = 0$.

semiprime differential ideal   A differential ideal which is semiprime. Let R be a commutative ring with an identity. If a mapping $\delta : R \to R$ is such that, for any $x, y \in R$, (i.) $\delta(x+y) = \delta x + \delta y$; (ii.) $\delta(xy) = \delta x \cdot y + x \cdot \delta y$, then δ is called a *differentiation*. A ring R provided with a finite number of mutually commutative differentiations in R is called a *differential ring*. Let $\delta_1, \ldots, \delta_m$ be the differentiations of a differential ring R. An ideal a of R with $\delta_j a \subset a$ ($i = 1, 2, \ldots, m$) is called a *differential ideal*. We say that a differential ideal a is *semiprime* if a contains all those elements x that satisfy $x^g \in a$ for some natural number g.

semiprime ideal   *See* semiprime ring.

semiprime ring   A ring such that 0 is the only nilpotent ideal.

semiprime ring   A ring R is semiprime if 0 is the only nilpotent ideal.

semiprimitive ring   A ring in which the radical is the set containing only the zero element. *See* radical.

semireductive action   A rational action defined by ρ such that $N = f_1 K + \cdots f_r K$ is a G-admissible module ($f_1, \ldots, f_r \in R$) and that if $f_0 \mod N$ ($f_0 \in R$) is G-invariant, then there is a homogeneous form h in $f_0, \ldots, f_r$ of positive degree with coefficients in K such that h is monic in f_0 and is G-invariant, where K is a commutative ring with identity and G is a matrix group over K (i.e., a subgroup of the general linear group $\mathrm{GL}(n, K)$). A homomorphism ρ of G into $\mathrm{GL}(m, K)$ is called a *rational representation* of G if there exist rational functions $\phi_{kl}, 1 \leq k, l \leq m$, in n^2 variables x_{ij}

$(1 \leq i, j \leq n)$ with coefficients in K such that $\rho((\sigma_{ij})) = (\phi_{kl}(\sigma_{ij}))$ for all $(\sigma_{ij}) \in G$. Assume that R is a commutative ring generated by $x_1, \ldots, x_n$ over K and that an action of the group G on R is defined such that $(\sigma x_1, \ldots, \sigma x_n)^t = \sigma^t(x_1, \ldots, x_n)$ where ^{t} means the transpose of a matrix. In this case we say that G acts on R as a matrix group. Assume that ρ is a rational representation of a matrix group G and $\rho(G)$ acts on a ring R as a matrix group. Then we have an action of G on R defined by $\sigma f = (\rho\sigma)f$ $(\sigma \in G, f \in R)$, called the *rational action* defined by ρ.

semisimple Banach algebra A Banach algebra A, such that its regular module A° is completely reducible. *See also* Banach algebra, regular module, completely reducible.

semisimple component The semisimple linear transformation ϕ_s of the Jordan decomposition of a linear transformation $\phi : L \to L$, where L is an n-dimensional linear space over a perfect field. By the Jordan decomposition, any such ϕ is represented as the sum of the ϕ_s and a nilpotent linear transformation ϕ_n.

semisimple Jordan algebra A Jordan algebra A such that its regular module A° is completely reducible. *See also* Jordan algebra, regular module, completely reducible.

semisimple Lie algebra A Lie algebra whose only Abelian ideal is the zero ideal.

semisimple Lie group A Lie group G is called *simple* if
(i.) $\dim(G)$ is greater than 1;
(ii.) G has only finitely many connected components; and
(iii.) any proper normal subgroup of the identity component of G is finite.
We say that G is *reductive* if it has finitely many connected components, and some finite cover of the identity component G_e is a product of simple and Abelian Lie groups. It is *semisimple* if there are no Abelian factors in this decomposition.

semisimple linear transformation A linear transformation T on a vector space V of dimension n over a field F such that V contains a basis of eigenvectors for T.

semisimple matrix A diagonalizable matrix. Hence a matrix $A = SMS^{-1}$, where S is nonsingular and M is diagonal. *See* diagonal matrix, nonsingular matrix.

semisimple module An R-module M of a ring R such that M is the sum of simple submodules. Equivalent conditions are
(i.) M is the direct sum of simple submodules and
(ii.) for every submodule E there exists a submodule F such that $M = E + F$.

semisimple part (1) A nonsingular matrix can be represented uniquely as a product (commutatively) of a diagonalizable matrix (the *semisimple part*) and a unipotent matrix (the unipotent part) of identical dimensions.
(2) The semisimple part of a group is the set of all semisimple elements in that group.

semisimple ring A ring R such that, as a module, R is semisimple.

Semistable Reduction Theorem A theorem proved by Grothendieck that asserts: Given an Abelian variety over F, there is a finite extension of F, over which it has a semistable reduction.

Let F be a field with a discrete valuation v, valuation ring o_v, and maximal ideal m_p. Let A_F be an Abelian variety over F. Let $\mathbf{A}$ be a Néron model, A^0 the connected Néron model, $k = k(v)$ and A_k^0 the special fiber over the residue class field k. We say that A_F has *semistable reduction* if A_k^0 is an extension of an Abelian variety by a torus (so that A_k^0 is a semi-Abelian variety).

sense of inequality The direction of an inequality is sometimes also referred to as the *sense of an inequality*. For example, $x > 5$ and $y > z$ have the same sense.

sensitivity analysis The analysis of response to imposed conditions, i.e., variations of solutions in response to variations in a problem's parameters. A particularly important topic in chaotic dynamics, wherein small changes in initial conditions or parameters (such as a butterfly flapping its wings in China) often produce dramatically different behavior at later times (such as the formation of a hurricane that proceeds to ravage much of Florida).

separable algebra A finite dimensional algebra A over a field K such that the tensor product $E \otimes_K A$ is semisimple over E for every extension field E of K.

separable element An algebraic element a over a field K which is a root of a separable polynomial over K.

separable extension An extension field F of a field K whose elements are all separable elements over K.

separable polynomial An irreducible polynomial with a nonzero formal derivative.

separable scheme *See* purely inseparable scheme.

separated morphism A morphism $f : X \to S$ such that the image of the diagonal morphism $\Delta_{X/S} : X \to X \times_S X$ is closed, where X and S are local-ringed spaces.

separated scheme A scheme X such that the diagonal subscheme $\Delta_{X/\mathrm{Spec}(A)}(X) = X \times_{\mathrm{Spec}(A)} X$ is closed.

separating transcendence basis A subset S of a field extension F of a field K such that S is algebraically independent and F is algebraic and separable over $K(S)$.

sequence of factor groups Let G be a group and let $G = G_0 \triangleright G_1 \triangleright \cdots \triangleright G_n$ be a chain of normal subgroups of G. Then the *sequence of factor groups* is given by $G_{i-1}/G_i, i = 1, 2, \ldots, n$.

sequence of Ulm factors Suppose R is a complete valuation ring with prime p and reduced R-module M. Define $\{M_\mu\}$ to be the descending transfinite series obtained by successive multiplication by p where intersections are taken at the limit ordinals. Define φ by $\varphi(x) = \mu$ if $x \in M_\mu$ but $x \notin M_{\mu+1}$; and de-

fine $\varphi(0) = \infty$ where ∞ exceeds any ordinal. Then associated to each x there is a sequence of ordinals $\{\mu_i : \mu_i = \varphi(p^i x), i \geq 0\}$ called the *sequence of Ulm factors. See also* complete valuation ring, reduced module, transfinite series, limit ordinal.

serial subgroup A subgroup H of a group G is *serial* in G, denoted H ser G, if there exists a totally ordered set Σ and a set

$$\{(\Lambda_\sigma, V_\sigma) : \sigma \in \Sigma\}$$

such that
(i.) $H \leq V_\sigma \equiv \Lambda_\sigma \leq G \ \forall \sigma \in \Sigma$;
(ii.) $\Lambda_\tau \leq V_\sigma$ if $\tau < \sigma$;
(iii.) $G \setminus H = \bigcup \{\Lambda_\sigma \setminus V_\sigma : \sigma \in \Sigma\}$.

series (1) The summation of a finite sequence of terms, written with the Greek letter sigma as follows:

$$\sum_{n=0}^{N} a_n = a_0 + a_1 + \cdots + a_N$$

where n is the index of summation, 0 and N are the limits of the summation, and each a_n is a term.

(2) An *infinite series* is a formal sum of the above type for which $N = \infty$. An infinite series is said to converge if $\lim_{N \to \infty}(a_1 + \cdots + a_N) = S$ for some finite S; it is said to diverge otherwise. There are many methods used to test the convergence or divergence of infinite series, such as the root test, ratio test, and comparison test. Infinite series are also particularly useful in analysis.

(3) *Power series,* or series of the form

$$\sum_{n=0}^{\infty} a_n x^n,$$

for complex a_n, can be used to represent some functions on intervals; these functions are called real analytic in the interval (as opposed to being singular at some point in the interval). Real analytic functions can be differentiated or integrated by differentiating or integrating their representative power series term by term. Moreover, the value of a function can be approximated at a point c by evaluating its truncated power series

$\Sigma_{n=0}^{N} a_n (x-c)^n$, with increasing accuracy in the approximation as N approaches infinity.

(4) Series appear in a variety of other contexts, including differential equations (in which solutions can sometimes be represented by power series) and Fourier analysis, in which series of trigonometric terms are used to represent otherwise complicated functions such as pulse functions. *See also* Fourier series.

Serre conjecture   Let $F[X_1, X_2, \ldots, X_n]$ be a ring of polynomials over a field F. Then every finitely generated projective module over $F[X_1, X_2, \ldots, X_n]$ is free. Serre's conjecture was proved by Quillen and Suslin and is also known as the Quillen-Suslin Theorem.

Serre's Theorem   For a positively graded, commutative algebra C, the category of coherent sheaves on the scheme $\mathrm{Proj}\, C$ is equivalent to a certain category which is completely defined in terms of C-gr.

set of antisymmetry   *See* antisymmetric decomposition.

Shaferevich's reciprocity law   The explicit reciprocity law which asserts that
(i.) $(\pi, E(\alpha, \pi')) = 1$ for $p \nmid i, \alpha \in \mathcal{D}_T$.
(ii.) $(\pi, E(\alpha)) = \zeta^{\mathrm{Tr}\alpha}$ if $\alpha \in \mathcal{D}_T$, $\mathrm{Tr}\alpha :=$ $\mathrm{Tr}_{T/\mathbf{Q}_p}\alpha$.
(iii.) $(\pi, \pi) = (\pi, -1)$.
(iv.) $(E(\alpha), \epsilon) = 1$ for $\alpha \in \mathcal{D}_T, \epsilon \in \mathcal{D}_K^{\times}$.
(v.) If $p \neq 2$,

$$\left(E\left(\alpha, \pi^i\right), E\left(\beta, \pi^j\right) \right)$$
$$= \left(\pi^j, E\left(\alpha\beta, \pi^{i+j}\right) \right) ;$$

if $p = 2$,

$$\left(E\left(\alpha, \pi^i\right), E\left(\alpha, \pi^j\right) \right)$$
$$= \left(-\pi^j, E\left(\alpha\beta, \pi^{i+j}\right) \right)$$
$$\prod_{s=1}^{\infty} \left(-1, E\left(\alpha F^s(\beta), \pi^{i+jp^s}\right) \right)$$
$$\cdot \prod_{r=1}^{\infty} \left(-1, E\left(F^r(\alpha)\beta, \pi^{ip^r+j}\right) \right)$$

for $\alpha, \beta \in \mathcal{D}_T$, $p \nmid i$, $p \nmid j$.

(vi.) If $p = 2$,

$$\left(-1, E\left(a, \pi^i\right)\right)$$
$$= \prod_{s=0}^{\infty} \left(\pi, E\left(i2^s F^{s+1}(a), \pi^{i2^{s+1}}\right) \right)$$

for $\alpha \in \mathcal{D}_T$, $p \nmid i$, where K is an algebraic number field, $p \in K$ is a prime ideal and π is a fixed prime element of K, T is the inertia field of K, $E(\alpha)$ is the Artin-Hasse function, and $E(\alpha, \chi)$ is the Shaferevich function with $\alpha \in \mathcal{D}_T$ and $\chi \in p$. The explicit reciprocity law is a generalization of Gauss's quadratic reciprocity law, which allows one to decide whether an integral number in a field is an nth power residue with respect to a prime ideal.

sheaf   A presheaf $\mathcal{F}$ on X which satisfies the following conditions for every open $U \subseteq X$ such that $\cup_i U_i$ is an open covering of U: (i.) For every i, if $s, s' \in \mathcal{F}(U)$ are such that $s|_{U_i} = s'|_{U_i}$, then $s = s'$; and (ii.) for every i and j, if $s_i \in \mathcal{F}(U_i)$ and $s_j \in \mathcal{F}(U_j)$ are such that $s_i|_{U_i \cap U_j} = s_j|_{U_i \cap U_j}$, then there is an $s \in \mathcal{F}(U)$ such that for every i, $s|_{U_i} = s_i$. *See* presheaf.

sheaf of germs of regular functions   A sheaf of rings $\mathcal{O}_V$ on a variety V where the stalk $\mathcal{O}_{V,x}$ ($x \in V$) is the local ring of the ring regular functions of open sets on V at x.

sheaf theory   A mathematical method for providing links between the local and global properties of topological spaces. *Sheaf theory* is also used to study problems in mathematical areas such as algebra, analysis, and geometry.

Shilov boundary   Let B be a Banach algebra, M the set of all multiplicative linear functionals λ ($\lambda : B \to \mathbf{C}$ linear and $\lambda(fg) = \lambda(f)\lambda(g)$), along with the weakest topology which makes the mapping $M \ni m \mapsto m(f) \in \mathbf{C}$ continuous for every $f \in B$, and $\hat{f}$ the Gel'fand transform of $f \in B$. Then the Shilov boundary S of M is the smallest closed subset of M such that $\sup_S |\hat{f}| = \sup_M |\hat{f}|$, $f \in B$. *See also* Banach algebra, Gel'fand transform.

short representation   A representation of an ideal I in a commutative ring R as an intersection of primary ideals, $I = Q_1 \cap Q_2 \cap \cdots \cap Q_k$, satisfying two additional conditions: (i.) The representation is *irredundant,* that is, it is not possible to omit one of the primary ideals Q_i from the intersection. (ii.) The prime ideals $P_1, P_2, \ldots, P_k$ belonging to the $Q_1, Q_2, \ldots, Q_k$ are all different. (P_i is the *prime ideal belonging to* Q_i if P_i contains Q_i, but there is no smaller prime ideal P_i' containing Q_i. Equivalently, P_i is the radical of Q_i, i.e., $P_i = \{p \in R : p^n \in I$ for some integer $n\}$.) *See also* isolated component, isolated primary component, irredundant, primary ideal, prime ideal, radical.

Siegel domain   An open subset S of complex N space $\mathbf{C}^N$ of the form $S = \{(z, w) \in \mathbf{C}^m \times \mathbf{C}^n : \operatorname{Im} z - \Phi(w, w) \in \Omega\}$. Here, Ω is an open convex cone in real m space $\mathbf{R}^m$ with vertex at the origin, $\Phi : \mathbf{C}^n \times \mathbf{C}^n \to \mathbf{C}^m$ is Ω Hermitian, $\operatorname{Im} z$ denotes the imaginary part of z, and $m+n = N$. In addition, the cone Ω is usually assumed to be acute; that is, Ω is usually assumed not to contain an entire line. Finally, Ω *Hermitian* means (i.) $\Phi(v, w)$ is linear with respect to v, (ii.) $\Phi(v, w) = \overline{\Phi(w, v)}$, (iii.) $\Phi(w, w) = 0$ if and only if $w = 0$, and (iv.) $\Phi(w, w) \in \overline{\Omega}$. In property (ii.), $\overline{Phi(w, v)}$ denotes the complex conjugate of $\Phi(w, v)$, whereas in property (iv.) $\overline{\Omega}$ denotes the topological closure of Ω.

An example of a Siegel domain is the unbounded domain $\mathcal{H}$ consisting of all $n \times n$ symmetric matrices with complex entries, and with positive definite imaginary parts. (A matrix M is *symmetric* if $M^t = M$, where M^t is the transpose of M. A matrix M with complex entries can be written uniquely as $M = A + iB$, where A and B have real entries. The matrix B is the *imaginary part* of M. An $n \times n$ matrix B with real entries is *positive definite* if $x^t Bx \geq 0$ for all n dimensional column vectors x with real entries. Here, x^t is just x, made into a row vector.) The Siegel domain $\mathcal{H}$ is called a *generalized Siegel half plane.* In the special case where $n = 1$, the generalized Siegel half plane is just the familiar upper half plane of elementary complex analysis.

A Siegel domain is *homogeneous* if it is homogeneous as a domain, that is, if it has a transitive group of analytic (holomorphic) automorphisms. The generalized Siegel half planes are homogeneous Siegel domains. *See* homogeneous domain.

Siegel domains are divided into three classes, *Siegel domains of the first kind, Siegel domains of the second kind,* and *Siegel domains of the third kind.* There is a deep connection between Siegel domains of the second kind and perfectly general bounded homogeneous domains. (A domain is bounded if it is contained in a ball of finite radius. A bounded homogeneous domain is a domain which is both bounded and homogeneous. An open ball of finite radius in complex N space is a bounded homogeneous domain. A generalized Siegel half plane is a homogeneous domain, but it is not bounded.) The connection between Siegel domains and bounded homogeneous domains is provided by a famous theorem of Vinberg, Gindikin, and Pyatetskii-Shapiro: Every bounded homogeneous domain in $\mathbf{C}^N$ is isomorphic to a homogeneous Siegel domain of the second kind. *See also* bounded homogeneous domain, irreducible homogeneous Siegel domain, Siegel domain of the first kind, Siegel domain of the second kind, Siegel domain of the third kind.

Siegel domain of the first kind   A domain $D(V) = \{x + iy \in R^c | y \in V\}$, where V is a regular cone in an n-dimensional vector space R and R^c is the complexification of R.

Siegel domain of the second kind   A domain $D(V, F) = \{(x + iy, u) \in R^c \times W | y - F(u, u) \in V\}$, where V is a regular cone in an n-dimensional vector space R, R^c is the complexification of R, W is a complex vector space, and $F : W \times W \to R^c$ is a V-Hermitian form. If $W = 0$, then a Siegel domain of the first kind is obtained. *See also* Siegel domain of the first kind.

Siegel domain of the third kind   A domain which is holomorphically equivalent to a bounded domain and can be written as $D(V, L, B) = \{(x + iy, u, p) \in R^c \times W \times X | y - \operatorname{Re} L_p(u, u) \in V, p \in B\}$, where V is a regular cone in an n-dimensional vector space R, R^c is the complexification of R, W and X are complex vector spaces, B is a bounded domain in X and $L, L_p : W \times W \to R^c$ are nondegenerate semi-Hermitian forms, for $p \in B$. *See also* Siegel domain of the first kind, Siegel domain of the second kind.

Siegel's Theorem Let χ be a real Dirichlet character of modulus k such that

$$\chi(n) \neq \begin{cases} 1 & \text{if } (n,k) = 1 \\ 0 & \text{if } (n,k) \neq 1 \end{cases}.$$

Siegel's Theorem states that, for any $\epsilon > 0$, there exists $c_\epsilon > 0$ such that

$$\sum_{n=1}^{\infty} \frac{\chi(n)}{n} > \frac{c_\epsilon}{k^\epsilon}.$$

Siegel zeta function A ζ-function attached to an indefinite, quadratic form, meromorphic on the whole complex plane and satisfying certain functional equations.

sigma field Consider an arbitrary non-empty subclass A of the power set of a set X such that if P and Q are in A, then $P \cap Q$, $P \cup bQ$, and ^cP are also in A. Such an A is called a *field*. If, in addition, A is closed under countable unions, then it is called a σ-*field*.

signature The ordered pair (p,q) corresponding to a real quadratic form Q of rank $p + q$, where Q is equivalent to

$$\sum_{i=1}^{p} x_i^2 - \sum_{j=p+1}^{q+p} x_j^2.$$

See also signature of Hermitian form.

signature of Hermitian form The ordered pair (p, q) corresponding to a Hermitian form H of rank $p + q$, where H is equivalent to

$$\sum_{i=1}^{p} \overline{x}_i x_i - \sum_{j=p+1}^{q+p} \overline{x}_j x_j.$$

signed numbers Numbers that are either positive or negative are said to be *signed* or *directed numbers*.

significant digits In a real number in decimal form, the digits to the left of the decimal point beginning with first non-zero digit, together with the digits to the right of the decimal point ending with the last non-zero digit.

sign-nonsingular matrix *See* sign pattern.

sign pattern The *sign pattern* of a real $m \times n$ matrix A is the $(0, 1, -1)$–matrix obtained from A when zero, positive, and negative entries are replaced by $0, 1, -1$, respectively. Thus, A determines a (qualitative) class of matrices, $Q(A)$, consisting of all matrices with the same sign pattern as A.

Sign patterns are one of the objects of study in combinatorial matrix analysis, concerned with the study of properties of matrices that are determined from its combinatorial structure (such as the *directed graph*) and other qualitative information. For example, A is called an L-matrix provided that every matrix in $Q(A)$ has linearly independent rows. The sign pattern

$$\begin{pmatrix} 1 & 1 & 1 & -1 \\ 1 & 1 & -1 & 1 \\ 1 & -1 & 1 & 1 \end{pmatrix}$$

is an example of an L-matrix. Such matrices first arose in the study of sign-solvable systems in economics models. A sign-nonsingular matrix A is a square L-matrix; that is, every matrix in $Q(A)$ is nonsingular.

similar (**1**) (Similar figures) Geometric figures such that the ratio of the distance between pairs of points in one figure and the distance between the corresponding pairs of points in another figure is constant for every pair of points in the first figure and corresponding pair of points in the second figure.

(**2**) (Similar terms) Terms in an expression which have the same unknowns and each unknown is raised to the same power in each term. For example, in the expression $5x^2y^3 + 8x^3 + 3y^3 + 7x^2y^3$, $5x^2y^3$ and $7x^2y^3$ are similar terms.

(**3**) (Similar triangles) Triangles which have proportional corresponding sides.

similar decimals Numbers that have the same number of decimal places are said to be similar. For example, 6.003, 2.232, and 100.000 are all similar decimals.

similar fractions Simple fractions in which the denominators are equal. *See* simple fraction, denominator.

similar isomorphism An isomorphism between two ordered fields, under which positive elements are mapped to positive elements.

similar matrices Two square matrices A and B such that $A = P^{-1}BP$ for some nonsingular matrix P. *See* nonsingular matrix.

similar permutation representations Two permutation representations, such that there exists a G-bijection between the corresponding G-sets, where G is a group. Let $\mathcal{S}_M$ be the group of all permutations of a set M. A *permutation representation* of a group G in M is a homomorphism $G \to \mathcal{S}_M$. In general, if the product $ax \in M$ of $a \in G$ and $x \in M$ is defined and satisfies $(ab)x = a(bx)$, $1x = x$, for all $a, b \in G$ and for all $x \in M$, with the identity element 1, then G is said to operate on M from the left and M is called a left G-set. We call a left G-set a G-set. A mapping $f : M \to M'$ of G-sets is called a G-mapping if $f(ax) = af(x)$ for any $a \in G$ and $x \in M$. G-injection, G-surjection, G-bijection are defined naturally.

similar terms *See* like terms.

simple Abelian variety A commutative algebraic group without Abelian subvarieties (except itself and its zero element). *See* algebraic group, Abelian subvariety.

simple algebra An algebra A such that the ring A is both a simple ring and a semisimple ring.

simple co-algebra A co-algebra which has no nontrival proper subco-algebras. *See* coalgebra.

simple component (1) Suppose R is a ring written as $R = R_1 \otimes \cdots \otimes R_n$ where $R_1, \dots, R_n$ are ideals satisfying the condition that if $R_i = R'_i \otimes R''_i$, where R'_i and R''_i are ideals, then either $R'_i = 0$ or $R''_i = 0$. Then the ideals $R_1, \dots, R_n$ are called the *simple components* of R.

(2) Suppose P is a polynomial in n variables over a field F. Write $P = \prod_{i=1}^{n} P_i^{\alpha_i}$ where the P_i are the distinct, irreducible (over F) factors of P. P_j is called a *simple component* of P if $\alpha_j = 1$.

simple extension An extension E of a field F such that $E = F(x)$, for some element $x \in E$, where $F(x)$ denotes the subfield of E over F, generated by x. *See also* extension, subfield.

simple fraction A fraction $\frac{a}{b}$, in which both a and b are integers, and $b \neq 0$. Also known as a *common fraction*.

simple group A group that has no normal subgroup other than itself and its identity $\{e\}$.

simple Lie algebra A Lie algebra whose only ideals are itself and the zero ideal.

simple Lie group A Lie group whose only proper Lie subgroup is the trivial subgroup.

simple module A module M, with more than one element, whose only submodules are itself and the zero submodule.

simple point A point of an irreducible variety V of dimension d over a field K such that $\left(\frac{\partial f_i}{\partial x'_j}\right)$ is of rank $n - d$, where $f_1(x_0, x_1, \dots, x_n) = 0$, $f_2(x_0, x_1, \dots, x_n) = 0, \dots, f_r(x_0, x_1, \dots, x_n) = 0$ form a basis for equations of V.

simple ring A ring R with more than one element whose only ideals are itself and the zero ideal.

simple root A root of an equation that is not repeated, i.e., a root of multiplicity 1. *See* multiplicity of root.

simplest alternating polynomial The polynomial $p(X_1, \dots, X_n) = (X_2 - X_1)(X_3 - X_2) \dots (X_n - X_{n-1})$. *See* alternating polynomial.

simplex (1) In geometry, the convex hull of an affine independent set of points, the vertices of the simplex.

(2) In topology, a homeomorphic image of a geometric simplex (as in (1)).

simplex method A method for optimizing a basic feasible solution of a primary linear programming problem in a finite number of iterations. Let I be the set of basic variables and J be the set of non-basic variables. Let **x** be a basic feasible solution of a primary linear programming problem, and write **x** in the basic form:

$$x_i + \sum_{j \in J} d_{ij} x_j = g_i \quad (i \in I)$$

$$z + \sum_{j \in J} f_j x_j = v.$$

If, for all $j \in J$, $f_j \geq 0$, then **x**, given by

$$x_\ell = \begin{cases} g_\ell & \text{if } \ell \in I \\ 0 & \text{if } \ell \in J \end{cases},$$

is an optimal solution. If there is a $j \in J$ such that $f_j < 0$, then let $r_{i*} = \min \frac{g_i}{d_{ij}}$, where the minimum is taken over all $i \in I$ such that $d_{ij} > 0$. Then, replace i^* in I with j, obtaining a new basis. Rewrite the basic feasible solution into a (new) basic form, and repeat the procedure.

simplex tableau   *See* simplex method.

simplification of an expression   Changing the form of an expression (but not the content) with the purpose of either making the expression more concise or readying the expression for the next step in a method or proof or solution.

simply connected covering Lie group   The simply connected Lie group G_0, among those Lie groups G which have g as their Lie algebra, where g is a finite-dimensional Lie algebra over **R**. Such G_0 is unique up to isomorphism.

simply connected group   A set G which has the structure of a topological group (making the group operation continuous) and which is simply connected.

simultaneous equations   Any system of two or more equations that must both/all be satisfied by the same solutions. For example,

$$\begin{cases} y = 3x + 5 \\ y = 2x + 6 \end{cases}$$

constitutes a system of two simultaneous equations satisfied by $x = 1$, $y = 8$;

$$\begin{cases} y = x - 1 \\ y = 3x - 3 \end{cases}$$

is a system of simultaneous equations with no solution.

simultaneous inequalities   Any system of two or more inequalities that must both/all be satisfied by the same solutions. For example,

$$\begin{cases} y^2 + x^2 \leq 1 \\ y \geq 0 \end{cases}$$

are simultaneous inequalities satisfied by the upper half of the unit circle.

sine function   One of the fundamental trigonometric functions, denoted $\sin x$. It is (i.) periodic, satisfying $\sin(x + 2\pi) = \sin x$; (ii.) bounded, satisfying $-1 \leq \sin x \leq 1$, for all real x; and (iii.) intimately related with the cosine function, $\cos x$, satisfying $\sin x = \cos(\frac{\pi}{2} - x)$, $\sin^2 x + \cos^2 x = 1$, and many others. It is related to the exponential function via the identity $\sin x = \frac{e^{ix} - e^{-ix}}{2i}$ ($i = \sqrt{-1}$), and has series expansion

$$\sin x = x - \frac{x^3}{3!} + \frac{x^5}{5!} - \frac{x^7}{7!} + \cdots$$

valid for all real values of x.
    See also sine of angle.

sine of angle   Written $\sin \alpha$, the y-coordinate of the point where the terminal ray of the angle α whose initial ray lies along the positive x-axis intersects the unit circle. If $0 < \alpha < \frac{\pi}{2}$ (α in radians) so that the angle is one of the angles in a right triangle with adjacent side a, opposite side b, and hypotenuse c, then $\sin \alpha = \frac{b}{c}$.

single step method for solving linear equations   *See* Gauss-Seidel method for solving linear equations.

single-valued function   A relation (a set R of pairs (x, y), x from a set X, and y from a set Y) such that $(x, y_1), (x, y_2) \in R$ implies $y_1 = y_2$. Thus, the function $f : X \rightarrow Y$ defined by $f(x) = y$ if $(x, y) \in R$ assigns only one y to a given x.

Indeed, every function has the above property. However, it is occasionally necessary to include multiple-valued "functions" (such as inverses of functions) in function theory and the term single-valued is required for emphasis.

singular locus   The subvariety of an irreducible variety V of dimension d over a field K consisting of all points V which are not simple points.

singular matrix   An $n \times n$, non-invertible matrix. A square matrix A is singular if and only if its null space contains at least one nonzero vector. Equivalently, A is singular if and only if $\det A = 0$ or if and only if at least one of its eigenvalues is 0.

singular point   A point at which a function ceases to be regular, even though the function remains regular at the points near the singular point. *See* regular function. The term is often used with "regular" replaced by "differentiable," "n times differentiable," "C^∞," etc.

singular value decomposition   Let A be an $n \times n$ real normal matrix. A *singular value decomposition* of A is the representation $A = USU^*$, where U is a unitary matrix, U^* is the adjoint of U, and S is a diagonal matrix of which the diagonal entries are the singular values of A.

singular values of a matrix   Eigenvalues of a matrix A; i.e., the values of λ such that $\lambda I - A$ is singular. *See* eigenvalue.

skew field   A skew field F is a ring in which the non-zero elements form a multiplicative group.

skew-Hermitian matrix   A matrix A of complex elements (necessarily square) such that $\bar{A}^T = -A$; i.e., the conjugate of its transpose is equal to its negation. Also called *anti-Hermitian*.

skew-symmetric matrix   A matrix A (necessarily square) such that $A^T = -A$, that is, its transpose is equal to its negation. Also called *antisymmetric*.

slack variable   The linear inequality $a_1 x_1 + a_2 x_2 + \cdots + a_n x_n \leq b$ may be converted to a linear equality by adding a nonnegative variable z, called a *slack variable*. Namely, $a_1 x_1 + a_2 x_2 + \cdots + a_n x_n \leq b$ becomes $a_1 x_1 + a_2 x_2 + \cdots + a_n x_n + z = b$. *See also* linear programming.

Slater's constraint qualification   The general nonlinear programming problem is the following: let X^0 be a connected closed set and let $\theta, g = (g_1, \ldots, g_m)$ be real-valued functions defined on X^0. Determine the set of $x \in X^0$ which minimizes (or maximizes) θ, subject to given constraints. If X^0 is convex and g_i are convex, then the convex vector function g is said to satisfy *Slater's constrait qualification* on C^0 provided there exists an $x \in X^0$ such that $g_i(x) < 0$, for all i.

slope-intercept form   An equation of a line in the plane of the form $y = mx + b$, where m is the slope of the line and b is its intercept on the y-axis. Example: $y = 3x - 2$ is a line with a slope of 3 and a y-intercept of -2 ($y = -2$, when $x = 0$).

small category   If the objects of a category form a set, then the category is called a small category or a diagram scheme.

small category   If the objects of a category form a set, then the category is called a *small category* or a *diagram scheme*.

smooth affine variety   Let k be a field and $X \subseteq k^n$. We call X a variety if it is the common zero set of a collection of polynomials. The variety is said to be *smooth* at a point $x \in X$ if the collection of polynomials has non-degenerate Jacobian at x.

smooth affine variety   Let k be a field and $X \subseteq k^n$. We call X a variety if it is the common zero set of a collection of polynomials. The variety is said to be smooth at a point $x \in X$ if the collection of polynomials has non-degenerate Jacobian at x.

Snapper polynomial   Let X be a k-complete scheme, F be a coherent $\mathcal{O}_X$-module, and L be an invertible sheaf. The *Snapper polynomial* is the polynomial in m given by $\chi(F \otimes L^{\otimes m})$, where $\chi(F) = \sum_q (-1)^q \dim H^q(X, F)$.

solution   Anything that satisfies a given set of constraints is called a solution to that set of constraints. A number may be a solution to an equation or a problem. A function may be a solution to a differential equation. A region on a plane that satisfies a set of inequalities is a solution of that set of inequalities.

solution by radicals   Any method of solving equations using arithmetic operations (including nth roots). For instance, the quadratic formula is a well-known method of solving second order equations. *See* quadratic formula. Some equations of fifth order and higher cannot be solved by radicals.

solution of an equation   (1) For an equation $F(x_1, \ldots, x_n) = 0$, with unknowns $x_1, \ldots, x_n$ in some set S, any n-tuple of numbers $(a_1, \ldots, a_n)$, $a_i \in S$ satisfying $F(a_1, \ldots, a_n) = 0$.

(2) For a functional equation $F(x, y) = 0$ with unknown function y, any function $f(x)$ such that $F(x, f(x)) = 0$.

(3) For a differential equation

$$F\left(t, x, x', \ldots, x^{(n)}\right) = 0 \, ,$$

where $x^{(m)} = \frac{d^m x}{dt^m}$, any n times differentiable function $f(t)$ such that

$$F\left(t, f(t), f'(t), \ldots, f^{(n)}(t)\right) = 0 \, .$$

Many other types of equations are possible, including differential equations with values of the function and certain derivatives specified at certain points.

solution of an inequality   For an inequality $F(x_1, \ldots, x_n) \geq 0$, with unknowns $x_1, \ldots, x_n$ in some ordered set S, any n-tuple of numbers $(u_1, \ldots, a_n)$, $a_i \in S$ satisfying $F(a_1, \ldots, a_n) \geq 0$. For instance, the region beneath the line $y = x + 2$ contains solutions to the inequality $y - x < 2$.

solution of oblique spherical triangle   A spherical triangle is formed by the intersection of the arcs of three great circles on the surface of a sphere. A spherical triangle has six parts, the three angles α, β, and γ, and three sides AC, AB, and BC. The angle α is specified by the dihedral angle between the plane containing the arc AB and the plane containing the arc AC, and similarly for β and γ. An oblique spherical triangle is one in which none of the three angles is a right angle, and solving such a triangle

consists of using three given parts (for example, the two sides and the included angle) to find the remaining parts. *See also* solution of right spherical triangle.

solution of plane triangle   A plane triangle has six principal parts: three angles and three sides. Given any three of these parts, which are sufficient to determine the triangle (e.g., two sides and the included angle), the solution of the triangle consists of determining the remaining three parts.

solution of right spherical triangle   A spherical triangle is formed by the intersection of the arcs of three great circles on the surface of a sphere. A spherical triangle has six parts, the three angles α, β, and γ, and three sides AC, AB, and BC. The angle α is specified by the dihedral angle between the plane containing the arc AB and the plane containing the arc AC, and similarly for β and γ. If one (or more) of the angles is a right angle, the triangle is called a right spherical triangle. Note that in contrast to a plane triangle, a spherical triangle may have one, two, or three right angles. The solution of a right spherical triangle consists of finding all the parts of the triangle, given the right angle and any two other elements. The many possible cases which can occur were all summarized in two formulas discovered by J. Napier (1550-1617), known as Napier's rules.

solvable algebra   Given an algebra $\mathcal{A}$, define the sequence of algebras $\mathcal{A}^{(1)}, \mathcal{A}^{(2)}, \mathcal{A}^{(3)}, \ldots$, by $\mathcal{A}^{(1)} = \mathcal{A}, \mathcal{A}^{(i+1)} = (\mathcal{A}^{(i)})^2, i = 1, 2, 3 \ldots$, where $(\mathcal{A}^{(i)})^2$ denotes the span of squares of elements of $\mathcal{A}^{(i)}$. Then $\mathcal{A}$ is *solvable* if $\mathcal{A}^{(j)} = 0$ for some j.

solvable by radicals   From antiquity, it has been known that the quadratic equation $ax^2 + bx + c = 0$ has solutions

$$x = \frac{-b \pm \sqrt{b^2 - 4ac}}{2a} \, .$$

In the Renaissance, analogous formulas for the solution of general cubic and quartic equations involving arithmetic combinations of the coefficients of the polynomial and their roots were discovered, and in the nineteenth century it was proved that no such formula exists for the roots of general polynomials of degree five or higher. In general one says that a polynomial $p(x)$, with coefficients in a field F (with characteristic 0) is *solvable by radicals* if there exists an algorithm for computing the roots of $p(x)$ which consists only of performing the field operations on the coefficients of p, together with extracting roots of such expressions. Since taking roots may take one to an extension field of F, a more precise formulation would say that there exists a radical extension of F which contains a splitting field of $p(x)$. *See also* Cardano's formula.

solvable group   Given a group G, a normal series is a finite set of subgroups, $\{N_j\}_{j=1}^n$, which satisfy: (i.) $\{e\} = N_n \subset N_{n-1} \subset \cdots \subset N_2 \subset N_1 = G$, where e is the identity element in G. (ii.) N_{j+1} is a normal subgroup of N_j for $j = 1 \ldots n - 1$. G is *solvable* if there exists a normal series $\{N_j\}_{j=1}^n$ for which each of the groups N_j/N_{j+1} is Abelian.

solvable ideal   An ideal in which the derived series becomes zero.

solvable ideal   An ideal in which the derived series becomes zero.

solvable Lie group   A Lie group G, which, ignoring its Lie group structure and considering it only as an abstract group, is solvable. *See* solvable group.

solving a triangle   The act of using the size of certain sides and angles of a triangle to determine the remaining sides and angles. *See also* solution of oblique spherical triangle, solution of plane triangle, solution of right spherical triangle.

space   (**1**) A set of objects, usually called points. Often equipped with some additional structure, e.g., a *topological* space, if equipped with a topology, or a *metric* space, if equipped with a metric.

(**2**) The unbounded three-dimensional region $\mathbf{R}^3$ in which all physical objects exist.

space coordinates   A set of three numbers sufficient to specify uniquely the location of a

point in space, according to some convention to establish such location. Examples of such coordinates are Cartesian coordinates, spherical coordinates, or cylindrical coordinates. Note that while the coordinates must specify a unique point in space, a given point in space may be represented by several (or even infinitely many) different sets of coordinates.

space curve   A one-dimensional variety in three-dimensional projective space.

space group   A discrete subgroup $\mathcal{G}$ of the Euclidean group, for which the subgroup of translations in $\mathcal{G}$ forms a three-dimensional lattice group. Synonyms are crystallographic group and crystallographic space group. *See also* lattice group.

span of vectors   The span of the vectors x_1, x_2, $\ldots$, x_k in a vector space V over a field F, is the set X of all possible linear combinations of $x_1, x_2, \ldots, x_k$. If $x_1, x_2, \ldots, x_k$ are linearly independent vectors, then they constitute a basis for their span, X.

spatial $*$-isomorphism   Suppose that H_i is a Hilbert space and that M_i is a von Neumann algebra contained in the set of bounded linear operators on H_i, for $i = 1, 2$, and consider a $*$-isomorphism $\pi : M_1 \to M_2$. If there exists a bijective isometric linear mapping $U : H_1 \to H_2$ such that $UAU^* = \pi A$, for each A in M_1 (with U^* denoting the adjoint of U), then π is called *spatial*.

spatial tensor product   The C^*-algebra resulting when the algebraic tensor product of two C^* algebras A and B is equipped with the spatial, or minimal, C^* cross-norm,

$$\|x\|_{\min} = \sup_{\rho, \eta} \|(\rho \otimes \eta)(x)\|$$

where $x \in A \otimes B$, and the supremum runs over all $*$-representations of A and B. There may be more than one norm on the algebraic tensor product $A \otimes B$ whose completion gives a C^*-algebra.

special Clifford group   If R is a ring with identity, M a free R-module, and Q a quadratic form, define the Clifford algebra $C(Q)$ as the quotient algebra $T(M)/\mathcal{I}(M)$ where $T(M) = R \oplus T^1 M \oplus T^2 M \oplus \ldots$, $T^j M$ is the j-fold tensor product of M with itself, and $\mathcal{I}(M)$ is the ideal generated by all elements of the form $x \otimes x - Q(x)\mathbf{1}$. Define $C^+(Q)$ to be the subalgebra of $C(Q)$ generated by an even number of elements of M. The *special Clifford group* $G^+(Q)$ consists of the set of invertible elements $g \in C^+(Q)$ for which $gMg^{-1} = g$.

special divisor   (**1**) A positive divisor a of a nonsingular, complete, irreducible curve C such that the specialty index of a is positive.

(**2**) A divisor D on an algebraic function field F in one variable such that the specialty index of D is positive.

See positive divisor, specialty index.

specialization   (**1**) Suppose that D and E are extension fields of a field F, and denote by D^n and E^n the n-dimensional affine spaces over D and E, respectively. Suppose that x is a point of D^n and that y is a point of E^n. If every polynomial p with coefficients in F that satisfies $p(x) = 0$ also satisfies $p(y) = 0$, then y is called a *specialization* of x over F.

(**2**) Suppose that S is a scheme, that s and t are geometric points of S, and that t is defined by an algebraic closure of the residue field of a point of the spectrum of the strict localization of S at s. Then s is called a *specialization* of t.

special Jordan algebra   Let $\mathcal{A}$ be an associative algebra over a field F of characteristic not equal to 2. Let $\mathcal{A}^+$ be the commutative algebra obtained by defining a new multiplication via $x * y = \frac{1}{2}(xy + yx)$. A special Jordan algebra J is a commutative algebra satisfying the Jordan identity

$$(xy)x^2 = x\left(yx^2\right)$$

for every x, y in J, which is isomorphic to a subalgebra of $\mathcal{A}^+$ for some $\mathcal{A}$.

special linear group   The multiplicative group of $n \times n$ matrices with determinant 1. The notation SL(n) is used.

special orthogonal group   The multiplicative group of $n \times n$ matrices M, with real entries and determinant 1, such that the transpose of M is equal to M^{-1}. The notation SO(n) is used.

special representation   If J is a Jordan algebra, $\mathcal{A}$ an associative algebra, and $\mathcal{A}^+$ the commutative algebra obtained from $\mathcal{A}$, by defining a new multiplication via $x * y = \frac{1}{2}(xy + yx)$, a special representation of J is a homomorphism $\rho : J \to \mathcal{A}^+$.

specialty index   (**1**) Of a divisor a of a nonsingular complete irreducible curve C, the number $d - n + g$, where d is the dimension of the complete linear system determined by a, n is the degree of a, and g is the genus of C.

(**2**) Of a divisor D of an algebraic function field F in one variable, the number $d - n + g - 1$, where D has dimension d and degree n, and F has genus g.

(**3**) Of a curve D on a nonsingular surface S, the dimension of $H^2(S, \mathcal{O}(D))$, where $\mathcal{O}(D)$ is the sheaf of germs of holomorphic cross-sections of D.

special unitary group   The multiplicative group of $n \times n$ matrices M such that the inverse M^{-1} is equal to the adjoint of M and the determinant of M is 1. The notation SU(n) is used.

special universal enveloping algebra   Given a Jordan algebra J, its *special universal enveloping algebra* is a pair (U, ρ) where U is an associative algebra, and $\rho : J \to U^+$ is a special representation of J with the property that for any other special representation $\tilde{\rho}$ of J, there exists a homomorphism h, such that $\tilde{\rho} = h\rho$.

special valuation   A valuation v of a field such that the rank of v is 1. Also called *exponential valuation*. *See* rank of valuation.

spectral radius   For a square ($n \times n$) matrix A, the real number

$$\rho(A) = \max\{|\lambda| : \lambda \text{ an eigenvalue of } A\}.$$

More generally, given a normed vector space X and a linear operator $T : X \to X$, the spectral radius of T coincides with

$$\lim_{k \to \infty} \left\| T^k \right\|^{1/k}.$$

This limit is independent of the choice of the norm $\| \cdot \|$.

spectral sequence   A sequence $M = \{M^k, d^k\}$, in which each M^k, $k = 2, 3, 4, \ldots$ is a **Z**-bigraded module and each d^k is a differential of bidegree $(-r, r - 1)$ mapping $M^k_{p,q}$ to $M^k_{p-r,q+r-1}$. Furthermore, one requires that $H(M^k, d^k) \cong M^{k+1}$, where $H(M^k, d^k)$ is the homology of M^k with respect to the differential d^k. An equivalent definition is obtained by considering an exact couple $\{D, E, \alpha, \beta, \gamma\}$ of Z-bigraded modules D and E, where the map α has degree $(1, -1)$, β has degree $(0, 0)$, and γ has degree $(-1, 0)$. One then associates to this exact couple a derived exact couple $\{D', E', \alpha', \beta', \gamma'\}$ in such a way that the differential object (E, d) (where $d = \beta \circ \gamma$) naturally associated to $\{D, E, \alpha, \beta, \gamma\}$, satisfies $H(E, d) = E'$. Iterating the construction of the derived exact couple then yields a spectral sequence isomorphic to that defined above.

spectral sequence functor   A map from the category of exact couples to the category of spectral sequences, which to an exact couple $\{D, E, \alpha, \beta, \gamma\}$, assigns a derived exact couple $\{D', E', \alpha', \beta', \gamma'\}$ in such a way that the differential object (E, d) (where $d = \beta \circ \gamma$) naturally associated to $\{D, E, \alpha, \beta, \gamma\}$, satisfies $H(E, d) = E'$. (Here, $H(E, d)$ is the homology of E with respect to the differential d.) Iterating the construction of the derived exact couple then yields a spectral sequence which is the image of $\{D, E, \alpha, \beta, \gamma\}$, under the *spectral sequence functor.*

spectral synthesis   If I is a closed ideal in the L^1-algebra of an Abelian group, then the problem of determining whether I is characterized by the set of common zeros of the Fourier transforms of elements of I is called the problem of *spectral synthesis*.

Spectral Theorem   Let A be a Hermitian operator on a Hilbert space H. Then there exists a unique, compact, complex spectral measure E such that $A = \int \lambda \, dE(\lambda)$.

spectrum   (1) Let L be a linear operator on a Banach space X. The *spectrum* of L is the complement of the set of complex numbers λ for which the operator $(L - \lambda I)^{-1}$ is bounded. In the case in which L is an $n \times n$ matrix, the spectrum is equal to the set of eigenvalues of L.

(2) Let x be an element of a commutative Banach algebra $\mathcal{A}$, with identity e. The *spectrum* of x is the set of complex numbers λ such that $x - \lambda e$ fails to have an inverse in $\mathcal{A}$.

spherical excess   The amount by which the sum of the three angles in a spherical triangle exceeds $180°$.

spherical Fourier transform   Let G be a connected Lie group, and K a compact subgroup. Let f be a continuous function on G, which is bi-invariant under K, by which one means that $f(g) = f(kgk')$, for every point g in G and every pair of points k and k' in K, and let dg be a left invariant measure on G. The *spherical Fourier transform* of f is

$$\hat{f}(\phi) = \int_G f(g)\phi(g^{-1})dg ,$$

where ϕ runs over the set of all positive definite spherical functions on G. One recovers the classical Fourier transform if one takes $G = \mathbf{R}^n$, $K = \{0\}$, and dg equal to Lebesgue measure.

spherical function   (1) Two linearly independent solutions, $P_{m,n}(z)$ and $Q_{m,n}(z)$ of the ordinary differential equation

$$\left(1 - z^2\right) \frac{d^2y}{dz^2} - 2z\frac{dy}{dz}$$
$$+ \left(n(n+1) - \frac{m^2}{1 - z^2}\right) y = 0$$

for n a positive integer, and m an integer between $-n$ and n. Their name arises because they are commonly encountered when applying the method of separation of variables to Laplace's equation in spherical coordinates. These functions, particularly when the range of m and n is extended to include arbitrary complex values, are also known as associated Legendre functions.

(2) Let G be a connected Lie group, and K a compact subgroup. A *spherical function* is a continuous function ϕ, on G, with the following properties:

(i.) ϕ is bi-invariant under K, by which one means that $\phi(g) = \phi(kgk')$, for every point g in G and every pair of points k and k' in K.

(ii.) ϕ is normalized so that $\phi(e) = 1$, where e is the identity element of G

(iii.) ϕ is an eigenfunction of every differential operator D which is invariant under the action of G, and also invariant under right translations of K.

Note that the classical spherical functions defined in (1) result from taking $G = SO(3)$, and $K = SO(2)$, while for $G = \mathbf{R}^n$, and $K = \{0\}$, the spherical functions are complex exponentials. *See also* spherical Fourier transform.

spherical representation (1) Let γ be a closed curve in three-dimensional Euclidean space, E. Fix an origin, 0, and to each point p on γ, associate the point p' on the unit sphere in E such that the ray $\overline{0p'}$ is the translate of the unit tangent vector to γ at p. Let γ' be the curve on the unit sphere traced out by p' as p varies over γ. Then γ' is the *spherical representation* of γ.

(2) Let G be a connected Lie group, and K a compact subgroup of G. A unitary representation U of G acting on a Hilbert space, $\mathcal{H}$, is a *spherical representation* of G with respect to K if there exists a nonzero vector v in $\mathcal{H}$ such that $U(k)v = v$, for every k in K. *See also* spherical function.

spherical triangle A geometrical figure on a sphere whose three sides are arcs of great circles. *See also* solution of right spherical triangle, solution of oblique spherical triangle.

spherical trigonometry The study of the properties and measurement of spherical triangles. *See also* solution of right spherical triangle, solution of oblique spherical triangle.

spinor Any element of the spin representation space of the group Spin$(n, \mathbf{C})$. *See* spin representation.

spinor group Let $C(Q)$ be the Clifford algebra obtained from a vector space V, with field of scalars, F, and inner product defined by the quadratic form Q. The *spinor group* Spin(Q) is the group of invertible elements of $C(Q)$ which

can be written as a product of an even number of vectors $v_j \in V$, with each v_j of norm 1. If V is $\mathbf{R}^n$ or $\mathbf{C}^n$, with the usual inner product, the spinor groups are denoted by Spin$(n, \mathbf{R})$, or Spin$(n, \mathbf{C})$, respectively.

spinorial norm Let $G^+(Q)$ be the special Clifford group obtained from a vector space V, with field of scalars, and quadratic form Q. *See* special Clifford group. Let β be the principal antiautomorphism (on the Clifford algebra C(Q)), defined by $\beta(v_1, v_2 \ldots v_n) = v_n v_{n-1} \ldots v_1$, for every $v_1, \ldots, v_n$ in V. Then the map $n : G^+ \to F$, defined by $n(g) = \beta(g) \cdot g$, for every g in G^+ is called the *spinorial norm* on C^+.

spin representation A linear representation of the spinor group Spin$(n, \mathbf{C})$. *See* spinor group. If n is odd, this is a faithful, irreducible representation of degree $2^{(n-1)/2}$. If n is even, the spin representation can be decomposed into a pair of inequivalent, irreducible representations, each of degree $2^{\frac{n}{2}-1}$. In this case, the two irreducible representations are referred to as half-spinor representations.

split extension Suppose that G, H, and K are groups, that 1 denotes the trivial group, and that

$$1 \to G \xrightarrow{\beta} H \xrightarrow{\alpha} K \to 1$$

is a short exact sequence. If there is a group homomorphism $\gamma : K \to H$ such that $\alpha \circ \gamma$ is the identity mapping on H, then H is called a *split extension* of G by K. *See also* Ext group.

splitting field (1) Let F be a field, E an extension field of F, and f a polynomial of positive degree and having all coefficients in F. If F contains all the roots of f, and E is generated over F by the roots of f, then E is called a *splitting* field of f over F. One says that f splits into linear factors over E.

(2) Suppose that P is a set of polynomials, each of positive degree and having all coefficients in F. If F contains all the roots of each polynomial in P, and E is generated over F by the roots of all the polynomials in P, then E is called a *splitting field* of P over F.

(3) Suppose that F is a field, and A an algebra over F. If every irreducible representation of A

over F is absolutely irreducible, then F is called a *splitting field* for A.

(**4**) Suppose that F is a field, and G a group. If F is a splitting field for the group ring $F[G]$, in the sense of (**3**) above, then F is called a *splitting field* for G.

(**5**) Suppose that A is a division algebra over a field F such that the center of A equals A, and that E is an extension field of F. If E effects a complete decomposition of A into simple left ideals so that $A \times F$ is isomorphic, for some positive integer i, to the ring of $i \times i$ matrices with elements in A, then E is called a *splitting field* of A over F. The number i is called the index of A.

(**6**) Suppose that F is a field, and denote by G the multiplicative group of nonzero elements in a universal domain that contains F. If an algebraic torus defined over F is isomorphic to the direct product G^n over an extension field E of F, then E is called a *splitting field* for the torus.

splitting ring   Suppose that R is a commutative ring, A an R-algebra, and S a commutative R-algebra. If there is some finitely generated faithful projective R-module P such that the S-algebra $S \otimes A$ is isomorphic to the ring of endomorphisms of P over S, then S is called a *splitting ring* of A.

square   (**1**) A geometrical figure with four sides of equal length which meet at right angles.

(**2**) To multiply a number by itself.

square integrable representation   Suppose that U is an irreducible unitary representation of a unimodular locally compact group G. Denote by dg the Haar measure of G. Suppose that, for some nonzero element x of the representation space of U, the function φ defined by $\varphi(g) = (U_g x, x)$ belongs to $L^2(G, dg)$. Then U is called *square integrable*.

square matrix   A matrix (rectangular array

$$\begin{bmatrix} a_{11} & a_{12} & \cdots & a_{1n} \\ a_{21} & a_{22} & \cdots & a_{2n} \\ & \cdots & \cdots & \\ a_{m1} & a_{m2} & \cdots & a_{mn} \end{bmatrix}$$

of real or complex numbers) in which the number of rows equals the number of columns; $m = n$.

square root   The *square root* of a number a is a number which, when multiplied by itself, gives a, denoted $\sqrt{a}$ or $a^{\frac{1}{2}}$. For example, a square root of 4 is 2 since 2×2 is 4. Note that a number may have more than one square root. In the present case, -2 is also a square root of 4 since $(-2) \times (-2)$ is also equal to 4. However, no number may have more than two distinct square roots.

stabilizer   If G is a group that acts on a set S and x is an element of S, then the *stabilizer* of x is the subgroup $\{g \in G : gx = x\}$ of G.

stabilizer   The subgroup of elements of a group of permutations of a non-empty set, under which the image of a given subset is itself.

stable matrix   Most commonly, an $n \times n$ matrix A with complex entries is called *stable* if its eigenvalues lie in the open left half plane, namely, if $\Re\lambda < 0$ for every eigenvalue λ of A. Matrices whose eigenvalues lie in the open right half plane are usually called positive stable. However, stability of a matrix appears in the literature with respect to a variety of regions of the complex plane. For example, a matrix whose eigenvalues lie in the open unit circle is sometimes referred to as Schur-stable.

stable reduction   (**1**) Suppose that E is an elliptic curve defined over a field F, complete with respect to a discrete valuation. Then E is said to have *stable reduction* over F if the reduced curve for a minimal Weierstrass equation is nonsingular.

(**2**) Suppose that R is a discrete valuation ring with quotient field F, and A is an Abelian variety over F. Denote by S the spectrum of R, and denote by M the Néron minimal model of A. Suppose that the connected component of the fiber of M over the closed point of S that contains 0 has no unipotent radical. Then A is said to have *stable reduction*. *See* Neron minimal model.

(**3**) Suppose that R is a discrete valuation ring with quotient field F, and C is a smooth connected curve over F. Then there is a finite separable algebraic extension field E of F such that the curve $C \times_F E$ extends to a flat family of stable curves over the spectrum of the integral closure R_E of R in E, and $C \times_F E$ is said to have *stable reduction* in R_E.

Stable Reduction Theorem If R is a discrete valuation ring with quotient field F, and A is an Abelian variety over F, then A has potential stable reduction at R. *See* stable reduction.

stable vector bundle A vector bundle E with the property that for any proper subbundle E' of E, the inequality

$$\text{rank}(E) \cdot \text{degree}(E') < \text{rank}(E') \cdot \text{degree}(E)$$

holds.

standard complex Suppose that R is a commutative ring, and that g is a Lie algebra over R that is R-free. Denote by U the enveloping algebra of g, and denote by $\bigwedge_R(g)$ the exterior algebra of the R-module g. The U-free resolution $U \otimes \bigwedge_R(g)$ of R is called the *standard complex* of g.
See enveloping algebra, exterior algebra.

standard form of difference equation If y is a function from the integers to the real or complex numbers, and Δ is the finite difference operator defined by

$$(\Delta y)(x) = y(x + 1) - y(x) \,,$$

a difference equation is an equation relating y, and its finite differences, i.e., an equation of the form $F(x, y, \Delta y, \ldots, \Delta^k y) = 0$. This standard form of this equation results by rewriting it in such a way that the finite difference operator does not explicitly occur, i.e., by choosing ℓ so that the equation becomes $G(x, y(x), y(x + 1), \ldots, y(x + \ell)) = 0$.

standard parabolic k-subgroup Suppose that k is a field, and G is a connected reductive group defined over k. Suppose that S is a maximal k-split torus, denote by Z the centralizer of S in G, and by P a minimal parabolic k-subgroup of G. Denote by r the set of all k-roots of G with respect to S, choose an ordering of r, and suppose that θ is a subset of the fundamental system of r with respect to that ordering.

Suppose that α is a linear combination of the roots of r, in which all roots not in θ occur with a nonnegative coefficient. Denote by ad the adjoint representation of g, and denote by P_α the unipotent k-subgroup of G normalized by S so that the Lie algebra of P_α is the set of all elements X in g such that $\text{ad}(s)X = \alpha(s)$ for all s in S. Denote by P_θ the subgroup generated by Z and P_α. Then P_θ is called a *standard parabolic k-subgroup* of G that contains P.

standard position of angle An angle with its vertex at the origin and its initial ray lying on the positive x-axis. The magnitude of the angle is then measured in a counter-clockwise direction with respect to the side lying on the x-axis.

star of element (1) (Of an element A of a complex Δ) The set of all elements of Δ that contain A.

(2) (Of an element x of a Banach algebra B with involution $*$) The image of x under $*$ (denoted x^*).

state (1) A positive linear functional of norm 1 on a C^*-algebra.

(2) The *state* of a system of N particles in classical mechanics is described by the instantaneous position and velocity of each of the particles. If these particles are moving in three-dimensional Euclidean space, the state is then described by a point in $\mathbf{R}^{6N}$. Given Newton's laws, the knowledge of the state of a system at any given time allows one to predict its state at any time in the future or past. These states of the system are sometimes referred to as the dynamical or microscopic state of the system.

In thermodynamics, the state of a system is often specified by a much smaller number of quantities, such as its density, volume, and temperature. In this case the number of quantities chosen to specify the state of the system are the minimum needed to predict its macroscopic properties, and these states correspond to a statistical average over a very great number of microscopic states of the system.

Steenrod algebra Suppose that p is a prime, and denote by $\mathbf{Z}_p$ the field of order p. Suppose that q is a nonnegative integer, and denote by A_q the Abelian group that consists of all stable cohomology operations of type $(\mathbf{Z}_p, \mathbf{Z}_p)$ and of degree q. The graded algebra $\sum_{q=0}^{\infty} A_q$ is called the *Steenrod algebra* mod p.

Steinberg's formula   Suppose that G is a compact, connected, semisimple, Lie group, and that (ρ_1, V_1) and (ρ_2, V_2) are irreducible representations of G. Denote by g the Lie algebra of G, and choose a Cartan subalgebra h of g. Decompose the tensor product mapping $\rho_1 \otimes \rho_2$ relative to h into a direct sum of irreducible constituent representations. For each weight μ of $\rho_1 \otimes \rho_2$, denote by ρ_μ the irreducible representation of G that has μ as its highest weight, and denote by $m(\mu)$ the multiplicity of ρ_μ in the decomposition of $\rho_1 \otimes \rho_2$. Denote by h^* the linear space of all complex-valued forms on h, and denote by $(h^C)_{\mathbf{R}}^*$ the real linear subspace of h^* that is spanned by the roots of g relative to h. Denote by W the Weyl group of g relative to h, by $P(\mu)$ the number of ways in which an integral form μ in $(h^C)_{\mathbf{R}}^*$ may be expressed as a sum of positive roots of g relative to h, and by Λ_1 and Λ_2 the highest weights of the representations of the differentials $d\rho_1$ and $d\rho_2$ of the complexification g^C over V_1 and V_2, respectively. Denote by s the sum of the positive roots of g relative to h. For w and w' in W, put

$$a(w) = w(\Lambda_1 + \delta) + w'(\Lambda_2 + \delta)$$

and put $b = \mu + s$. Then, if $\rho_i = \rho_{\Lambda_i}$ for $i = 1$ and 2,

$$m(\mu) = \sum_{w \in W} \sum_{w' \in W} \det(ww') P(a(w) - b).$$

See Weyl group, highest weight.

Steinberg type group   A connected, semisimple, algebraic group G, defined over a field F, such that G has a maximal, connected, solvable, closed subgroup defined over F.

stereographic projection   A particular method of associating to each point on a sphere a point in the plane. Let N be the north pole of the sphere, and Π the equatorial plane. Through each point, p, of the sphere, draw the line passing through N. The stereographic projection of p is the point q at which this line passes through Π. Note that this is a one-to-one (and hence invertible) mapping of the sphere minus the north

pole onto the plane. This mapping is often used in complex analysis since in addition to being invertible, it is conformal. The plane Π is sometimes chosen to be the plane tangent to the south pole, rather than the equatorial plane.

Using the inverse of this mapping, the plane can be thought of as embedded in the (Riemann) sphere.

Stiefel-Whitney class   Let ξ be a q-sphere bundle with base space B, and let $U_\xi \in H^{q+1}(E_\xi, \dot{E}_\xi; \mathbf{Z}_2)$ be its orientation class over $\mathbf{Z}_2$. The ith *Stiefel-Whitney class* $w_i(\xi) \in H^i(B; \mathbf{Z}_2)$ for $i \geq 0$ is defined by

$$\Phi_\xi^*(w_i(\xi)) = \mathrm{Sq}^i(U_\xi).$$

Stiemke Theorem   Suppose that A is an $m \times n$ real matrix. Then either
(i.) there is a vector x in $\mathbf{R}^n$ such that $Ax = 0$ and each coordinate of x is positive, or
(ii.) there is a vector u in $\mathbf{R}^m$ such that each coordinate of $u^T A$ is nonnegative.

stochastic programming   The theory, methods and techniques of incorporating stochastic information into mathematical programming. Typically, one wishes to optimize one or more functions subject to some set of constraints. In classical mathematical programming, it is assumed that both the function to be maximized, and the constraint functions are known with certainty. In stochastic programming, any of these functions may vary randomly.

Stoker multiplier   A concept from the theory of water waves.

Stokes multiplier   With respect to a given formal fundamental solution $\widehat{Y}$ of

$$\widehat{Y}(x) = \widehat{H}(x) \times^\ell e^{Q(x)},$$

the Stokes multiplier (or Stokes matrix) $S_\alpha \in \mathrm{GL}(n, \mathbf{C})$ corresponding to the singular line α of Δ at 0 is defined by $Y_{\alpha-} = Y_{\alpha+} + S_\alpha$ on α.

Stone space   The space of all uniform ultrafilters on ω with the natural topology. Equivalently, the remainder $\beta\omega \setminus \omega$ in the Cech-Stone compactification of the integers. This construction is denoted $\mathrm{St}(\mathcal{P}(\omega)/\mathrm{fin})$, where $\mathcal{P}(\omega)/\mathrm{fin}$

is the power set of ω modulo the ideal of finite sets.

strict Albanese variety   Suppose that V is an irreducible variety, A is an Abelian variety, and $f : V \to A$ is a morphism such that
(i.) the image of f generates A, and
(ii.) for every Abelian variety B and every morphism $g : V \to B$, there is a homomorphism $h : A \to B$ and a point b of B such that $g = h \circ f + b$.

Then the pair (A, f) is called a *strict Albanese variety* of V.

strict localization   Suppose that S is a scheme, and s is a geometric point of S. The stalk of the sheaf of étale neighborhoods of s is called the *strict localization* of S at s.

stronger equivalence relation   Suppose that R and $\tilde{R}$ are equivalence relations on a set S. Suppose that for all elements s and t of S, if $s R t$ holds, then $s \tilde{R} t$ holds. Then R is called stronger than $\tilde{R}$.

Strong Lefschetz Theorem   Suppose that V is a projective nonsingular irreducible variety of dimension n, defined over the complex field $\mathbf{C}$, and suppose that V is embedded in the complex projective space $\mathbf{P}^N$ for some N. Suppose that W_0 and W_∞ are elements of a Lefschetz pencil obtained from a generic linear pencil of hyperplanes in $\mathbf{P}^N$. Suppose that $\tilde{V}$ is a smooth variety and $\pi : \tilde{V} \to \mathbf{P}^1$ is a surjective morphism obtained from blowing up V at $W_0 \cap W_\infty$. Denote by $\mathbf{Q}$ the ring of rational numbers, by c the cohomology class of $H^2(V, \mathbf{Q})$ that corresponds to the hyperplane section $\pi^{-1}(0)$, and by h the homomorphism from $H^*(V, \mathbf{Q})$ to $H^{*+2}(V, \mathbf{Q})$ that is defined by the cup product with c. Then, for each $i \leq n$, the homomorphism $L^{n-1} : H^i(V, \mathbf{Q}) \to H^{2n-1}(V, \mathbf{Q})$ is an isomorphism.

See Lefschetz pencil, blowing up.

strongly closed subgroup   A subgroup H of a finite group G such that $g^{-1}Hg \cap N(H)$ is a subgroup of H for each g in G, where $N(H)$ is the normalizer of H in G.

strongly continuous homomorphism   Suppose that G is a topological group, H is a Hilbert space with at least two elements, and U is a homomorphism of G into the group of unitary operators on H. For each g in G, denote by U_g the unitary operator $U(g)$ on H that is associated with U. Consider the mapping $\sigma : G \to H$ defined by $\sigma(g) = U_g(x)$. If σ is a continuous mapping, then U is called *strongly continuous*.

strongly embedded subgroup   A proper subgroup H of a finite group G such that H is of even order and $H \cap g^{-1}Hg$ is of odd order for every g in $G \backslash H$.

strongly normal extension field   Suppose that U is a universal differential field of characteristic zero, and denote by K the field of constants of U. Suppose that F is a differential subfield of U such that U is universal over F, and E is a finitely generated differential extension field of F. Denote by C the field of constants of E. Suppose that every isomorphism σ of E over F satisfies the following conditions:
(i.) σ leaves invariant every element of C;
(ii.) each element of σE is an element of EK; and
(iii.) each element of E is an element of $\sigma E \cdot K$.

Then E is called a strongly normal extension field of F.

strong minimality   The property of a complete, nonsingular, irreducible variety V over a field that every rational mapping from a nonsingular variety to V is a morphism.

Also called *absolutely minimality*.

structural constant   (1) If V is a vector space over the field of complex numbers, then multiplication in V is a tensor. The entries of the matrix of that tensor relative to fixed bases of the underlying spaces are called the *structural constants* of the corresponding tensor algebra.

(2) Suppose that G is a local Lie group and D is a differentiable system of coordinates in G. Suppose that x and y are elements of G that are sufficiently close to the identity, and put $f(x, y) = xy$. If the second-order term of the ith coordinate function f^i of f is $a^i_{jk} x^j y^k$, then the *structural constants* of G in D are the numbers c^i_{jk} defined by $c^i_{jk} = a^i_{jk} - a^i_{kj}$.

(3) If G is an infinitesimal group (that is, a vector space together with a commutation op-

eration defined on it), and if the commutation operation in G is expressed in terms of coordinates, then the coefficients of the corresponding matrix are called the *structural constants* of G in the given system of coordinates. *See* commutator.

(4) Suppose that F is a field of characteristic 0, that g is a Lie algebra over F of dimension n, and that $X_1, \ldots, X_n$ is a basis of g over F. If $[X_i, X_j] = \sum_{k=1}^{n} c_{ij}^k X_k$ for $i = 1, \ldots, n$ and $j = 1, \ldots, n$, then the elements c_{ij}^k of F are called the *structural constants* of g relative to $X_1, \ldots, X_n$.

structure   A description of a mathematical object in terms of sets and relations, subject to a set of axioms is called a *mathematical structure*. For example, a group, defined in terms of a set and an operation, subject to the usual axioms for a group, is a mathematical structure.

structure sheaf   (1) Suppose that V is an affine variety. Denote by A_U the ring of regular functions on the open set U. The sheaf obtained from the assignment of A_U to U for each open set U is called the *structure sheaf* of V (or the *sheaf of germs of regular functions* on V).

(2) Suppose that V is a topological space, F is a field, and $\mathcal{O}$ is a sheaf of germs of mappings from V to F. Suppose that there is a finite open covering $U_1, \ldots, U_n$ of V such that each pair $(U_i, \mathcal{O}_{U_i})$ is isomorphic to an affine variety. Then $\mathcal{O}$ is called the *structure sheaf* of the pair $(V, \mathcal{O})$.

(3) Suppose that X is a topological space, and $\mathcal{O}$ is a sheaf on X of commutative rings, each with a unit, such that the stalk of $\mathcal{O}$ at each point x of X does not equal $\{0\}$. Then $\mathcal{O}$ is called the *structure sheaf* of the pair $(X, \mathcal{O})$.

structure space   The set of kernels of algebraically irreducible representations of a Banach algebra.

Structure Theorem for Type-III von Neumann Algebras   Suppose that M is a von Neumann algebra of type III. Then there exist a von Neumann algebra N of type II$_\infty$, a faithful normal trace τ, and a one-parameter group G of $*$-automorphisms θ_t with $\tau \circ \theta_t = e^{-t}\tau$, such that M is a crossed product of N with G.

See type-III von Neumann algebra, crossed product.

Structure Theorem of Complete Local Rings   Suppose that R is a semilocal ring with unique maximal ideal m, that I is a coefficient ring of R, and that m is generated by n elements. Denote by S the ring of formal power series in n variables over I. Then R is a homomorphic image of S.

Sturm method of solving algebraic equations   Suppose that f is a real-valued polynomial of degree d in one variable, and that f has no multiple zero. For real numbers a and b with $a < b$, denote by $N(a, b)$ the number of real roots of the equation $f = 0$ in the interval (a, b). By applying the division algorithm, one may construct polynomials $f_0, f_1, \ldots, f_d$ and $q_1, \ldots, q_{d-1}$ such that $f_0 = f$, $f_{i-1} = f_i q_i + f_{i+1}$ for $1 \leq i \leq d - 1$, and f_d is constant. Denote by $V(a)$ the number of changes in sign of the terms of the finite sequence $f_0(a), \ldots, f_d(a)$, and denote by $V(b)$ the number of changes in sign of the terms of the finite sequence $f_0(b), \ldots, f_d(b)$; in each case we ignore zero terms. Then $N(a, b) = V(a) - V(b)$. Sturm's method uses this result to approximate the real roots of the equation $f = 0$.

Sturm's Theorem   A theorem which relates the position of zeros of solutions of two linear, second order, ordinary differential equations. Suppose that $y_1(x)$ and $y_2(x)$ are nonzero solutions of

$$
\begin{aligned}
y_1'' + q(x)y_1 &= 0 \\
y_2'' + p(x)y_2 &= 0
\end{aligned}
$$

on some interval I. If $q(x) > p(x)$ for all x in some interval I, then between any two zeros of y_2, there must be a zero of y_1.

subalgebra   A subset S of an algebra $\mathcal{A}$, with the property that S is itself an algebra with respect to the operations of addition and multiplication of $\mathcal{A}$.

subbialgebra   A subspace of a bialgebra $\mathcal{B}$, which is both a subalgebra and a subco-algebra of $\mathcal{B}$.

subbundle   If E and F are vector bundles over a common base space X, then E is a subbundle of F if there exists a vector bundle morphism such that for each $x \in X$, the restriction map $E_x \to F_x$ has rank one.

subco-algebra   Let C' be a subspace of a co-algebra C with comultiplication Δ and counit ϵ. C' is a subco-algebra, if $\Delta(C') \subset C \otimes C$, and $(C', \Delta|_{C'}, \epsilon|_{C'})$ is a co-algebra.

subcomplex   (1) If C is a simplicial complex, a subset C' of C which is itself a simplicial complex is called a *subcomplex.*

(2) If C is a chain complex, a subset C' of C which is itself a chain complex and for which C'_j is a subset of C_j for each j, and for which the boundary operator ∂'_j is the restriction of the boundary operator ∂_j to C'_j for each j is called a *subcomplex.*

subdirect sum of rings   A subring T of a direct sum of rings S for which each projection map from T into the coordinate rings S_i is surjective. In more detail, let $S_1, S_2, S_3, \ldots$ be a finite or countable family of rings. The ring S, consisting of all sequences $s = (s_1, s_2, s_3, \ldots)$, where each $s_i \in S_i$ and the operations are taken coordinate-wise, is called the *direct sum* of the rings S_i. Let $p_i : S \to S_i$ be the function defined by $p_i(s) = s_i$. p_i is a ring homomorphism from S onto S_i called the ith *projection map.* A ring T is a *subdirect sum of the rings S_i* if two things are true: (i.) T is a subring of S, and (ii.) each projection map p_i is surjective from T to S_i, that is, $p_i(T) = S_i$. *See also* direct sum, ring homomorphism.

The notions of direct sum and subdirect sum extend without difficulty to arbitrary, possibly uncountable, families of rings. *See* countable set, uncountable set.

subfield   A subset S of a field F, with the property that S is itself a field with respect to the operations of addition and multiplication of F.

subgroup   Let G be a group and H a subset of G. We call H a *subgroup* of G if H contains the identity element, H is closed under multi-

plication, and H contains the inverse of each of its elements.

submatrix   *See* principal submatrix.

subnormal subgroup   Given a group G, a normal series is a finite set of subgroups, $\{N_j\}_{j=1}^n$, which satisfy:
(i.) $\{e\} = N_n \subset N_{n-1} \subset \cdots \subset N_2 \subset N_1 = G$, where e is the identity element in G.
(ii.) N_{j+1} is a normal subgroup of N_j for $j = 1 \ldots n - 1$.

If H is a subgroup of G such that $H = N_i$, for N_i an element of some normal series of G, then H is called a *subnormal subgroup* of G.

Note that if H is a normal subgroup of G, we can take the normal series $N_1 = G$, $N_2 = H$, $N_3 = \{e\}$, so every normal subgroup is also subnormal.

subordinate subalgebra   Suppose that G is a simply connected nilpotent Lie group, g is the Lie algebra of G, and g^* is the dual space of g. Suppose that h is a subalgebra of g and f is an element of g^* such that $(f, [X, Y]) = 0$ for every X and Y in h. Then h is called *subordinate* to f.

subrepresentation   Let ρ be a representation of a group G, with representation space V, and suppose the vector space V can be written as the direct sum of two subspaces, S and T, each left invariant by the representation. That is, let $\rho(g)$ be the image in $GL(V)$ of a point $g \in G$ and suppose $\rho(g)S \subset S$, and $\rho(g)T \subset T$, for all $g \in G$. The maps ρ_S and ρ_T obtained by restricting the range of the representation to $GL(S)$ and $GL(T)$, respectively, are themselves representations, and are called *subrepresentations* of ρ. *See* representation, representation space.

subring   A subset S of a ring R, with the property that S is itself a ring with respect to the operations of addition and multiplication of R.

substitution   The act of replacing one mathematical expression by another equivalent one. For instance, if one wishes to solve the system of equations

$$x + y = 1$$
$$x - y = 2$$

one can conclude from the first equation that $x = 1 - y$, and then substitute this value of x into the second equation; that is, replace the x in the second equation by the equivalent expression $1 - y$. One then obtains the equation $1 - 2y = 2$, or $y = -\frac{1}{2}$, from which one concludes that $x = \frac{3}{2}$.

subtraction   The process of finding the difference of two numbers or quantities. For instance, the number $z = x - y$ is that number which if added to y gives x.

subtraction formulas in trigonometry   Formulas that express the sine, cosine, or tangent of the difference of two angles α and β, in terms of trigonometric functions of α and β. For example,

$$\sin(\alpha - \beta) = \sin(\alpha)\cos(\beta) - \sin(\beta)\cos(\alpha)$$
$$\cos(\alpha - \beta) = \sin(\alpha)\sin(\beta) + \cos(\alpha)\cos(\beta)$$
$$\tan(\alpha - \beta) = \frac{\tan(\alpha) - \tan(\beta)}{1 + \tan(\alpha)\tan(\beta)}$$

subvariety   Let W be a variety, that is, the set of common zeros of some collection of polynomials. If W is a (proper) subset of some other variety V, then W is called a subvariety of V.

subvariety   Let W be a variety, that is, the set of common zeros of some collection of polynomials. If W is a (proper) subset of some other variety V, then W is called a *subvariety* of V.

successive substitutions   A method of solving systems of equations in the variables $x_1, \ldots, x_n$ in which the first equation is solved for x_1 as a function of $x_2, \ldots, x_n$. This result is then substituted into the remaining equations yielding a system of equations in the variables $x_2, \ldots, x_n$. This process is then repeated until only a single equation remains. For example, given the system of equations

$$x_1 + x_2 + x_3^2 = 0$$
$$2x_1 - x_2 + x_3 = 0$$
$$x_1 + 2x_2 - x_3^2 = 0$$

one can rewrite the first equation as $x_1 = -x_3^2 - x_2$. Substituting this into the second and third equations leads to

$$-3x_2 + x_3 - 2x_3^2 = 0$$
$$x_2 - 2x_3^2 = 0$$

Solving the second of these equations for x_2, gives $x_2 = 2x_3^2$, which when substituted into the first equation gives

$$x_3\left(1 - 8x_3^2\right) = 0$$

so that $x_3 = 0$, or $x_3 = \pm 1/(2\sqrt{2})$. One can then obtain x_1 and x_2 for each of these possible choices of x_3, by substituting the chosen value of x_3 into the previous equations. Note that this method may fail to yield a solution if it is impossible to solve for any of the variables at some stage in the procedure, or if the final equation one obtains has no solution.

sum   **(1)** The result of adding two or more numbers, or other mathematical objects (e.g., vectors).
    (2) To add.

summation   The act of adding numbers or other mathematical objects (e.g., vectors).

summation notation   Notation which uses the Greek letter Σ to denote a sum. Typically the symbol Σ is followed by an expression (the "summand") which depends on a "summation variable," say j, and Σ will have a subscript indicating the first value of j to be included in the sum, and a superscript indicating the final value of j to be included. For example, the symbol $\sum_{j=1}^{5} \frac{j(j-1)}{2}$ denotes the sum of the quantity $\frac{j(j-1)}{2}$ from $j = 1$ to $j = 5$, i.e., $0 + 1 + 3 + 6 + 10$.

sum of like powers   A sum in which the summands are all expressions raised to a common power, e.g., $15^5 + 20^5 + 25^5 + 30^5$, or $x^r + y^r + z^r$.

sum of perfect powers of integers   A sum in which the summands are integers, each raised to the same positive integral power, e.g., $15^5 + 20^5 + 25^5 + 30^5$, or $1 + 8 + 27 + 64 + 125$. (Note that in the second example, each summand is the cube of an integer.)

sum of vectors   *See* addition of vectors.

superabundance   Suppose that S is a nonsingular surface, and D is a curve on S. Denote by $\mathcal{O}(D)$ the sheaf of germs of holomorphic cross-sections of D. The dimension of $H^1(S, \mathcal{O}(D))$ is called the *superabundance* of D.

supersolvable group   A finite group G such that there are subgroups $G_0, \dots, G_n$ of G such that G_0 is trivial, G_i is a normal subgroup of G_{i+1} for $i = 0, 1, \dots, n-1$, each G_i is normal in G, and each quotient group G_{i+1}/G_i is cyclic and of prime order.

supplemented algebra   Let R be a commutative ring with a unit. If A is an R-algebra and $\varepsilon : A \to R$ is an R-algebra homomorphism, then (A, ε) is called a *supplemented algebra* (or *augmented algebra*).

support   (**1**) The support of a complex-valued function is the closure of the set of points at which the function is nonzero. Equivalently, it is the complement of the largest open set on which the function vanishes.

(**2**) The support of a measure is the complement of the largest open set of zero measure.

supremum   The least upper bound or join of a set of elements of a lattice. The term is most frequently used with regard to sets of real numbers. If S is a set of real numbers, the *supremum* of S is the unique real number $B = \sup S$ defined by the following two conditions: (i.) $x \leq B$, for all $x \in S$; (ii.) if $x \leq C$, for all $x \in S$, then $B \leq C$.

See also infimum.

surd   An irrational root of a positive integer, e.g., $\sqrt[3]{2}$, or $5^{2/5}$, or a sum of such expressions.

surjection   A function e from a set X to a set Y such that $e(X) = Y$, that is, such that every element $y \in Y$ is the image $e(x)$ of an element $x \in X$. Surjections are also called *surjective functions, surjective mappings,* and *onto functions*. The notion of a surjection generalizes to

the notion of an *epimorphism* or *epic morphism* in a category. *See also* epimorphism, injection.

Suzuki group   A finite simple group of order $q^2(q-1)(q^2+1)$, where n is a positive integer, and $q = 2^{2n+1}$.

sweepable bounded domain   A bounded domain D in $\mathbf{C}^n$ such that there is a compact subset K of D and a subgroup Γ of the group of holomorphic automorphisms of D with $\Gamma K = D$.

Sylow's Theorems   Suppose that G is a finite group.

(**1**) If m and n are relatively prime and the order of G is $p^n m$, then G contains a subgroup of order p^i for each $i = 1, 2, \dots, n$. If $1 \leq i \leq n$ and S is a subgroup of G of order p^i, then S is a normal subgroup of some subgroup of G of order p^{i+1}.

(**2**) If H is a p-subgroup of G and S is a Sylow p-subgroup of G, then there is some element g of G such that H is a subgroup of gSg^{-1}. In particular, every two Sylow p-subgroups of G are conjugate in G.

(**3**) Denote by s the number of distinct Sylow p-subgroups of G. Then k is a factor of the order of G and $s \equiv 1 (\bmod\ p)$. If H is a p-subgroup of G, denote by t the number of distinct Sylow p-subgroups of G that contain H. Then $t \equiv 1\ pmod p$.

See Sylow subgroup.

Sylow subgroup   A maximal p-subgroup of a group G, for a prime number p. Also called *Sylow p-subgroup* of G.

See Sylow's Theorems.

Sylvester's elimination method   A method used in elimination theory for obtaining the resultant of two polynomials. Suppose that I is a unique factorization domain and $a_0, \dots, a_m$ and $b_0, \dots, b_n$ are elements of I. Put $f(x) = \sum_{i=0}^m a_i x^{m-i}$ and $g(x) = \sum_{i=0}^n b_i x^{n-i}$. Define an $m \times n$ matrix $C = (c_{ij})$ as follows. For $i = 1, \dots, n$ and $j = i, \dots, i+m$, put $c_{ij} = a_{j-i}$. For $i = n+1, \dots, n+m$ and $j = i-n, \dots, i$, put $c_{ij} = b_{j+n-i}$. Put $c_{ij} = 0$ for all other appropriate values of i and j. If a_0 and b_0 have no common factor, then the determinant of C equals the resultant of f and g.

See resultant.

Sylvester's Theorem (**1**) If F is an ordered field and V is a vector space over F with a non-degenerate bilinear form, then there is a non-negative integer r such that if $\{v_1, v_2, \ldots, v_n\}$ is an orthogonal basis of V, then exactly r of the squares $v_1^2, v_2^2, \ldots, v_n^2$ are positive and $n - r$ of these squares are negative.

(**2**) Suppose that $A = (a_{ij})$ is an $n \times n$ matrix, and r is a positive integer less than n. For i and k in the set $\{1, \ldots, n - r\}$, denote by B_{ik} the $(r + 1) \times (r + 1)$ submatrix of A consisting of entries contained in one of rows $1, \ldots, r, r + i$ of A and in one of columns $1, \ldots, r, r + k$ of A. Denote by C the $r \times r$ submatrix of A consisting of entries contained in one of rows $1, \ldots, r$ of A and one of columns $1, \ldots, r$ of A. Then

$$\det B_{ik} = (\det A)(\det C)^{n-r-1} .$$

symmetric algebra (**1**) If V is a vector space, T is the tensor algebra over V, and I is the ideal generated by

$$\{x \otimes y - y \otimes x : x, y \in V\} ,$$

then T/I is called the *symmetric algebra* of (or over) V. This algebra is also the direct sum of all the symmetric powers of V.

(**2**) If R is a commutative ring, M is an R-module, T is the tensor algebra of M, and I is the ideal generated by

$$\left\{x \otimes y - (-1)^{(\deg x)(\deg y)} y \otimes x : x, y \in T\right\} ,$$

then T/I is called the *symmetric algebra* of (or over) M.

(**3**) Suppose that A is a finite-dimensional algebra over a field F and that the left regular module of A is isomorphic to the group of homomorphisms from the right regular module of A to F. If A has a nondegenerate bilinear form that is symmetric and associative, then A is called a *symmetric algebra*.

symmetric bilinear form Let V be a vector space, and F a field. Let $h : V \times V \to F$ be a bilinear form. Then h is *symmetric* if $h(v, v') = h(v', v)$ for any two vectors v and v' in V.

symmetric equation An equation in two or more unknowns $x_1, \ldots, x_n$ which is unchanged by any permutation of the unknowns.

symmetric form A matrix M is said to be *symmetric* if $^t M = M$. The quadratic form induced by M is then called a *symmetric form*.

symmetric group The set of permutations of a set $\mathcal{N}$ of n elements, together with the operation of composition, forms the *symmetric group*. The notation S_n is used.

symmetric matrix A real matrix S which is equal to its adjoint, i.e., if the elements of S are S_{ij}, then $S_{ij} = S_{ji}$.

symmetric points Two points a and b are symmetric with respect to a third point c if c is the midpoint of the line segment $\overline{ab}$. Two points a and b are symmetric with respect to a line if that line is the perpendicular bisector of $\overline{ab}$. Finally, a and b are symmetric with respect to a plane if that plane passes through the midpoint of $\overline{ab}$, and is perpendicular to this line segment.

symmetric polynomial A polynomial, $p(x_1, \ldots, x_n)$, which is transformed into itself by any permutation of the variables $x_1, \ldots, x_n$.

symmetric property The property of a relation R on a set S:

$$(s, t) \in R \text{ implies } (t, s) \in R \text{ for all } s, t \in S .$$

symmetric relation A relation R (on a single set S) such that $(a, b) \in R$ implies $(b, a) \in R$. *See* relation. An example of a *symmetric relation* is equality: if $a = b$, then $b = a$. An example of a relation that is not symmetric is the "less than" relation. If $a < b$, then it is not true that $b < a$.

symmorphism Suppose that T is an n-dimensional lattice in n-dimensional Euclidean space, and that K is a finite subgroup of the orthogonal group. Denote by A and by S the sets of arithmetic crystal classes and of space groups of (T, K), respectively. There is a surjective mapping $s : S \to A$ and there is an injective mapping $i : A \to S$ such that $s \circ i$ is the identity mapping on A. A space group that belongs to the image of the mapping i is called *symmorphic*.

See arithmetic crystal class, space group.

symplectic group   The set of all symplectic transformations from a symplectic vector space (V, ω) to itself form the symplectic group of (V, ω). If V is $\mathbf{R}^{2n}$, with the standard symplectic structure, the notations $\mathrm{Sp}(2n)$ and $\mathrm{Sp}(2n, \mathbf{R})$ are used. *See* symplectic transformation.

symplectic manifold   A manifold equipped with a distinguished closed 2-form of maximal rank.

symplectic transformation   A linear transformation T from a symplectic vector space (V_1, ω_1) to a symplectic vector space (V_2, ω_2), which preserves the symplectic form; i.e.,

$$\omega_2\left(T\left(v_1\right), T\left(v_2\right)\right) = \omega_1\left(v_1, v_2\right)$$

for any vectors v_1 and v_2 in V_1.

synthetic division   A compact notation for representing the division of a polynomial $p(x)$ by a linear factor of the form $x - a$. For instance, to divide $p(x) = 3x^2 - 2x - 1$ by $x - 1$, first write down the coefficients of $p(x)$, and then to the left, the constant term in the linear factor:

$$-1 \quad | \quad 3 \quad -2 \quad -1$$

The coefficient of the first term in the quotient will always be the same as the coefficient of the leading term in the original polynomial (3 in this example), so that is not written. Since $3x(x - 1) = 3x^2 - 3x$, we must subtract $-3x$ from the $-2x$ in the original polynomial. This is represented by changing the sign of -3, and writing it under the -2 in the first line, and adding:

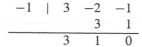

The next term in the factorization is $1 \cdot (x - 1) = x - 1$, and repeating the above procedure, we find:

$$
\begin{array}{ccccc}
-1 & | & 3 & -2 & -1 \\
   &   &   & 3 & 1 \\
\hline
   &   & 3 & 1 & 0
\end{array}
$$

The bottom row of this diagram gives the coefficients of the quotient polynomial — in this case $3x + 1$, and the final entry in the last row gives the remainder — in this case 0.

system of equations   A set of equations in the variables $x_1, x_2, \ldots, x_n$, which are all required to be satisfied simultaneously.

system of fundamental solutions   (1) Suppose that F is a field, and consider a system S of linear homogeneous equations with coefficients in F. Denote by V the vector space over F of solutions of S. If the dimension of V is positive, then a basis for V is called a *system of fundamental solutions* of S.

(2) Suppose that D is a division ring, and consider a system S of linear homogeneous equations with left coefficients in D. Denote by M the unitary right D-module of solutions of S. If the rank of M is positive, then a basis for M is called a *system of fundamental solutions* of S.

system of fundamental units   A minimal set of units in terms of which the units of all other physical quantities can be defined. For example, in classical mechanics, if we choose the meter as the unit of length, the kilogram as the unit of mass, and the second as the unit of time (the mks-system), then the units of other quantities such as velocity (meters/second) or force (kilogram-meters/second^2) can be expressed in terms of these three. Note that what constitutes a system of fundamental units may depend on the physical phenomena under consideration. For instance, if one wished to consider electrodynamics in addition to classical mechanics, one would need to add to the list of fundamental units a unit for current.

system of generators   A set of elements S, of a group G, such that every element in G can be written as a product of elements of S (possibly with repetitions).

system of inequalities   A set of inequalities in the variables $x_1, x_2, \ldots, x_n$, all of which are required to be satisfied simultaneously. For example the set of values (x_1, x_2) which satisfy the system of inequalities

$$\begin{aligned} x_1 + x_2 &< 0 \\ 2x_1 + x_2 &> -1 \\ x_1 + 3x_2 &> -2 \end{aligned}$$

is represented by the set of points in the x_1x_2-plane, lying between, and not on, the lines $x_2 = -x_1$, $x_2 = -2x_1 - 1$ and $x_2 = -\frac{1}{3}x_1 - \frac{2}{3}$.

system of linear equations   A system of m linear equations in n variables $x_1, x_2, \ldots, x_n$, is a set of equations of the form

$$\sum_{k=1}^{n} a_{jk} x_k = c_j; \quad j = 1, 2, \ldots, m,$$

where a_{jk} and c_j are constants. For example, if n and m are both 2, one could have

$$\begin{aligned} x_1 + 2x_2 &= 1 \\ 3x_1 - x_2 &= 1. \end{aligned}$$

The important point is that in order for the system to be linear, none of the variables x_j may appear with a power higher than 1, nor can there be any products of x_i with x_j.

The above system can be written as a single matrix equation

$$AX = C,$$

where

$$A = \begin{bmatrix} a_{11} & a_{12} & \cdots & a_{1n} \\ a_{21} & a_{22} & \cdots & a_{2n} \\ & \cdots & \cdots & \\ a_{m1} & a_{m2} & \cdots & a_{mn} \end{bmatrix}$$

$$X = \begin{bmatrix} x_1 \\ x_2 \\ \vdots \\ x_n \end{bmatrix}, \quad C = \begin{bmatrix} c_1 \\ c_2 \\ \vdots \\ c_m \end{bmatrix}.$$

system of linear homogeneous equations
A system of linear equations in which the constants, c_j, appearing on the right-hand side of the equations are all zero. For example,

$$\begin{aligned} x_1 + 2x_2 &= 0 \\ 3x_1 - x_2 &= 0. \end{aligned}$$

See also system of linear equations.

system of parameters   Let R be a ring of dimension n, and let $\mathcal{I}$ be its ideal of non-units. A *system of parameters* is a set of n elements of R which generates a primary ideal of $\mathcal{I}$.

system of resultants   Suppose that I is an integral domain, and $f_1, \ldots, f_n$ elements of a polynomial ring over I. Denote by a the set of resultants of $f_1, \ldots, f_n$. Suppose that G is a finite subset of a such that a equals the radical of the ideal generated by G. Then G is called a *system of resultants*.
    See resultant.

syzygy   (1) A point on the orbit of a planet (or the moon) at which that planet, the sun, and the earth all lie on a straight line.
    (2) Given a ring R, and an R-module M, a *syzygy* of $(m_1, \ldots, m_n) \in M^n$ is an n-tuple $(r_1, \ldots, r_n) \in R^n$, such that

$$\sum_{j=1}^{n} r_j m_j = 0.$$

syzygy theory   The study of syzygies and related concepts such as chains of syzygies and modules of syzygies. *See* syzygy. *See also* Hilbert's Syzygy Theorem.

T

Tamagawa number Suppose that F is an algebraic number field of finite degree and G is a connected algebraic group defined over F. Denote by G_F the group of all F-rational points on G, by G_A the adele group of A, by I the idele group of G, by M the module of all F-rational characters of G, and by $|\cdot|$ the standard norm in I. Denote by G_A^0 the set of all elements of G_A such that $|\chi_A(g)| = 1$ for each element χ of M. The volume of G_A^0/G_F with respect to the normalized invariant measure is called the *Tamagawa number* of G.

Tamagawa zeta function Suppose that F is a field, and A a division algebra over F. Denote by A^* the group of invertible elements of A, and by G the idele group of A. Suppose that ω is a positive definite zonal spherical function that belongs to the spectrum of A^*. Denote by P the set of prime divisors of F. For p in P, denote by F_p the completion of F with respect to p, by A_p the algebra obtained from A by the scalar extension F_p over F, and by T_p the reduced trace of $A_p/\mathbf{R}$. For p in P and p finite, put $\varphi_p(g_p)$ equal to the characteristic function of a maximal order of A_p. Suppose that $*$ is a positive involution. For p in P and p infinite, put $\varphi_p(g_p) = \exp(-\pi T_p(g_p g_p^*))$. For g in G, put $\varphi(g) = \prod_{p \in P} \varphi_p(g_p)$, and denote by $||g||$ the volume of g. For s a complex number, put

$$\zeta(s, \omega) = \int_G \varphi(g)\omega\left(g^{-1}\right) ||g||^s \, dg \ .$$

The function ζ is called the *Tamagawa zeta function* with character ω.

See zonal spherical function.

Tanaka embedding The embedding induced by $\pi \exp$ (as an open submanifold) of a Siegel domain D into the quotient group G_C/B. Here, G_C is the connected complex Lie group generated by the complexification of the Lie algebra g_h of the full holomorphic automorphisms of D, B is the normalizer in G_C of $b = g_-^{-1} + g_C^0 +$

$g_C^1 + g_C^2$ where C denotes complexification, g_-^{-1} is the $-i$-eigenspace in the complexification g_C^{-1} of g^{-1} under ad$I$, and g^λ is the λ-eigenspace of ad$E = 2\sum_k z_k \frac{\partial}{\partial z_k} + \sum_\alpha i u_\alpha \frac{\partial}{\partial u_\alpha}$ in g_h.

tangent function One of the fundamental trigonometric functions, denoted $\tan x$. It is (i.) periodic, satisfying $\tan(x + \pi) = \tan x$; (ii.) undefined at multiples of π, satisfying

$$\lim_{x \to k\pi-} \tan x = -\infty, \quad \lim_{x \to k\pi+} \tan x = \infty \ ,$$

for every integer k; and (iii.) intimately related to the sine and cosine functions, satisfying $\tan x = \frac{\sin x}{\cos x}$. The tangent function satisfies many important identities, including $1 + \tan^2 x = \sec^2 x$, $\tan x = -i\frac{e^{ix}-e^{-ix}}{e^{ix}+e^{-ix}}$, etc.

See also tangent of angle.

tangent line Let P be a point on a curve C parameterized by $\mathbf{x} = f(t)$. The tangent line to C at P is the line passing through P with direction vector $df(\mathbf{x})$. Thus, if C is the graph of a function of one variable, $y = f(x)$, then the tangent line to C at (x, y) is the line passing through (x, y) with slope $f'(x)$.

tangent of angle Written $\tan \alpha$, the x-coordinate of the point where the tangent line to the unit circle at $(1, 1)$ intersects the infinite line, along which lies the terminal ray of the angle α, whose initial ray lies along the positive x-axis. If $0 < \alpha < \frac{\pi}{2}$ (α in radians) so that the angle is one of the angles in a right triangle with adjacent side a, opposite side b, and hypotenuse c, the $\tan \alpha = \frac{b}{a}$.

tangent plane Let S be a surface in $\mathbf{R}^2$, defined by the vector equation $\mathbf{x} = \mathbf{x}(s, t)$. The tangent plane to S at the point $\mathbf{x}_0 = \mathbf{x}(s_0, t_0)$ is the plane passing through $\mathbf{x}_0$ and with direction vectors $\mathbf{x}_s(s_0, t_0)$ and $\mathbf{x}_t(s_0, t_0)$.

tangent space The space of all tangent vectors to a manifold or other analytic object. Equivalently, the collection of equivalence classes of curves in the object, where two curves are equivalent if they induce the same directional derivative at a specified base point.

Taniyama-Weil conjecture   An important conjecture in the study of ζ-functions. Let $L(s, E)$ be the L-function of an elliptic curve E over the rational number field $\mathbf{Q}$. The *Taniyama-Weil conjecture* states that if $L(s, E) = \sum_{n=1}^{\infty} a_n n^{-s}$, then $f(\tau) = \sum_{n=1}^{\infty} a_n e^{2\pi i n \tau}$ is a cusp of weight 2 for the congruence subgroup $\Gamma_0(N)$ which is an eigenfunction for Hecke operators.

Tannaka Duality Theorem   A compact group is isomorphic to the group of representations of its group of representations.

Tate cohomology   A variation on the cohomology of finite groups. It forms a free resolution of the finite group Z as a ZG module. *See also* free resolution.

Tate's conjecture   For a projective nonsingular variety V over a finite algebraic number field K, let $\mathcal{U}^r(\bar{V})$ denote the group of algebraic cycles of codimension r on $\bar{V} = V \otimes_K \mathbf{C}$ modulo homological equivalence and let $\mathcal{U}^r(V)$ be the subgroup of $\mathcal{U}^r(\bar{V})$ generated by the algebraic cycles rational over K. Tate conjectured that the rank of $\mathcal{U}^r(V)$ is equal to the order of the pole of the Hasse ζ-function $\zeta_{2r}(s, V)$ at $s = r + 1$.

Tate-Shafarevich group   The set of elements of the Weil-Châtelet group that are everywhere locally trivial.

Tate's Theorem   Let K/k be a Galois extension with Galois group G. Tate's Theorem states that the $n - 2$ cohomology group of G with coefficients in $\mathbf{Z}$ (denoted $\hat{H}(G, \mathbf{Z})$) is isomorphic to the nth cohomology group of G with coefficients in the idele class group C_k of K (denoted $\hat{H}(G, C_k)$). The isomorphism is given by $\Phi_n(\alpha) = \zeta_{K/k} \smile \alpha$, where $\smile$ denotes the cross product and $\zeta_{K/k}$ is the canonical cohomology class for K/k.

tautological line bundle   The subspace of $P_n X R^{n+1}$ (where P_n is projective space) that consists of the pairs (L, x) where $L \in P_n$ and $x \in L$.

tensor product   Given two linear spaces, M and N, the tensor product, $M \otimes N$, is the unique linear space with the property that, given a bi-linear form b on $M \otimes N$, there is a unique linear mapping, ϕ, on $M \otimes N$ so that $b(x, y) = \phi(x \otimes y)$. If $L(= M)$ is the space of transformations of a linear space and A and B are transformations, then there is a unique linear transformation $A \otimes B$ such that $A \otimes B(x \otimes y) = A(x) \otimes B(y)$.

term   One of the ordered elements to which an operation is applied. Examples would be summands in a polynomial or any sum, finite or infinite.

terminating decimal   A real number that can be expressed as $m + \frac{p}{10^n}$ for integers p, m and n. Rational numbers can always be expressed as either a terminating or repeating decimal. *See also* repeating decimal.

term of polynomial   An individual summand in the polynomial expression.

tetrahedral group   The alternating group of degree 4, denoted A_4. It is called the tetrahedral group because it can be realized as the group of rigid motions that preserve a tetrahedron.

theory of equations   The study of polynomial equations, the most general of which is

$$x^n + a_1 x^{n-1} + \cdots + a_{n-1} x + a_n = 0 .$$

Some of the questions studied in the theory of equations are the existence of roots, methods of solutions, and the possibility of solutions by radicals.

theory of moduli   Given a class of objects, such as "projective surfaces of general type" or "rank two vector bundles on a given Riemann surface," one may classify them (up to isomorphism, say, or some other similar equivalence relation) in two steps:

First, identify numerical invariants (invariants under the chosen equivalence). Then, for the objects having a fixed set of numerical invariants, describe an algebraic variety (or an analytic variety) which parametrizes these. The parametrization should satisfy certain naturality conditions; most often, the parameter space (called *moduli space,* since the parameters are called *moduli*) must represent some Hom functor.

A well-known example is the moduli space of Riemann surfaces (smooth algebraic curves) of genus g: this is a point, for $g = 0$; a one-dimensional variety, for $g = 1$; and a $3g - 3$-dimensional variety, if $g \geq 2$. Here g is the numerical invariant, which must be chosen first.

Of course, in each given context, a moduli space may or may not exist (or may exist only after changing the class of objects slightly); when the moduli space does exist, we say that the class of objects has a *theory of moduli*.

Theta formula   A formula relating the sum of particular values of a continuous complex function ϕ on the adele ring of a finite degree algebraic number field k to the sum of certain values of the Fourier transform of ϕ. The formula states that, for each idele a of k,

$$\sum_{\alpha \in k} \phi(a\alpha) = V(a)^{-1} \sum_{\alpha \in k} \hat{\phi}\left(a^{-1}\alpha\right) ,$$

where $V(a)$ is the volume of a.

theta function   Any one of the four functions

$$\Theta_1(\nu, \tau) = 2 \sum_{n=0}^{\infty} (-1)^n q^{(n+1/2)^2} \sin(2n+1)\pi\nu$$

$$\Theta_2(\nu, \tau) = 2 \sum_{n=0}^{\infty} q^{(n+1/2)^2} \cos(2n+1)\pi\nu$$

$$\Theta_3(\nu, \tau) = 1 + 2 \sum_{n=1}^{\infty} q^{n^2} \cos 2n\pi\nu$$

$$\Theta_4(\nu, \tau) = 1 + 2 \sum_{n=1}^{\infty} (-1)^n q^{n^2} \cos 2n\pi\nu,$$

where $q = e^{in\tau}$ and $\Im\tau > 0$.

Any elliptic function can be expressed as a quotient of theta functions.

theta series   A series of the form

$$\sum_{(x_1,...,x_2) \in \mathbf{Z}^n} e^{2\pi i Q(x_1,...,x_n)z}$$

where $Q(x_1, \ldots, x_n)$ is a positive definite form with integral coefficients and z is a complex number.

If $\Im z > 0$ the series converges and is an entire function of z.

Thom class   Let ξ be a q-sphere bundle with base space B and let Γ_ξ be the local system on B such that $\Gamma_\xi(b) = H_{q+1}(E_b, \dot{E}_b)$. Let $p_\xi^*(\Gamma_\xi)$ be the local system on E_ξ induced from Γ_ξ by $p_\xi : E_\xi \to B$. A *Thom class* of ξ is an element of $U_\xi \in H^{q+1}(E_\xi, \dot{E}_\xi; p_\xi^*(\Gamma_\xi))$ such that for every $b \in B$ the element

$$U_\xi \bigg|(E_b, \dot{E}_b) \in H^{q+1}(E_b, \dot{E}_b; p_\xi^*(\Gamma_\xi) \mid E_b)$$

$$= H^{q+1}(E_b, \dot{E}_b; H_{q+1}(E_b, \dot{E}_b))$$

corresponds to the identity map of $H_{q+1}(E_b, \dot{E}_b)$ under the universal-coefficient isomorphism

$$H^{q+1}(E_b, \dot{E}_b; H_{q+1}(E_b, \dot{E}_b))$$

$$\cong \operatorname{Hom}(H_{q+1}(E_b, \dot{E}_b), H_{q+1}(E_b, \dot{E}_b)) .$$

Thue's Theorem   Let $n > 2$, $a \in \mathbf{Z}$, $a \neq 0$ and let $a_1, \ldots, a_n \in \mathbf{Z}$. If $f(x) = \sum_{\nu=0}^{n} a_\nu x^\nu$ has distinct roots, then the number of rational integral solutions of the binary form $\sum_{\nu=0}^{n} a_\nu x^\nu y^{n-\nu} = a$ is finite.

Tometa-Takesaki theory   A generalization of the theory of Hilbert algebras.

topological vector space   A vector space V over a field F, equipped with topology such that (i.) each point is closed and (ii.) the vector space operations are continuous (as maps from $V \times V$ to V and $F \times V$ to V).

Tor $\mathbf{Z}$-modules $\operatorname{Tor}_n^A(M, N)(n = 0, 1, 2, \ldots)$ that are defined from given right and left A-modules. They are defined as the homology modules $H_n(M \bigotimes_A Y)$ where Y is a projective resolution of N.

Torelli's Theorem   Two curves are birationally equivalent if their canonically polarized Jacobians are isomorphic.

toroidal embedding   The observation (by Kempf, Knudsen, and Mumford) that a normal algebraic variety Y and a nonsingular Zariski open subset U such that $Y \supset U$ is formally isomorphic at each point to a torus embedding.

toroidal subgroup   Given a compact, connected, semisimple Lie group G, the connected Lie subgroup H that is associated to a Cartan subalgebra of the Lie algebra of G is called a *toroidal subgroup*. *See also* maximal torus.

torsion element   An element a of an A-module for which there is a non-zero divisor λ of A such that $\lambda a = 0$.

torsion-free group   An Abelian group with no elements of finite order except for the identity.

torsion group   An Abelian group with the property that all of its elements are of finite order.

torsion product   A left-derived functor of the tensor product of modules. *See also* Tor.

torsion subgroup   The subgroup of an Abelian group consisting of the set of all elements of finite order.

torus embedding   A normal scheme X, locally of finite type, over a closed field k on which an algebraic torus T acts with a dense open orbit isomorphic to T.

total boundary operator   For the associated chain complex, (X_n, ∂), ∂ is called the *total boundary operator.*

total degree   *See* filtration degree, complementary degree.

total differential   Let f be a real-valued function of the real variables, $x_1, x_2, \ldots, x_n$. The total differential is a generalization of the differential for a function of one variable. Its existence for a function of two variables is equivalent to the existence of a unique tangent plane to the surface at a point. When it does exist, the total differential of f at a point $(a_1, a_2, \ldots, a_n)$ is the function

$$df(dx_1, dx_2, \ldots, dx_n)$$
$$= f_{x_1}\, dx_1 + f_{x_2}\, dx_2 + \cdots + f_{x_n}\, dx_n \ .$$

where the partials are evaluated at $(a_1, a_2, \ldots, a_n)$. Note, though, that the existence of the partials $f_{x_1}, f_{x_2}, \ldots, f_{x_n}$ is *not* sufficient for the existence of the total differential when $n > 1$.

totally imaginary field   A field with no infinite prime divisors.

totally isotropic   *See* isotropic.

totally isotropic subspace   A subspace W of a linear space with the property that the symmetric bilinear form B (associated with a quadratic form Q) is equal to zero for all pairs $x, y \in W$.

totally real field   A number field with no imaginary infinite prime divisors.

total matrix algebra   The set of all $n \times n$ matrices over a field K. It is usually denoted $M_n(K)$.

total step method for solving linear equations   *See* Jacobi method for solving linear equations.

total transform   Let T be a rational mapping from the algebraic variety V to the algebraic variety W and let T' be an irreducible subvariety of T whose projections have the closed images V' and W' which are irreducible subvarieties of V and W, respectively. The total transform of V' is the set of points of W that correspond to V' by T.

trace   The trace of a square matrix, $A = (a_{ij})$ is the sum of the elements along the main diagonal,

$$\mathrm{Tr}(A) = \sum_{i=1}^{n} a_{ii} \ .$$

trace formula   Let T be the regular representation of a connected semisimple Lie group G on $\Gamma\, G$, where Γ is a discrete subgroup of G, and let χ_k be the characters of the irreducible unitary representations in the irreducible decomposition of T. The trace formula is an identity between two ways of calculating the trace of the integral operator with kernel $\sum_{y \in \Gamma} f(x^{-1}\lambda y)$:

$$\sum_{k=1}^{\infty} \int_G \chi_k(g)\, dg = \sum_{\gamma} f\left(x^{-1}\gamma x\right) dx \ ,$$

where γ is the conjugate class of γ in Γ and D_γ is the quotient space of the centralizer G_γ of γ in G by the centralizer Γ_γ of γ in Γ.

trace of surface (**1**) The image of a surface (a map from $\mathbf{R}^n$ into $\mathbf{R}^m$).

(**2**) The set of C-divisors $C \cdot C'$ where C is a generic component of an irreducible linear system and C' is a member of the system different from C.

transcendence In the theory of fields, the term has two meanings:

(**1**) In a field extension, $k \subset K$, an element of K is transcendental over k if it is not algebraic over k.

(**2**) The extension itself is transcendental if all elements of K except those in k are transcendental over k. *See also* transcendence degree.

transcendence basis A collection of elements S of an extension field K over k that is algebraically independent over k and is algebraic over the smallest intermediate extension field of K over k containing S.

transcendence degree For an extension field K over k, the cardinal number of an algebraically independent basis.

transcendence equation A field element is transcendental if it is not algebraic, that is, does not satisfy an algebraic equation. A transcendental number has a transcendence degree. A *transcendence equation* is an equation that one would solve to determine the transcendence degree. *See* transcendence degree.

transcendence equation A number is transcendental over a field k if it is not algebraic over k; that is, it does not satisfy an algebraic equation with coefficients in k. A transcendental number has a transcendence degree. A *transcendence equation* is an equation that one would solve to determine the transcendence degree.

transcendental A number is called *transcendental* if it is not the solution of a polynomial equation with rational coefficients. *See also* algebraic number.

transcendental curve An analytic curve that is not algebraic.

transcendental element An element α of an extension field K over k that is not algebraic over k. That is, α is not a zero of a polynomial with coefficients in k. For example, π is not transcendental over the rationals.

transcendental extension A field extension K over k that contains at least one element that is not algebraic over k.

transcendental function A multiple valued analytic function $w = w(z)$ that is not algebraic; that is, it does not satisfy an irreducible polynomial equation, $P(z, w) = 0$.

transcendental number A complex number that is not algebraic over the rationals.

transfer Let H be a subgroup of G with finite index n and let g_i $(i = 1, \ldots, n)$ be representatives of the right cosets of H. Let H' be the commutator subgroup of H and let G' be the commutator subgroup of G. The transfer from G/G' to H/H' is the homomorphism ϕ from G/G' to H/H' that is defined by $\phi(G'(x)) = H' \Pi_{i=1}^n g_i g_{x(g_i x)^{-1}}$, where g_y is g_i when $y \in H g_i$.

transfinite series Let Ω be the first uncountable ordinal and $\beta \leq \Omega$. A transfinite sequence of real numbers $\{a_\xi\}_{\xi < \beta}$ is said to be convergent and to have a limit a, provided that to any $\varepsilon > 0$ there exists $\xi_0 < \beta$ such that $|a_\xi - a| < \varepsilon$ for $\xi \geq \xi_0$. Given a transfinite sequence $\{f_\xi\}_{\xi < \beta}$, we define a sequence $\{s_\xi\}_{\xi < \beta}$ of partial sums belonging to a transfinite series $\sum_{\xi < \beta} f_\xi$ in the following way: $s_0 = f_0, s_\lambda = s_{\lambda-1} + f_\lambda$ if λ is not a limit ordinal and $s_\lambda = \lim_{\xi < \lambda} s_\xi$ if λ is a limit ordinal and all s_ξ exist for $\xi < \lambda$. We say that a transfinite series $\sum_{\xi < \beta} f_\xi$ converges (uniformly) to a function s if the sequence $\{s_\xi\}_{\xi < \beta}$ exists and converges (uniformly) to s.

transfinite series Let Ω be the first uncountable ordinal and $\beta \leq \Omega$. A transfinite sequence of real numbers $\{a_\xi\}_{\xi < \beta}$ is said to be convergent and to have a limit a, provided that to any $\varepsilon > 0$ there exists $\xi_0 < \beta$ such that $|a_\xi - a| < \varepsilon$ for $\xi \geq \xi_0$. Given a transfinite sequence $\{f_\xi\}_{\xi < \beta}$, we define a sequence $\{s_\xi\}_{\xi < \beta}$ of partial sums belonging to a transfinite series $\sum_{\xi < \beta} f_\xi$ in the following way: $s_0 = f_0, s_\lambda = s_{\lambda-1} + f_\lambda$ if λ is not a limit ordinal and $s_\lambda = \lim_{\xi < \lambda} s_\xi$ if λ is a limit ordinal and all s_ξ exist for $\xi < \lambda$. We say that a transfinite series $\sum_{\xi < \beta} f_\xi$ converges (uniformly) to a function s if the sequence $\{s_\xi\}_{\xi < \beta}$ exists and converges (uniformly) to s.

transformation A function or mapping. The term *transformation* usually implies that the mapping is from a set to itself rather than to another set and it often implies linearity.

transformation equation An equation or set of equations that preserve the structure under study. For example, in special relativity a change of variables that preserves the form $x^2 + y^2 + z^2 - c^2 t^2$ would be called a *transformation equation*.

transformation formula A formula that gives the evaluation of a function at an element in its domain in terms of the value of the function at another related element in the domain.

transformation function In quantum mechanics, the function given by

$$U(\sigma, \sigma_0) = P \cdot \exp\left[i \int_{\sigma_0}^{\sigma} dx\, j_\mu A_\mu\right],$$

where $j_\mu = e \psi^+ \gamma_\mu \psi$ and P is Dyson's chronological operator.

The name *transformation function* is also give to the Heaviside function in real-variable contexts and to the Cauchy kernel in complex variable contexts.

transformation of coordinates A one-to-one and onto linear transformation. It transforms vectors of coordinates in one basis to vectors of coordinates in a second basis.

transformation problem The problem of determining a general procedure that will decide, in a finite number of steps, whether two given words interpreted as elements of a finitely presented group can be transformed into each other by an inner automorphism of the group.

transformation space An algebraic variety that is acted on by an algebraic group. The ac-

tion is defined by an everywhere regular rational mapping $f(g, v)$ from $G \times V$ to V.

transgression A *transgression* in a fiber space is a certain special correspondence between the cohomology classes of the fiber and the base.

transitive extension Given a permutation group H on a set Ω, a transitive extension of H is the transitive permutation group on the set $\Omega \cup \infty$ (where $\infty \notin \Omega$) in which the stabilizer group of ∞ is H.

transitive permutation group A permutation group G on a set Ω with the property that whenever $a, b \in \Omega$ there is a $\pi \in G$ such that $\pi(a) = b$.

transitive property For a relation R on a set X, the property that if both $x R y$ and $y R z$ then $x R z$.

transitive relation A relation R on a set X with the property that if $x R y$ and $y R z$ then $x R z$. *See* transitive property. Here $x R y$ means that $(x, y) \in R$.

Translation Theorem One of the basic theorems of class field theory. It states that if K/k is the class field for an ideal group and Ω is an arbitrary finite extension of k, then $K\Omega/\Omega$ is the class field for the ideal group of Ω consisting of the ideals in Ω with their relative norms in the original ideal group.

transportation problem A linear programming problem which has several simple algorithmic solutions. The problem is to minimize $\sum_{i,j} c_{ij} x_{ij}$ under the condition that $\sum_j x_{ij} \geq a_i, \sum_i x_{ij} \leq b_j$ and $x_{ij} \geq 0$.

transpose Given a linear mapping ϕ from the linear space L to the linear space M, the transpose of ϕ (denoted $^t\phi$) is the linear map from the dual of M to the dual of L given by $^t\phi(f) = f \circ \phi$. In the case where $L = R^n$ and $M = R^m$, the transpose of the matrix of ϕ is the ordinary matrix transpose: if the matrix of ϕ is (a_{ij}), then the matrix of $^t\phi$ is (a_{ji}).

transposed matrix   The matrix M^t that is obtained from M by exchanging the rows and columns of M. If the ij entry of M is a_{ij}, then the ij entry of M^t is a_{ji}. *See also* transpose.

transposed representation   *See* adjoint representation.

transposition   A permutation of a set that exchanges exactly two elements. Any permutation can be written as a product of transpositions.

transvection   An element, other than the identity, of the general linear group of a right linear space V over a noncommutative field with the properties that it leaves a subspace of dimension $(n-1)$ fixed and that it acts as the identity on the remainder of V.

triangular factorization   The factorization of a matrix A into the product of a lower triangular matrix L, with ones on its diagonal, and an upper triangular matrix U: $A = LU$. The solution to the matrix equation $Ax = b$ can be determined by first solving the equation $Ly = b$ and then solving $Ux = y$.

trichotomy   Literally, *division into three parts*. The *trichotomy axiom* in a totally ordered set S, states that for $x, y \in S$, one of the assertions $x > y$, $x < y$ or $x = y$ must hold.

trigonometric curve   The graph of any of the trigonometric functions in rectangular coordinates.

trigonometric equation   An equation that involves trigonometric functions and constants. The solution set for such an equation consists of angles that satisfy it.

trigonometric expression   An expression that contains one or more of the trigonometric functions.

trigonometric form of a complex number
The representation of the complex number z as $z = r[\sin \theta + i \cos \theta](= re^{i\theta})$ where r is the modulus of z (distance from the origin to z) and θ is an argument of z (angle that the line $\overline{0z}$ makes with the positive x-axis).

trigonometric function   One of the six functions: sine, cosine, tangent, cotangent, secant, cosecant. *See* sine function, cosine function, tangent function, cotangent function, secant function, cosecant function.

trigonometric functions of a sum   Formulas of the type $\sin(a+b) = \sin a \cos b + \sin b \cos a$. (Similar formulas hold for the other trigonometric functions.)

trigonometric identity   An equation between two trigonometric expressions that is satisfied by any angle.

trigonometric series   Any series of the form

$$\frac{1}{2}a_0 + \sum_{k=1}^{\infty} (a_k \cos kx + b_k \sin kx)$$

or

$$\sum_{k=-\infty}^{\infty} a_k e^{ikx} .$$

See also Fourier series.

trigonometry   The study of the trigonometric functions of angles (sine, cosine, tangent, cotangent, secant, and cosecant) and the relationships among them. Of special importance is the use of these methods to solve triangles. Fields of application include surveying and navigation.

trinomial   An algebraic expression with exactly three summands.

triple   (**1**) An element of a triple Cartesian product of sets $A \times B \times C$. Usually written (a, b, c), where $a \in A$, $b \in B$ and $c \in C$.
(**2**) To multiply by 3.

trivial solution   The solution $x = 0$ to a system of linear homogeneous equations.

trivial valuation   A valuation that maps the non-zero elements of a field onto one.

truncation of number   Removing significant digits from a number without rounding.

Tsen's Theorem   A normal simple algebra over a field K of algebraic functions of one variable over an algebraically closed field is a total matrix algebra over K.

Tucker's Theorem on Complementary Slackness   For any matrix A, the inequalities $A\mathbf{x} = \mathbf{0}$, and $\mathbf{u}'A \geq \mathbf{0}$ have solutions $\mathbf{x}, \mathbf{u}$ that satisfy $A'\mathbf{u} + \mathbf{x} > \mathbf{0}$.

two-phase simplex method   A method used to introduce new artificial variables so that there is a solution to start with. This usually happens if $(0, 0, \ldots, 0)$ is not a feasible solution to the linear programming problem. The first phase is to use the new variables to have a solution to start with, then after performing some row transformations, we can discard all the artificial variables. This reduces the original question to an ordinary linear programming problem. Solving this problem is the second phase of this algorithm.

type-I factor   A factor that is isomorphic to the algebra of bounded operators on an n-dimensional Hilbert space.

type-II factor   A type-II von Neumann algebra that is a factor. *See* factor, type-II von Neumann algebra, von Neumann algebra.

type-III factor   *See* factor, type-III von Neumann algebra, von Neumann algebra.

type-I von Neumann algebra   A von Neumann algebra that contains an Abelian projection E for which I is the only central projection covering E. Also called a discrete von Neumann algebra.

type-II von Neumann algebra   A von Neumann algebra that is semifinite and contains no Abelian projection.

type-III von Neumann algebra   A von Neumann algebra for which a semi-finite normal trace does not exist. Also called a purely infinite von Neumann algebra.

type of a group   A generic phrase that could refer to (i.) the homotopy type of a group, (ii.) the property of the group being of Lie type, or (iii.) the property of the group being of Schottky type, or any of several other commonly used phrases.

U

U-invariant closed subspace   A subspace of the representation space of a unitary representation U that is invariant under each of the unitary operators U_g.

Ulm factor   A quotient group, $G^\alpha / G^{\alpha-1}$, of an Abelian p-group G, where α is less than the type of G and the G^α are defined by transfinite induction with G^1 equal to the elements of G of infinite height. *See* type of a group.

unconditional inequality   An inequality which is true for all values of the variables (or contains no variables). Examples are inequalities such as $x + 3 > x$, $x^2 \geq 0$, $3 > 1$.

uncountable set   A set S such that there does not exist a one-to-one mapping $f : S \to \mathbf{N}$ from S onto the set of natural numbers. *See also* cardinality.

undefined term   The initial objects about which the basic hypotheses of an axiom system are written. An example from set theory of an undefined term is *element*.

undetermined coefficients   Referring to a method of solving a system of n linear differential equations, when the matrix of the system has unique distinct real eigenvalues and the initial conditions are given. The method uses the fact that each solution x_i of such a system is of the form

$$x_i(t) = c_{i1}e^{t\lambda_1} + \cdots + c_{in}e^{t\lambda_n} .$$

From this fact, the initial conditions, and the original system of differential equations, a system of linear equations in the unknown coefficients c_{ij} can be derived and solved, yielding the final solution.

unipotent component   The unipotent linear transformation that is one of the products of the multiplicative Jordan decomposition of a nonsingular linear transformation.

unipotent group   An algebraic group that has no semisimple elements.

unipotent linear transformation   A linear transformation with the property that the semisimple component of its Jordan decomposition is the identity.

unipotent matrix   A square matrix whose characteristic roots are all equal to one.

unipotent radical of a group   The unipotent part (in the multiplicative Jordan decomposition) of the largest connected solvable closed normal subgroup of an algebraic group.

unique factorization domain   An integral domain with the property that each nonzero element is a product of prime elements (up to invertible factors).

Unique Factorization Theorem   Every positive integer can be written uniquely as a product of primes. *See also* unique factorization domain.

uniqueness theorem   In general, a theorem that states that an object that satisfies certain specified conditions must be unique. Examples are theorems that state that certain differential equations have unique solutions that satisfy initial values and theorems that state that functions (especially analytic functions) that satisfy certain conditions must be unique.

unirational variety   An irreducible algebraic variety over a field k that has a finite algebraic extension which is purely transcendental over k.

uniserial algebra   An algebra that can be decomposed into a direct sum of ideals that are primary rings.

unit   (**1**) An invertible element, especially in a ring.

(**2**) The multiplicative identity element.

unitary algebra   An algebra with a unitary element.

unitary equivalence   Let S and T be operators on a Hilbert space. We say that S and T are unitarily equivalent if there exists a unitary operator U such that $S = U^{-1} \circ T \circ U$.

unitary group   The group of all unitary matrices with complex coefficients, with matrix multiplication as the operation. It is denoted $U(n)$.

unitary homomorphism   A homomorphism between unitary rings that maps the unitary element of the first group onto the unitary element of the second group.

unitary matrix   A matrix of a unitary transformation. The adjoints of these matrices are their inverses and this property characterizes the unitary matrices. They are also characterized by the fact that they leave the inner product invariant.

unitary representation   Let G be a group. A homomorphism $\varphi : G \to U(n)$ of G into some unitary group is called a *unitary representation*.

unitary restriction   The Lie algebra

$$g_u = \sum \mathbf{R}\sqrt{-1} H_j$$
$$+ \sum \mathbf{R}(E_\alpha + E_{-\alpha})$$
$$+ \sum \mathbf{R}\left(\sqrt{-1}\,(E_\alpha - E_{-\alpha})\right),$$

where $\{H_i, E_\alpha\}$ is Weyl's canonical basis for g.

unitary ring   A ring with a unitary element.

unitary transformation   A linear transformation that leaves the inner product $\sum_{i=1}^{n} \overline{x}_i x_i$ invariant. *See also* unitary matrix.

unit circle   The set of all points in the plane that are unit distance from the origin.

$$\mathbf{T} = \left\{ z = e^{it} : 0 \le t < 2\pi \right\}.$$

unit element   In a group G or ring R, the element e that is the multiplicative identity. That is, either $e \cdot g = g$ for all $g \in G$ or $e \cdot r = r$ for all $r \in R$.

unit group   The set of units from an algebraic number field.

unit mapping   A mapping ν from a field k to a vector space A over k that, along with another mapping μ (called multiplication), form an algebra (A, μ, ν). *See* algebra.

unit matrix   A diagonal matrix whose elements on the main diagonal are all unity. Also called *identity matrix,* since it acts as the identity under matrix multiplication.

unit representation   The one-dimensional representation of a group G that maps each element of G to the number 1. The *unit representation* is sometimes also called the trivial representation.

univalent function   *See* injection. The term is usually applied to *holomorphic* or *analytic* functions which are injections. *See also* analytic function, holomorphic function.

Universal Coefficient Theorem for Homology
Let $A \subset X$ be a pair of topological spaces and let G be a group. *The Universal Coefficient Theorem for Homology* is the fact that the exact sequence

$$0 \to H_n(X, A) \otimes G \to H_n(X, A; G)$$
$$\to \mathrm{Tor}\,(H_{n-1}(X, A), G) \to 0$$

splits (non-naturally) in (X, A).

universal domain   An algebraically closed field K that has infinite transcendence degree over a given field k.

universal enveloping algebra   A *universal enveloping algebra* of a Lie algebra $\mathcal{G}$ over a commutative ring R with a unit element is an associative R-algebra $U(\mathcal{G})$ with a unit element, together with a mapping $\sigma : \mathcal{G} \to U(\mathcal{G})$ for which the following properties hold:
(i.) σ is a homomorphism of Lie algebras, i.e., σ is R-linear and

$$\sigma([X, Y]) = \sigma(X)\sigma(Y) - \sigma(Y)\sigma(X),$$

for all $X, Y \in \mathcal{G}$;
(ii.) For every associative R-algebra $\mathcal{A}$ with a

unit element and every $\mathcal{R}$-algebra mapping $\phi : \mathcal{G} \to \mathcal{A}$ such that

$$\phi([X, y]) = \phi(X)\phi(Y) - \phi(Y)\phi(X) \,,$$

for all $X, Y \in \mathcal{G}$, there exists a unique homomorphism of associative algebras $\psi : (\mathcal{G}) \to \mathcal{A}$, mapping the unit element to the unit element such that $\phi = \psi \circ \sigma$.

universal enveloping bialgebra The cocommutative bialgebra $(U(L), \mu, \eta, \Delta, \epsilon)$ over a field k, where L is a Lie algebra over k, $U(L)$ is the universal enveloping algebra of L with multiplication μ and unit mapping η, the algebraic homomorphism $\Delta : U(L) \to U(L \oplus L) \simeq U(L) \otimes U(L)$ and $\epsilon : U(L) \to k$ are the maps induced by the Lie algebraic homomorphisms $L \to L \oplus L, x \mapsto x \oplus x$ and $L \to \{0\}, x \mapsto 0$, respectively.

universal quantifier An operator containing a variable written "(x)" or "$(\forall x)$," that indicates that the open sentence that follows is true of every member of the relevant domain. More precisely, every replacement of the variable x by a name yields a true statement.

unknown A variable, or the quantity it represents, whose value is to be discovered by solving an equation.

unmixed ideal An ideal which is not mixed. *See* mixed ideal.

Unmixedness Theorem If R is a locally Macaulay ring, and an ideal a of the polynomial ring $R[x_1, \ldots, x_n]$ over R is generated by r elements, with height $a = r$, then a is unmixed. *See* local Macaulay ring, unmixed ideal.

unramified covering A triple (X, Y, π) where X and Y are connected complex spaces and $\pi : X \to Y$ a surjective, holomorphic map such that any point $y \in Y$ has a connected neighborhood V_y with the property that $\pi^{-1}(V_y)$ consists of the union of disjoint open subsets of X, each of which is mapped isomorphically onto V_y by π. *See also* ramified covering.

unramified extension Let H and K be two fields. H is an unramified extension of K if all prime ideals of K are unramified in H.

unramified ideal A prime ideal $\mathcal{B}$ of an algebraic number field F lying over a prime number p such that the principal ideal (p) has in F a product decomposition into prime ideals of the form

$$(p) = \mathcal{B}_1^{\alpha_1} \mathcal{B}_2^{\alpha_2} \cdots \mathcal{B}_m^{\alpha_m} \,,$$

where $\mathcal{B}_1 = \mathcal{B}$ and $\mathcal{B}_2, \ldots, \mathcal{B}_m \neq \mathcal{B}$ and $\alpha_1 = 1$.

upper bound A value greater than or equal to each element of an ordered (or partially ordered) set S is called an *upper bound* of S.

upper central series A sequence of subgroups $Z_0 = \{e\} \subset Z_1 \subset Z_2 \subset \ldots$, where Z_1 is the center of a group G, Z_2/Z_1 is the center of G/Z_1, etc.

upper semi-continuous function Let X be a topological space and $f : X \to \mathbf{R}$ a function. Then f is said to be *upper semi-continuous* if $f^{-1}((-\infty, \alpha))$ is open for every real α.

upper triangular matrix A square matrix having only zero entries below the main diagonal.

V

one of n given numbers $x_1, \ldots, x_n$:

$$\det \begin{bmatrix} 1 & x_1 & \cdots & x_1^{n-1} \\ 1 & x_2 & \cdots & x_2^{n-1} \\ & \cdot & \cdot & \cdot \\ 1 & x_n & \cdots & x_n^{n-1} \end{bmatrix} = \prod_{j<k} (x_j - x_k).$$

valuation   A map Φ from a field F to an ordered ring R such that for all $x, y \in F$,
(i.) $\Phi(x) \geq 0$ and $\Phi(x) = 0$ if and only if $x = 0$;
(ii.) $\Phi(x \cdot y) = \Phi(x) \cdot \Phi(y)$;
(iii.) $\Phi(x + y) \leq \Phi(x) + \Phi(y)$.

valuation ideal   Let K be a number field and let v be an additive valuation of the field K into an ordered additive group G. It follows that the set $R_v = \{a \in K : v(a) \geq 0\}$ is a subring of K and this ring has only one maximal ideal $\{a : v(a) = 0\}$ which is the *valuation ideal* of v (or of R_v).

valuation ring   If $A \subseteq B \subseteq K$ are two local rings with maximal ideals a and b, respectively, then one says that B dominates A if $a \subset b$. Dominance is a partial order relation on the set of subrings of K. The maximal elements of this set are exactly the valuation rings of K.

value   When a variable stands in place of constants, then any one of these constants represents a *value* of the variable.

value group   The submodule $\{v(a) : a \in K \backslash \{0\}\}$ of G, where G is a totally ordered group (i.e., G is a lattice-ordered commutative group that is totally ordered, with the group operation expressed by addition), K is a field, and v is an additive valuation $v : K \to G \cup \{\infty\}$ of the field K (i.e., a mapping satisfying (i.) $v(a) = \infty$ if and only if $a = 0$; (ii.) $v(ab) = v(a) + v(b)$ for all $a, b \neq 0$; and (iii.) $v(a + b) \geq \min\{v(a), v(b)\}$).

Vandermonde determinant   The determinant of an $n \times n$ square matrix of which each row consists of the 0th to $(n-1)$st powers of

Vandiver's Conjecture   Let G be the ideal class group of $\mathbf{Q}(\zeta_\ell)$, the ℓ-th cyclotomic field. Let $\mathcal{A}$ be the subgroup of G containing all elements whose order divides ℓ. Define $\mathcal{A}_j = \{A \in \mathcal{A} : A^{\sigma_t} = A^{t^j}, 1 \leq t < \ell\}$. It can be shown that $\mathcal{A}$ is the direct product of the $\mathcal{A}_j$, i.e.,

$$\mathcal{A} = \mathcal{A}_1 \mathcal{A}_2 \cdots \mathcal{A}_{j-1}$$

and $\mathcal{A}_j \cap \mathcal{A}_k = \{e\}$, if $j \neq k$. If ℓ is odd, *Vandiver's conjecture* states that the group $\mathcal{A}_2 \mathcal{A}_4 \mathcal{A}_6 \cdots \mathcal{A}_{\ell-1}$ is trivial.

vanish   To equal zero, e.g., a function f from a set Ω to the real or complex numbers is said to *vanish* on a subset B of Ω if it maps every element of B to zero.

vanishing cocycle   A cocycle which is equal to some coboundary in a cochain complex.

variable   A quantity that is allowed to represent any element of a given set. Any number in the set is called a value of the variable and the set itself is called the domain of the variable.

variable component   A term that stands in contrast to the *fixed component* of a linear system. See fixed component. The precise meaning of variable component, however, may vary depending upon the context.

variation   (1) A term, introduced by Lagrange, used to denote a small displacement of an independent variable or of a functional. If f is a functional defined on a space Ω, a *variation of the argument* $\omega_0 \in \Omega$ is a curve $\omega(t, v)$, where $a \leq t \leq b, a \leq 0, b \geq 0$, passing through ω_0. As v ranges through V (space of parameters), the variations range through a family of curves where, on each curve, $t = 0$ corresponds to position ω_0. In the case where $V = \Omega$, v is referred to as variation and is called the *directional variation*.

(2) In statistics, the quantity $100\sigma/\bar{x}$ is called the *coefficient of variation* of the variable x, where σ is the standard deviation of x and $\bar{x}$ is the mean.

(3) A term used to describe how one quantity changes with respect to another quantity (quantities). If the relation between two variables is such that their ratio is a constant, one is said to *vary directly* as the other. *See also* inverse variation, joint variation.

(4) If f is a complex valued function defined on an interval $[a, b] \subset \mathbf{R}$, the *variation of* f is

$$V(f) = \sup \sum_{k=1}^{\infty} |f(x_k) - f(x_{k-1})|,$$

where the supremum is taken over all partitions $a = x_0, x_1, \ldots, x_n = b$ of the interval $[a, b]$. If $V(f) < \infty$, then f is said to be of *bounded variation*.

(5) Assume μ is a finitely additive measure defined on an algebra $\mathcal{A}$ of subsets of a set Ω with values in a normed linear space $(X, \|\cdot\|)$. If $E \in \mathcal{A}$, the *variation of* μ *over* E is defined as

$$|\mu|(E) = \sup_{\pi} \sum_{A \in \pi} \|\mu(A)\|,$$

where the supremum is taken over all finite partitions π of E into disjoint subsets of E, each element of π belonging to $\mathcal{A}$. If $|\mu|(\Omega) < \infty$, then μ is said to be of *bounded variation*.

variety If $\mathbf{A}^n$ denotes the affine n-space over a field k and $k[x_1, x_2, \ldots, x_n]$ the polynomial ring in n variables over k, an *affine algebraic variety* (or simply *affine variety*) is a subset of $\mathbf{A}^n$ of the form

$$\{p \in \mathbf{A}^n : f(p) = 0 \text{ for all } f \in T\},$$

where T is some subset of $\mathbf{K}[x_1, x_2, \ldots, x_n]$. If $\mathbf{P}^n$ denotes the projective n-space over a field k and $k[x_0, x_1, \ldots, x_n]$ the polynomial ring in $n + 1$ variables over k, a *projective algebraic variety* (or simply *projective variety*) is a subset of $\mathbf{P}^n$ of the form

$$\{p \in \mathbf{P}^n : f(p) = 0 \text{ for all } f \in T\},$$

where T is a subset of homogeneous polynomials in $\mathbf{K}[x_0, x_1, \ldots, x_n]$.

A variety which is a subset of another variety is called a *subvariety*. A variety V is called *reducible* (resp. *irreducible*) if it can (resp. cannot) be written as a union of two proper subvarieties.

The modern, and more general, definition of algebraic varieties is as follows: Consider a pair $(V, \mathcal{O})$ of a topological space V and its structure sheaf $\mathcal{O}$ of mappings $V \longrightarrow k$, where k is an algebraically closed field. This pair is called a *prealgebraic variety over* k, and is denoted simply by V, if V has a finite open covering (V_i) and there exists a homomorphism $V_i \longrightarrow U_i$ from V_i to an affine variety U_i that transforms $\mathcal{O}|V_i$ to the structure sheaf of U_i. A prealgebraic variety V is then called an *algebraic variety* if the image of the diagonal mapping $V \longrightarrow V \times V$ is closed in the Zariski topology of the product variety $V \times V$.

An algebraic variety V is called an algebraic group if it has a group structure and the mapping $V \times V \longrightarrow V$, $(a, b) \mapsto ab^{-1}$ is a morphism. In the case V is irreducible, then V is called a *group variety*. A complete group variety is called an *Abelian variety*.

vector **(1)** An entity having both magnitude and direction. Commonly represented by a *directed* line segment whose length and orientation in space give the magnitude and direction, respectively. Two directed line segments represent the same vector if they are equivalent; that is, if they have the same length and orientation.

(2) An element of a vector space. *See* vector space.

vector bundle A mapping π of a vector space E onto a vector space M such that for each $p \in M$ the inverse image $\pi^{-1}(p) = \{e \in E : \pi(e) = p\}$ is a real vector space. More specifically, a vector bundle is a five-tuple $(E, M, V^k, W(k), \pi)$ where (i.) E is the *bundle space*, (ii.) M is the *base space*, (iii.) V^k is a k-dimensional vector space, called the *fiber space*, (iv.) π is a mapping of E onto M, called the *projection*, such that $\pi^{-1}(p)$ is homeomorphic to V^k for all $p \in M$, and (v.) $W(k)$ is the general linear group of all non-singular $k \times k$ matrices acting on V^k in such a way that if $Av = v$ for all $v \in V^k$ then $A = I$, the identity matrix.

vector invariant An invariant of a matrix group when each matrix representation of the group is either the identity or the contragredient map.

vector of shadow prices Linear Programming is a method of finding extremal values of certain linear functions satisfying conditions in terms of a system of linear equations, inequalities, or both. Suppose that A is an $m \times n$ matrix, $B \in \mathbf{R}^m$, and $P \in \mathbf{R}^{n^*}$. The Problem P is the following: to find $X \in \mathbf{R}^n$ for which $x_0 = P'X$ is maximal under the condition (i.) $AX=B$; (ii.) $X \geq 0$. An X satisfying (i.) is called a *solution;* an X satisfying both (i.) and (ii.) is called a *feasible solution.* We denote by S the set of all feasible solutions. A necessary and sufficient condition on $X_0 \in S$ to attain a maximal value at $X = X_0$ is the existence of a $U = U_0 \in \mathbf{R}^{m^*}$ such that (a.) $U_0'A \geq P'$ and (b.) $(U_0'A - P')X_0 = 0$. U_0 is called the *vector of shadow prices.*

vector quantity Any mathematical or physical quantity having a direction (usually in 2 or 3 dimensions) as well as magnitude, and hence representible as a vector.

vector representation (of a Clifford group) If G is the Clifford group of a quadratic form q on an n-dimensional vector space V, a homomorphism ϕ of G into the orthogonal group of V relative to q is called a *vector representation* of G. *See* Clifford group.

vector space A pair (V, F), where V is an Abelian additive group of elements called vectors, and F is a field such that for each $\alpha \in F$ and every $\mathbf{v} \in V$ the product $\alpha \mathbf{v} = \mathbf{v}\alpha$ is defined as an element of V, and for all $\mathbf{v}, \mathbf{w} \in V$ and $\alpha, \beta \in F$, the following hold:
(i.) $(\alpha + \beta)\mathbf{v} = \alpha\mathbf{v} + \beta\mathbf{w}$,
(ii.) $\alpha(\mathbf{v} + \mathbf{w}) = \alpha\mathbf{v} + \alpha\mathbf{w}$,
(iii.) $(\alpha\beta)\mathbf{v} = \alpha(\beta\mathbf{v})$, and
(iv.) $1\,\mathbf{v} = \mathbf{v}$.
The pair (V, F) is usually referred to by saying that V is a *vector space over* F.

versine The function $f(x) = 1 - \cos x$.

very ample (1) A linear system S of divisors on a complete irreducible variety V is said to be *very ample* if it has no fixed components and the rational mapping ϕ_S defined by S is a closed immersion.

(2) If D is a divisor, the set of positive divisors that are linearly equivalent to D is a linear system. If that linear system is very ample, the divisor D is said to be *very ample.*

virtual arithmetic genus The integer $\chi_V(D) := \chi(V) - \chi(L(-D))$, with respect to D over V, where V is a normal variety, D is a divisor on V, $\chi(V)$ is the arithmetic genus of V, $L(D)$ is a coherent algebraic sheaf, and

$$\chi(F) = \sum_q (-1)^q \dim H^q(V, F)$$

for each coherent sheaf F.

volume A numerical characteristic of the extent of a body in three-dimensional space. In the simplest case, the body is a rectangular parallelepiped with sides of length a, b, c and volume abc. The volume (of bodies in three-dimensional space) has the following four properties: (i.) volume is non-negative; (ii.) volume is additive, that is, if two bodies U and V, with no points in common, have volumes $v(U)$ and $v(V)$, then the volume of their union is the sum of their volumes, $v(U \cup V) = v(U) + v(V)$; (iii.) volume is invariant under translation, that is if two bodies U and V have volumes $v(U)$ and $v(V)$, and the two bodies are congruent, then $v(U) = v(V)$; (iv.) volume of the unit cube is 1.

The concept of volume is extended, while preserving properties (i.) through (iv.), to a wider class of bodies by a process called the method of exhaustion. Let G be a body whose surface satisfies certain conditions. Find a set $K \subset G$ that can be expressed as a disjoint union of finitely many parallelepipeds (and whose volume can be calculated by additivity, property (ii.)). Also find a set $O \supset G$ that can be expressed as a finite disjoint union of parallelepipeds. If G is to have a volume, we must have

$$v(K) \leq v(G) \leq v(O).$$

If

$$\sup_{K \subset G} v(K) = \inf_{O \supset G} v(O),$$

317

this common value is called the *volume of G*, $v(G)$.

The concept of volume of a body in three-dimensional space can be extended to sets in n-dimensional space while preserving properties (i.) through (iv.). This is done by replacing rectangular parallelepipeds with cells. An n-dimensional cell is a set of the form

$$I = \{ (x_1, x_2, \ldots, x_n) : a_i \leq x_i \leq b_i ,$$

$$i = 1, 2, \ldots, n \}.$$

The volume of the cell I is $v(I) = (b_1 - a_1) \cdot \cdots \cdot (b_n - a_n)$.

von Neumann algebra A subalgebra, **A**, of the algebra $\mathcal{B}$ of bounded linear operators on a Hilbert space that is self-adjoint and coincides with its bicommutant. That is, (i.) $A \in \mathbf{A}$ implies $A^* \in \mathbf{A}$ and (ii.) $\mathbf{A} = (\mathbf{A}')'$, where $'$ means commutant:

$$\mathbf{A}' = \{T \in \mathcal{B} : TA = AT, \text{ for } A \in \mathbf{A}\} .$$

von Neumann Density Theorem If **A** is a subalgebra of the algebra $\mathcal{B}$ of bounded linear operators on a Hilbert space, the following statements are equivalent:
(i.) **A** is self-adjoint, contains the identity operator, and is closed in the weak operator topology.
(ii.) **A** is self-adjoint, contains the identity operator, and is closed in the strong operator topology.
(iii.) **A** is the commutant of some subset of $\mathcal{B}$.
The strong operator topology is defined by

$$T_n \to T \iff T_n x \to Tx, \text{ for all } x .$$

See weak convergence, von Neumann algebra.

von Neumann reduction theory A theory which reduces the study of von Neumann algebras on a Hilbert space to the study of factors, that is, von Neumann algebras whose centers consist of scalar multiples of the identity operator.

W

Waring's Second Theorem   Let N, p, q, n, and d be as in Waring's Theorem. If $N > 0$ then $N \geq q^{n-d}$. *See* Waring's Theorem.

Waring's Theorem   Let $\mathbf{K}$ be a field of characteristic p consisting of q elements. Let f_1, $\ldots$, f_m be polynomials in n variables with coefficients in $\mathbf{K}$ of degrees d_1, $\ldots$, d_m, respectively, and suppose that $d = d_1 + \cdots + d_m < n$. If N is the number of common zeros of f_1, $\ldots$, f_m, then $N \equiv 0 \pmod{p}$. *See also* Waring's Second Theorem.

weak (inductive) dimension   The empty set $\emptyset$ is declared to have dimension -1. Inductively, we proceed as follows: If R is a set and if, for every neighborhood $U(p)$ of each point $p \in R$, there exists an open neighborhood V such that $p \in V \subset U(p)$ and the dimension of the boundary of V is $n - 1$, then we say that R has *weak inductive dimension* $\leq n$.

If it is the case that the dimension of R is less than or equal to n and it is *not* the case that the dimension of R is less than or equal to $n - 1$, then the dimension of R is n.

weak (inductive) dimension   The empty set $\emptyset$ is declared to have dimension -1. Proceed inductively as follows. If R is a set and if for every neighborhood $U(p)$ of each point $p \in R$ there exists an open neighborhood V such that $p \in V \subset U(p)$ and the dimension of the boundary of V is $n - 1$, then we say that R has weak inductive dimension $\leq n$.

If it is the case that the dimension of R is less than or equal to n and it is *not* the case that the dimension of R is less than or equal to $n - 1$, then the dimension of R is n.

weak convergence   (1) Convergence in the weak topology. A sequence $\{x_n\}$ in a linear topological space V is said to converge weakly to an element x in V if, for every continuous linear functional f on V, $\lim_{n \to \infty} f(x_n) = f(x)$.

(2) Weak convergence of a sequence of linear operators $\{L_n : V \to W\}$ to L means weak convergence of $L_n x$ to Lx, for every $x \in V$.

weak Dirichlet algebra   A closed subalgebra $A \subset C(X)$ such that $A + \bar{A}$ is weakly dense in $L_\infty(m)$, where $C(X)$ is a Banach algebra of continuous complex-valued functions on a compact Hausdorff space X, $\mathcal{M}(A)$ is the maximal ideal space of A, $\varphi \in \mathcal{M}(A)$, M_φ is the set of representing measures for φ and $m \in M_\varphi$.

weaker equivalence   An equivalence relation R is said to be *weaker* than an equivalence relation S if equivalence in the sense of S implies equivalence in the sense of R.

weak global dimension   If M is a left R-module over some ring R, then M is isomorphic to the quotient of a free R-module F by some submodule S. So M is isomorphic to F/S. S is called the first syzygy of M. Taking the first syzygy of S one gets the second syzygy of M, etc. The flat dimension of M is the least number of times the procedure has to be repeated to get a flat module, i.e., it is the smallest n such that the nth syzygy is flat. If there is a bound on such n, as M varies, then the least upper bound of all such n is called the *weak global dimension* of R. More precisely, it is the *left* weak global dimension. But, in fact, the number is the same, whether left or right modules are used.

Weak Lefschetz Theorem   Let X be an algebraic subvariety of complex dimension n in the complex projective space $C\mathbf{P}^N$, let $P \subset C\mathbf{P}^N$ be a hyperplane passing through all singular points of X (if any) and let $Y = X \cap P$ be a hyperplane section of X. Then the relative homology groups $H_i(X, Y; \mathbf{Z})$ vanish for $i < n$. This implies that the natural homomorphism

$$H_i(Y; \mathbf{Z}) \longrightarrow H_i(X; \mathbf{Z})$$

is an isomorphism for $i < n - 2$ and surjective for $i = n - 1$.

Weak Mordell-Weil Theorem   If A is an Abelian variety of dimension n, defined over an algebraic number field k of finite degree, and A_k the group of all k-rational points on A, then

$A_k / m A_k$ is a finite group for any rational integer m.

web group   A Kleinian group whose region of discontinuity consists only of Jordan domains.

Wedderburn-Mal'tsev Theorem   Let A be a finite-dimensional associative algebra over a field k, and let $R(A)$ denote the radical of A. If the quotient algebra $A/R(A)$ is separable, then there exists a subalgebra S such that $A = S + R(A)$, $S \cap R(A) = \{0\}$ and S is uniquely determined up to an inner automorphism.

Wedderburn's Theorem   Every finite division ring is commutative and, hence, a field.

wedge product   (**1**) If X and Y are topological spaces and $x_0 \in X$, $y_0 \in Y$ fixed points, the subspace $X \times \{y_0\} \cup \{x_0\} \times Y$ of the product $X \times Y$ is called a *wedge product* of X and Y.

(**2**) A multiplication operation defined on the alternating multilinear forms on a K-module V, where K is a commutative ring with identity. The wedge product is also called *exterior product* and is defined to be associative. Assume L and M are alternating multilinear forms on V of degree r and s, respectively. Let S_{r+s} be the set of permutations of $\{1, 2, \ldots, r+s\}$ and $G \subset S$ the set of all permutations σ which permute the sets $\{1, 2, \ldots, r\}$ and $\{r+1, r+2, \ldots, r+s\}$ within themselves. Let S be any set of permutations of $\{1, 2, \ldots, r+s\}$ which contains exactly one element from each left coset of G. The *wedge product* of L and M is defined by the equation

$$L \wedge M = \sum_{\sigma \in S} (\operatorname{sgn} \sigma)(L \otimes M)_\sigma \, ,$$

where

$$(L \otimes M)_\sigma \, (\alpha_1, \ldots, \alpha_{r+s})$$
$$= (L \otimes M) \left(\alpha_{\sigma 1}, \ldots, \alpha_{\sigma(r+s)} \right) \, .$$

Weierstrass canonical form   (**1**) An equation representing the reciprocal of the gamma function as an infinite product,

$$\frac{1}{\Gamma(x)} = x e^{Cx} \prod_{n=1}^{\infty} \left(1 + \frac{x}{n} \right) e^{-x/n} \, .$$

Here, C is Euler's constant

$$C = \lim_{n \to \infty} \left(1 + \frac{1}{2} + \frac{1}{3} + \cdots + \frac{1}{n} - \log(n) \right) \, .$$

(**2**) An equation of the form $y^2 = 4x^3 - g_2 x - g_3$ defining a curve which serves as a normal model of an elliptic function field $\mathbf{K}$ (over a field k) of dimension 1 that has a prime divisor of degree 1.

Weierstrass point   A point p, on a Riemann surface R of genus g, such that there exists a nonconstant rational function on R which has no singularities except a pole at p of order not exceeding g.

weight   (**1**) The smallest cardinal number which is the cardinality of an open base of a topological space X is called the *weight* of the space X.

(**2**) Let G be a complex semisimple Lie algebra and fix a Cartan subalgebra H of G. Furthermore, fix a lexicographic linear ordering on $H_{\mathbf{R}}^*$, the linear space of all $\mathbf{C}$-valued forms on H. Let (ρ, V) be a representation of G. For each $\lambda \in H^*$ let $V_\lambda = \{v \in V : \rho(H)v = \lambda(H)v\}$. Then V_λ is a subspace of V and if $V_\lambda \neq \{0\}$, then λ is called the *weight of the representation* ρ (with respect to H). The set of weights is finite and its maximum element with respect to the ordering on $H_{\mathbf{R}}^*$ is called the *highest weight* of ρ.

(**3**) Let R be a root system in a Euclidean space E. Let Λ be the set of all elements λ in E for which $2(\lambda, r)/(r, r)$ is an integer, for all $r \in R$. The elements of Λ are called *weights*.

weight function   (**1**) If p and f are nonnegative functions defined and integrable on Ω, let $M_r(f)$ be defined as

$$M_r(f) = \left(\frac{\int_\Omega p f^r \, dx}{\int_\Omega p \, dx} \right)^{1/r} , \quad r \neq 0$$

and if $M_r(f) > 0$ for some $r > 0$, define $M_0(f) = \lim_{r \to 0+} M_r(f)$. The quantity $M_r(f)$ is called the mean of degree r of f with respect to the *weight function* p. If $p = 1$, then $M_1(f)$, $M_0(f)$, and $M_{-1}(f)$ are the arithmetic, geometric, and harmonic means of f, respectively.

(2) Let (X, μ) be a measure space, where X is a subspace of Euclidean space and the measure μ is absolutely continuous with Radon-Nikodym derivative ϕ with respect to Lebesgue measure. Then $L^2(X, \mu)$, with norm

$$\|f\| = \left(\int_X |f(x)|^2 \phi(x) \, dx \right)^{\frac{1}{2}}$$

is referred to as the L^2 space with *weight function* ϕ.

weight group   Let (T, K) be an element of an arithmetic crystal group; that is, T is an n-dimensional lattice in an n-dimensional Euclidean space V, K a finite subgroup of the orthogonal group $\mathcal{O}(V)$ ((T_1, K_1) is equivalent to (T_2, K_2) (arithmetically equivalent) if there exists an invertible linear transformation $g \in$ GL(v) so that $T_2 = gT_1$ and $K_2 = gK_1g^{-1}$). The *weight lattice* or *weight group* of (T, K) is defined as the set

$$\{v \in V : v - kv \in T, \text{ for all } k \in K\} \, .$$

weight lattice   *See* weight group.

Weil-Chatelet group   The group of principal homogeneous spaces over an Abelian variety.

Weil cohomology   A cohomology theory for algebraic varieties over a field of arbitrary characteristic. Let k be an algebraically closed field and K a field of characteristic zero, called the coefficient field. A contravariant functor $V \longrightarrow H^*(V)$ from the category of complete connected smooth varieties over k to the category of augmented $\mathbf{Z}^+$-graded finite-dimensional anticommutative K-algebras is called a *Weil cohomology* with coefficients in K if the following hold:
(i.) If dim $V = n$, then a canonical isomorphism $H^{2n}(V) \simeq K$ exists and the cup product $H^j(V) \times H^{2n-j}(V) \longrightarrow H^{2n}(V)$ induces a perfect mapping.
(ii.) For any V_1 and V_2, the mapping $H^*(V_1) \otimes H^*(V_1) \longrightarrow H^*(V_1 \times V_2)$ defined by $a \otimes b \longrightarrow \mathrm{Proj}_1^*(a) \cdot \mathrm{Proj}_2^*(b)$ is an isomorphism.
(iii.) Let $C^j(V)$ be the group of algebraic cycles of codimension j on V. There exists a fundamental-class homomorphism $C^j(V) \longrightarrow$ $H^{2j}(V)$ for all j which is functorial in V, compatible with products via the Kunneth formula, has compatibility of the intersection with the cup product, and maps a 0-cycle in $C^n(V)$ to its degree as an element of $K \simeq H^{2n}(V)$.

Weil conjecture   A conjecture, first posed in 1949 by A. Weil, regarding the properties of the zeta function of some special varieties. Let X be an n-dimensional complete nonsingular variety over a field $\mathbf{F}_q$ of q elements. Let $Z(u, X) = Z(t)$ be the zeta function of X. (*See* congruence zeta function.) Then
(i.) $Z(t)$ is a rational function of t; that is, a quotient of polynomials with rational coefficients.
(ii.) $Z(t)$ satisfies the functional equation

$$Z\left(\left(q^n t\right)^{-1}\right) = \pm q^{n\chi/2} t^\chi Z(t) \, ,$$

where the integer χ is the intersection number of the diagonal subvariety Δ_X with itself in the product $X \times X$.
(iii.) We can write $Z(t)$ in the form

$$Z(t) = \prod_{i=1}^{2n-1} P_i(t)^{(-1)^i} \, ,$$

where $P_0(t) = 1 - t$, $P_{2n}(t) = 1 - q^n t$, and for each i, $1 \leq i \leq 2n-1$, $P_i(t) = \prod_{j=1}^{B_i}(1 - \alpha_{ij}t)$ is a polynomial with integer coefficients and α_{ij} are algebraic integers of absolute value $q^{i/2}$.
(iv.) When X is the reduction modulo p of a complete nonsingular variety Y of characteristic zero, then the degree, B_i, of P_i is the ith Betti number of the topological space Y considered as a complex manifold.

Weil group   Let K be a finite Galois extension of an algebraic number field k, let C_K be the idele class group of K, and let

$$\alpha_{K/k} \in H^2(\mathrm{Gal}(K/k), C_K)$$

be the canonical cohomology class of field theory. Then $\alpha_{K/k}$ determines an extension $W_{K/k}$ of $\mathrm{Gal}(K/k)$ and if L is a Galois extension of k containing K, there is a canonical homomorphism $W_{L/k} \longrightarrow W_{K/k}$. The *Weil group* W_k for $\bar{k}/k$ is defined as the projective limit group $\mathrm{proj}_K \lim W_{K/k}$ of the $W_{K/k}$. If k_v is a local field, the Weil group W_{k_v} for $\bar{k}_v/k_v$ is defined by replacing the idele class group C_K with the multiplicative group $K_w^\times$ in the above definition, where K_w is a Galois extension of k_v.

Weil L-function   Let W_k be a Weil group of an algebraic number field k, and let $\rho : W_k \longrightarrow GL(V)$ be a continuous representation of W_k on a complex vector space V. (*See* Weil group.) Let p be a finite prime of k or an Archimedean prime of k, let ρ_p be the representation of W_{k_p} induced by ρ and let

$$\phi : W_{k_p} \longrightarrow Gal(\bar{k}_p/k_p)$$

be a surjective homomorphism. Let Φ be an element of W_{k_p} such that $\phi(\Phi)$ is the inverse Frobenius element of p in $Gal(\bar{k}_p/k_p)$. Let I be the subgroup of W_{k_p} consisting of elements w such that $\phi(w)$ belongs to the inertia group of p in $Gal(\bar{k}_p/k_p)$ and let V^I be the subspace of V consisting of elements that are fixed by $\rho_p(I)$. Let $N(p)$ denote the norm of p and define the function

$$L(V, s) = L(s)$$
$$= \prod_p \det(1 - (N(p))^{-s}\rho_p(\Phi)|V^I)^{-1}.$$

Then this product converges for s in some right half-plane and is called the *Weil L-function* for the representation $\rho : W_k \longrightarrow GL(V)$.

Weil number   An algebraic integer of a finite field K of q elements is called a *Weil number* for q if each of its conjugates has absolute value $\sqrt{q}$.

well-behaved   A *-derivation δ of a C^*-algebra such that, for every self-adjoint x in its domain $D(\delta)$, there exists a state φ with $|\varphi(x)| = \|x\|$ and $\varphi((x)) = 0$.

Weyl chamber   (1) *See* Weyl group, for the Weyl chambers of a root system in a finite dimensional Euclidian space.

(2) Let G be a connected affine algebraic group, T a maximal torus in G, and B a Borel subgroup of G containing T. The set of all semi-regular one-parameter subgroups in T whose associated Borel groups equal B is called a *Weyl chamber* of B with respect to T in G.

Weyl group   (1) If R is a root system in a finite dimensional Euclidean space E, an element r_α of the general linear group $GL(E)$ is called a *reflection* with respect to $\alpha \in R$ if $r_\alpha(\alpha) = -\alpha$ and r_α fixes the points of a hyperplane H_α in E. The subgroup $W(R)$ of $GL(E)$, generated by the reflections r_α, $\alpha \in R$, is called the *Weyl group* of R. The connected components of $E \setminus \cup_\alpha H_\alpha$ are called the *Weyl chambers* of R.

(2) If G is a connected affine algebraic group and T, a torus in G, the quotient group $W(T, G) = N_G(T)/Z_G(T)$, where $N_G(T)$ is the normalizer of T in G and $Z_G(T)$ is the centralizer of T in G, is called the Weyl group of G relative to T. Weyl groups of maximal tori are isomorphic and are referred to as the Weyl groups of the group G.

Weyl's canonical basis   Let G be a semisimple Lie algebra over $\mathbf{C}$. Let H be a Cartan subalgebra of G and α a $\mathbf{C}$-valued form on H. Let $G_\alpha = \{x \in G : ad(h)x = \alpha(h)x \text{ for all } h \in H\}$. Let Δ be the set of all nonzero linear forms α on H for which $G_\alpha \neq \{0\}$. Then Δ is a finite set whose elements are the roots of G relative to H, and for each $\alpha \in \Delta$, G_α is of dimension one. Write G as a sum $G = H + \sum_{\alpha \in \Delta} G_\alpha$.

If $\{h_1, \ldots, h_n\}$ is a basis for H and $\{e_\alpha\}$ a basis for G_α, then $\{h_i, e_\alpha\}$ is a basis for G called *Weyl's canonical basis* if the following three conditions are satisfied:

(i.) $\alpha(h_i) \in \mathbf{R}$, $\alpha \in \Delta$, $1 \leq i \leq n$.
(ii.) The Killing form B of G satisfies $B(e_\alpha, e_{-\alpha}) = -1$ for every $\alpha \in \Delta$.
(iii.) If $\alpha, \beta, \alpha+\beta \in \Delta$ and $[e_\alpha, e_\beta] = n_{\alpha,\beta}e_{\alpha+\beta}$ ($n_{\alpha,\beta} \in \mathbf{C}$), then $n_{\alpha,\beta} \in \mathbf{R}$ and $n_{\alpha,\beta} = n_{-\alpha,-\beta}$.

Weyl's character formula   The formula

$$\chi_\rho(h) = \frac{\xi_{\Lambda+\delta}(h)}{\xi_\delta(h)}, \quad h \in H,$$

where G is a compact, connected, semisimple Lie group, (ρ, V) is an irreducible representation of G, ξ_ρ is the character of ρ ($\chi_\rho(g) = tr\rho(g)$), Λ is the highest weight of ρ, $\lambda \in P$, $\xi_\rho(h) = \sum_{w \in W} \det(w)e^{(w(\lambda))(X)}$, $h = \exp X$, and $\delta = \frac{1}{2}\sum_{\alpha \in \Delta^+} \alpha$.

Weyl's Theorem   A theorem of H. Weyl stating that any finite dimensional representation of a semisimple Lie algebra is completely reducible.

Whittaker model   Let G be a group with a subgroup N (as in the Iwasawa decomposition). Let χ be a unitary character of N on a space V of one complex dimension. Then the action of N on V via χ induces a line bundle L on G/N. The *Whittaker model* for G *defined by* χ is the space of smooth sections of L (namely C^∞ functions from G to V such that $f(gn) = \chi(n)^{-1} f(g)$).

whole number   **(1)** One of the natural numbers, $1, 2, 3, \ldots$.
     (2) An integer that is positive, negative, or zero.

Witt decomposition   **(1)** If V is a vector space over a field of characteristic different from 2, equipped with a metric structure induced by a symmetric or skew-symmetric bilinear form f, a decomposition $V = V_1 + V_2 + V_3$ of V is called a *Witt decomposition* if V_1 and V_2 are isotropic, V_3 is anisotropic, and V_3 is perpendicular to $V_1 + V_2$ with respect to the bilinear form f.
     (2) Let F be a field of characteristic different from 2. A decomposition $q = q_1 \oplus q_2$ of a quadratic form, where $q(x_1, \ldots, q_m) = q_1(x_1, \ldots, x_{2n}) + q_2(x_{2n+1}, \ldots, x_m)$, is called a *Witt decomposition* of q if $q_1 = x_1 x_2 + \cdots + x_{2n-1} x_{2n}$ and $q_2(x_{2n+1}, \ldots, x_m) = 0$ only if $x_{2n+1} = x_{2n+2} = \cdots = x_m = 0$.

Witt's Theorem   If V is a finite dimensional vector space over a field of characteristic different from 2, equipped with a metric structure induced by a non-degenerate symmetric or skew-symmetric bilinear form, then any isometry between two subspaces of V may be extended to a metric automorphism of the entire space.

Witt vector   An infinite sequence $x = (x_0, x_1, \ldots)$ with components in an associative and commutative ring with a unit. Witt vectors are added and multiplied as follows:

$$(x_0, x_1, x_2, \ldots) + (y_0, y_1, y_2, \ldots)$$
$$= (p_0 (x_0, y_0), p_1 (x_0, x_1, y_0, y_1),$$
$$p_2 (x_0, x_1, x_2, y_0, y_1, y_2), \ldots)$$

$$(x_0, x_1, x_2, \ldots) \cdot (y_0, y_1, y_2, \ldots)$$
$$= (q_0 (x_0, y_0), q_1 (x_0, x_1, y_0, y_1),$$
$$q_2 (x_0, x_1, x_2, y_0, y_1, y_2), \ldots),$$

where p_n and q_n are polynomials in variables $v_0, \ldots, v_n$ and $w_0, \ldots, w_n$, respectively, with integer coefficients such that

$$f_n(p_0, \ldots, p_n) = f_n(v_0, \ldots, v_n) + f_n(w_0, \ldots, w_n)$$

and

$$f_n(q_0, \ldots, q_n) = f_n(v_0, \ldots, v_n) \cdot f_n(w_0, \ldots, w_n),$$

where $f_n = z_0^{p^n} + p z_1^{p^{n-1}} + \cdots + p^n z_n$ are polynomials, n a natural number and p a prime number.

word   A finite string w of syllables, written in juxtaposition, where a syllable is any symbol of the form a_i^n, n any integer, where a_i, called a letter, is an element of some set A called alphabet. If the alphabet is $A = \{a_1, a_2, a_3, a_4\}$, then $a_1 a_2^3 a_4^{-1} a_3^2$ is a word. Every word can be changed to a reduced word by replacing an occurrence of $a_i^n a_i^m$ by a_i^{n+m} and by replacing an occurrence of a_i^0 by 1, that is, dropping it out of the word. The reduced form of the word $a_1 a_2 a_2^{-7} a_4^2 a_4 a_3^0$ is $a_1 a_2^{-6} a_4^3$.

word problem   A problem of determining if a given word w is a consequence of a given set of words $\{w_i\} \subset F(A)$, where $F(A)$ is the free group generated by the alphabet A; that is, to determine if w is an element of the least normal subgroup N of $F(A)$ containing $\{w_i\}$.

X

x-axis   The first of the coordinate axes used in locating points in two- or three-dimensional space.

(**1**) In the plane (two-dimensional space), the coordinate axes are two straight lines intersecting in a point O called the origin and denoted by the pair $(0, 0)$. By indicating which of the axes is the first one, we introduce an orientation to the plane. The location of any point P in the plane is given by an ordered pair (a, b), which means that the point P can be reached from the origin by moving a units along the x-axis (the first axis) and then b units parallel to the second axis, which is called the y-axis.

(**2**) In three-dimensional space, the coordinate axes are the three lines of intersection of three planes which intersect in a single point O called the origin and denoted by the triple $(0, 0, 0)$. By labeling the axes as the first, second, and third we introduce an orientation to the three-dimensional space. The location of any point P is given by an ordered triple (a, b, c), which means that the point P can be reached from the origin by moving a units along the x-axis (first axis), then b units parallel to the second axis, called the y-axis, and finally c units parallel to the third axis which is called the z-axis

Y

y-axis   *See* x-axis.

Young diagram   A graphical representation of a partition of a natural number. If $\alpha = (\alpha_1, \alpha_2, \ldots, \alpha_k)$ is a partition of a natural number n, that is, $n = \sum \alpha_i$ with each $\alpha_i \in \mathbf{N}$, the *Young diagram* (of order n) of the partition consists of n blocks arranged in rows and columns in such a way that the ith row has α_i blocks. The first block in each row lies in the first column. The transposed Young diagram α' corresponds to the partition $\alpha' = (\alpha'_1, \alpha'_2, \ldots, \alpha'_m)$, where

α'_j is the number of cells in the jth column of the Young diagram. If the numbers $1, 2, \ldots, n$ are inserted in the blocks of a Young diagram in some order, the diagram is called a *Young tableau*.

Young symmetrizer   An element e_α of the group ring of the symmetric group on n elements defined by a Young tableau α of order n. *See* Young diagram. The symmetrizer is determined by the following rule: Let R_α and C_α be the subgroup of the symmetric group consisting of all permutations that permute $1, \ldots, m$ in each row and each column, respectively, of α. Let

$$r_\alpha = \sum_{\sigma \in R_\alpha} \sigma, \qquad c_\alpha = \sum_{\sigma \in C_\alpha} \mathrm{sgn}_n(\sigma)\sigma \,,$$

where $\mathrm{sgn}_n(\sigma) = \pm 1$ is the sign of σ. Then $e_\alpha = c_\alpha r_\alpha$.

Z

Zariski closed set A set which is closed in the Zariski topology. *See* Zariski topology.

Zariski dense Dense in the Zariski topology. *See* Zariski topology.

Zariski open set A set which is open in the Zariski topology. *See* Zariski topology.

Zariski ring A Noetherian ring R, equipped with the I-adic topology, satisfying the following condition: every element $b \in R$ such that $1 - b \in I$ has an inverse in R if and only if every ideal of R is a closed subset of R in the I-adic topology. (The I-adic topology determined by an ideal I of R is defined by taking $\{I^n\}_{n \in \mathbf{N}}$ as a fundamental system of neighborhoods of zero.)

Zariski's Connectedness Theorem (1) If $f : X \longrightarrow Y$ is a proper surjective morphism of irreducible varieties, the field $K(Y)$ of rational functions separably algebraically closed in $K(X)$ and $y \in Y$ a normal point, then $f^{-1}(y)$ is connected.

(2) For a proper morphism $f : X \longrightarrow Y$ of locally Noetherian schemes with $f_*(\mathcal{O}_X) = \mathcal{O}_Y$, every fiber $f^{-1}(y)$ of f is connected and nonempty for $y \in Y$.

Zariski's Main Theorem (1) If $f : X \longrightarrow Y$ is a birational transformation of projective varieties, where X is normal, then for any fundamental point p of f, the total transform $f(p)$ is connected and of dimension ≥ 1.

(2) A birational morphism $f : X \longrightarrow Y$ of algebraic varieties is an open imbedding into a neighborhood of a normal point $y \in Y$ if $f^{-1}(y)$ is a finite set.

Zariski topology (1) A topology defined on $\mathbf{A}^n$, the affine space of all n-tuples of elements of a field F. This topology is defined by taking the closed sets to be the algebraic subsets of $\mathbf{A}^n$.

(2) A topology defined on the affine scheme $\mathrm{Spec}(A)$ of a ring A by taking the closed sets to be the sets $\{p \in \mathrm{Spec}(A) : I \subset p\}$, where I is an ideal of A.

Zassenhaus group A subgroup H of S_A, the group of permutations of a finite set A, satisfying the following three conditions:
(i.) H is 2-transitive, that is, for every $a, b, c, d \in A$ there exists $\sigma \in H$ such that $\sigma(a) = b$ and $\sigma(c) = d$,
(ii.) the only element of H that leaves more than two elements of A fixed is the identity, and
(iii.) for any pair $a, b \in A$, the subgroup $\{h \in H : h(a) = a, h(b) = b\}$ is nontrivial.

z-axis The third (and last) of the coordinate axes used to locate points in three-dimensional space. *See* x-axis.

Z-basis Let $\mathbf{Z}$ be the ring of integers. Let S be a ring containing $\mathbf{Z}$. A left S module V has a $\mathbf{Z}$ basis if V is freely generated as a $\mathbf{Z}$ module (Abelian group) by a family of generators v_α, $a \in A$. This means (i.) each $v_\alpha \in V$; (ii.) if a finite sum $n_1 v_{\alpha_1} + \cdots + n_k v_{\alpha_k} = 0$, where $n_1, \ldots, n_k \in \mathbf{Z}$, then $n_i = 0$ for $i = 1, \ldots, k$; (iii.) the set of all finite sums $n_1 v_{\alpha_1} + \cdots + n_k v_{\alpha_k}$, where $n_1, \ldots, n_k \in \mathbf{Z}$, is equal to V. The family $v_\alpha, \alpha \in A$, is called a $\mathbf{Z}$-*basis for* V.

This generalizes to the case where $\mathbf{Z}$ is replaced by an arbitrary subring R of S, in which case the family $v_\alpha, \alpha \in A$, is called an R-*basis for* V. *See also* R-basis.

Z-equivalence (1) Let $\mathbf{Z}$ be the ring of integers, let S be an arbitrary ring, and let V and W be S submodules with finite $\mathbf{Z}$ bases, B_V and B_W. The modules V and W are $\mathbf{Z}$-equivalent if there is an S module isomorphism between V and W which, relative to the bases B_V and B_W, has a matrix U with integer entries, and such that the inverse matrix U^{-1} also has integer entries. It is a theorem that $\mathbf{Z}$-equivalence is independent of the particular choice of the bases B_V and B_W. Furthermore, $\mathbf{Z}$-equivalence is related to the classical notion of *integral equivalence* for matrices. *See also* integral equivalence, isomorphism, Z-basis.

The two most important applications of $\mathbf{Z}$-equivalence arise in (i.) the Jordan-Zassenhaus

Theorem, and (ii.) the theory of integral representations of groups. *See* integral representation, Jordan-Zassenhaus theorem.

The notion of **Z**-equivalence generalizes to the case where **Z** is replaced by an arbitrary subring R of S, in which case the notion is called R-equivalence. *See also* R-equivalence.

(**2**) Two integral representations T_1 and T_2 of a finite group G are **Z**-equivalent if there is an invertible matrix U, with integer entries, and such that the inverse matrix U^{-1} also has integer entries, which intertwines T_1 and T_2. This means that $T_2(g) = U^{-1}T_1(g)U$ for all $g \in G$. *See also* integral representation.

zero algebra   An algebra having only one element, 0 (zero), which then is also the unit of the algebra.

zero divisor   If a and b are two nonzero elements of a ring R such the $a \cdot b = 0$, then a and b are *divisors of zero* or *zero divisors*. More specifically, a is a left divisor of zero and b is a right divisor of zero. In a commutative ring, every left divisor of zero is also a right divisor of zero and conversely.

zero element   The identity element of a group where the group operation is denoted by $+$ (addition). Depending on what additional structures the group has, the element may be called the zero element of a field, of a linear space, of a ring.

zero exponent   The integer 0, as a power to raise elements of a multiplicative group. A consequence of one of the laws of exponents is $a^m/a^n = a^{m-n}$. Thus, when the base a is different from zero and the two exponents are equal, $n = m$ the equation becomes $1 = a^m/a^n = a^{m-n} = a^0$.

zero homomorphism   A homomorphism $\phi : G \longrightarrow G'$ that maps every element of G to the identity in G. A zero homomorphism is also called the *trivial homomorphism.*

zero matrix   A matrix in which every entry is zero.

zero of a function   A value of the argument for which the function is zero. (A value of x such that $f(x) = 0$.)

zero point   (**1**) If f is a nonzero holomorphic function and $f(a) = 0$, then a is a *zero point* of f. (*See* zero of a function.) If a is a zero point of f, then there exists a unique natural number n and a holomorphic function f_n such that $f(z) = (z-a)^n f_n(z)$, where $f_n(a) \neq 0$. In that case, n is called the *order* of the zero point a and a is a zero point of order n.

(**2**) If P is a polynomial in n variables over a ring R and $P(\alpha_1, \alpha_2, \ldots, \alpha_n) = 0 \in R$, then $(\alpha_1, \alpha_2, \ldots, \alpha_n)$ is a *zero point* of P. (*See* zero of a function.) If $n = 1$, the zero points are called *roots.*

(**3**) Consider a polynomial ring $K[x_1, \ldots, x_n]$ in n variables over a field K and a field Ω containing K. A point $(\alpha_1, \alpha_2, \ldots, \alpha_n)$ in $\Omega^n = \{(\omega_1, \omega_2, \ldots, \omega_n) : \omega_i \in \Omega\}$ is a *zero point* of a subset S of $K[x_1, \ldots, x_n]$ if $f(\alpha_1, \alpha_2, \ldots, \alpha_n) = 0$ for all $f \in S$.

zero representation   If K is a commutative ring with a unit and A an associative algebra over K, an algebra homomorphism $A \longrightarrow E_K(M)$, where $E_K(M)$ is the associative algebra over K of all K-endomorphism of a K-module M, is called the *zero representation* of A if the module $M = \{0\}$.

zero ring   A ring having only one element, the zero element (the identity with respect to addition).

zeta function   (**1**) The series

$$\zeta(s) = 1 + \frac{1}{2^s} + \frac{1}{3^s} + \frac{1}{4^s} + \cdots$$

which converges for all real numbers $s > 1$ is known as the *Riemann ζ-function.* B. Riemann was the first to treat $\zeta(s)$ as a function of a complex variable and showed that $\zeta(s)$ is holomorphic and has no zeros in $\Re s > 1$. Furthermore, Riemann proved that $\zeta(s)$ can be extended to a meromorphic function in the whole complex plane whose only pole is a simple pole a $s = 1$. *See also* Riemann hypothesis.

(**2**) Any of several special functions called ζ-functions that (i.) are meromorphic on the whole complex plane, (ii.) have Dirichlet series expansion, (iii.) have Euler product expansion, and (iv.) satisfy certain functional equations. *See also* L-function.

zeta function defined by Hecke operators
Let T_n ($n = 1, 2, 3, \ldots$) be a Hecke operator defined with respect to a discontinuous group Γ. Let M be the representation space of the Hecke operator ring K and denote by (T_n) the matrix of the operation of $T_n \in K$ on M. The matrix-valued function $\sum_n (T_n) n^{-s}$ is called the *ζ-function defined by Hecke operators*.

ZG-lattice   Let G be a finite group. A finitely generated $\mathbf{Z}G$-module which is $\mathbf{Z}$-torsion free is called a *$\mathbf{Z}G$-lattice*.

zonal spherical function   (1) A spherical function of degree n (a harmonic polynomial in three variables of degree n restricted to the unit sphere S^2 of Euclidean space E^3), $Z_{\mathbf{u}}^{(n)}(x_1, x_2, x_3)$ ($\mathbf{u} \in S^2$), which has a constant value on any circle in S^2 whose plane is perpendicular to the vector $\mathbf{u}$.

(2) Let G be a locally compact unimodular group and K a compact subgroup of G.

Denote by $C(G; K)$ the set of all continuous complex-valued functions f on G that are invariant under every left translation by elements in K, by $C(G, K)$ the set of elements in $C(G; K)$ that are two-sided K-invariant, and by L the subset of $C(G, K)$ consisting of all functions with compact support. Then L is an algebra over $\mathbf{C}$ if the product of two elements is defined by convolution. If λ is an algebra homomorphism from L into $\mathbf{C}$, and the eigenspace $\{g \in C : f * g = \lambda(f)g$ for all $f \in L\}$ contains a nonzero element, then it contains a two-sided K-invariant element ω normalized by $w(e) = 1$. This element ω is unique and is called the *zonal spherical function* associated with λ.

Z-order   Let $\mathbf{Z}$ be the ring of integers. Let A be a finite dimensional algebra with unit element e over a field F. A subring G of A is called a *$\mathbf{Z}$-order* in A if it satisfies (i.) $e \in G$, (ii.) G contains a basis of A as a vector space over the field F (an F-basis of A), and (iii.) G is a finitely generated $\mathbf{Z}$ module (a finitely generated Abelian group).

This generalizes to the case where $\mathbf{Z}$ is replaced by an arbitrary ring R, in which case G is called an *R-order*. *See also* R-basis, R-order.